D0892086

THE EXCITED STATE IN CHEMICAL PHYSICS

ADVANCES IN CHEMICAL PHYSICS

VOLUME XXVIII

ADVANCES IN CHEMICAL PHYSICS

EDITORS

I. Prigogine

University of Brussels, Brussels, Belgium

S. Rice

Department of Chemistry and The James Franck Institute,
University of Chicago, Chicago, Illinois

EDITORIAL BOARD

THE EXCITED STATE IN CHEMICAL PHYSICS

Edited by J. Wm. McGOWAN
Department of Physics and the Centre for Interdisciplinary Studies in Chemical Physics
The University of Ontario
London, Ontario, Canada

Volume XXVIII

AN INTERSCIENCE® PUBLICATION
JOHN WILEY AND SONS
New York · London · Sydney · Toronto

An Interscience ® Publication

Library of Congress Cataloging in Publication Data:

McGowan, James William, 1931–
The excited state in chemical physics.

(Advances in chemical physics, v. 28)
"An Interscience publication."
1. Excited state chemistry I. Title. II. Series. QD453.A27
vol. 28 [QD461.5] 541'.08s [539.7'54]
ISBN 0-471-58425-8 74-6240

Printed in the United States of America

10 9 8 7 6 5 4 3 2 1

INTRODUCTION

From time to time the broad front of advance in any field is pierced by significantly greater and more important developments in some subareas. Recent developments in laser technology, mass spectrometry, and molecular beam studies have made that the case for the properties and reactions of excited states of simple atoms and molecules. This volume of the Advances in Chemical Physics is, therefore, devoted to a collection of contributions that are relevant to aspects of the physics and chemistry of excited species. The articles cover topics as diverse as theoretical estimation of potential energy, surface properties, and upper atmosphere chemistry, but all are tied together by the common denominator of the need to understand the properties of the excited states of molecules. It is hoped that this and succeeding volumes will supplement the rather broadly scattered literature, and provide an introduction for both the interested student and the working scientist.

Stuart A. Rice

Chicago, Illinois
January 1975

CONTRIBUTORS TO VOLUME XXVIII

ROBERT C. AMME, Department of Physics, University of Denver, Denver, Colorado

ROBERT H. BULLIS, United Aircraft Research Laboratories, East Hartford, Connecticut

FORREST R. GILMORE, R and D Associates, Santa Monica, California

JOYCE J. KAUFMAN, Department of Chemistry, The Johns Hopkins University, and Division of Anesthescology, Department of Surgery, The John Hopkins School of Medicine, Baltimore, Maryland

L. KRAUSE, Department of Physics, University of Windsor, Windsor, Ontario, Canada

RALPH H. KUMMLER, Department of Chemical Engineering and Material Sciences, Wayne State University, Detroit, Michigan

J. WM. MCGOWAN, Department of Physics, and the Centre for Interdisciplinary Studies in Chemical Physics, The University of Western Ontario, London, Ontario, Canada

E. E. NIKITIN, Institute of Chemical Physics, Academy of Sciences, Moscow, U.S.S.R.

IAN W. M. SMITH, Department of Physical Chemistry, University Chemical Laboratories, Lensfield Road, Cambridge, England

PREFACE

For the chemical physicist the rapid growth of studies associated with the formation and destruction of excited atoms, ions, and molecules has created an awkward situation. Much of the literature lies buried in many unexpected places and is very difficult to locate. Furthermore, when and if it is reviewed, it has largely been prepared for specialist groups such as those interested in lasers, biophysics, chemical kinetics, or perhaps the physics and chemistry of the atmosphere. Consequently the nonspecialist and to a large extent even the specialist in one of the above-mentioned areas finds it difficult to keep reasonably in touch with the various aspects of excited-state research.

It is the primary purpose of this project to help alleviate this problem. A study of excited-state production and destruction quite naturally constitutes a large measure of chemical physics. In order to try to cover the field, two volumes are planned. In this volume, emphasis is placed on neutral–neutral collisions, while in the following volume, ion–electron collisions or collisions leading to ionization plus excitation will constitute the bulk of the material.

In this volume, the first chapter focuses upon some chemical reactions discussed in sufficient detail so that the excited reaction products can be definitely identified. In the second chapter, some of the general rules are considered that govern the development of the potential-energy surfaces associated with the intermediate collision complex. The third chapter deals with the theoretical and experimental aspects of nonreactive interchange of energy among kinetic, rotational, and vibrational channels, while the fourth and fifth chapters focus upon some aspects of electronic energy transfer primarily between electronic and vibrational modes. Two short specialized chapters follow which deal with some of the important excited-state reactions in atmospheric and laser studies.

There is always a measure of artificiality associated with abstracting the excited-state material from the total literature of reaction physics and chemistry. However, the particular emphasis brought to the excited state today through its special applications tends to vindicate our approach. As we began this review, it was our hope that it could be a critique as well as a summary, but the recent literature, though vast, still lacks depth.

This project has received the support of many groups and has been completed largely because of the assistance granted by the Office of Standard Reference Data, National Bureau of Standards. To this office, the information centers at JILA and at Oakridge, and to many others, we owe much. Included in those who have helped me are Ernest Bauer, William Bayliss, Y. F. Bow, R. J. Cvetahovic, Donald Gallagher, Radha Govindarajan, J. K. McGowan and Marion Wilson.

As editor of this volume I must express my most sincere appreciation to those who have worked hard on the various chapters and who have been patient with me as this volume has slowly come together.

J. Wm. McGowan

January, 1975

CONTENTS

CHAPTER ONE

THE PRODUCTION OF EXCITED SPECIES IN SIMPLE CHEMICAL REACTIONS

Ian W. M. Smith

Department of Physical Chemistry, University Chemical Laboratories, Cambridge, England

Contents

I. INTRODUCTION

Until recently, two major objectives of chemical kineticists were to explain overall chemical change in terms of elementary reactions and to determine the rates of these individual steps over a wide range of temperature. Within the last twenty years, the development of experimental techniques, such as flash photolysis, shock tubes, and gaseous fast-flow systems, have made it possible to observe the rapid changes that accompany many elementary reactions. The rate of reaction is defined in terms of the appearance or disappearance of a particular chemical entity. For example, for a bimolecular atom-exchange reaction like

$$\mathrm{A + BC \rightarrow AB + C}, \tag{1}$$

the rate is $-dN_{\mathrm{A}}/dt$ $(= -dN_{\mathrm{BC}}/dt = dN_{\mathrm{AB}}/dt = dN_{\mathrm{C}}/dt)$ and is proportional to the product of N_{A} and N_{BC}, that is, the concentrations of the reactants. The constant of proportionality is termed the rate constant or *rate coefficient* (k) and its variation with temperature is normally presented in terms of the Arrhenius parameters, defined by

$$k(T) = A \exp(-E_a/RT) \tag{2}$$

The *preexponential factor*, A, has the same dimensions as k, for bimolecular reactions we shall use cm³/molecule-sec, and E_a is the *activation energy*. It has long been recognized that this approach cannot lead to an understanding of reactions at the molecular level, since a conventional experiment determines a rate that is an average over an enormous number of collisions with a wide spectrum of collisional parameters.

During the last few years, a growing proportion of research in chemical kinetics has been directed at achieving an understanding of the detailed molecular dynamics of reactive collisions. Broadly speaking, an attempt can be made either to study a reaction under conditions where some of the collisional parameters are controlled or to observe the distribution of energy among the products of the reaction. There are two possible ways to ensure that collisions do not occur with a complete distribution of parameters, particularly relative kinetic energy. First, one can make use of "hot-atom" effects by producing one of the reagents with such a high translational energy that the random thermal motions of the second reagent are insignificant in comparison [1]. Alternatively, one can study reactive scattering from two crossed molecular beams [2]. Both the translational energy and the internal quantum states can, in principle, be selected; also, the distribution of relative translational energy among the products can either be inferred from the measured angular distribution or determined directly. To obtain information about the discrete distribution of reaction products among their quantized levels, spectroscopic methods are needed [3]. So far, observations of this kind have been restricted to reactions between reagents at equilibrium, although there are indications that the application of molecular-beam resonance spectroscopy [4] may fuse these powerful techniques together.

This chapter concentrates on the experimental determination of product energy distributions and their interpretation in terms of the dynamics of collisions. In a volume concerned with excited states, it seemed appropriate to bias the article in favor of the spectroscopic methods and results, at the expense of the crossed-beam studies. Electronically adiabatic reactions, which pass smoothly from reactants to products in their electronic ground states, are emphasized, since the results of such processes may be compared with trajectories computed with the equations of classical, rather than wave, mechanics, and the effects of kinematic factors on the sharing of energy can be explored.

The ultimate experiment, where the complete distribution of energy among the degrees of freedom of the products of a reaction between species that have been completely selected in regard to relative translational energy, internal quantum states, orientation, and so on, is a long way off. That very limited data can add considerably to our understanding of a particular reaction can be clearly illustrated by reference to the first experiments of this kind, those of Polanyi and co-workers [5]. They observed the sodium lines from a diffusion flame containing Na vapor and Cl_2, and showed that $Na(^2P)$ was excited by energy transfer from vibrationally excited NaCl formed in the reaction[1]

$$Cl + Na_2 \rightarrow NaCl^{\dagger} + Na, \qquad \Delta H_0^0 = -81 \text{ kcal/mole} \quad (-3.51 \text{ eV}) \quad (3)$$

$$NaCl^{\dagger} + Na \rightarrow NaCl + Na^*(^2P), \quad \Delta H_0^0 = 48.5 \text{ kcal/mole} \quad (2.10 \text{ eV}) \quad (4)$$

Their results showed that a considerable fraction of the NaCl from reaction (3) could excite atoms to the ^2P state, indicating that the NaCl vibration absorbed a large part of the energy released in reaction (3). It was clear that the primary reaction between Na and Cl_2 was remarkably fast, and Ogg and Polanyi [6], and later Magee [7], proposed that an electron switched from the approaching Na atom to the halogen molecule at long range, and thereafter the Na^+ ion and a Cl^- ion attracted one another Coulombically, so that

[1] Throughout this chapter * denotes electronic excitation and † vibrational excitation. Unless an excited state is actually specified, the heats of reaction quoted are ΔH_0^0, which, for a reaction like (3), is equal to ΔD_0, and the thermodynamic data are taken from refs. 328 and 329. (Author's ΔD_0, originally in kcal/mole, have been converted into electron volts—Editor.)

most of the heat released became concentrated in the vibration of this newly formed bond [8].

Recent advances have resulted from the development of more powerful experimental methods and because the classical collision dynamics can now be calculated fully using high-speed computers. By applying Monte Carlo techniques to the selection of starting conditions for trajectory calculations, a reaction can be simulated with a sample very much smaller than the number of reactive encounters that must necessarily occur in any kinetic experiment, and models for reaction can therefore be tested. The remainder of this introduction is devoted to a simple explanation of the classical dynamics of collisions, a description of the parameters needed to define them, and the relationship between these and the rate coefficient for a reaction [9].

A. Classical Collision Dynamics

1. Two-body Systems

A collision between two particles X and Y, with masses m_x and m_y and velocities $\mathbf{v}_x$ and $\mathbf{v}_y$, can be reduced to the motion of a single representative point (P), with mass $\mu = m_x m_y/(m_x + m_y)$, moving with *relative velocity* $\mathbf{v} = \mathbf{v}_x - \mathbf{v}_y$ in a potential $V(r)$, which is a function only of the distance of P from a scattering center (O). This one-body, two-dimensional problem is represented in Figure 1.1. For a particular interaction potential, a collision

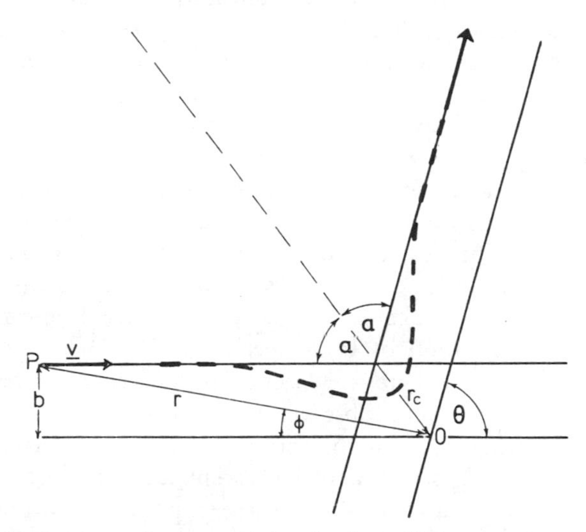

Figure 1.1 Reduction of a two-body elastic collision to the equivalent one-particle problem.

is completely characterized by $\mathbf{v}$ and b, the *impact parameter*. In an elastic collision, E, the collision energy, is unchanged, the trajectory is symmetrical about a line joining O to the distance of closest approach and the classical deflection function $\chi = \pi - 2\alpha$. For particular values of $\mathbf{v}$ and b, χ is uniquely defined, and since the azimuthal orientation of b cannot be selected, the scattering is cylindrically symmetric about $b = 0$, or on any experimentally attainable scale about $\mathbf{v}$. However, even for monoenergetic collisions, the scattering at some values of the *scattering angle*, θ, need not arise from collisions with a single value of b.

We now consider a stream of X particles all approaching a collection of Y with the same relative velocity $\mathbf{v}$. The number of particles per second of X crossing a unit area perpendicular to $\mathbf{v}$ is $I_x = vN_x$, and this current density will be attenuated due to the scattering of X because of their interaction, that is, collision, with particles of Y. The decrease in I_x on the beam passing through a length dl of the absorbing medium is given by

$$dI_x = -\sigma_{xy}(v)I_x(l)N_y\,dl, \tag{5}$$

which defines a *scattering cross section*, $\sigma_{xy}(v)$, which is a scattering analog to the Beer–Lambert optical extinction coefficient [10]. For a given v, the total cross section is

$$\int_{b_{\min}}^{b_{\max}} 2\pi b\,db = \pi b_{\max}^2,$$

since $b_{\min} = 0$, that is, for "head-on" collisions. For hard spheres, where $V(r) = 0$ for $r > (r_x + r_y)$, and $V(r) = \infty$ for $0 \leq r < (r_x + r_y)$, the cross section $\sigma_{xy} = \pi(r_x + r_y)^2$ and is independent of v. However, for more realistic potentials describing the interaction between two nonreactive atoms, the problem is much more complex, and for any potential where the attraction between X and Y only becomes zero at $r = \infty$, the calculated cross section only remains finite if a quantum-mechanical analysis is made.

Classically, $V(r)$ has been derived from measurements of the coefficients of bulk transport properties, or from virial coefficients [10, 11], but these methods suffer from the same drawbacks as conventional measurements on reaction rates: collisions occur with a wide distribution of collisional parameters. The application of beam techniques can provide much more detailed information, since there is a closer connection between the interaction potential and the experimental observations. Two kinds of experiment may be distinguished. In the first, a high-energy neutral beam is passed through a container of scattering gas at low pressure. The attenuation of the beam is observed at different beam energies. These experiments probe the high-energy, repulsive region of the intermolecular potential, which is quite inaccessible to conventional methods. In the second type of experiment,

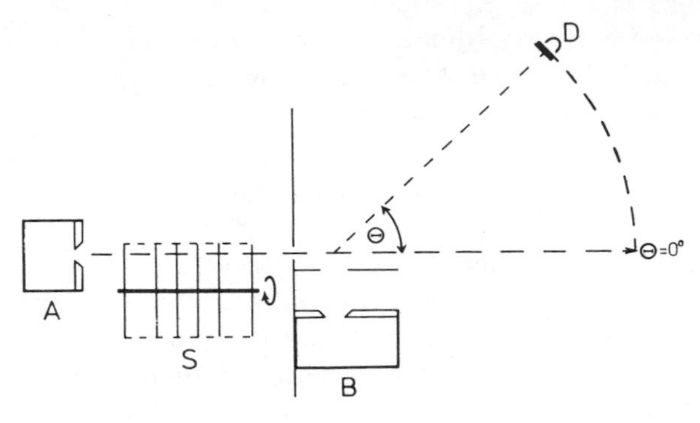

Figure 1.2 Representation of a simple crossed-molecular-beam source [16]. The primary beam effusing from an oven source (A) is velocity selected (S) and then crosses the thermal beam issuing from a second source (B). This diagram shows the detector (D) positioned at the lab angle Θ.

low-energy, monochromatic beams are formed by collimating a gas effusing from an "oven" and passing the resultant beam through a series of slotted discs that rotate on a common shaft and block all molecules except those in a narrow band of velocities. The scattering gas is replaced by a second beam. The experimental arrangement, which is also used for reactive scattering measurements, is shown in Figure 1.2. Besides determining the reduction in the incident-beam intensity, the angular distribution of scattered particles can also be observed, and since the scattering must be cylindrically symmetric about $\mathbf{v}$ in the center-of-mass (c.m.) system, it is only necessary to scan over a fairly limited range of laboratory angles (lab) to reconstruct the entire c.m. scattering pattern. The results are normally presented in terms of the differential scattering cross section, $I_v(\theta) = d\sigma/d\omega$, where $d\omega$ is an element of solid angle in the c.m. system and equals $2\pi \sin\theta\, d\theta$, so that

$$I_v(\theta) = \frac{d\sigma}{2\pi \sin\theta\, d\theta} = \frac{b\, db}{\sin\theta\, d\theta}. \tag{6}$$

In collisions at thermal energies, the scattering is sensitive to the attractive portion of $V(r)$, and any realistic analysis must involve the use of wave mechanics [13, 14]. Here, just two effects will be mentioned, because of their relevance to studies of reactive systems.

Application of the conservation laws [10] yields the equations

$$\mu b v = \mu r^2 \dot{\phi} \tag{7}$$

and

$$\tfrac{1}{2}\mu v^2 = \tfrac{1}{2}\mu(\dot{r}^2 + r^2\dot{\phi}^2) + V(r), \tag{8}$$

where r and ϕ are the coordinates defined in Figure 1.1, v is the initial relative speed, and $V(r)$ is the centrosymmetric scattering potential. Elimination of $\dot{\phi}$ from (7) and (8) yields the expression

$$\tfrac{1}{2}\mu v^2 = \tfrac{1}{2}\mu \dot{r}^2 + \tfrac{1}{2}\mu v^2 \frac{b^2}{r^2} + V(r), \tag{9}$$

which represents the one-dimensional motion of a particle (mass μ, total energy $\frac{1}{2}\mu v^2$) in an *effective potential* equal to $V(r) + \frac{1}{2}\mu v^2(b^2/r^2)$. The second term is called the *centrifugal potential*, and the whole expression gives rise to a family of curves. A few of these are shown in Figure 1.3; they have been calculated [15] for specific values of $q = \mu v^2 b^2/2\varepsilon$ and an exp-6 potential,

$$V\left(\frac{r}{r_m}\right) = \frac{\varepsilon}{1 - 6/\alpha}\left\{\frac{6}{\alpha} \exp\left[\alpha\left(1 - \frac{r}{r_m}\right)\right] - \left(\frac{r_m}{r}\right)^6\right\}, \tag{10}$$

where ε is the depth of the minimum in $V(r)$ located at r_m, and α is an additional parameter.

If the classical equations of motion are solved for a particular value of E, then χ is obtained as a function of b. In general, χ goes through a minimum, and here the classical value of $I_v(\theta)$ becomes infinite; at lower E, when the initial relative energy exactly equals the effective potential at the *centrifugal barrier*, that is, the maximum in $V_{\text{eff}}(r)$, χ itself becomes infinite and "orbiting" occurs. Wave-mechanical considerations remove these singularities and predict a slightly different value for the *rainbow angle*, where $I_v(\theta)$ is a maximum. Smaller maxima, called supernumary rainbows, have been found inside the main band, arising from the interference between scattered waves when the De Broglie wavelength of the collision is about the same as

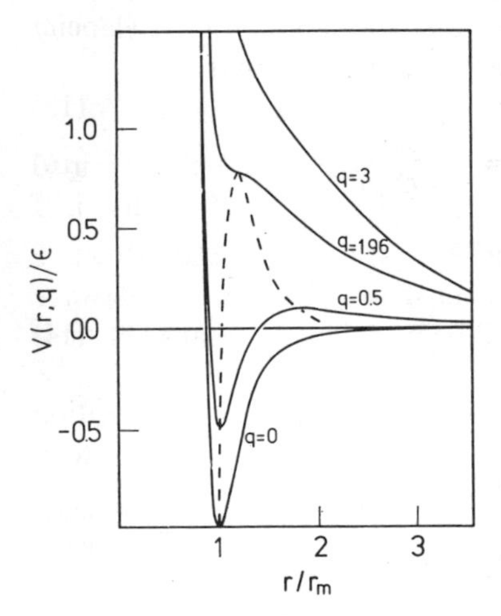

Figure 1.3 The effective potential [15], calculated with the exp-6 function given in equation (10) and for different values of $q = \mu v^2 b^2/2\varepsilon$. The positions of extrema for different q are shown by the dashed line.

the range of the intermolecular forces. The quenching of these effects for systems capable of reaction can yield useful information about the potentials operative in such systems [15].

2. Reactive Systems

In a collision between an atom (A) and a diatomic molecule (BC) new dynamical features arise. The potential is now a function of three, rather than one, internuclear distances: r_{AB}, r_{BC}, and r_{CA}. Furthermore, to characterize an individual collision, it is necessary to define the initial vibrational and rotational states of BC and, even if reaction does not occur, these may change as a result of *inelastic scattering*.

If reaction (1) occurs, the total energy of the products of a reactive collision exceeds that of the reactants by the difference in the dissociation energies of AB and BC, so

$$E' + W' = E + W + \Delta D_0, \tag{11}$$

so where E' and W' are the relative kinetic and internal energies of AB and C after reaction, and E and W are the relative kinetic and internal energies of A and BC before reaction. The energy released must be shared between (a) relative translational energy of AB and C, (b) the vibrational and rotational degrees of freedom of AB, and (c) electronically excited states of AB and C if any are energetically accessible.

Where reaction is possible, the total scattering cross section is made up of reactive and nonreactive contributions. The *reaction cross section* σ_R is defined by the expression

$$-\frac{dN_A}{dt} = \sigma_R(v)vN_A N_{BC}, \tag{12}$$

and k_v, a rate coefficient associated with collisions at the relative velocity $\mathbf{v} = \mathbf{v}_A - \mathbf{v}_{BC}$, is given by

$$k_v = \sigma_R(v)v. \tag{13}$$

The "Boltzmann" rate coefficient, $k(T)$, at temperature T is then obtained by averaging this expression over the Maxwell–Boltzmann distribution of relative velocities,[1] or relative energies. In terms of E,

$$k(T) = \left(\frac{1}{\pi\mu}\right)^{1/2}\left(\frac{2}{kT}\right)^{3/2}\int_0^\infty \sigma_R(E)Ee^{-E/kT}\,dE. \tag{14}$$

In arriving at equation (14), no account has been taken of the distribution of A and BC among internal quantum states. To derive $k(T)$ completely,

[1] $k(T)$ (cm^3/molecule-sec) and $\bar{\sigma}_R(T)$, the averaged reaction cross section, can be interconverted using the approximate relationship $k(T) = \bar{\sigma}_R(T)\bar{v}(T)$, where $\bar{v}(T)$ is the mean relative velocity of colliding reagents at temperature T.

expressions like (14) for each pair of quantum states should be summed, each term being weighted according to its statistical probability.

Reaction cross sections can be estimated by crossed-beam techniques in two ways. The more direct way is to detect one or other of the scattered chemical products [2] over a sufficiently wide range of lab angles, to permit the integration of the observed differential reaction cross section over the whole of c.m. space. The second method is to measure the elastic scattering of one of the reagents in the reactive system and compare plots of the differential cross section against θ, with the angular scattering when the constituent of the second beam is replaced by a similar, but nonreactive, species [16]. Total cross sections may again be determined by observing the attenuation of one incident beam intensity by the second beam.

Classically, σ_R becomes zero below E_0, a critical value of the relative energy known as the *threshold energy*, which is typical of the reaction and will be approximately equal in magnitude to the activation energy, although these two quantities should not be confused since they are measured and defined in quite different ways. Crossed-beam determinations of $\sigma_R(E)$ have until now been limited almost entirely to reactions with threshold energies below about 2 kcal/mole (0.09 eV), because of the difficulties of (a) producing intense beams with energies much greater than this, and (b) detecting low-intensity scattered species unless they consist of alkali metals or their salts. Values of E_0 for the reactions

$$K + HX \quad (X \equiv Cl, Br, \text{ or } I) \rightarrow KX + H \tag{15}$$

have been tabulated by Greene, Moursund, and Ross [15].

Kuppermann and White have reported an approximate determination of the form of $\sigma_R(E)$ for the important reaction

$$D + H_2 \rightarrow DH + H, \tag{16}$$

using "hot" atoms that were generated photochemically [18]. DI was photolyzed with various line sources between 3360 and 3030 Å, in the presence of a large excess of H_2. Because momentum must be conserved and the D and I atoms have such disparate masses, the D atom carries away 98.5% of that part of the energy which is supplied by the photon in excess of that required for dissociation. Collision of these "hot" atoms (**D**) with H_2 can either result in reaction,

$$\mathbf{D} + H_2 \rightarrow HD + H, \tag{17}$$

or in removal of energy,

$$\mathbf{D} + H_2 \rightarrow D + H_2, \tag{18}$$

while thermalized atoms can react with DI:

$$D + DI \rightarrow D_2 + I, \tag{19}$$

$$H + DI \rightarrow HD + I, \tag{20}$$

$$H + DI \rightarrow HI + D. \tag{21}$$

The relative yields of HD and D_2 are related to the relative cross sections for reaction and for elastic scattering, and this ratio was examined as a function of E, by altering the wavelength of the photolyzing source. Their results were extended by replacing DI by DBr and photolyzing it between 1829 and 2537 Å [17]. A preliminary analysis indicates $E_0 = 7.6\ (\pm 0.5)$ kcal/mole (0.33 eV) and that $\sigma_R(E)$ rises monotonically to the highest energies observed [$\approx$33 kcal/mole (1.43 eV)].

These experiments are important because they are performed on a reaction for which *a priori* calculations of $V(r_{AB}, r_{BC}, r_{CA})$ are likely to have their best chance of success as only three electrons are involved. Even here the accurate computation of V, frequently termed the *potential-energy hypersurface*, is extremely difficult. Porter and Karplus [19] have determined a semiempirical hypersurface, and Karplus, Porter, and Sharma [20] have calculated classical trajectories across it. This type of computer experiment has been mentioned before and will be described in greater detail later. The objective of Karplus et al. was to calculate $\sigma_R(E)$ and E_0. Collisions were therefore simulated at selected values of E, with other collision parameters selected by Monte Carlo procedures, and the subsequent trajectories were calculated using the classical equations of motion. Above E_0, σ_R was found to rise to a maximum value, of the same order of magnitude as the gas-kinetic cross section, and then gradually to decrease to greater energies.

II STUDIES OF REACTION DYNAMICS USING MOLECULAR BEAMS

The field of molecular-beam kinetics is in an era of extremely rapid growth, and the topic has been frequently reviewed in the past few years [2, 16, 21–25]. In this chapter, only a simple description will be provided of the instrumentation and interpretation of crossed-beam experiments. We shall concentrate on those results that provide information concerning the disposal of reaction energy among the separating products and shall pay particular attention to the data that have been reported since Herschbach's excellent review on reactive scattering appeared in 1966.

A. Instrumentation

At its simplest, the apparatus for a crossed-beam experiment consists of (a) sources for two molecular beams arranged so that the beams intersect, (b) a detector for at least one of the two incident reagent beams or for one of the scattered reaction products, and (c) a means of moving the position of the detector, relative to the direction of the incident beams, so that scattering can be determined at various lab angles. The apparatus must be maintained at very high vacuum (certainly less than 10^{-7} torr); this is greatly facilitated, if the beam molecules are condensable, by cooling the walls of the vacuum chamber with liquid N_2.

The problems associated with the formation and detection of molecular beams have already been referred to. They are interrelated and have largely determined which reactions have been studied with this technique. The simplest method to form a beam is to collimate the effusive flow occurring from a low-pressure source, conventionally called an "oven," although its temperature may be subambient. Unfortunately, this yields low beam intensities, and the velocities in the beam are thermally distributed. As a result, even for the accurate assessment of the incident-beam intensity, a highly sensitive detector is required. Moreover, the relatively low beam temperature requires that the reaction has a small threshold energy so that an appreciable proportion of the scattering is reactive.

Despite its limitations, considerable success has been achieved with the basic apparatus, although nearly all of this work has been carried out on reactions of alkali metal atoms (M) with halogen-containing compounds to produce alkali halides (MX), since then the differential surface-ionization detector, which was first developed by Taylor and Datz [26], can be used. It consists of two filaments: the first of tungsten, or rhenium, when heated to about 1900°K, produces M^+ ions from both M and MX with roughly the same efficiency, but the second filament of platinum–8%-tungsten alloy, after suitable pretreatment, is sensitive only to M. The performance of this type of two-mode detector [2] has been checked by Herm, Gordon, and Herschbach [27], who mounted a surface-ionization detector behind a magnet capable of producing strong, inhomogeneous fields, so that the paramagnetic M atoms could be deflected and the flux of diamagnetic MX determined directly, virtually in the absence of M. Results for three reactive systems ($K + CH_3I$, $K + Br_2$, and $K + ICl$) confirmed those which had been obtained earlier with a two-filament detector, and this technique, which is particularly useful at small scattering angles where the flux of M is much greater than that of MX and for studying the reactions of Li atoms, has now been applied [28, 29] in experiments on several other reactions.

The stage has now been reached where mass spectrometers are increasingly replacing the highly specific surface-ionization device. The efficiency of ionization by electron bombardment is much lower [24] than surface ionization of alkali atoms or halides, but it can be improved by special techniques, and the vital problem [30] is then to reduce the concentration of background species that interfere with the detection of product molecules. Modulating one of the beams and using phase-sensitive detection provides further improvements in sensitivity, and by observing the phase difference between the signal and that from an unscattered beam of a known temperature, the mean velocity of product molecules can be estimated.

Problems of detection are eased with more intense beams, but this requires that effusive sources be abandoned since they have to be operated at pressures low enough to ensure that the molecular mean free path is greater than the dimensions of the exit slit. If the pressure is about 20 times that needed to satisfy the requirement for effusive flow and if the exit slits are designed to meet hydrodynamic criteria, the emergent beam can be at least ten times more intense, although greater pumping capacity is needed to cope with the increased input to the scattering chamber.

An apparatus incorporating these improvements has been described by Lee et al. [31]. Similar machines are now operating in several laboratories.

The most important item of auxiliary equipment is the mechanical velocity selector, which can be mounted in the path of the incident beam (or beams) to narrow the spread of relative velocities in collisions in the scattering zone or positioned directly in front of the detector so that the distribution of lab *translational* velocities can be directly analyzed. In both positions, velocity selection greatly reduces the flux of molecules reaching the detector. Internal energy distributions are more difficult to measure in beam experiments, since the concentrations of excited products are too low to be observed by conventional spectroscopy. On the other hand, the absence of secondary collisions favors the determination of initial *rotational* distributions, which in experiments at higher pressures are rapidly destroyed by collisional relaxation. The rotational excitation in product molecules that are polar can be estimated by making use of electric deflection prior to detection. Using a two-pole field [34, 35], average rotational energies have been determined by observing the variation of the signal (a) as the detector was displaced from its central position behind the field, and (b) at the central position, as the voltage applied to the electrodes was increased. Particular rotational states of polar molecules can be selected with multipole fields, and Grice et al. [36] have recently reported the preliminary results of some elegant experiments in which RbBr, produced in the reaction

$$\mathrm{Rb} + \mathrm{Br_2} \rightarrow \mathrm{RbBr} + \mathrm{Br}, \qquad \Delta H_0^0 = -46\ \mathrm{kcal/mole}\quad (-1.99\ \mathrm{eV}), \tag{22}$$

was passed through a quadrupole field and the rotational state *distribution* measured directly.

The degree of *vibrational* excitation has generally been estimated from equation (11), using the observed value of E', and either the rotational energy yield was determined experimentally or the conclusion was drawn from trajectory studies that the vibrational contribution to W' is much greater than the rotational part, even where the latter is not obviously limited by kinematic factors. When $E' \ll \Delta D_0$, then much of the energy from reaction must enter the AB vibration, although only Moulton and Herschbach [37] have directly confirmed this. KBr scattered from the reaction

$$K + Br_2 \rightarrow KBr + Br, \qquad \Delta H_0^0 = -45 \text{ kcal/mole} \quad (-1.95 \text{ eV}) \quad (23)$$

was collimated and crossed with a beam of Na atoms in a second scattering chamber. Chemiluminescence from K atoms was excited in the reaction

$$KBr^{\dagger} + Na \rightarrow K^*(^2P_{3/2,1/2}) + NaBr, \qquad \Delta H_0^0 = 40.5, \quad 40.2 \text{ kcal/mole} \quad (1.76, 1.74 \text{ eV}). \quad (24)$$

σ_R for reaction (24) was about 10 Å^2, and clearly KBr retained sufficient energy to make this reaction effectively exothermic in a high proportion of collisions. Direct determinations of vibrational *distributions* may result from the application of two recent developments. Some experiments by Herschbach Klemperer, and their co-workers [4] have shown that it is possible to determine the populations in individual *vibration–rotation* states by passing reactively scattered polar products through a molecular-beam resonance spectrometer. Gillen and Bernstein [38] have described the preparation and calibration of a surface-ionization detector for alkali halides with a sensitivity that depends on the internal energy of the molecule.

Most reactions that have been studied in molecular-beam experiments are insufficiently exothermic to cause true *electronic* excitation of either of the products. Ottinger and Zare [39] have shown that where such excitation does occur, it is not difficult to observe chemiluminescent spectra from the reaction zone using a scanning monochromator and photomultiplier. Because the radiative lifetime of the excited state is short and the total pressure is very low, the vibration–rotation distribution observed within the excited state will be unrelaxed, but it is less easy to determine the proportion of reactive encounters leading to electronically excited products.

B. Interpretation

The nature of reactive collisions and of the reactive interaction potential can be examined by observing the angular distribution of both the elastically scattered reagents [15] and the reactively scattered products [2] in crossed-beam experiments. In both instances, the objective is to discover the spectrum of recoil velocity vectors that carry the scattered particles away from the center of mass, which proceeds at a velocity undisturbed by the collision. As a result any distribution observed in lab space is distorted from that required, because for any lab velocity vector

$$\mathbf{v} = \mathbf{u} + \mathbf{C}, \tag{25}$$

where $\mathbf{u}$ is the equivalent c.m. velocity and $\mathbf{C}$ is the constant vector describing the motion of the center of mass. The kinematic analysis, which is needed to reconstruct the c.m. scattering pattern from that observed, is similar for both elastic and reactive scattering. Elastic scattering is considered first, because the analytical problems are simpler, and the treatment is an extension of that given in the Introduction, but since little can be learned about product energy distributions, only a few of the results and conclusions from these experiments are given here. More details can be found in the review by Greene, Moursund, and Ross [15].

1. Elastic Scattering

The broad features of the lab–c.m. transformation can best be described by reference to a Newton vector diagram, like that shown in Figure 1.4, which represents a collision between A and BC with lab velocity vectors $\mathbf{v}_{\mathrm{A}}$ and $\mathbf{v}_{\mathrm{BC}}$ at right angles and a relative translational velocity $\mathbf{v} = \mathbf{v}_{\mathrm{A}} - \mathbf{v}_{\mathrm{BC}}$. The center-of-mass vector is given by

$$\mathbf{C} = \frac{m_{\mathrm{A}}\mathbf{v}_{\mathrm{A}} + m_{\mathrm{BC}}\mathbf{v}_{\mathrm{BC}}}{m_{\mathrm{A}} + m_{\mathrm{BC}}}, \tag{26}$$

and divides $\mathbf{v}$ in accordance with momentum balance into the individual c.m. vectors $\mathbf{u}_{\mathrm{A}}$ and $\mathbf{u}_{\mathrm{BC}}$,
with

$$\mathbf{u}_{\mathrm{A}} = \mathbf{v}_{\mathrm{A}} - \mathbf{C} = \frac{m_{\mathrm{BC}}\mathbf{v}}{(m_{\mathrm{A}} + m_{\mathrm{BC}})} \tag{27}$$

and

$$\mathbf{u}_{\mathrm{BC}} = \mathbf{v}_{\mathrm{BC}} - \mathbf{C} = \frac{-m_{\mathrm{A}}\mathbf{v}}{(m_{\mathrm{A}} + m_{\mathrm{BC}})}. \tag{28}$$

In an elastic collision, the direction of $\mathbf{v}$, but not its size (or u_{A} and u_{BC}), is altered. Since the scattering in collisions that have the same impact parameter must be cylindrically symmetric in the c.m. system about $\mathbf{v}$, the vectors, like $\mathbf{v}'$, representing the relative velocity after collision must terminate on the perimeters of the bases of the two cones, drawn with $u_{\mathrm{A}} = u'_{\mathrm{A}}$ and $u_{\mathrm{BC}} = u'_{\mathrm{BC}}$, shown in Figure 1.4. Because of this symmetry imposed by conservation

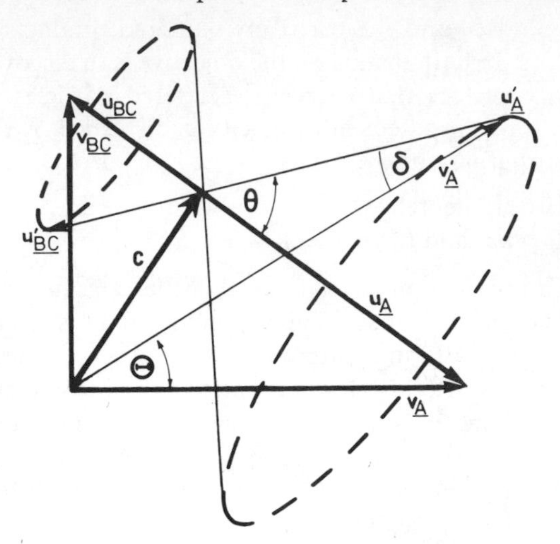

Figure 1.4 Newton vector diagram for *elastic* scattering. It is assumed that $M_A = M_B = M_C$ and that $\mathbf{v}_A/\mathbf{v}_{BC} = (M_{BC}/M_A)^{1/2}$. For elastic collisions with lab velocities initially $\mathbf{v}_A$ and $\mathbf{v}_{BC}$ and a scattering angle θ, A and BC are found at the lab coordinates corresponding to the perimeters of the bases of the two cones.

requirements, it is unnecessary to make observations out of the plane containing the beams, or even over a complete range of "in-plane" angles.

However, it is important to realize that although the differential cross sections in the c.m. system at θ and $-\theta$ are identical, the same is not true of the corresponding lab quantities. The lab differential cross section, which is actually determined in a scattering experiment, is defined by an expression exactly analogous to that for $I_{\text{c.m.}}(\theta)$, that is,

$$I_{\text{lab}}(\Theta) = \frac{d\sigma}{d\Omega}, \tag{29}$$

where $d\Omega$ is an element of the lab solid angle. Consequently,

$$I_{\text{c.m.}}(\theta) = I_{\text{lab}}(\Theta)\frac{d\Omega}{d\omega}, \tag{30}$$

and $d\Omega/d\omega$ can be shown to equal $(u/v)^2 \cos \delta$ [16]. For precisely defined elastic collisions, where $u^2 \cos \delta$ is constant, the interconversion of $I_{\text{c.m.}}(\theta)$ and $I_{\text{lab}}(\Theta)$ is clearly trivial, and even in practice, if *one* beam is velocity selected, the blurring resulting from the spread in the **v** and **C** vectors introduces no great difficulties.

The variation of $I_v(\theta)$ for two-body collisions controlled by a central potential $V(r)$ has been described previously. By observing $I_v(\theta)$, values can be obtained for the characteristic parameters in potentials such as the Lennard–Jones and exp-6 functions. The interpretation of elastic scattering in reactive systems [15] is based on the assumption that the long-range interaction between the reagents can be described in terms of a central potential of this type, and the method is inappropriate for any system where this clearly is not so. The low-angle scattering that results from collisions with large impact parameters is often virtually identical to that in non-reactive systems, and this similarity may often extend past the rainbow maximum. This portion of the $I_v(\theta)$-against-θ curve permits one to evaluate parameters for a potential of the exp-6 type and hence to compute a complete curve, assuming only nonreactive scattering [15]. However, at larger values of θ, corresponding to closer approach of the colliding reagents, reaction becomes more likely and the observed elastic differential cross section, $I_E(\theta)$, falls below that calculated, $I_E(\theta)_{\text{calc}}$. The probability of reaction can be defined by

$$P(E, \theta) = \frac{I_E(\theta)_{\text{calc}} - I_E(\theta)}{I_E(\theta)_{\text{calc}}}. \tag{31}$$

It is more useful to map P as a function of r_c, the distance of closest approach (see Figure 1.1), or of $V(r_c)$, the potential at that distance, both of which can be calculated once the parameters of the assumed potential are known. These representations yield values for the threshold energy and the threshold distance where P becomes greater than zero, and total reactive cross sections, at any relative energy E, can be estimated using the relationship

$$\sigma_{R,E} = 2\pi \int_0^{b\ \text{threshold}} P_E(b) b\, db. \tag{32}$$

Greene, Ross, and their co-workers [15, 40] have found the following total reaction cross sections for reactions of K atoms with hydrogen halides:

$$\text{K} + \text{HCl},\quad 4\ \text{Å}^2; \qquad \text{K} + \text{HBr},\quad 31\ \text{Å}^2,\ 35\ \text{Å}^2; \qquad \text{K} + \text{HI},\quad 24\ \text{Å}^2.$$

No anomalies appeared in the results which could be attributed to the assumption of a centrosymmetric potential $V(r)$.

For systems with larger values of σ_R, rainbow scattering may be completely quenched [2, 15, 41, 42]. This occurs, for example, in K + Br_2 and K + CBr_4. The long-range potential parameters must then be calculated in a different manner [15], based on the supposition that the reaction has a zero threshold energy, so that elastic scattering occurs only from the portion of $V_{eff}(r)$ outside the centrifugal barrier (see Figure 1.3). By comparing the observed scattering with the distribution calculated on this basis, ε can be estimated; a rough value of r_m can be obtained from data on similar, but nonreactive systems, and σ_R is typically ≥ 100 Å^2 and only weakly temperature dependent.

2. Reactive Scattering

Reactive collisions [2, 23, 43, 44] may be represented by a Newton diagram similar to that already shown for elastic scattering, but now $\mathbf{v}'$, the final relative velocity, may differ both in size and magnitude from the initial vector $\mathbf{v}$ and, although $\mathbf{v}'$ is again pivoted at the tip of $\mathbf{C}$, the relative length of its segments is determined by momentum balance between the masses of the products, not those of the reactants. Consequently,

$$\mathbf{v}' = \mathbf{u}'_{AB} - \mathbf{u}'_{C}, \tag{33}$$

where

$$\mathbf{u}'_{AB} = \mathbf{v}'_{AB} - \mathbf{C} = \frac{m_C \mathbf{v}'}{m_{AB} + m_C}, \tag{34}$$

and

$$\mathbf{u}'_{C} = \mathbf{v}'_{C} - \mathbf{C} = \frac{-m_{AB} \mathbf{v}'}{m_{AB} + m_C}. \tag{35}$$

Since $\mathbf{v}' = (2E'/\mu')^{1/2}$, the in-plane scattering of AB at various values of E' can be represented by the concentric circles with radius u'_{AB}, which are shown in Figure 1.5. If $u'_{AB} > C$, then the complete lab distribution of AB ranges over 4π steradians and is easy to resolve. On the other hand, if $u'_{AB} < C$, the scattering is confined to a cone defined by a small angle about $\mathbf{C}$. This may occur either because E' is small, or because $m_{AB} \gg m_C$.

If the reaction is appreciably exothermic, then v', and therefore u'_{AB} and u'_C, can have a wide distribution of values, even when $\mathbf{v}$ and $\mathbf{C}$ are precisely defined. A detector at a laboratory angle Θ observes products with a spread of lab velocities each corresponding to a different pair of values for u' and θ. Consequently, the lab–c.m. transformation of the differential cross sections, which is now represented [50] by the expression

$$\frac{I_{lab}(v', \Theta)}{I_{c.m.}(u', \theta)} = \left(\frac{v'}{u'}\right)^2, \tag{36}$$

cannot be carried out in the straightforward manner that is possible in elastic scattering, since the Jacobian factor, $(v'/u')^2$, possesses a spectrum of values for scattering at any particular angle Θ in the lab (or θ in the c.m.

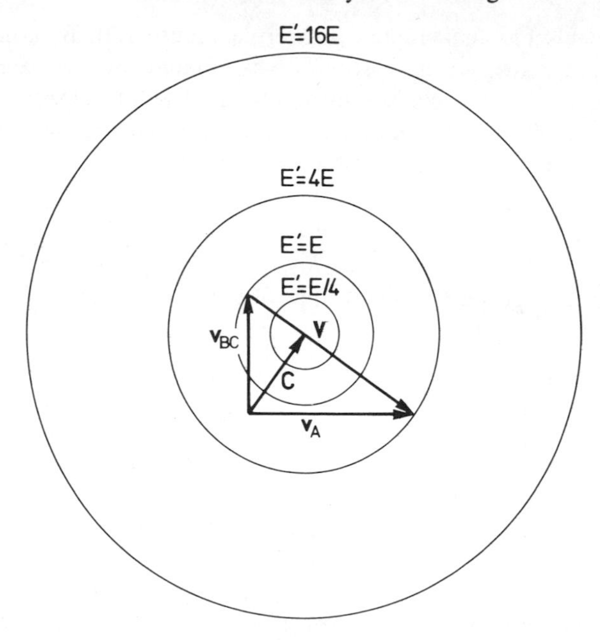

Figure 1.5 Example of a Newton diagram for reactive scattering from A + BC → AB + C, assuming that $M_A = M_B = M_C$ and $V_A/V_{BC} = (M_{BC}/M_A)^{1/2}$. The in-plane μ'_{AB} vectors must terminate on a circle of radius μ'_{AB} and the length of μ'_{AB} will be directly related to E', the final relative translational energy.

system). Because of the distortions introduced by the Jacobian, one must be extremely careful when attempting to translate features observed in the lab distribution into those in the c.m. system [45], particularly since, in general, it strongly weights those contributions to $I_{lab}(v', \Theta)$ which have low relative velocity. Nevertheless, peaks observed in the lab intensity distribution do provide vital evidence concerning the reaction mechanism. The kinematic analysis is much easier when $m_C \ll m_A$ and m_B, and C (rather than AB) is detected [46]. Then except for values of $E' \ll E$, u'_C and v' are both much greater than the initial relative velocity vector **v**; v/u is therefore close to unity and the lab–c.m. distortion is minimal.

Wolfgang and Cross [47] have suggested that the difficulties and ambiguities in the transformation of I_{lab} to $I_{c.m.}$ could be avoided if results were presented as probability functions in Cartesian coordinates defined by

$$P_c(v'_x, v'_y, v'_z) = P_c(u'_x, u'_y, u'_z) = \frac{I_{lab}}{(v')^2} = \frac{I_{c.m.}}{(u')^2}.$$

It is likely that more data will be presented in this way in the future.

In order to understand reactive scattering more fully, it is necessary to consider the features which arise because angular momentum must be conserved. In a two-particle system, only the *orbital angular momentum* associated with the relative motion of the two particles is present. The magnitude of **L**, the vector associated with this momentum before collision, is given by

$$L = \mu v b, \tag{37}$$

and its direction is perpendicular to that of **v**. Conservation requires that

$$\mathbf{L} = \mathbf{L}', \tag{38}$$

where **L**′ is the orbital angular momentum after the collision. This must be aligned at right angles to **v**′ and so the possible directions of **v**′ must all fall in the plane perpendicular to **L**. Since $\mu = \mu'$ and $v = v'$ in elastic scattering, $b = b'$ and the trajectory is symmetric, as shown in Figure 1.1.

If elastic scattering, in collisions defined by **v** and **L**, occurs with an equal probability at any angle in the plane containing **v** and **v**′, it can be represented by the diagram in Figure 1.6a. However, in any real scattering experiment, all azimuthal orientations of **L** relative to **v** are equally probable and the corresponding scattering pattern can be generated by rotating the disc in Figure 1.6a about **v**. This clearly leads to a density distribution of final velocity vectors per unit solid angle which is cylindrically symmetric about **v** but also symmetric about $\theta = 90°$ and strongly peaked at $\theta = 0°$ and 180°.

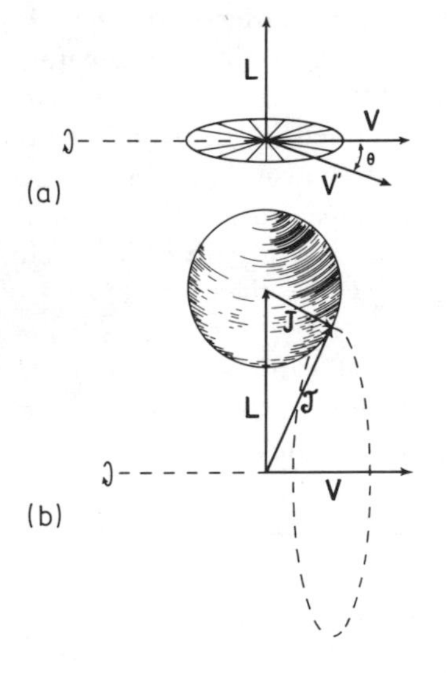

Figure 1.6 (a) Isotropic, in-plane scattering in collisions with initial relative velocity **v** and angular momentum **L**. The complete scattering distribution, including all, equally probable, orientations of **L**, is obtained by rotating this simple diagram around **v** [44].

(b) represents the case where one collision partner can possess rotational momentum. Since **J** will be randomly oriented relative to **L**, the total angular momentum vectors ($\mathscr{J}$) for a particular magnitude of **L** will terminate on the surface of the sphere. However, as all azimuthal orientations of **L** are again equally probable, the $\mathscr{J}$ vectors are distributed with cylindrical symmetry around **v** [44].

The same conclusions are reached using the expression

$$I_v(\theta) = (2\pi \sin \theta)^{-1} \frac{d\sigma}{d\theta} \tag{6}$$

for the differential scattering cross section, and supposing $d\sigma/d\theta$ is constant. This assumption that $\mathbf{v}'$ is oriented to $\mathbf{v}$ at any angle θ with an equal probability is characteristic of the scattering from a long-lived complex [44] that can survive for several rotational periods and hence "forget" the initial direction of $\mathbf{v}$. When the complex dissociates, the $\mathbf{v}'$ vectors are sprayed out randomly at right angles to $\mathbf{L}$, which remains well defined. Such behavior is not, of course, characteristic of a two-atom system, since the atoms must separate far too rapidly. For a more direct or impulsive collision, the scattering remains cylindrically symmetric about $\mathbf{v}$ but not about $\theta = 90°$. However, equation (6) indicates that unless $d\sigma/d\theta \rightarrow 0$ as $\theta \rightarrow 0$ or $180°$, strong peaking may be observed at the polar positions (the singularity, that is, $I_v(\theta) \rightarrow \infty$, predicted by a classical analysis is again removed by quantum mechanics). Isotropic scattering, that is, $I_v(\theta) =$ constant, is only observed when $d\sigma/d\theta$ is proportional to $\sin \theta$ which is so for "hard spheres" scattered from a specular potential.

For any three-atom reactive system, like

$$\mathrm{A} + \mathrm{BC} \rightarrow \mathrm{AB} + \mathrm{C}, \tag{1}$$

$\mathbf{J}$ and $\mathbf{J}'$, the rotational angular momenta of BC and AB, must be included in any statement of the conservation law. Then

$$\mathscr{J} = \mathbf{L} + \mathbf{J} = \mathbf{L}' + \mathbf{J}' = \mathscr{J}', \tag{39}$$

where $\mathscr{J}$ and $\mathscr{J}'$, the vector sums of the individual orbital and rotational angular momenta, are the total angular momenta of the reagents and products, respectively, and must be equal. $\mathbf{L}$ and $\mathbf{L}'$ must again be aligned perpendicular to $\mathbf{v}$ and $\mathbf{v}'$, but the $\mathbf{J}$ (and $\mathbf{J}'$) vectors can be oriented randomly, as indicated in Figure 1.6b. $\mathbf{L}$ and $\mathscr{J}$ are distributed with cylindrical symmetry around $\mathbf{v}$, and once again $I_v(\theta)$ is cylindrically symmetric about $\mathbf{v}$.

The magnitudes of the rotational momentum vectors are given by

$$J = (2I_{\mathrm{BC}}W_{\mathrm{rot}})^{1/2} = (2\mu_{\mathrm{BC}}W_{\mathrm{rot}})^{1/2} r_{\mathrm{BC}} \tag{40}$$

and

$$J' = (2I_{\mathrm{AB}}W'_{\mathrm{rot}})^{1/2}, \tag{41}$$

where I_{BC}, W_{rot}, μ_{BC}, and r_{BC} are respectively the moment of inertia, rotational energy, reduced mass, and internuclear distance of the diatomic BC. Now $L = \mu v b = (2\mu E)^{1/2} b$ and $\langle b \rangle \approx \frac{2}{3}(\sigma_R/\pi)^{1/2}$. Typically b is several times larger than r_{BC}, so that $L \gg J$ and $\mathscr{J} \approx \mathbf{L}$. The magnitudes of L' and J' cannot be estimated in this direct and simple fashion, and are generally only weakly constrained by the conservation laws: (a) since $|\mathbf{L}' + \mathbf{J}'|$ may be much smaller than either L' or J' individually, and (b) because any restriction due to energy conservation is slight, if ΔD_0 is appreciable. If $\mathbf{J}'$ contributes to the total angular momentum after reaction, since it can be randomly orientated, the $\mathbf{v}'$ vectors (perpendicular to $\mathbf{L}'$) are no longer limited to the plane perpendicular to $\mathbf{L}$. Nevertheless, because all azimuthal orientations of $\mathcal{J}$ to $\mathbf{v}$ are equally probable, any reactive scattering defined by particular values of $\mathbf{v}'$, $\mathbf{L}'$, and $\mathbf{J}'$ is equally likely at any azimuthal angle and the cylindrical symmetry of the c.m. scattering is maintained.

The symmetry of the $I_v(\theta)$ distribution about the $\theta = 90°$ plane remains a diagnostic test for complex formation. Moreover, the anisotropy of the $I_v(\theta)$ distribution can be used [44] to determine the partitioning of final angular momentum between $\mathbf{L}'$ and $\mathbf{J}'$ and therefore the rotational energy of the diatomic product AB. Unsymmetric scattering of the product about $\theta = 90°$ indicates a *direct* reactive mechanism, and observations of the preferred direction of recoil provide important evidence concerning the character of the reactive collisions.

Finally, we should note that L, J, L', or J' may have unusually low values if μ, μ_{BC}, μ', or μ_{AB} is small, particularly if A, B, or C is an H atom, and special kinematic effects may ensue. An example is provided by the reaction

$$K + HBr \rightarrow KBr + H. \tag{42}$$

Because $\mu \gg \mu_{HBr}$, $L \gg J$, and since $\mu' \ll \mu_{KBr'}$, J' is likely to be much larger than L', so that

$$\mathbf{L} \approx \mathbf{J}'. \tag{43}$$

This leads to two important results [2]. First, the rotation of AB will be strongly polarized, which may have a marked effect in electric deflection experiments. Secondly, if the rotational distribution of AB could be observed, the distribution of reactive $\mathbf{L}$ vectors and hence of $P(b)$ could be determined. Monte Carlo calculations [48] of reactive trajectories suggest that the $\mathbf{L} \approx \mathbf{J}'$ condition may hold in reactions where it is not required by conservation. Extended electric deflection experiments [35], carried out recently on a number of systems, indicate that although the AB rotation may absorb a large part of the initial orbital momentum, $J'/|\mathcal{J}|$ is appreciably less in reactions where the equality represented by equation (43) need not be satisfied. It does not appear that $\mathbf{L}'$ and $\mathbf{J}'$ can both have a high value (i.e., $\gg L$) and opposite signs in any significant number of encounters.

C. Measurements and Mechanisms

Studies of reactive scattering have been almost entirely restricted until recently to the reactions of alkali atoms with halogen-containing compounds.

Nevertheless, for bimolecular atom-exchange reactions it has been possible [49] to identify three categories of reactive dynamics, together with intermediate cases. On the other hand, very little is known about "four-center" reactions [50, 51], and no results have been reported for association reactions.

The following four families of reactions are now considered in some detail:

$$M + X_2(XY) \rightarrow MX + X(Y), \tag{44}$$

$$M + HX \rightarrow MX + H, \tag{45}$$

$$M + RX \rightarrow MX + R \tag{46}$$

$$M + M'X \rightarrow MX + M', \tag{47}$$

Here M and M′ are alkali metal atoms, X and Y are halogen atoms, and R is an alkyl group. Subsequently, the results for other reactions are reviewed.

1. Reactions $M + X_2 \rightarrow MX + X$

Table 1.1 gives data for reactions considered in this section.

Table 1.1

$-\Delta H_0^0$[kcal/mole (eV) for $M + X_2(XY) \rightarrow MX + X(Y)$ reactions

M =	Li	Na	K	Rb	Cs
$X_2 = Cl_2$	$55._4$	$40._9$	44	44	50
	(2.40)	(1.77)	(1.91)	(1.91)	(2.17)
Br_2	54	42	45	46	54
	(2.34)	(1.82)	(1.95)	(1.99_5)	(2.34)
I_2	48	35	41	46	48
	(2.08)	(1.52)	(1.78)	(1.99_5)	(2.08)
XY = ICl	$52._7$	$48._2$	$51._4$	51	57
	(2.28)	(2.09)	(2.23)	(2.21)	(2.47)
IBr	58	46	49	50	58
	(2.51)	(1.99_5)	(2.12)	(2.17)	(2.51)

[a] Thermochemical data has been taken from ref. 329. For Rb and Cs halides, the estimates of D_0 from flame equilibria are preferred. For M + XY, $-\Delta H_0^0$ is quoted for the more exothermic reaction.

Reference has already been made to the diffusion flame experiments of Polanyi and his co-workers. Some less well-known results [2, 52, 53] obtained with the same method demonstrated that KX is formed from $K + X_2$

with considerable internal excitation, since a high proportion of the product subsequently excited K atoms by energy transfer:

$$KX^{\dagger} + K \rightarrow KX + K^{*}({}^2P_{3/2'1/2}),$$
$$\Delta H_0^0 = +37.4, 37.1 \text{ kcal/mole} \quad (1.62, 1.61 \text{ eV}). \tag{48}$$

In these experiments, the role of $KX^{\dagger}$ produced in the secondary reaction of X with K_2 was minimized by raising the temperature in order to reduce the concentration of K_2.

Results of molecular-beam experiment on these reactions were first reported in 1964 [54, 55]. Since then, they have been investigated in great detail in several laboratories [33, 36, 56–63], and it is probably no exaggeration to assert that at present they are understood more fully at the molecular level than any other reaction.

Except at lab angles within 10° of the incident M beam, the intensity of MX is greater than that of elastically scattered atoms, and the absence of a rainbow maximum associated with scattering from a repulsive core confirms that the σ_R must be unusually large. For all the reactions of K, Rb, and Cs with Cl_2, Br_2, and I_2 (except Rb + I_2, which has not been studied), there is apparently zero threshold, and σ_R lies between about 200 and 400 Å^2 [59, 60], with no trend discernible along either the sequence of alkali metals or that of the halogens. The effective total scattering cross sections, including reactive and nonreactive contributions, are approximately 500–700 Å^2. A recent measurement [64] by kinetic absorption spectroscopy of $\bar{\sigma}_R$ for the Cs and I_2 reaction gave a value of 180 (±25) Å^2 at about 1200°K.

The outstanding feature in the product scattering is a strong maximum in $I_{\text{lab}}(\Theta)$ at an angle Θ_{peak} much less than that defining the direction of **C** and therefore corresponding to preferential scattering of MX into the *forward* hemisphere in the c.m. system ($0° < \theta < 90°$) defined relative to the direction of $\mathbf{v} = \mathbf{v}_M - \mathbf{v}_{X_2}$. Datz and Minturn [54] reported that the CsBr peak, from Cs + Br_2, was shifted to smaller lab angles as the selected velocity of the atomic beam was increased from 3.0×10^4 to 4.8×10^4 cm/sec (v_{Br_2} was thermally distributed, $T = 407°$K). The size of this shift was consistent with the change in the direction of **C** predicted by the Newton diagrams for the most probable velocities of Cs and Br_2. As most of the CsBr was scattered within quite a small range of lab angles, it was concluded (see Section II.B.2) that the yield of relative translational energy was small. In early papers on these as on other reactions, E'_{mp}, the most probable value of E' was roughly estimated by assuming that all the product recoiled along **v**, the initial relative velocity vector [2, 43, 65]. E'_{mp} was then taken as approximately equal to the value of E' at the point where a line drawn at Θ_{peak} cuts **v** on the Newton diagram (see Figure 1.7). For the M + X_2 reactions, the

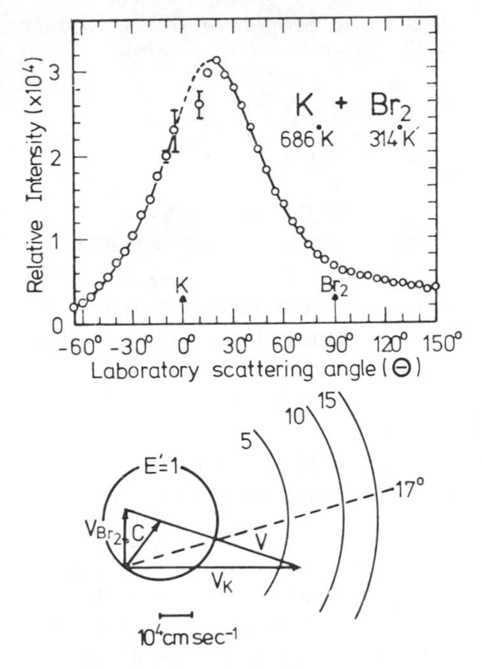

Figure 1.7 The in-plane lab angular distribution of KBr from K + Br_2 [2]. The Newton diagram is given for the most probable beam velocities (both beams are unselected and their temperature is given), and the circles indicate the length of $\mathbf{u}'_{KBr}$ vectors corresponding to various values of E' (kcal/mole). The simple interpretation of these results is to equate the lab peak at $\Theta = 17°$ with a c.m. peak at $\theta = 0°$ (direct forward scattering) and hence estimate $E'_{mp} \approx 1.2$ kcal/mole.

resultant conclusion that only a small fraction of the total energy $(E + W + \Delta D_0)$ is converted to relative translation of the products was confirmed by velocity analysis.

The experimental results on K + Br_2 of Grosser and Bernstein [33], who also selected the velocity of the K beam, and Birely and Herschbach [57] agree remarkably well. However, the more detailed information provided by velocity analysis deserves and requires a more sophisticated data analysis. Birely and Herschbach computed a probability distribution function $P(u'_{KBr})$ for each lab angle at which $P(v'_{KBr})$ was observed, and transformed these curves to scattering distributions at different values of E'. The Wisconsin group adopted an entirely different approach [58]. $P(\theta)$ and $P(f_{int})$, the probabilities of scattering at θ and of retaining a fraction f_{int} of $(E + W + \Delta D_0)$ as internal excitation of KBr, were assumed to be uncorrelated and given adjustable, analytical forms. About 10^5 points in velocity space were then computed by randomly combining E' and θ, and the lab distributions were compared to the experimental observations. $P(\theta)$ and $P(f_{int})$ were then adjusted until a "best fit" was achieved. $P(\theta)$ and $P(f_{int})$ were respectively narrower and broader than the corresponding functions calculated by the Harvard group, who estimated that as much as 20% of the MX from K, Rb, and Cs with Br_2 and I_2, and approximately 5–15% from K, Rb, and Cs with Cl_2, might be scattered into the backward hemisphere.

These experiments show conclusively that the available energy is released almost entirely as internal excitation of the products. The observation that the diatomic product can subsequently excite M atoms electronically demands a degree of excitation which precludes the formation of $^2P_{1/2}$ halogen atoms, which requires 21.7 kcal/mole (0.94 eV) for I and 10.5 kcal/mole (0.46 eV) for Br. Where electric deflection analysis has been performed [34–36], the averaged rotational energy yield, $\langle W'_{rot}\rangle$, is about 5 kcal/mole (0.22 eV), so a high proportion of the energy released must enter the MX vibration. For K + Br_2, it is then possible to estimate the proportion of KBr molecules that possess more than a certain amount of vibrational energy. The values shown in Table 1.2 are calculated from the results of Birely and Herschbach,

Table 1.2

Yield of KBr† from K + Br_2 reaction[a]

Range	(kcal/mole)	>27	>32	>37	>42
	(eV)	(1.17)	(1.39)	(1.60)	(1.82)
Yield	(%)	90	75	52	9

[a] This table differs slightly from that given by Birely and Herschbach (57), because a revised (36) value of $\langle W_{rot}\rangle$ has been used to calculate the yield of vibrational energy.

assuming that $(E + W + \Delta D_0) = 48$ kcal/mole (2.08 eV) and that $\langle W'_{rot}\rangle =$ 5.5 kcal/mole (0.24 eV). Warnock et al. [58] estimate that a higher, but still small, proportion of MX is formed with $f_{int} < 0.5$, although they also predict a maximum in $P(f_{int})$ close to $f_{int} = 1$. There are indications [33, 57] that the internal excitation is greatest in the product scattered directly forward.

The yield of recoil energy from the M + X_2 reaction appears to be independent of the nature of X_2 but does increase up the series from Cs to Li. This is paralleled by a decrease in σ_R and an increase in the amount of backward scattered product. Parrish and Herm [63] have speculated that this may simply arise from a mass effect, although they have indicated that other explanations are possible.

2. Reactions M + HX → MX + H

The reaction

$$\text{K} + \text{HBr} \rightarrow \text{KBr} + \text{H}, \qquad \Delta H_0^0 = -4 \text{ kcal/mole} \quad (-0.17 \text{ eV}) \qquad (49)$$

was the first investigated by the crossed-molecular-beam method [26]. The small reduced mass of the products (μ') and the limited amount of energy available to them requires that the KBr molecules scatter close to the direction of **C**, producing a pronounced peak in the lab distribution. However, because of this effect, it is important that experiments be performed with high resolution if the dynamics are to be satisfactorily analyzed. Martin and Kinsey [46] adopted an ingenious method of avoiding the problems resulting from the unusually low value of μ' and, indeed, made it work for them. They used the β-activity of tritium atoms to observe their angular scattering from the reaction of K + TBr. Since u'_T, $v'_T \gg v$ except for very small values of E', the lab and c.m. distributions virtually correspond. Their conclusion, which seemed to conflict with the earlier findings of Grosser et al. [32] for K + HBr, was that KBr mainly recoiled backwards, although there was a significant forward contribution. Bernstein's group [66, 67] have now performed a high-resolution study of the velocity and angular distributions from K + HBr and K + DBr using a velocity-selected K beam and collision energy of 2.8 kcal/mole (0.12 eV). These experiments showed that the angular distributions were broad in both cases, with K + DBr showing rather more forward scattering. It appears that isotopic substitution does not leave the reaction dynamics unaltered in this series of reactions, and Roach [68] has suggested that these changes may be explained if the products have to surmount a small activation barrier as they separate.

Gillen et al. [67] estimated the distribution of c.m. recoil energies from the velocity analysis. No significant differences were observed in scattering from K + HBr and from K + DBr, and E'_{mp} was approximately 1.5 kcal/mole (0.06 eV). Electric deflection analysis [34, 35] on MBr from K (and Cs) + HBr indicates that $\langle W'_{\text{rot}} \rangle \approx 1.5$ kcal/mole (0.06 eV) and confirms that the rotational momentum of KBr is approximately equal to the orbital momentum of the reactive collisions [see equation (43)]. These measurements suggest that a considerable fraction of the small amount of energy available in this reaction enters the KBr vibration.

3. Reactions M + RX → MX + R

Approximate σ_R for these systems—assuming a central potential—can be estimated from elastic scattering measurements: for $E = 2$ kcal/mole (0.09 eV), $\sigma_R = 23$ Å² for K + CH_3Br (15) and 30 Å² for K + CH_3I [41]. The latter value agrees with that estimated from measurements of the total reactive scattering [69].

As in M + X_2, the reactive scattering is strongly anisotropic, but now the diatomic product is found predominantly in the *backward* hemisphere. The early experiments [2, 43, 70], performed without velocity selection or velocity analysis, were difficult to interpret quantitatively because of the unfavorable kinematics, which arose because: (a) $m_{\text{KI}} \gg m_{\text{CH}_3}$, (b) considerable "blurring" was introduced by the spread in the incident velocities, and (c) the $(v'/u')^2$ factor in the Jacobian distorts the lab distribution and cannot be accurately accounted for in any simple treatment. Estimates of the recoil energy based on an analysis like that shown in Figure 1.7 suggested a low yield, but velocity analysis of the products of the reactions

$$K + CH_3I \rightarrow KI + CH_3, \qquad \Delta H_0^0 = -24 \text{ kcal/mole} \quad (-1.04 \text{ eV}), \quad (50)$$

$$Cs + CH_3I \rightarrow CsI + CH_3, \qquad \Delta H_0^0 = -31 \text{ kcal/mole} \quad (-1.34 \text{ eV}) \quad (51)$$

has since shown [45] that this conclusion was erroneous. In both cases, the flux density of E' is broadly distributed with E'_{mp} being approximately 10 (0.43) and 20 kcal/mole (0.87 eV) for reactions (50) and (51), respectively. A more detailed analysis [69] of the angular scattering from reaction (50) (and other reactions with larger alkyl groups) leads to an approximate estimate of E'_{mp} in agreement with that determined from the velocity analysis.

The rotational energy yield for reaction (51) corresponds to $\langle W'_{\text{rot}} \rangle \approx$ 1.3 kcal/mole (0.06 eV) [35] and that for reaction (50) is likely to be similar. Clearly, the vibrational excitation from these reactions, although still considerable, is appreciably less than from $M + X_2$ systems, and the internal excitation from $Cs + CH_3I$ is significantly less than that from $K + CH_3I$. It has generally been assumed that most of this energy will excite the MX vibration, although some excitation of the alkyl group might be expected [69], particularly if, as in the case of a CH_3 group, it undergoes a change in its equilibrium configuration during the course of the reaction. Maximum excitation would result from an instantaneous change.

The anisotropy of scattering from $M + X_2$ and $M + RX$ shows that these reactions occur in *direct* encounters that do not involve the formation of an intermediate complex. However, the difference in the preferred direction of recoil implies that the detailed molecular dynamics are not the same in these two systems. The $M + X_2$ reactions are prototypes of sharp forward scattering in which the product containing the new bond is scattered predominantly in the forward direction defined in the c.m. system. The large reactive cross sections are consistent with an electron switching from the incoming M atom to the X_2 molecule at a separation of about 7 Å. The instantaneous transition of $X_2 \rightarrow X_2^-$ produces a diatomic ion which can be easily dissociated by the perturbation of the M^+ ion. Subsequently, oppositely charged M^+ and X^- ions attract one another strongly, and much of the released energy is finally stored in the MX (or M^+X^-) vibration. It was customarily thought that the X atom exerted little influence on the other species and was left a virtual "spectator," but recent trajectory calculations [71–74], which are discussed in greater detail later, indicate that the experimental results can only be reproduced if the electron is allowed to migrate between the two X atoms. Frequently the MX product contains the X atom that was initially farthest from the approaching M atom. In contrast, the $M + RX$ reactions constitute examples of *backward* scattering. Here, the electron switch mechanism may be operative but migration is impossible, and a significant proportion of the energy may be released as repulsion between the separating R and X species. Whenever the degree of internal excitation is relatively small, a major part of the energy liberated in the reaction is released as repulsion between the products and the newly bound product is likely to "rebound" into the backward hemisphere.

Wilson and Herschbach [75] have discovered several systems, involving polyhalide molecules, where the peak intensity of the scattered product corresponded to wide c.m. angles and the dynamics is intermediate between the limiting cases represented by rebound and stripping. The preference for forward scattering could be correlated with large values for σ_R. E' was estimated to lie between 10 and 30% of the total energy.

Polanyi [49] has suggested that all reactions will exhibit an increasing preference for "stripping" behavior as the relative collision energy is raised, although the "stripping threshold" may not always be accessible and a proportion of collisions, namely, those with $b \approx 0$, must always result in rebound. If a minimum or "well" occurs on the potential hypersurface when all the atoms are in close proximity, reaction may occur through a complex, and its formation will be favored by low collision energies. Recently, this type of dynamics has been observed in beam studies [44] of alkali metal atoms with alkali halides; these systems are considered next.

4. Reactions $M + M'X \rightarrow MX + M'$

By combining mass analysis with surface ionization, Miller, Safron, and Herschbach [44] have been able to observe the distribution of all four constituents of the systems:

$$Cs + RbCl \rightarrow CsCl + Rb, \qquad \Delta H_0^0 = -5.4 \text{ kcal/mole} \quad (-0.23 \text{ eV}), \tag{52}$$

$$K + RbCl \rightarrow KCl + Rb, \qquad \Delta H_0^0 = -0.5 \text{ kcal/mole} \quad (-0.22 \text{ eV}). \tag{53}$$

The products of the endothermic as well as the exothermic reactions are widely scattered, and their distribution is clearly consistent with the formation of a complex capable of surviving several rotations before dissociating to produce a symmetric (about $\theta = 90°$) c.m. distribution that is distorted in the lab system by the $(v'/u')^2$ Jacobian factor. This conclusion was confirmed by the presence at wider lab angles of a strong, "sticky-collision" peak in the distribution of M, resulting from the break-up of the complex to reform the original reagents rather than new products.

The probability functions $P(\theta)$ and $P(E')$ were assumed to be separable and each to depend on a single adjustable parameter, θ^* and T^*, respectively. A thermal distribution with a temperature T^* was assumed for $P(E')$, while

$P(\theta)$ was approximately of the form predicted for a simple statistical model with the value of θ^* controlling the degree of anisotropy in the c.m. scattering. θ^* and T^* were then adjusted until the "best fit" with the lab distribution was found. The resultant value of T^* indicated that the relative translational energy associated with the dissociation of the complex—to reactants as well as to products—was appreciably greater than that with which the reactants collided, $E' \approx 3$ kcal/mole (0.13 eV) compared to $E \approx 1.4$ kcal/mole (0.06 eV). Preliminary experiments [75] employing velocity analysis indicate that E' is greater and the "superelasticity" is less than at first thought. The strong peaking of the c.m. angular distribution at $\theta = 0°$ and $180°$, as indicated by the "best" value of θ^*, is consistent with the formation and dissociation of the complex with considerably greater orbital than rotational momentum, and the complex was apparently formed with high probability in collisions with impact parameters as large as 8–9 Å.

Internal excitation of the products of these reactions is limited by the small amount of energy which is available for redistribution, and, so far, no details of the energy partitioning have been reported. As Miller, Safron, and Herschbach point out, it should be possible to apply unimolecular theories [76] or statistical theories such as that invoking phase space, which has been developed by Light and his co-workers [77] and requires "strong coupling" in the reactive collisions.

Complexes are also formed between (a) alkali metals and NO_2, with which reaction occurs, and SO_2, CO_2, and NO, which do not react [78–81], and (b) two alkali halides [75].

Very few reactions that do not involve alkali metals have yet been studied in any detail, although considerable progress is likely within the next few years. Three groups [82–86] have investigated the reactive scattering from

$$\mathrm{Cl} + \mathrm{Br_2} \rightarrow \mathrm{ClBr} + \mathrm{Br}, \qquad \Delta H_0^0 = -6.1 \text{ kcal/mole} \quad (-0.26 \text{ eV}). \tag{54}$$

Despite the relatively small reactive cross section ($\sigma_R \approx 5$–20 Å^2), the diatomic product is scattered predominantly forward and the angular distribution is not unlike that from the reaction of K + Br_2. For

$$\mathrm{Br} + \mathrm{I_2} \rightarrow \mathrm{BrI} + \mathrm{I}, \qquad \Delta H_0^0 = -6.3 \text{ kcal/mole} \quad (-0.27 \text{ eV}), \tag{55}$$

a greater scattering intensity was found in the backward hemisphere, and the distribution was closer to that expected from complex formation. Herschbach's group interpreted [82] these results in terms of short-range attraction between the reagents, stronger for Br + I_2 than for Cl + Br_2, giving rise to an "*osculating complex*" capable of surviving for about a rotational period. Similar results have been obtained for the reaction

$$\mathrm{Cs} + \mathrm{TlX} \rightarrow \mathrm{CsX} + \mathrm{Tl} \tag{56}$$

(where X = Cl or I), and some features of the "osculating-complex" model have been discussed [86].

A preliminary estimate [82] of the recoil energy from reaction (54) gave $E'_{mp} = 2.8$ kcal/mole (0.12 eV) for $E_{mp} = 2.9$ kcal/mole (0.13 eV). The conclusion that the energy is largely channeled into internal excitation has been supported by experiments by Blais and Cross [86], in which the products were velocity analyzed. In view of the relatively small reaction cross section, and hence the low incident orbital momentum, it is likely that almost all the balance of the available energy excites the ClBr vibration. Blais and Cross found that the degree of internal excitation is quite markedly dependent on the scattering angle, with the most excited distribution being that along the forward direction of the initial velocity vector. They proposed that their results could be interpreted in terms of a stripping model, although it is not clear whether migration and secondary encounters (see Section IV.D.1) could play any role in the dynamics if this is the case.

Two short articles [88, 89] have appeared on crossed-beam experiments on the reaction

$$D + H_2 \rightarrow DH + H. \tag{57}$$

The reported laboratory distributions differ. The more recent measurements of Geddes, Krause, and Fite indicate that DH is scattered mainly into the backward hemisphere. Excited vibrational levels of HD are not energetically accessible in the majority of reactive encounters, which will occur at energies fairly close to threshold [$E_0 = 7.6$ kcal/mole (0.33 eV) (17)], and consequently the energy must be released mainly in relative translation of the products, generating H atoms which are quite "hot" [90].

A study by Lee and his co-workers (91) of the reaction

$$F + D_2 \rightarrow DF + D, \qquad \Delta H_0^0 = -30.4 \text{ kcal/mole} \quad (-1.32 \text{ eV}) \tag{58}$$

provides the clearest demonstration of the potential of the molecular-beam method. As a consequence of (a) the limited amount of energy which can be released as rotational energy, (b) the wide spacing of the DF vibrational levels, and (c) the predominantly backward scattering, DF molecules in different vibrational states scatter at different lab angles and the relative reaction rates into these levels could be estimated from the observed angular distribution. Because this reaction [92] and that between F and H_2 [92–94] have been studied by spectroscopic methods, this represents the first case in which the two main techniques for studying reactive molecular dynamics have overlapped.

We close this section by mentioning one important "negative" result. Using nozzle expansion, Jaffe and Anderson [95] produced HI beams with translational energies of 40–215 kcal/mole (1.73–9.32 eV), and passed them into a reaction chamber containing DI. Despite the large collision energies [E = 20–113 kcal/mole (0.87–4.90 eV)], no reaction

$$HI + DI \rightarrow HD + I_2, \tag{59}$$

could be detected. These reactions showed that appreciable internal excitation may be required to promote four-center reactions of this kind, and this conclusion is supported by the results of trajectory calculations that are mentioned later in this chapter.

III. SPECTROSCOPIC OBSERVATIONS OF CHEMICALLY EXCITED SPECIES

The results of molecular-beam experiments suggest that the distribution of energy among the various degrees of freedom of the products of a reaction is perhaps controlled by the way in which energy is released during the course of reactive collisions, and this conclusion is supported by the results of Monte Carlo trajectory studies. The information that is provided by crossed-beam experiments is necessarily inexact since the distribution functions are continuous, and to discover the distributions among the quantized—rotational, vibrational, and possibly electronic—states, spectroscopic measurements are needed. Observations have been confined to "bulb" reactions, but the results of beam experiments and trajectory calculations indicate that the product distributions are almost independent of relative collision energy for energies close to threshold. A more serious drawback is that collisional relaxation tends to destroy the distribution initially produced in the reaction, and efforts must be made to eliminate or allow for this effect.

Most of the useful data which have been obtained relate to the two classes of reaction:

(i) Metathetical reactions like

$$A + BC \rightarrow AB + C, \tag{1}$$

where A and B are atoms but C may be slightly more complex.

(ii) Association reactions like

$$A + B(+M) \rightarrow AB^{*(\dagger)}(+M) \rightarrow AB + h\nu, \tag{60}$$

particularly where A and B are either both atoms or an atom and a diatomic molecule. More complex reactions, such as four-center reactions

(iii) $$AB + CD \rightarrow AC + BD, \tag{61}$$

and displacement reactions

(iv) $$A + BCD \rightarrow ACD + B, \tag{62}$$

generally have higher activation energies, so that relaxation of the excited product dominates over its formation and observations are difficult.

Electronic excitation can occur in atom-exchange reactions, although apparently only occasionally. Any realistic, theoretical explanation of such excitation must involve at least an approximate quantum-mechanical formulation of the energy surfaces that lead to excited- and ground-state products and also of the probability that the system may cross from one surface to another, either during the course of the reactive collision or subsequently. Such processes are quite well understood in a number of association reactions. Consequently, although the emphasis in this chapter is placed on atom-exchange reactions, we begin by examining mechanisms of recombination. Chemiluminescence arising from the production of electronically excited states in simple association reactions is an important topic, which has been reviewed several times recently [96–99]; here, a brief description is provided of the processes which can lead to excitation, and these are illustrated by reference to just two selected examples.

A. Association Reactions

If a collision between two species that can form a stable molecule is to result in recombination, there must be some mechanism by which the intermediate can lose energy—either transfer to a "third body" or emission of light. For association directly into the ground electronic state, only the collisional mechanism is possible. It may be represented by the reaction sequence

$$\mathrm{R} + \mathrm{R} \underset{k_{-1}}{\overset{k_1}{\rightleftarrows}} \mathrm{R}_2^{\dagger} \xrightarrow[+\mathrm{M}]{k_2} \mathrm{R}_2 + \mathrm{M}, \tag{63}$$

and the rate is given by

$$\frac{d[\mathrm{R}]}{dt} = \frac{k_1 k_2 [\mathrm{R}]^2\,[\mathrm{M}]}{k_{-1} + k_2[\mathrm{M}]}\,. \tag{64}$$

For two atoms colliding with relative velocity v, the first part of this process can be reduced to that of a single representative particle moving in the one-dimensional, effective potential given by

$$V_{\mathrm{eff}}(r) = V(r) + \tfrac{1}{2}\mu v^2 \frac{b^2}{r^2}, \tag{65}$$

where the symbols retain the meanings defined earlier, but $V(r)$ is now the central potential for the rotationless, chemically bound diatomic molecule. The function of M is to remove energy from the colliding system and thus stabilize R_2 within the potential well. If R is polyatomic, the process is more complex, since energy may be redistributed within internal degrees of freedom of R and a complex will form; this complex will not redissociate until sufficient energy accumulates in one bond to break it.

Porter [100] has suggested that atomic recombination may occur via an alternative mechanism involving the formation of "atom-chaperon" complexes. Weakly bound RM species, formed in the much more common collisions of R + M and stabilized by a second M, are thought to exist in a small equilibrium concentration. R_2 is then produced in bimolecular collisions of R with RM,

$$R + RM \rightarrow R_2 + M. \tag{66}$$

One would expect both these models of three-body recombination to result in stabilized molecules that are initially highly vibrationally excited. Unfortunately, experimental information is almost nonexistent.

1. Recombination into the Electronic Ground State

Polanyi and his co-workers have observed infrared chemiluminescence from vibrationally excited molecules formed in simple chemical reactions. In some cases [101–103], excitation was thought to occur in recombination reactions. The highest vibrational level observed in these experiments was always considerably below the dissociation energy of the excited molecule, but no firm conclusion can be drawn from this fact because there is little doubt that the observed distribution was considerably relaxed from the first stabilizing collisions.

Besides vibrational excitation of the chemical product, the exothermicity of a recombination reaction may excite the third-body in what, in the "energy-transfer" theory, is a physical process or in Porter's mechanism would correspond to excitation in an exchange reaction. Again, information is very scarce. Cashion and Polanyi [101] proposed that DCl could be excited to its first vibrational level, but no further, when it serves as third body in the recombination of H + Cl. They also suggested that Br atoms might be promoted to their $^2P_{1/2}$ state when acting as third body for H + H and H + Br recombinations, but later results [104] from the same laboratory indicate that a binary process may be responsible for exciting Br atoms in these experiments. Oxygen atoms are excited to the ^1S state in mixtures containing N and O atoms, and the three reactions

$$O + O + O \rightarrow O_2 + O(^1S), \tag{67}$$

$$N + N + O \rightarrow N_2 + O(^1S), \tag{68}$$

$$N + O + O \rightarrow NO + O(^1S) \tag{69}$$

have been proposed [105–107] as possible sources of excitation. Since the rate of reaction (68) depends on the nature of the carrier gas, it is believed to occur in two steps: the formation of an excited intermediate, followed by collisional energy transfer. The rate of reaction (67), on the other hand, was independent of the carrier gas, so it might occur directly. Setser and Thrush [108] have shown that CN can be excited to the $A\ ^2\Pi$ and $B\ ^2\Sigma^+$ states in the overall reaction

$$O + O + CN \rightarrow O_2 + CN(A\ ^2\Pi,\ B\ ^2\Sigma^+)\ \Delta H_0^0 = -92.0,\ 44.3\ \text{kcal/mole} \quad (3.99,\ 1.92\ \text{eV}). \quad (70)$$

2. Recombination via Electronically Excited States

Several quite different mechanisms can lead to the formation of electronically excited molecules when atoms, or radicals, themselves in their ground states, recombine.

When more than one state correlates with the electronic states of the separated species, collisions populate the various molecular states at statistically controlled relative rates. If more than one such state is bound, then it may be stabilized in a third-order process. For complex species, the rate of predissociation of the energy-rich complex, that is, $k_{-1}[R^\dagger]$, depends on its dissociation energy, so the ground-state complex will survive longest and have the highest chance of being collisionally stabilized. For diatomic molecules, for example, N_2, O_2, and NO, dipole transitions from these excited states to the ground state are not fully allowed and the excited species are almost certainly quenched in collisions.

Recombination may also proceed via an electronically excited state if during the course of a bimolecular collision the system may transfer from the nonquantized part of the potential curve associated with one electronic state to a second state from which emission is allowed. This process is called *preassociation* or *inverse predissociation*, and the selection rules that control the probability of crossing in both directions are well known [109]. In such encounters total angular momentum must be conserved. For diatomic molecules, the system can pass only into the rotational level of the excited bound state which corresponds to the initial orbital angular momentum in the collision.

The kinetics of the chemiluminescence associated with simple association reactions is frequently found to be third order. The mechanisms that can lead to such a dependence have been discussed in detail by Thrush [98]. One special case is when crossing can take place between high levels in the ground electronic state and levels in an excited state. This process is analogous to that of "internal conversion," which is important in the photochemical behavior of large molecules [110]. Two possibilities should be distinguished.

If levels in two electronic states are mutually perturbed, the electronic wave function associated with each level is essentially a mixture of the wave functions for the two "pure" states, and it becomes of doubtful value to assign these levels to either electronic state. If these levels are formed as a result of recombination, they may be able to radiate to much lower levels in the ground state and hence to stabilize. Moreover, collisions will tend to repopulate these levels from those where this mechanism is less probable, and an efficient relaxation pathway may result. Alternatively, in cases where the levels in different states lie close together, but do not perturb one another, it may be possible for collisions to induce transitions to the state from which radiation can occur. Both mechanisms become more likely with complex molecules since the manifold of levels in each state grows accordingly.

The recombinations of N + O atoms and of NO + O are now discussed to provide examples of the processes which can lead to the production of electronically excited molecules in the association reactions of simple species.

a. Radiative Recombination of N + O

The potential diagram for NO is shown in Figure 1.8. Baulch et al. [112] have recently reviewed the rate data and recommend a value for the third-order rate constant for recombination at 298°K with N_2 as the third body of $1.0_3 \times 10^{-32}$ cm^6/molecule^2-sec. The β ($B\,^2\Pi$–$X\,^2\Pi$), γ ($A\,^2\Sigma^+$–$X\,^2\Pi$), δ ($C\,^2\Pi$–$X\,^2\Pi$), and Ogawa ($b\,^4\Sigma^-$–$a\,^4\Pi$) bands have all been identified in the complex chemiluminescence that accompanies the recombination. Young and Sharpless [113, 114] determined the total intensities of the first three of these systems at room temperature, and the temperature dependences of these processes have since been measured by Gross and Cohen [115].

Young and Sharpless concluded that the $a\,^4\Pi$ state played a decisive part in the formation of excited species in this reaction. They proposed the

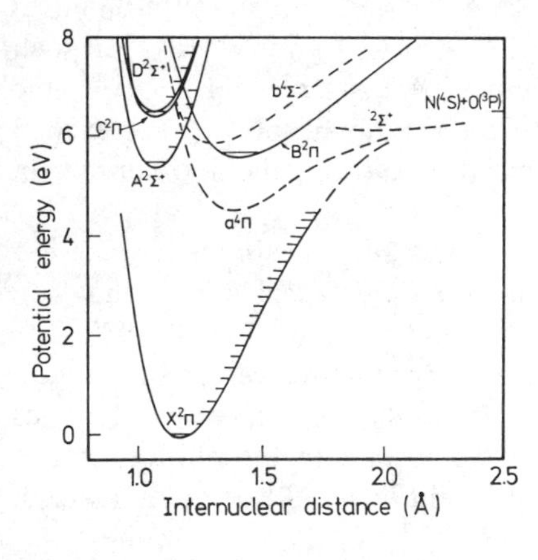

Figure 1.8 The potential-energy curves of NO [117].

following: (a) that NO $C\,^2\Pi(v = 0)$ is produced as the result of a spin-forbidden preassociation via the continuum of the $a\,^4\Pi$ state; (b) that NO $A\,^2\Sigma^+$ is produced by cascade from $C\,^2\Pi(v = 0)$, that is,

$$\text{NO } C\,^2\Pi(v = 0) \rightarrow \text{NO } A\,^2\Sigma^+(v = 0) + h\nu, \qquad (71)$$

and by an energy-transfer with N_2, [116–122], which apparently occurs by two mechanisms, one direct and the other involving $N_2\ A\,^3\Sigma_u^+$ as an intermediate; (c) that collision-induced transitions transfer molecules into the $B\,^2\Pi$ state and higher vibrational levels of $A\,^2\Sigma^+$.

The proposals relating to the γ and δ bands have been supported by the results of a detailed investigation of the NO fluorescence [117, 118, 123–125], illustrating the generally close relationship between these two kinds of experiment. At high total pressures, 95% of the molecules excited to the $C\,^2\Pi(v = 0)$ level predissociate; the rest radiate either to the ground state or to $A\,^2\Sigma^+$. Spontaneous emission can compete, albeit unsuccessfully, with predissociation because the latter is both spin-forbidden and likely to have a low overlap factor [117]. At pressures of about 1 torr or less, a higher proportion of $C\,^2\Pi(v = 0)$ molecules reemit, and it appears that a few rotational levels lie below the dissociation limit [125].

Callear and Smith [117] discussed δ-band emission during preassociation at high total pressures in terms of the reaction scheme

$$\text{N}(^4S) + \text{O}(^3P) \underset{k_{-72}}{\rightleftharpoons} \text{NO}\,(a\,^4\Pi) \xrightarrow{k_{72}} \text{NO } C\,^2\Pi(v = 0) \qquad (72)$$

$$\downarrow k_{-7}$$

$$\text{NO X}\,^2\Pi + h\nu. \qquad (73)$$

The rate coefficient for δ-band emission is $k_{73}k_{72}/(k_{73} + k_{72})$ which is approximately equal to $k_{73}(k_{72}/k_{-72})$ since $k_{-72} \gg k_{73}$. The overall rate is therefore controlled simply by the spontaneous emission rate and an equilibrium constant. This result is quite general and has been applied to S_2 radiative recombination by Fair and Thrush [126]. The magnitude of k_{72} is only revealed if the rate of predissociation is less than that of spontaneous emission, which is unusual, or if the rate of predissociation has been determined. The latter has been done for NO $C\,^2\Pi(v = 0)$ ($k_{73} = 5.1 \times 10^7$ sec^{-1}; $k_{-72} = 1.65 \times 10^9$ sec^{-1}; ref. 125) and k_{72} can then be calculated either from the observed rate constant for the emission accompanying recombination or by calculating k_{72}/k_{-72} using the methods of statistical mechanics and assuming $\Delta E_0^0 = 0$. This yields $k_{72} = 1.3 \times 10^{-15}$ cm^3/molecule sec. Finally, it should be noted that two-body radiative recombination can only predominate over third-order recombination at pressures below 10^{-2} torr.

b. Radiative Recombination of NO + O

The third-order rate coefficient for this recombination reaction has been determined many times. Baulch et al. [127] recommend a value of 6.8×10^{-32} cm^6/molecule^2 sec (M = O_2, Ar) at 298°K. The higher value, compared to that for N + O + M, reflects the formation of a triatomic, short-lived complex.

Recombination is accompanied by the familiar air afterglow that extends from 390 to 1400 nm and is so similar to the fluorescence [128] that it is reasonable to suppose that only the $\tilde{A}\,^2B_1$ excited state is involved in both cases. At total pressures of about 1 torr, the intensity is certainly independent of the pressure of the carrier gas, that is, $I = I_0[O][NO]$, but Clyne and Thrush [129] showed that I_0 possessed a negative temperature coefficient and also depended on the nature of M, the diluent gas. Consequently, they proposed that M is involved in the formation and quenching of NO_2^* with quenching predominating over spontaneous emission until pressures as low as about 10^{-2} torr were reached. Despite some evidence to the contrary [130, 131], it seems now to be established [132–134] that a decrease in emission intensity does occur at pressures between 3×10^{-3} and 2×10^{-2} torr. Using quenching data obtained from a careful investigation of the fluorescence [128], Kaufman and Kelso [135] estimated that 50–90% of the total reaction occurs through the excited state, although they were unable to decide whether this takes place by direct, three-body recombination into this state, or as a result of crossing from high vibrational levels of the ground state, a possibility which had been suggested by Hartley and Thrush [136]. The NO_2 spectrum is very complex, but detailed spectroscopic evidence may help to decide whether the $\tilde{A}\,^2B_1$ state is produced preferentially in a third-order recombination process, which seems unlikely in view of the smaller dissociation energy of the excited state, or in transitions from high vibrational levels of the $\tilde{A}\,^2A_1$ ground state. Recently Becker et al. [137] have shown that although the chemiluminescent intensity does decrease at total pressures below about 10^{-2} torr, there is a significant contribution at lower pressures from two-body radiative recombination.

B. Atom-Exchange Reactions

The exothermicity of simple atom-exchange reactions is frequently sufficient for *electronically* excited states of the products to be energetically accessible. However, appreciable excitation is found only rarely since the reaction producing excited species is usually spin-forbidden or possesses a greater activation energy than that leading to ground-state products. When excited molecules are produced, unless the emission is very strongly forbidden, the reaction will be accompanied by chemiluminescence. Moreover, if the radiative lifetime of the upper state is short, the vibrational, and possibly rotational, distribution within that state may be unrelaxed, so that the relative spectral intensities can provide valuable evidence about its exact mode of formation. As in association reactions, one must be careful to establish that a product is not excited in a secondary process, subsequent to its chemical formation.

The formation of *vibrationally* excited products is nearly always energetically possible in an exothermic reaction, and these products can be detected by observing either an electronic banded system in absorption or the vibration–rotation bands in emission. In principle, *rotational* level distributions may be determined by resolving the fine structure of these spectra, but rotational energy is redistributed at almost every collision, so that any non-Boltzmann distribution is rapidly destroyed and difficult to observe. In contrast, simple, vibrationally excited species are much more stable to gas-phase deactivation and the effects of relaxation are less difficult to eliminate or allow for.

1. Electronic Excitation of Atoms

Excitation of the atomic product of an exchange reaction involving three atoms is certainly rare. When excited states are accessible energetically, their formation is often precluded by the spin correlation rules. For example, the important reaction

$$N(^4S) + NO(X\,^2\Pi) \rightarrow N_2(^1\Sigma_g^+) + O(^3P), \quad \Delta H_0^0 = -75.0 \text{ kcal/mole} \quad (-3.25 \text{ eV}) \quad (74)$$

is spin-allowed and rapid [138], but the reaction

$$N(^4S) + NO(X\,^2\Pi) \rightarrow N_2(^1\Sigma_g^+) + O(^1D), \quad \Delta H_0^0 = -29.7 \text{ kcal/mole} \quad (-1.29 \text{ eV}) \quad (75)$$

is spin-forbidden and apparently cannot compete with reaction (74) despite its exothermicity.

Where an atom has a multiplet ground state, reaction may populate these sublevels with a non-Boltzmann distribution. This is difficult to observe since for light atoms the spin-orbit splitting is small and relaxation is rapid, and also because optical transitions between the components of the multiplet are strongly forbidden. Absorption measurements are possible but have scarcely been applied at all to this particular problem.

The forbidden emission lines $Br(4\,^2P_{1/2}–4\,^2P_{3/2})$ and $I(5\,^2P_{1/2}–5\,^2P_{3/2})$ have been observed [102, 104, 139, 140] from systems where the reactions

$$H + HBr \rightarrow H_2 + Br, \quad \Delta H_0^0 = -16.6 \text{ kcal/mole} \quad (-0.72 \text{ eV}), \quad (76)$$

$$H + HI \rightarrow H_2 + I, \quad \Delta H_0^0 = -32.8 \text{ kcal/mole} \quad (-1.42 \text{ eV}), \quad (77)$$

were taking place. Reaction (77) has been studied by photolyzing HI [139] and by using H atoms from a discharge [140]. Cadman and Polanyi estimated that $>98\ (\pm 1.5)\%$ of the I atoms produced in reaction (77) are in the $5\ ^2P_{3/2}$ ground state. This conclusion that reaction between H and HI only produces $5\ ^2P_{1/2}$ atoms in low yield (although it could be much greater than that corresponding to a thermal distribution) is supported by the results of flash photolysis [141] and photochemical [142] experiments. Airey, Pacey, and Polanyi [104] have discussed the processes which might excite Br ($4\ ^2P_{1/2}$) in the $H + Br_2$ reaction system.

2. Electronic Excitation of Molecules

Most simple atom-exchange reactions that are sufficiently exothermic to produce electronically excited species cannot be studied in isolation. Banded spectra are emitted from a number of complex reaction systems, but invariably there are difficulties in establishing the mechanism responsible for causing the observed excitation. There are three general possibilities: (a) direct chemical formation, (b) crossing into the emitting state from another electronic state, and (c) promotion of ground-state molecules by collisional energy transfer.

Perhaps the best-known example of chemiluminescence is the emission of the red ($A\ ^2\Pi$–$X\ ^2\Sigma^+$) and violet ($B^2\Sigma^+$–$X^2\Sigma^+$) systems of CN from mixtures of active nitrogen with hydrocarbons, halogenated hydrocarbons, or molecules containing a CN group. Almost certainly, all three types of excitation mechanism operate.

Potential curves for the X, A, and B states have been determined by the RKR procedure [143] and are shown in Figure 1.9. The unusual crossing of the ground state by the $A\ ^2\Pi$ curve occurs because the promoted electron in a $\sigma_g 2p$ orbital is more strongly bonding at large internuclear distances than a $\pi_u 2p$ electron [116]. Rotational perturbations have been observed [144] in the levels $A\ ^2\Pi(v = 7)$ and $X\ ^2\Sigma^+(v = 11)$, and also [109] in the levels $A\ ^2\Pi(v \geq 10)$ and $B\ ^2\Sigma^+(v \geq 0)$.

The relative intensities of the CN bands depend strongly on the nature of the species added to active nitrogen [145, 146]. With halogenated hydrocarbons, red emission from the $A\ ^2\Pi$ state predominates; this was classified by Bayes [146] as the "P_2 emission." The v' distribution is broad, peaking at $v' = 7$, and the red system is accompanied by bands from $B\ ^2\Sigma^+\ (v = 0)$ that is populated via $A\ ^2\Pi\ (v = 10)$ because of perturbations and collision-induced transitions. The "P_1 emission" from mixtures of active nitrogen with halogen-containing compounds is quite different, and the relative band intensities in the red system show that $v' = 0$ is now the most populated level.

Setser and Thrush [116, 147] extended Bayes' classification by observing bands from $B\ ^2\Sigma^+\ (5 \leq v' \leq 15)$ and suggesting that these formed part of the

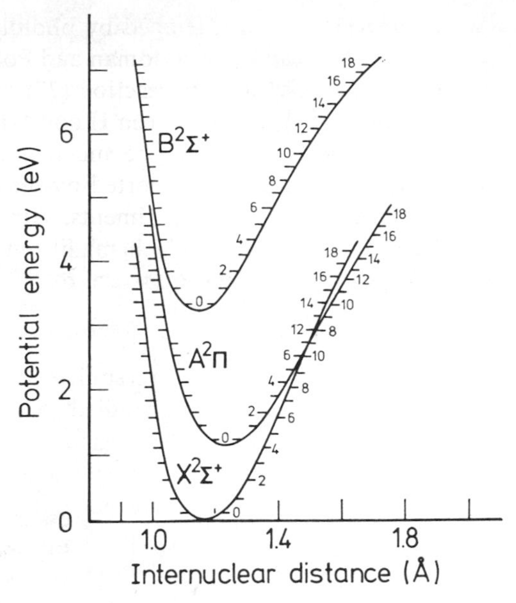

Figure 1.9 The potential-energy curves of CN [147].

P_1 emission. They proposed that the reaction between active nitrogen and molecules like CX_4 (X = H, Cl, Br, or I) proceeded by a chain mechanism and that CN A $^2\Pi$ was a product of the reaction

$$N + CX \rightarrow CN + X. \tag{78}$$

The spin correlation rules forbid the intermediate formation of a $NCX^\dagger$ complex, at least in the singlet ground state, and reaction must presumably proceed across one or more triplet surfaces. Setser and Thrush suggested that the reaction surface leading to CN $A\,^2\Pi$ may possess a lower barrier height than that correlating with ground-state products, since the $A\,^2\Pi$ state is more stable than $X\,^2\Sigma^+$ at large internuclear distances. They attributed the P_1 emission to promotion of $X\,^2\Sigma^+$ molecules, either by transfer of energy from an excited, metastable species, or during action as a third body for atomic recombination, for example,

$$N_2^* + CN \rightarrow N_2 + CN^* \tag{79}$$

or

$$N + N + CN \rightarrow N_2 + CN^*. \tag{80}$$

Broida and his colleagues [148–150] have concentrated on the interaction between the $A\,^2\Pi(v = 10)$ and $B\,^2\Sigma^+(v = 0)$ levels. By irradiation with the appropriate microwave frequency, a single level in the $B\,^2\Sigma^+(v = 0)$ state was populated [150], and rotational relaxation rates were determined by observing the increase in intensity of other lines as the pressure was increased. Besides inducing transitions between rotational levels within the same state, it was found [149] that collisions can cause transitions between unperturbed levels in different states with about 1% efficiency.

From the results of a high-resolution study of the CN chemiluminescence, Iwai, Savadatti, and Broida [151] proposed that reactions (79) and (80) should be reserved to explain the high-energy "tail" bands, which extend to $A\,^2\Pi(v' = 24)$ and $B\,^2\Sigma^+(v' = 15)$, and that Bayes' P_1 emission from low levels in $A\,^2\Pi$ probably results from excitation in the reactions

$$N + N + YCN \rightarrow N_2 + CN^* + Y \tag{81}$$

and

$$N_2^* + YCN \rightarrow N_2 + CN^* + Y. \tag{82}$$

As the potential curves suggest, the $A\,^2\Pi$ and $B\,^2\Sigma^+$ can mix considerably above about $B\,^2\Sigma^+(v = 5)$, and since the $B\,^2\Sigma^+$ state has a shorter radiative lifetime, the $B\,^2\Sigma^+$–$X\,^2\Sigma^+$ system predominates in the tail bands. The relative intensities of the perturbed lines at low pressure indicated that $A\,^2\Pi(v = 10)$ was populated in CH_2Cl_2 flames about 30 times more rapidly than $B\,^2\Sigma^+(v = 0)$, and that $A\,^2\Pi(v = 7)$ was formed 4 times faster than $X\,^2\Sigma^+(v = 11)$.

The increase of rotational temperature in the red bands at low pressure and the absence of any marked intensity anomalies appear to preclude collision-induced transitions from $X\,^2\Sigma^+$ to $A\,^2\Pi$ as a mechanism for excitation in the cross-over region. Iwai et al. suggested that the excited state might be favored because of its higher density of states resulting from λ doubling and its smaller rotational constant. This would be particularly valid if the reaction occurs through a strongly coupled complex.

The $B\,^2\Sigma$–$X\,^2\Sigma$ and C–$A\,^2\Pi$ systems of SiN are emitted when volatile silicon compounds are added to active nitrogen [152]. The $B\,^2\Sigma$ vibrational distribution shows a pronounced peak.

The $A\,^3\Pi_g$ $X\,^3\Pi_u$ Swan bands of C_2 are emitted from several complex reaction systems. The potential curves for C_2 [153] are shown in Figure 1.10.

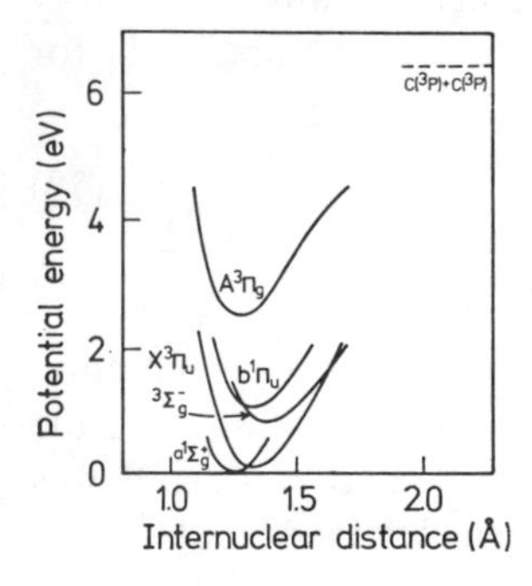

Figure 1.10 The potential-energy curves of C_2 [153].

The $^1\Sigma_g^+$ state is taken as the ground state [154], and the dissociation energy is given the value 144 kcal/mole (6.24 eV) [155, 156]. The $^1\Sigma_g^+$ potential curve crosses those of the excited $3\Sigma_g^-$ and $X\,^3\Pi_u$ states for the same reason as the $X\,^2\Sigma^+$ and $A\,^2\Pi$ curves of CN cross.

Remarkably selective excitation of the $A\,^3\Pi_g(v = 6)$ level has been observed in a number of reaction systems. Some years ago, Herzberg [157] suggested that an inverse predissociation might be responsible, but this proposal is incompatible with the modern value of D_0 (C_2) and has been replaced by the proposal [153] that crossing from the $^3\Sigma_g^-$ state takes place. The band progression from $A\,^3\Pi_g(v = 6)$, often called the "high-pressure bands," is observed when $<0.01\%$ of CO in He is irradiated with α particles [158]. The reaction

$$C + C_2O \rightarrow C_2 + CO \tag{83}$$

was proposed to explain this result, and it probably occurs also in condensed discharges through ≥ 10 torr of CO, where the high-pressure bands were first observed. When the products from a carbon arc are mixed with active nitrogen, the high-pressure bands are again emitted [159].

Palmer and his co-workers have observed more complex C_2 emission from diffusion flames of alkali metals (Na or K) in haloforms and carbon tetrahalides [160–165]. The Swan bands show a much broader v' distribution than those emitted from other systems, and although the $v' = 6$ level is sometimes excited preferentially, it is clear that more than one process excites the $A\,^3\Pi_g$ state in the diffusion flame reactions. Detailed interpretation is difficult and is hampered by a lack of precise thermochemical data for some of the species involved. Tewanson, Naegeli, and Palmer [165] have suggested that three quite distinct mechanisms may cause excitation: (i) the association of C atoms, either by three-body recombination into the $A\,^3\Pi_g$ state or via the successive, fast bimolecular steps

$$C + K_2 \rightarrow CK^\dagger + K, \tag{84}$$

and

$$C + CK^\dagger \rightarrow C_2 + K; \tag{85}$$

(ii) the reaction

$$C + CX \rightarrow C_2 + X, \tag{86}$$

where X is a halogen atom, and (iii) $A\,^3\Pi_g(v = 6)$ may be selectively populated by the curve-crossing route from $^3\Sigma_g^-$.

Palmer and his co-workers have also reported on the chemiluminescence from diffusion flames of K atoms with a wide variety of volatile inorganic halides [166–168].

Several diatomic species are electronically excited in the reaction of O atoms with C_2H_2, and their spectra have been observed in flame [169, 170], shock-tube [171, 172] and flow-discharge [173–177] experiments. Williams and Smith [178] have recently reviewed the combustion and oxidation of acetylene and have considered what processes might give rise to the complex chemiluminescence that is observed.

CO is formed in three excited electronic states, namely, $A\,^1\Pi(v \leq 6)$, $e\,^3\Sigma^-(v \leq 6)$, and $d\,^3\Delta(v \leq 9)$, with a maximum excitation energy corresponding to about 205 kcal/mole (8.89 eV) [175, 176]. Becker and Bayes substituted C_3O_2 for C_2H_2, and since the emission was unchanged and behaved identically as the flame conditions were altered, they proposed that the reaction

$$O(^3P) + C_2O \rightarrow CO(A\,^1\Pi, d\,^3\Delta, e\,^3\Sigma^-) + CO \qquad (87)$$

occurred in both systems. When two CO molecules in their ground electronic states are produced, reaction (87) is about 205 kcal/mole (8.89 eV) exothermic; Palmer and Cross' [179] low value for ΔH_f (C_2O), which is referred to by Becker and Bayes [176], is now thought to be incorrect [180]. Bayes [177] has studied the reaction between ^{18}O atoms and $C_2^{16}O$. $C^{18}O$ $A\,^1\Pi$ and $C^{16}O$ $A\,^1\Pi$ emit with comparable intensities, and Bayes concludes that reaction must occur via an intermediate OCCO complex which exists for at least several vibrations. The production of several excited states in a single process indicates that the system can cross between different potential surfaces during the course of the reaction.

Addition of H atoms to the $O + C_3O_2$ flame gives rise to the CH, C_2, and OH emission [176] that is associated with the $O + C_2H_2$ reaction. A number of processes could excite CH $A\,^2\Delta$; the $v' = 0$ level, which is predominantly filled, lies 66.0 kcal/mole (2.86 eV) above the $X\,^2\Pi$ ground state. The most likely [170, 178, 181] excitation mechanism is the reaction

$$C_2 + OH \rightarrow CH + CO, \qquad \Delta H_0^0 = -91.7 \text{ kcal/mole} \quad (-3.98 \text{ eV}), \qquad (88)$$

although the reaction

$$O + C_2H \rightarrow CH + CO, \qquad \Delta H_0^0 = -60.5 \text{ kcal/mole} \quad (-2.62 \text{ eV}), \qquad (89)$$

can populate the $A\,^2\Delta$ state of CH if C_2H is itself internally excited, and Brennen and Carrington [182] have reported that this is not inconsistent with their results. Finally, CH might be excited by transfer of electronic energy from CO $a\,^3\Pi$. OH also appears to be excited in a four-center reaction [183, 184]:

$$CH + O_2 \rightarrow CO + OH, \qquad \Delta H_0^0 = -159.5 \text{ kcal/mole} \quad (-6.92 \text{ eV}). \qquad (90)$$

According to Becker et al. [183], this leads to a "hot" rotational distribution in the excited $A\,^2\Sigma^+$ state, while a second unidentified mechanism yields excited OH with an almost thermal rotational distribution.

The uncertainty of the conclusions associated with these systems only emphasizes that the energy of the highest level observed only rarely correlates with the exothermicity of the process leading to formation of an excited state—although the excitation of CO in reaction (87) may be one such case. Moreover, particularly in flames, energetic species produced in other reactions may react or transfer energy to molecules in their ground state causing important, and frequently confusing, effects.

Besides vibrational excitation of OH, which is considered later, there have been reports that the reaction

$$H + O_3 \rightarrow OH + O_2, \quad \Delta H_0^0 = -77.1 \text{ kcal/mole} \quad (-3.43 \text{ eV}) \quad (91)$$

can result in electronic excitation of OH [185] and O_2 [186]. The $A\ {}^2\Sigma^+$ state of OH cannot be excited directly, and Broida [185] has concluded promotion to the excited electronic state may occur when two vibrationally excited OH molecules collide. Reaction (91) can produce $O_2\ b\ {}^1\Sigma_g^+$ directly, because it lies only 36.6 kcal/mole (1.59 eV) above the ground state. However, the ground-state vibrational distribution of OH shows two maxima that are separated by roughly the excitation energy of $O_2\ b\ {}^1\Sigma_g^+$ and that disappear as the total pressure and concentration of O_2 are reduced [187]. This suggests that O_2 may also be excited by vibrational–electronic energy transfer rather than by direct excitation.

There are only two metathetical reactions for which accurate rate parameters have been determined for formation of both the ground and electronically excited state of a reaction product. Perhaps surprisingly, these both involve five atoms and a triatomic product:

$$NO(X\ {}^2\Pi) + O_3(X\ {}^1A_1) \rightarrow NO_2(\tilde{X}\ {}^2A_1, \tilde{A}\ {}^2B_1) + O_2(X\ {}^3\Sigma_g^-),$$
$$\Delta H_0^0 = -47.6, -4.8 \text{ kcal/mole}$$
$$(-2.06, -0.21 \text{ eV}) \qquad (92a,b)$$

$$SO(X\ {}^3\Sigma_g^-) + O_3(X\ {}^1A_1) \rightarrow SO_2(\tilde{X}\ {}^1A_1, \tilde{a}\ {}^3B_1, \tilde{A}\ {}^1B_1) + O_2(X\ {}^3\Sigma_g^-),$$
$$\Delta H_0^0 = -106.6, -33.0, -22.0 \text{ kcal/mole}$$
$$(-4.62, -1.43, -0.95 \text{ eV}) \qquad (93a,b,c)$$

(Energies of excited NO_2 and SO_2 are taken from ref. 188.) The reagents are relatively stable and the reactions can be studied in the absence of complicated chain processes.

The chemiluminescence from reaction (92b) was discovered by Tanaka and Shimayi [189], identified as emission from NO_2 by Greaves and Garvin [190], and studied by Thrush and his co-workers in flow systems [191, 192]. From absolute intensity measurements, Clough and Thrust [192] were able to determine rate parameters for the formation of $NO_2(\tilde{X}\,^2A_1)$ and of $NO_2(\tilde{A}\,^2B_1)$:

$$k_{92a} = 1.26(\pm 0.25) \times 10^{-12} \exp\left(-4180 \pm \frac{300}{RT}\right) \text{ cm}^3\text{/molecule sec},$$

$$k_{92b} = 7.1(\pm 1.6) \times 10^{-13} \exp\left(-2330 \pm \frac{300}{RT}\right) \text{ cm}^3\text{/molecule sec}.$$

The different activation energies indicate that reaction proceeds across two quite distinct potential surfaces and that the excited state is not populated via the ground state by crossing. This conclusion does not conflict with the mechanism proposed for the recombination of O and NO (see Section III.A.2.b), since much higher levels are excited in that reaction. For the lower levels involved here, the vibrational overlap factors will be much smaller since the equilibrium geometry of the two states is very different [193]. Probably, only a few low-lying levels in the $^2B_1$ state are populated and the spread of emitted wavelengths arises from the large number of ground-state levels to which transitions occur.

Superficially, reactions (92) and (93) are similar, but quite different electronic states are involved [97, 194]. The rate coefficients for reaction (93) are given by

$$k_{93a} = 2.5 \times 10^{-12} \exp\left(\frac{-2100}{RT}\right) \text{ cm}^3\text{/molecule sec},$$

$$k_{93b} \ngtr 5 \times 10^{-14} \exp\left(\frac{-3900}{RT}\right) \text{ cm}^3\text{/molecule sec},$$

$$k_{93c} = 1.7 \times 10^{-13} \exp\left(\frac{-4200}{RT}\right) \text{ cm}^3\text{/molecule sec}.$$

There is now a marked difference between the preexponential factors for the reactions to excited- and ground-state products that is not found for the $NO + O_3$ reaction, and the difference is explained in terms of the potential surfaces which can correlate with the reactants and possible products. With NO, two closely spaced surfaces arise because of the degeneracy associated with the orbital angular momentum of NO $X\,^2\Pi$, but with SO only a single triplet surface correlates with ground-state reagents. Accordingly, Halstead and Thrush [194] conclude that excited SO_2 is produced when the system crosses from this surface to one correlating with excited products at a point near the potential barrier. This is probably assisted by the states having fairly similar equilibrium geometry [188].

3. Vibrational and Rotational Excitation

Although atom-exchange reactions rarely result in electronic excitation, the internal degrees of freedom of a polyatomic product—certainly of a product containing a new bond—are likely to absorb some part of the energy of reaction. The distribution of molecules among the vibrational and rotational levels can, in principle, be determined from absorption, spontaneous emission, or stimulated emission spectra. In order to calculate from these results the initial distribution formed in the reaction, two steps must be taken. First, a distribution must be derived from the relative intensities of the observed spectral features, and then this distribution must be related to the distribution being sought, by taking account of relaxation processes before, or during, the period of observation. Rotational excitation is equilibrated very rapidly in collisions, and only recently have any worthwhile results on rotational distributions resulting from chemical reaction been obtained. On the other hand, vibrationally excited simple molecules can frequently survive very many collisions, and initial vibrational distributions are consequently easier to determine.

The relaxation of vibrationally excited species can frequently be observed directly by kinetic absorption spectroscopy, and it is rather surprising that this technique has provided so little detailed information about how energy is partitioned in chemical reactions, particularly when one remembers that experiments by Norrish and his co-workers [195] did much to initiate the present interest in this subject. Unfortunately, several conditions must be simultaneously fulfilled if the experiments are to succeed. These requirements are as follows:

a. a species, say ABC, which can be photodissociated to AB, which should be unreactive, and C;
b. a reagent (XYZ) to react with C, which must either be unexcited by the flash, or can be undissociated ABC;
c. diatomic products of $C + XYZ \rightarrow CX + YZ$, which are chemically relatively stable and possess banded spectra unobscured by other absorptions, and in a convenient wavelength region for observation;
d. if ABC and XYZ are different, they should not react while being mixed.

The method has been largely restricted to reactions of O atoms. Early experiments were reviewed by Basco and Norrish [196] in 1960.

The absorption spectra of vibrationally excited products were first detected following flash photolysis of ClO_2 and NO_2 [195] and of O_3 [197]. O_2 was vibrationally excited in the chemical reaction following photodissociation:

$$XO_2 + h\nu \rightarrow XO + O,$$

$$O + XO_2 \rightarrow O_2 + XO.$$

The interpretation of the experiments is complicated because both atomic and molecular oxygen possess low-lying electronic states.

When O_3 is flash photolyzed, absorption below 300 nm produces $O(^1D)$ atoms, probably by the process

$$O_3 + h\nu \rightarrow O_2(^1\Delta_g) + O(^1D),$$

which is consistent with spin conservation, the chemical properties of the atoms produced and the long-wavelength limit to the absorption. The reactions of both $O(^3P)$ and $O(^1D)$ atoms with O_3 are highly exothermic:

$$O(^3P) + O_3 \rightarrow 2\,O_2, \qquad \Delta H_0^0 = -93.7\ \text{kcal/mole} \quad (-4.06\ \text{eV}), \tag{95}$$

$$O(^1D) + O_3 \rightarrow 2\,O_2, \qquad \Delta H_0^0 = -139.0\ \text{kcal/mole} \quad (-6.03\ \text{eV}), \tag{96}$$

but reaction (95) has an activation energy of a few kcal/mole, and at 298°K, $k_{95} = 1.5 \times 10^{-14}$ cm³/molecule sec [198], whereas at the same temperature $k_{96} = 6.7 \times 10^{-11}$ cm³/molecule sec [199]. Consequently, the lifetime of an atom produced initially in the ^1D state depends on whether it is quenched.

McGrath and Norrish [197] detected $O_2^\dagger\,(v \leq 17)$ on flashing O_3 in a 40:1 excess of N_2. At that time, it was not appreciated that N_2 quenches $O(^1D)$ rapidly [199, 200], but a high proportion of the atoms must have been de-excited before they eventually reacted with O_3. The formation of $O(^3P)$ atoms is consistent with the relatively slow consumption of O_3 that was observed once the flash was over, and these free atoms doubtless played a major role in relaxing $O_2^\dagger$ [201].

The flash photolysis of O_3 has been very thoroughly studied by Bair and his co-workers [199, 202–204]. Fitzsimmons and Bair [202] photolyzed mixtures of 1 torr O_3 + 16 torr Ar, which does not quench $O(^1D)$ as efficiently as N_2 [200], in a multipass absorption cell. Bands with $v'' \leq 23$ were "moderately strong" and the formation and decay of $O_2(13 \leq v \leq 19)$ were monitored photoelectrically. Despite the concentration of O_3 being lower than in McGrath and Norrish's experiments, $O_2^\dagger$ again relaxed rapidly, either by vibrational energy exchange with O_3 or "cold" O_2, vibrational–electronic energy transfer with O_2, or relaxation by free O atoms whose concentration built up as O_3 was destroyed. The variation of vibrational level populations was analyzed assuming that $O_2^\dagger$ relaxed by single quantum steps, and it was estimated that 0.14% of reactive events produced $O_2(v = 18)$. A reestimate based on Hebert and Nicholls' [205] more accurate absorption data approximately doubles this figure. If the levels are assumed to be populated according to a Boltzmann distribution, this concentration in

$v = 18$ corresponds to a temperature of $\sim$7000°K and $\langle W'_{vib} \rangle \sim -0.1\,\Delta H$ for reaction (96).[1]

Lipscomb, Norrish, and Thrush [195] observed $O_2(v \leq 8)$ on flashing NO_2. $O(^1D)$ atoms were unimportant since most of their experiments were performed with light filtered through sodaglass, so that the short-wavelength photodissociation did not occur. Relaxation of $O_2^\dagger$ was clearly rapid since the absorption followed the decay of the photoflash. Excitation of O_2 occurred in the reaction.

$$O(^3P) + NO_2 \rightarrow O_2 + NO, \qquad \Delta H_0^0 = -46.1 \text{ kcal/mole } (-1.99 \text{ eV}), \qquad (97)$$

which has a rate coefficient at 298°K = 6.0×10^{-13} cm³/molecule sec [206]. Two attempts have been made to determine the vibrational distribution from the reaction. Kane, McGarvey, and McGrath [207] estimated that $\langle W'_{\text{vib}} \rangle \sim -4 \times 10^{-3}\,\Delta H$. Bass and Garvin [208], on the other hand, concluded that O_2 was produced in $v = 2$ to $v = 6$ at approximately equal rates, which agrees with a rough estimate by Lipscomb et al., and that molecules in these levels account for about 15% of the total O_2 produced. It seems likely that the higher resolution in Bass and Garvin's experiments enabled them to distinguish NO bands which could be mistaken for O_2 bands with low values of v'', and their conclusion is probably correct.

Lipscomb, Norrish, and Thrush [195] also observed $O_2^\dagger$ when ClO_2 was flash photolyzed. ClO_2 possesses a number of strong absorption bands between 300 and 400 nm, which can yield only $O(^3P)$ atoms, and, on flashing, was completely destroyed within a few μsec by photolysis and the rapid [209] reaction

$$O(^3P) + ClO_2 \rightarrow O_2 + ClO, \qquad H_0^0 = -59.7 \text{ kcal/mole } (-2.59 \text{ eV}). \qquad (98)$$

Observations of $O_2^\dagger$ were limited to levels above $v = 4$. The concentrations in $v = 5$, 6, and 7 were roughly equal and significantly greater than that in $v = 8$.

The reaction

$$O(^3P) + CS_2 \rightarrow SO + CS, \qquad \Delta H_0^0 = -20 \text{ kcal/mole } (1.30 \text{ eV}) \qquad (99)$$

has been studied [210, 211], using the long-wavelength ($\lambda \geq 300$ nm) photolysis of NO_2 to generate $O(^3P)$ atoms. At 298°K, $k_{99} = 3.5 \times 10^{-12}$

[1] Bair and Fitzsimmons compared the observed concentration in $v = 18$ with its equilibrium concentration at 9800°K, which would be the vibrational temperature (again assuming a Boltzmann distribution) if the reaction energy was shared equally between the degrees of freedom of the products. However, their figure for $(N_{v=18}/N)_{T=9800°K}$ should read 5.3×10^{-3} and not 5.3×10^{-4}.

cm³/molecule-sec [210, 212, 213], and the products are almost certainly in their ground electronic state, because no CS excited states are accessible and the formation of $SO(^1\Sigma)$ and $SO(^1\Delta)$ is spin-forbidden.

Spectra of both CS and SO could be observed. Surprisingly, for it is the "old" bond in this reaction, the CS vibration was appreciably excited. Relaxation of CS† was quite slow [214, 215] and the initial distribution could be determined; it was roughly "Boltzmann" with T_{vib}-1775°K, corresponding to 9% of the reaction energy entering this mode. It was more difficult to measure the relative vibrational concentrations of SO, since the reaction

$$SO + NO_2 \rightarrow SO_2 + NO, \qquad \Delta H_0^0 = -59 \text{ kcal/mole } (-2.56 \text{ eV}) \quad (100)$$

is fast [216], and consequently the SO absorption was never strong. During the short period the SO bands could be observed, the vibrational distribution was constant, and it seems that reaction (100) proceeds at a rate independent of the vibrational state of SO and was fast enough to eliminate relaxation. No SO bands with $v'' \geq 5$ were detected despite the strongest bands being in a favorable region for their observation. This strongly indicates that less than 40% of the available energy enters the SO vibration, and an estimate based on the relative concentrations in $v = 1$ to $v = 4$ gave $\langle W'_{vib}(SO) \rangle \sim -0.21\ \Delta H$.[1]

Table 1.3

Miscellaneous Examples of Vibrational Excitation from Absorption Measurements Following Flash Photolysis

REACTION	REF.
$O(^1D) + HR \rightarrow OH\ (v \leq 2) + R, \quad (R = H, Cl, OH, NH_2, CH_3)$	217
$I(^2P_{1/2}) + Br_2 \rightarrow IBr\ (v \leq 3) + Br$	218
$X + O_3 \rightarrow XO\ (v \leq 5) + O_2 \quad (X = Cl, Br)$	219
$S + S_2Cl_2 \rightarrow S_2\ (v \leq 12) + SCl_2$	220
$S + S_2Cl$ (or SCl) $\rightarrow S_2\ (v \leq 15) + SCl$ (or Cl)	

The absorption spectra of vibrationally excited diatomic molecules have been observed in several other flash photolysis experiments. Table 1.3 lists the systems where it has been proposed that these species are produced by chemical reaction.

The insensitivity of spectrographic absorption measurements constitutes a serious disadvantage in flash photolysis experiments. In order to obtain a

[1] Figures in this paragraph have been adjusted to take account of new thermochemical data.

sufficiently intense background continuum in a single shot from the spectroscopic flash lamp, normally a spectrograph with only moderate resolution is used, and optical densities are measured at the unresolved band heads. As a result rather large concentrations of reagent or precursor must be used, and unwelcome effects such as fast relaxation, secondary reactions, and overlapping absorptions arise frequently. Resonance absorption techniques, which have been applied to the study of atomic reactions, would provide appreciably greater sensitivity, although it may be difficult to develop sufficiently intense resonance emission lamps.

Del Greco and Kaufman [221] used resonance absorption measurements in an attempt to observe the formation of $OH^\dagger$ in the reaction

$$H + NO_2 \rightarrow OH + NO, \qquad \Delta H_0^0 = -29.5 \text{ kcal/mole} \quad (-1.28 \text{ eV}) \qquad (101)$$

in a flow-discharge system. They failed to do so, and concluded that less than 2% of the OH produced in reaction was in levels above $v = 0$. This unusually low yield might result from the reaction taking place via a HONO complex that releases very little energy into the OH bond on fragmenting, although the reported yield seems much less than would be expected if the available energy was statistically shared among the degrees of freedom of the products.

Since it has been shown that in many exothermic reactions a large part of the energy liberated enters the vibration of the newly formed molecule, there has been considerable interest in developing vibrational lasers pumped by chemical reaction. Although a total inversion between vibrational states is not required [223], it is difficult to reach the threshold condition for lasing action if relaxation is faster than excitation by reaction. These effects can be reduced where a light flash (or pulsed discharge) initiates reaction, and stimulated emission has now been observed from a number of systems [224].

The large Einstein radiative coefficients [225] and the widely spaced vibration–rotation quantum states make $HF^\dagger$ peculiarly prone to stimulated emission, and a large proportion of the chemical lasers which have been reported operate on lines in the infrared bands of this molecule [224]. H-atom abstraction reactions by F and F-atom abstraction by H are both normally exothermic, and HF is quite generally produced in a vibrational distribution giving rise to oscillation. However, the systems are complex; frequently both types of reaction occur, and the details of the vibrational distribution resulting from chemical reaction are difficult to evaluate.

An ingenious approach to solving this problem was adopted by Parker and Pimentel [226]. Experiments were carried out on the reactions

$$F + H_2 \rightarrow HF + H, \qquad \Delta H_0^0 = -31.5 \text{ kcal/mole} \quad (-1.37 \text{ eV}), \qquad (102)$$

$$F + D_2 \rightarrow DF + D, \qquad \Delta H_0^0 = -31.3 \text{ kcal/mole} \quad (-1.36 \text{ eV}), \qquad (58)$$

$$F + CH_4 \rightarrow HF + CH_3, \qquad \Delta H_0^0 = -33.2 \text{ kcal/mole} \quad (-1.44 \text{ eV}) \qquad (103)$$

using flash photolysis of UF_6 to provide F atoms. The gain on any individual transition in the vibration–rotation spectrum depends—among other factors—on the rotational distributions in the upper and low vibrational levels and on the translational temperature. Parker and Pimentel kept the rotational distributions in equilibrium with the translational temperature by adding buffer gas and adjusted the ambient temperature of the system until lasing occurred simultaneously on two lines in the same band. The theoretical expressions for gain on these lines were then equated yielding N_v/N_{v-1}, the ratio of the concentrations in the upper and lower vibrational levels. One drawback to the method is that the "equal-gain temperature" is insensitive to N_v/N_{v-1} once this ratio becomes much greater than unity [224]. Parker and Pimentel's results are compared with those from infrared chemiluminescence experiments in Table 1.4.

Table 1.4

Comparison of Relative Rate Coefficients for Reactions Producing HF and DF

REACTION	$k_{v=2}/k_{v=1}$	$k_{v=3}/k_{v=2}$	TECHNIQUE	REF.
$F + H_2 \rightarrow HF + H$	5.5	—	Chemical laser	226
	3.2	0.48	Infrared emission	227, 228
	3.4	0.55	Infrared emission	229, 230
$F + D_2 \rightarrow DF + D$	—	1.6	Chemical laser	226
	—	~2.0	Infrared emission	228
$F + CH_4 \rightarrow HF + CH_3$	3.0	—	Chemical laser	
	3.0	0.23	Infrared emission	229

HCl is excited to laser action when mixtures of $Cl_2 + H_2$ (231, 232), $Cl_2 + HBr$ [233], or $Cl_2 + HI$ [234, 235] are flash photolyzed in a suitable laser cavity. In $Cl_2 + H_2$ mixtures, excitation occurs through the chain reaction

$$Cl_2 + h\nu \rightarrow 2Cl, \tag{104}$$

$$Cl + H_2 \rightarrow HCl + H, \quad \Delta H_0^0 = +0.9 \text{ kcal/mole} \quad (0.04 \text{ eV}), \tag{105}$$

$$H + Cl_2 \rightarrow HCl + Cl, \quad \Delta H_0^0 = -45.0 \text{ kcal/mole} \quad (-1.95 \text{ eV}), \tag{106}$$

whereas in the other two systems, $HCl^\dagger$ is formed when Cl atoms react with the hydrogen halide:

$$Cl + HBr \rightarrow HCl + Br, \quad \Delta H_0^0 = -15.6 \text{ kcal/mole} \quad (-0.68 \text{ eV}), \tag{107}$$

$$Cl + HI \rightarrow HCI + I, \quad \Delta H_0^0 = -31.9 \text{ kcal/mole} \quad (-1.38 \text{ eV}). \tag{108}$$

In none of these systems are absorbed ultraviolet photons converted to infrared laser photons efficiently, nor have detailed rate coefficients been obtained from these experiments.

The only nonhydride molecule to have been excited directly to chemical laser action is CO. When mixtures of $CS_2 + O_2$ are flash photolyzed [236, 237], or subjected to a pulsed electrical discharge [238], oscillation occurs on vibration–rotation lines in a wide range of CO fundamental bands. Infrared emission experiments [239, 240] have established that the reaction mechanism is

$$CS_2 + h\nu \rightarrow CS + S, \tag{109}$$

$$S + O_2 \rightarrow SO + O, \tag{110}$$

$$O + CS_2 \rightarrow SO + CS, \tag{99}$$

$$O + CS \rightarrow CO + S, \ \Delta H_0^0 = 85 \text{ kcal/mole (3.68 eV)} \tag{111}$$

Levels above about $v = 15$ are populated by vibration–vibration energy exchange and, because CS is formed vibrationally excited [241] in reaction (109).

Table 1.5

Comparison of Relative Rate Coefficients (k_v) for Reaction $O + CS \rightarrow CO_{(v)} + S$

v	(i)	(ii)	(iii)
4			
5		0	
6		0.05	
7	~0.06	0.17	
8	0.27	0.32	
9	0.61	0.41	
10	0.66	0.55	
11	0.80	0.65	~0.6
12	0.87	0.85	0.87
13	1.0	1.0	1.0
14	0.64	0.90	0.72
15	~0.2	0.58	0.28
16	0	0.32	0
17	0	0.18	0
18	0	0	0

Some recent experiments [242, 243] in the author's laboratory confirm these proposals. The formation of CO† on flash photolyzing CS_2 in the presence of O_2 and buffer gas has been observed by measuring the gain or absorption when single lines from a continuous-wave (cw) CO vibrational laser [244] were passed through the reaction cell. Detailed rate coefficients for reaction (111) were determined and are compared in Table 1.5 to those derived from an analysis of the infrared chemiluminescence from steady-state flow-discharge experiments [242, 245]. All results are quoted relative to $k_{13} = 1.0$. Results in column (i) are derived from steady-state infrared chemiluminescence experiments where CS is formed in the reaction $O + CS_2 \rightarrow SO + CS$ and therefore $300 < T_{vib} < 1775°K$ [210]. Data in column (ii) are from time-resolved experiments where CS is formed by photodissociation of CS_2 and therefore in a much more excited distribution. The larger values of k_{17} to k_{14} indicate that vibrational energy in the CS reagent is efficiently converted to vibrational excitation of the CO product. The results in column (iii) are also from time-resolved experiments, but N_2O was added to completely relax $CS^†$.

Several factors work against the efficient conversion of chemically produced vibrational excitation into infrared laser photons. Two related techniques can effect significant improvements. In the first, and most widely used, the excitation energy of the chemical reaction product is transferred by near-resonant vibration–vibration energy exchange to a particular mode of a triatomic molecule, usually the asymmetric stretching vibration of CO_2. This level is populated selectively and oscillation can then take place on the transition between this state (00^01) and a lower level (10^00) that is initially only thermally populated and that relaxes faster than the upper state. Chen, Stephenson, and Moore [246] have shown that the photon output can be increased by an order of magnitude if CO_2 is added to a $Cl_2 + HI$ flash photolysis laser, and a similar improvement has been achieved in a DF laser [247]. Cool and his co-workers have applied this technique to a number of chemical laser systems where the reactions produce excited hydrogen (or deuterium) halides [248–253], and they compared the output powers to those which can be obtained directly from the chemical product [254].

The second method has been applied to cw CO lasers; it relies on the fact that $CO(v = 0)$ and a number of other "vibrationally cold" gases preferentially deexcite lower vibrational levels of CO, where vibration–vibration energy exchange is closest to resonance [256, 257]. Consequently, controlled addition of these gases can make the populations in neighboring vibrational levels more suitable for laser action and can enhance the power output [258, 259].

Most of the chemical lasers mentioned above have also been operated continuously [224, 254]. A major problem is to supply and mix the reagent gases sufficiently quickly for the pumping reaction to keep pace with the relaxation processes—including the stimulated emission itself—which continuously tend to destroy the population inversion. The highest powers are generally attained by flowing the gases across the laser cavity at near sonic or supersonic speeds.

Two cw chemical lasers have been reported that operate without any external source of power to produce atomic reagents. Cool, Shirley, and Stephens [250–252] achieved oscillation on the $(00^01) \rightarrow (10^00)$ transition of CO_2. The mechanism was comprised of three important stages: (a) production of F atoms by the reaction $NO + F_2 \rightarrow NOF + F$; (b) reaction between F and D_2 to produce $DF^\dagger$; (c) vibrational–vibrational energy exchange from $DF^\dagger$ to CO_2 to produce CO_2 (00^01). Meinzer [259] was able to obtain stimulated emission from $DF^\dagger$ itself using F atoms generated in an H_2-F_2 flame.

Research into chemical lasers is new, fascinating, and in a period of tremendous growth. The object of most of the experiments up to the present time has been the improvement of output powers and efficiencies, but although little quantitative kinetic data have been derived so far from studies of laser systems, it seems unlikely that this will be so for very long.

Detailed rate constants for the production of chemical reaction products in specific internal quantum states have been determined most successfully from infrared spontaneous emission spectra. The experiments are basically quite straightforward. Flows of the reagent gases are mixed in a vessel which forms part of a discharge-flow system and which is equipped to collect as much of the infrared chemiluminescence as possible. The radiation is passed through a spectrometer and the spectrum is recorded. Photoconductive detectors have been used extensively. Their response is strongly dependent on wavelength: PbS (at 77°K) is sensitive at wavelengths $\lesssim 3.5\ \mu m$, and cooled PbSe and InSb cells can be used out to approximately $6\ \mu m$. The study of chemiluminescence in the instrumental infrared was pioneered by J. C. Polanyi, and has been brought to a sophisticated level in his laboratory. Recent improvements include the replacement of the conventional infrared spectrometer by a Fourier transform spectrometer [260].

Vibrationally excited diatomic molecules will only emit if they are polar, and most of the available results are for reactions which produce diatomic hydrides. Because of their unusually small reduced mass, these molecules have high frequency and very anharmonic vibrations, and their rotational levels are widely spaced. Consequently, their spectra can be resolved more easily than those of nonhydrides, where there are many more individual lines in the vibration–rotation spectrum. Furthermore, the molecular dynamics of these reactions are particularly interesting because of the special kinematic features that arise when an H atom is involved in a reactive collision and because these reactions provide the most stringent test of the assumption of classical behavior that is usually made in analyzing the dynamics. Both these features can be studied further by substituting D atoms for H atoms in these systems.

Relative state populations ($N_{v,j}$ or N_v) are derived from the observed spectrum in two stages. First, the spectrometer-detector unit must be calibrated with a standard blackbody source to allow for changes in sensitivity with wavelength. Then the corrected relative intensities are converted to the $N_{v,j}$ (or N_v) using values of the spontaneous emission coefficients. This procedure is quite simple when individual rotational lines can be resolved [101, 102]. Karl et al. have described a computational technique for analyzing the overlapped first overtone ($\Delta v = 1$) spectra of CO [261] and NO [262] when the rotational distribution is known to be equilibrated, and Hancock and Smith [256] have extended this method.

The $N_{v,j}$ or N_v may be quite different from the relative rates ($R_{v,j}$ or R_v) at which the reaction populates the levels, as a result of molecules relaxing during the time they remain within the volume observed by the spectrometer. There are three ways to take account of this, and these methods will be described briefly in order of increasing sophistication (and precision).

The first method is to attempt to *allow for* relaxation. This can be done most accurately if molecules relax predominantly by spontaneous emission, and effects due to collisions with other gas molecules and the vessel walls, and the removal of excited species by pumping them out of the observation zone are minimized [242, 245]. To ensure that surface deactivation is unimportant, a large excess of inert gas can be added to reduce diffusion, as long as it does not cause significant vibrational deexcitation. This maintains the rotational distribution in equilibrium with the translational temperature, and there is no chance of discovering anything about rotational excitation in the reaction being investigated. The reaction rate into a particular vibrational level is calculated as the difference between the total rate of loss of molecules from that level by spontaneous emission, gas-phase collisions, diffusion-controlled surface loss, and pumping, and the rate of population from higher levels by "cascade" processes [263, 245].

The second technique (method I of ref. 264) is to *measure* the relaxation. Here the infrared emission is observed from different points downstream from where the reagents are mixed in a fast-flow system. Even at the shortest times, rotational relaxation is complete, but the relaxation of the vibrational states can be followed and the distributions extrapolated back to yield a set of R_v. Pacey and Polanyi [265] have found small, but significant, differences between the R_v derived from a simple extrapolation and those determined using an analysis that allowed for the concurrent processes of reaction, diffusion, flow, radiation, and deactivation. Using a large-capacity sorption pump [265], Jonathan et al. have carried out fast-flow experiments [229] at higher total pressures where the effects of diffusion should be reduced.

The most successful technique for obtaining information regarding the energy distribution in the products of simple reactions has proved to be Polanyi's method II [264]. Experiments are performed at very low background pressures ($\sim 10^{-4}$ torr) where the mean free path of molecules are of the same order of magnitude as the dimensions of the reaction vessel. The walls of this vessel are cooled (normally to 77°K, but sometimes to lower temperatures), and it appears that excited molecules reaching the cold surface are adsorbed sufficiently strongly for their energy to be removed completely before they reenter the gas phase. The observed rotational distributions are markedly non-Boltzmann, and because rotational relaxation is much more likely in collisions than vibrational relaxation, it is quite safe to assume that the latter is completely *arrested.* There are apparently sufficient collisions in the zone where the two streams of reagent gases cross to equilibrate the translational energy and therefore to ensure that product molecules in different (v, j) states (and thus initially with different translational energies) take the same time, on the average, to reach the walls. As a result the N_v correspond to the R_v, and the $N_{v,j}$ correspond very nearly to the $R_{v,j}$.

Polanyi and his colleagues have studied the infrared chemiluminescence from vibrationally excited hydrogen halides produced in three families of reaction of the type A + BC → AB + C: (a) hydrogen atom plus diatomic halogen, (b) halogen atom plus hydrogen halide, and (c) fluorine atom plus molecular hydrogen. The method of "arrested relaxation," described in the previous paragraph, has now been applied in each case, and the results have been summarized in a single paper [92]. As was mentioned earlier, reactions in all of these classes have formed the basis of chemical lasers.

Early work on the reaction

$$H + Cl_2 \rightarrow HCl + Cl, \qquad \Delta H_0^0 = -45.0 \text{ kcal/mole} \quad (-1.95 \text{ eV}) \qquad (106)$$

was reported in 1960 [101, 102]. Two years later, Charters and Polanyi [263] derived values of R_v allowing for relaxation according to various models. They concluded that the rates into levels with $v \geq 3$ rose steeply as v decreased ($R_{v=3} \approx R_{v=5} \times 30$) and that these values were quite insensitive to the choice of relaxation mechanism. However, below $v = 3$ the rates became increasingly uncertain.

Within the last few years, the problems arising from relaxation and the resultant difficulty of estimating values of R_v for the lower vibrational levels have been resolved by carrying out experiments using both the "arrested-relaxation" and the "measured-relaxation" techniques. The R_v that have been derived in these two sets of experiments are plotted in Figure 1.11 against f_v, the fraction of available energy in the HCl vibration. The relative rates

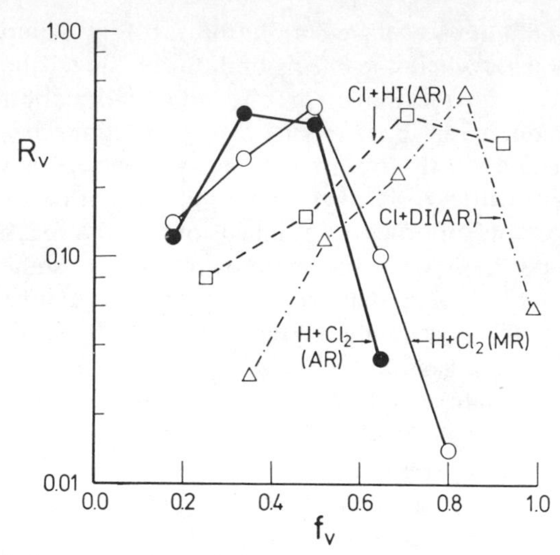

Figure 1.11 Relative rates of reaction into individual vibrational levels by arrested relaxation (AR) and measured relaxation (MR) techniques: $H + Cl_2 \rightarrow HCl + Cl$ from ref. 265; $Cl + HI \rightarrow HCl + I$ and $Cl + DI \rightarrow DCl + I$ from ref. 267. R_v are plotted against f_v, the fraction of available energy which goes into product vibration.

into $v \geq 3$ are similar to those determined in the early experiments, but at $v = 2$ or $v = 3$ the distribution turns through a maximum and R_v decreases with decreasing v. The mean vibrational energy of HCl produced in reaction (106) is approximately 45% of the total available energy, that is, $Q \approx -\Delta H + E_a$. A rather similar vibrational distribution is found in HBr produced in the reaction

$$H + Br_2 \rightarrow HBr + Br, \qquad \Delta H_0^0 = -41.2 \text{ kcal/mole} \quad (-1.79 \text{ eV}), \qquad (112)$$

the fractional vibrational energy yield increases to 55%.

The distribution of HCl produced in the reaction

$$Cl + HI \rightarrow HCl + I, \qquad \Delta H_0^0 = -31.9 \text{ kcal/mole} \quad (-1.38 \text{ eV}), \qquad (108)$$

has also been determined in experiments of the "arrested-relaxation" type [264, 267]. A substantial proportion of the HCl is formed in the highest accessible level ($v = 4$), and a greater proportion of the total energy (~65%) is channeled into the HCl vibration than in the $H + Cl_2$ reaction.

Experiments on the reactions between $D + Cl_2$ and $Cl + DI$ reveal that the plots of R_v against f_v are scarcely changed by isotopic substitution, indicating that a classical analysis of the dynamics is useful even when the actual quantized vibrational levels are as widely spaced as in HCl. The greater degree of vibrational excitation from Cl + HI compared to that from H + Cl_2 may arise solely from kinematic factors, and the potential hypersurfaces for the two reactions could be very similar. Because of the small reduced mass for H + Cl_2 collisions, reactants approach unusually rapidly, so that the HCl bond distance is reached before the repulsion between the heavy Cl atoms can cause much separation. On the other hand, with Cl + HI, the reactants approach more slowly while the repulsion between H and I pushes the light H atom towards the approaching Cl atom. The broader vibrational distribution may be the result of energy being abstracted from HCl during secondary encounters before the newly formed HCl molecule and I atom have properly separated [267].

The rotational level distributions from reactions (106) and (108) have also been determined in the low-pressure experiments [264] and those for HCl(v = 2) are shown in Figure 1.12. The (unrelaxed) high J distribution corresponds to that produced by reaction and is clearly non-Boltzmann. Hydrides are particularly well suited to absorbing large amounts of energy as rotation because of their low moments of inertia. However, the orbital angular momentum associated with H + Cl_2 collisions is unusually low, because the collisional reduced mass is small, and this apparently causes the lower "cutoff" in the rotational distribution from this reaction. For Cl + HI,

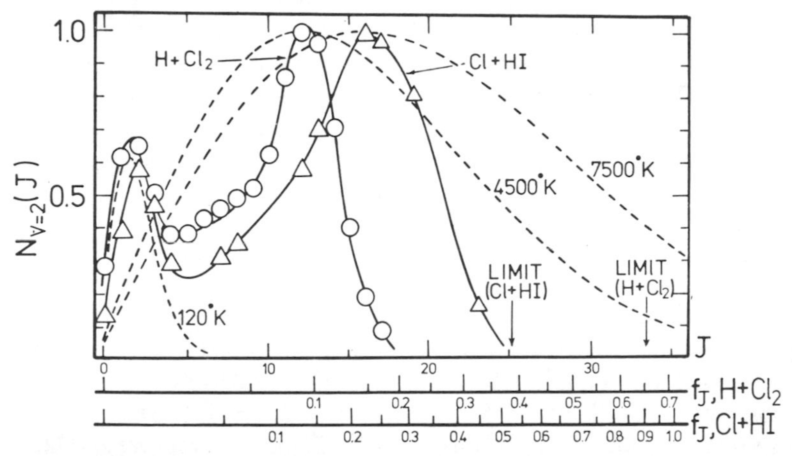

Figure 1.12 Rotational level distributions in HCl (v = 2) from the H + Cl_2 and Cl + HI reactions [264]. The abscissa records both J, the rotational quantum number, and F_J, the fraction of available energy present as rotation. The arrows indicate the limits of excitation determined by the fact that E (v = 2) + $E(J)$cannot exceed Q, the total available energy. The broken lines indicate "best fit" Boltzmann distributions and show that the majority of the rotators are in a highly non-Boltzmann distribution. The subsidiary peaks at low J conform to a low-temperature Boltzmann distribution.

rotational excitation occurs right up to the thermochemical limit; averaging the rotational and vibrational distribution for this reaction gives $\langle W' \rangle \sim 0.8Q$.

The "arrested-relaxation" method has been applied [227, 228] to the reactions of $F + H_2$ and $F + D_2$, and the "measured-relaxation" technique to $F + H_2$ [229, 230]. These values of $R_{v,j}$ and R_v are particularly important since they can be compared with the results of molecular beam and chemical laser experiments (see Table 1.4), and the agreement is satisfactory. The HF vibrational and rotational degrees of freedom absorb approximately 67% and 6% of the total energy and once again there is a marked parallelism between the results for a reaction and its isotopic analog. Preliminary measurements on other reactions producing $HF^\dagger$ have been reported by Jonathan et al. [230].

Setser and co-workers [268, 269] have compared the infrared emission from the reactions of H (and D) atoms with SCl_2, S_2Cl_2, $SOCl_2$, SO_2Cl_2, and Cl_2O with that from the H (or D) + Cl_2 reaction. The reactions with SCl_2 and S_2Cl_2, which have similar exothermicities to the $H + Cl_2$ reaction, were studied most thoroughly, and the conclusion was reached that the HCl vibrational distribution from all three reactions was similar and the additional degrees of freedom made little difference to the energy partitioning. However, more recently Heydtmann and Polanyi [270] have observed the chemiluminescence from $H + SCl_2$ at very low pressures, that is, under conditions of arrested relaxation, and found that although the average amount of energy entering the HCl vibration is the same as from $H + Cl_2$, the details of the energy distributions are quite different. The broader vibrational distribution and the double-peaked rotational distributions from $H + SCl_2$ suggest that two distinct kinds of reaction dynamics are involved.

Other reactions of the type $A + BCD \rightarrow AB + CD$ have been studied in experiments in Polanyi's laboratory. There is special interest in the reaction

$$H + O_3 \rightarrow OH + O_2, \qquad \Delta H_0^0 = -77.1 \text{ kcal/mole} \quad (-3.34 \text{ eV}) \quad (91)$$

since it is the source in the night sky airglow of the Meinel bands, which have been identified as high-overtone transitions of $OH^\dagger$. McKinley et al. [271] first observed these bands in the laboratory, and in 1960 the results of a detailed investigation of the spectrum out to 3 μm were reported [272, 273]. The vibrational distribution could be described by a temperature of 9250°K, but it was evident that considerable relaxation was taking place. A set of R_v was calculated allowing for collisional and radiative deexcitation in single quantum transitions. The resultant distribution possessed maxima at $v = 4$ and $v = 9$ and suggested that a substantial proportion of the reaction might yield $O_2(b\,^1\Sigma_g^+)$, since the excitation energy of this species is roughly equivalent to five OH vibrational quanta. However, because the total pressure and concentration of O_2 is reduced, the proportion of the emission from high v levels increases [187], and once the conditions for arrested relaxation are reached [260] the observed N_v distribution rises monotonically from $v = 4$ to $v = 9$, indicating that reaction is most probable into the highest level accessible ($v = 9$) and becomes progressively less likely into lower levels. When O_2 is present, efficient vibrational–electronic energy transfer,

$$OH(v) + O_2(X\,^3\Sigma_g^-) \rightarrow OH\,(v = 5) + O_2(b\,^1\Sigma_g^+), \qquad (113)$$

may take place giving rise to the double-peaked N_v distribution and the weak emission of the O_2 ($b\,^1\Sigma_g^+ - X\,^3\Sigma_g^-$) bands detected by Cawthorn and McKinley [186].

Emission from $OH^\dagger(v \leq 3)$ has also been observed when atomic hydrogen was mixed with O_2 [274]. The reaction

$$H + HO_2 \rightarrow 2\,OH, \qquad \Delta H_0^0 = -39\text{ kcal/mole} \quad (-1.34\text{ eV}) \qquad (114)$$

was thought to cause this excitation, after HO_2 had been formed in the three-body association,

$$H + O_2 + M \rightarrow HO_2 + M. \qquad (115)$$

The total pressure in these experiments was 4 torr, and it will not be easy to extend these measurements to very low pressures, since OH is excited in a secondary reaction and reaction (115) is a third-order process.

The reaction

$$H + NOCl \rightarrow HCl + NO, \qquad \Delta H_0^0 = -65.0\text{ kcal/mole} \quad (-2.82\text{ eV}) \qquad (116)$$

was studied some time ago at total pressures of 1 torr [275] and 10^{-2} torr [276], and more recently at 10^{-3} torr [277]. As both products are infrared active, conclusions can be reached about the vibrational yields in HCl and NO, although the interpretation of the experiments is complicated because of rapid transfer of vibrational energy from HCl to NOCl. Anlauf [277] estimates that the HCl vibration absorbs at least one-third of the total energy but that less than 10% of NO are formed in levels with $v > 0$. The observation that the "old-bond" vibration is not highly excited agrees with experimental data on other four-atom reactions of this type [196, 210], and indicates that the reaction does not proceed via an intermediate complex with energy eventually being shared statistically between the degrees of freedom of the products.

Hancock et al. [239, 240, 245] have measured emission from $CO^\dagger$ when atomic oxygen and CS_2 are mixed, and this constitutes the only detailed study of infrared chemiluminescence from a nonhydride diatomic molecule. CO is excited in a secondary reaction

$$O + CS_2 \rightarrow SO + CS, \qquad (99)$$

$$O + CS \rightarrow CO + S, \qquad \Delta H_0^0 = -85 \text{ kcal/mole} \quad (-3.68 \text{ eV}). \qquad (111)$$

Experiments have been performed at high total pressures (1–12 torr) and relaxation allowed for by the usual steady-state analysis. The R_v distribution, shown in Table 1.5, reaches a maximum close to the highest accessible level, and about 85% of the total energy enters the CO vibration. Reaction presumably occurs across the lowest triplet surface and not via the OCS ground state, which would require a breakdown of the spin conservation rule and would correspond to complex formation in the O + CS reaction. If the triplet state is neither strongly bonding nor strongly antibonding relative to separated CO + S, then it will be ideal for forming CO highly excited.

C. Four-Center, Displacement, and Elimination Reactions

There are no examples of *four-center reactions* where the extent of excitation of the products has been well characterized. Reactions (88) and (90) have been suggested as sources of electronically excited CH and OH in the O + C_2H_2 reaction. The reactions

$$PH + PH \rightarrow P_2 + H_2, \qquad \Delta H_0^0 \sim -70 \text{ kcal/mole} \quad (-3.04 \text{ eV}) \qquad (117)$$

and

$$CN + O_2 \rightarrow CO + NO, \qquad \Delta H = -114 \text{ kcal/mole} \quad (-4.94 \text{ eV}) \qquad (118)$$

may produce $P_2^\dagger$ and $NO^\dagger$, which has been observed in absorption in flash photolysis systems [278, 279], but alternative mechanisms have not been eliminated.

Clough and Thrush [280, 281] have studied the infrared emission from $N_2O^\dagger$ formed in the *displacement reaction*

$$N + NO_2 \rightarrow N_2O + O, \qquad \Delta H_0^0 = -41.7 \text{ kcal/mole} \quad (-1.81 \text{ eV}). \qquad (119)$$

N_v were estimated by comparing the band intensities with the continuous chemiluminescence from the NO + O_3 reaction (see Section III.B.2). The ν_1 and ν_3 modes were each excited separately with a number of quanta in ν_2 but were rarely excited simultaneously. The vibrational energy yield was estimated to be approximately 37% of the total energy.

Recently measurements on chemical lasers [282–285] and of infrared chemiluminescence [286] have provided information about the partitioning of energy in unimolecular *elimination reactions*, following production of energetically excited alkyl halides by radical recombination, energy transfer, or photochemical action, for example,

$$CH_3 + CF_3 \rightarrow CH_3{-}CF_3^{\dagger} \rightarrow CH_2{-}CHF + HF^{\dagger}, \tag{120}$$

$$CH_2 = CF_2 + Hg(^3P_1) \rightarrow CH \equiv CF + HF^{\dagger} + Hg(^1S_0), \tag{121}$$

$$CHCl = CHCl + h\nu \rightarrow CH \equiv CCl + HCl^{\dagger}. \tag{122}$$

Reaction (120) has been investigated by both techniques. From their observation that the $P_1(4)$ line was the first to reach the threshold for laser oscillation, Berry and Pimentel [282] were able to calculate that $1.11 < (N_0 : N_1) < 1.72$. Clough, Polanyi, and Taguchi [286] estimated that the relative rate coefficients for reaction into $v = 1$, 2, 3, and 4 are 1.0, 0.43, 0.13, and 0.033. A number of other elimination reactions have been shown to give laser emission [283–285], and the equal-gain temperature method (see Section III.B.3) has been applied to determine relative detailed rate coefficients. All these lasers operated on partial inversions, indicating that reaction occurs fastest into $v = 0$ and that the vibrational energy yield in the newly formed diatomic is lower than those generally found in atom-exchange reactions.

Infrared emission has been observed [287] from $CO^{\dagger}$ formed in mixtures of O atoms with C_2H_2. Their excitation was attributed to the reaction

$$O + CH_2 \rightarrow CO + 2H, \qquad \Delta H_0^0 = -76.8 \text{ kcal/mole} \quad (-3.33 \text{ eV}), \tag{123}$$

but the relative R_v were not determined, and it is not known whether reaction occurs via an $H_2CO^{\dagger}$ complex, which simultaneously or successively loses H atoms, or in direct encounters.

D. Vibrationally Excited Molecules as Chain Carriers

Schiff and co-workers [288–293] have used mass-spectrometric sampling in kinetic studies of a number of simple reactions in low-pressure flow systems. There was evidence indicating that energy chains were important, and they estimated the degree of excitation of a chain carrier by observing the extent of the decomposition which it subsequently induced.

The method was first applied [288–290] to the reaction

$$N + NO \rightarrow N_2 + O, \qquad \Delta H_0^0 = -75.0 \text{ kcal/mole} \quad (-3.25 \text{ eV}). \tag{74}$$

When O_3 was added to the products downstream from their formation it was consumed by reaction with vibrationally excited N_2,

$$N_2^{\dagger} + O_3 \rightarrow N_2 + O_2 + O, \tag{124}$$

as well as by

$$O + O_3 \rightarrow 2\, O_2. \tag{95}$$

Phillips and Schiff obtained evidence that O_2 produced in reaction (95) could not carry the chain, and concluded that either 75% of the N_2 possessed sufficient energy [>24 kcal/mole (1.04 eV)] to decompose one molecule of O_3, or that less N_2 possessed this amount of energy but some could decompose more than one O_3 molecule. They also determined the average yield of vibrational energy with an isothermal calorimetric probe and obtained $\langle W'_{vib} \rangle = 21\ (\pm 5)$ kcal/mole (0.91 eV).

The reaction

$$H + O_3 \rightarrow OH + O_2, \qquad \Delta H_0^0 = -77.1 \text{ kcal/mole} \quad (-3.34 \text{ eV}) \qquad (91)$$

was also studied by Phillips and Schiff [291]. For each H atom originally present, 3.1 (±0.2) O_3 molecules were decomposed. This yield was reduced if N_2O or H_2 was added because these species deactivated $OH^{\dagger}$ molecules and prevented them from dissociating a further molecule of O_3. 80% of the OH was thought to be sufficiently energetic to act as a chain carrier. OH produced in the less exothermic reaction,

$$H + NO_2 \rightarrow OH + NO, \qquad \Delta H_0^0 = -29.5 \text{ kcal/mole} \quad (-1.28 \text{ eV}), \qquad (101)$$

could not act as a chain carrier.

The estimates of the concentrations of vibrationally excited chain carrier were based on the assumption that the transfer of energy to O_3 is 100% effective. If the transfer efficiency is less, then the estimates of $\langle W'_{vib} \rangle$ must be increased. For $H + O_3$, this would bring the results into agreement with the latest spectroscopic measurements [187, 260], and the similarity of the results of Phillips and Schiff for $OH^{\dagger} + O_3$ and $N_2^{\ddagger} + O_3$ suggests that reaction (74) may give a higher vibrational energy yield than that indicated by the calorimetric probe measurements.

IV. MONTE CARLO CALCULATIONS OF REACTIVE DYNAMICS

In this article, particular attention has been paid to the results of laboratory experiments that have provided information about the molecular dynamics of electronically adiabatic atom-exchange reactions. For a reaction like $A + BC \rightarrow AB + C$, where the possibility of crossing between different electronic states can safely be ignored, a theoretical study comprises three distinct stages [20]: (a) the determination of the potential-energy hypersurface, (b) the solution of the equations that describe the motion of A, B, and C under the influence of V, and (c) the averaging of the results of these trajectory calculations so that they reflect the distribution of collisional parameters—for example, of v and b—that are present in a particular experiment.

In order to handle the equations of motion, whether they be classical or quantum mechanical, it is desirable for the potential energy to be expressed as a relatively simple function of some "distance" coordinates, for example, the three internuclear distances, r_{AB}, r_{BC}, and r_{CA}. Consequently, despite the increasing accuracy of *ab initio* calculations, at least for the simplest systems, trajectory calculations have been carried out using semiempirical expressions for V. These generally contain one or more adjustable parameters that allow the nature of the surface to be altered either to match known data such as the experimental activation energy or points on the *ab initio* surface, or to test the effect of any change on the dynamics of the trajectories. The assumption of the Born–Oppenheimer approximation, which is implied when V and the motions of A, B, and C are calculated separately, is reasonable and certainly leads to smaller errors than other approximations. It is less easy to justify not using quantum dynamics to describe the behavior of A, B, and C, although the De Broglie wavelengths associated with the momenta do seem sufficiently small for classical mechanics to provide a reasonably accurate description of the motion. The results of some quantum-dynamical calculations for linear systems of three particles have now been compared with those from the corresponding classical trajectories [294–299], and the conclusions from this work will be mentioned later. In comparison to these first two stages, the averaging is not difficult—although it may be mathematically lengthy—since the collisional parameters are distributed according to well-known statistical laws.

The results of a complete Monte Carlo trajectory study are those properties of a reaction and its chemical products that have already been mentioned at various points in this chapter:

a. the Arrhenius rate parameters A and E_a;
b. the threshold energy, E_0, and the variation of the reactive cross section (σ_R) with relative collision energy;
c. the differential reactive cross section, that is, $\sigma_R(\theta)$;
d. the vibrational and rotational energy distributions of AB;
e. the orbital and rotational angular momenta of the products, $\mathbf{L}'$ and $\mathbf{J}'$.

All this information can be obtained if a sufficiently large sample of trajectories is chosen, whereas, up to the present time, laboratory experiments on any single reaction have not provided all these data. The computer "experiments" have two main aims. The first is to discover how the results are altered by making selective changes either in the nature of the reaction—for example, in the exothermicity or the form of V—or in the starting conditions—for example, by suppressing the natural distribution of impact parameters.

The second, and more ambitious, objective is to find at least the general form of the potential function for a particular reaction, by successively adjusting V, until the observed and simulated results coincide. In essence, this is the same approach as that which is adopted to interpret elastic scattering measurements in terms of the central potential acting between two atoms (Section I.A.1), but the problem is now considerably more complex.

Wall, Hiller, and Mazur [300, 301] first used a computer to integrate the classical motion equations for a system of three atoms, and in the 1960s the technique was developed by Blais and Bunker [48, 302–306], and by Karplus [19, 20, 72, 307–311] and Polanyi [71, 73, 74, 267, 312, 313] and their co-workers. Recently calculations have been performed on systems simulating abstraction reactions involving more than three atoms [314] and four-center reactions involving four atoms, that is, $AB + CD \rightarrow AC + BD$ [315–317]. Here we present first a general survey of the Monte Carlo calculations of classical trajectories and then a brief review of some of the results of these calculations. Emphasis is placed on data for reactions that have been studied experimentally and have been mentioned earlier in this chapter.

A. Equations of Motion

The classical Hamiltonian for motion of three particles in three dimensions is the sum of the potential energy (V), which can be expressed in terms of nine position coordinates (q_i) and the kinetic energy (T), involving the conjugate momenta (p_i):

$$H = T + V$$
$$= (\tfrac{1}{2}M_A)\sum_{i=1}^{3} p_i^2 + (\tfrac{1}{2}M_B)\sum_{i=4}^{6} p_i^2 + (\tfrac{1}{2}M_C)\sum_{i=7}^{9} p_i^2 + V(q_1, q_2, \ldots, q_9). \quad (125)$$

The motion is then described by 18 simultaneous differential equations:

$$\frac{\partial H}{\partial p_i} = \frac{\partial T}{\partial p_i} = \dot{q}_i$$

and

$$-\frac{\partial H}{\partial q_i} = -\frac{\partial V}{\partial q_i} = \dot{p}_i.$$

This number can be reduced to 12 by transforming the coordinates [20] so that terms in the momentum components of the center of mass are separated. As they must remain constant, these terms can be ignored. If account were also taken of the constancy of total angular momentum, the number of equations could be decreased still further, but the algebra required is formidable and conservation of angular momentum throughout the calculated trajectory can be used to check the accuracy of the computation.

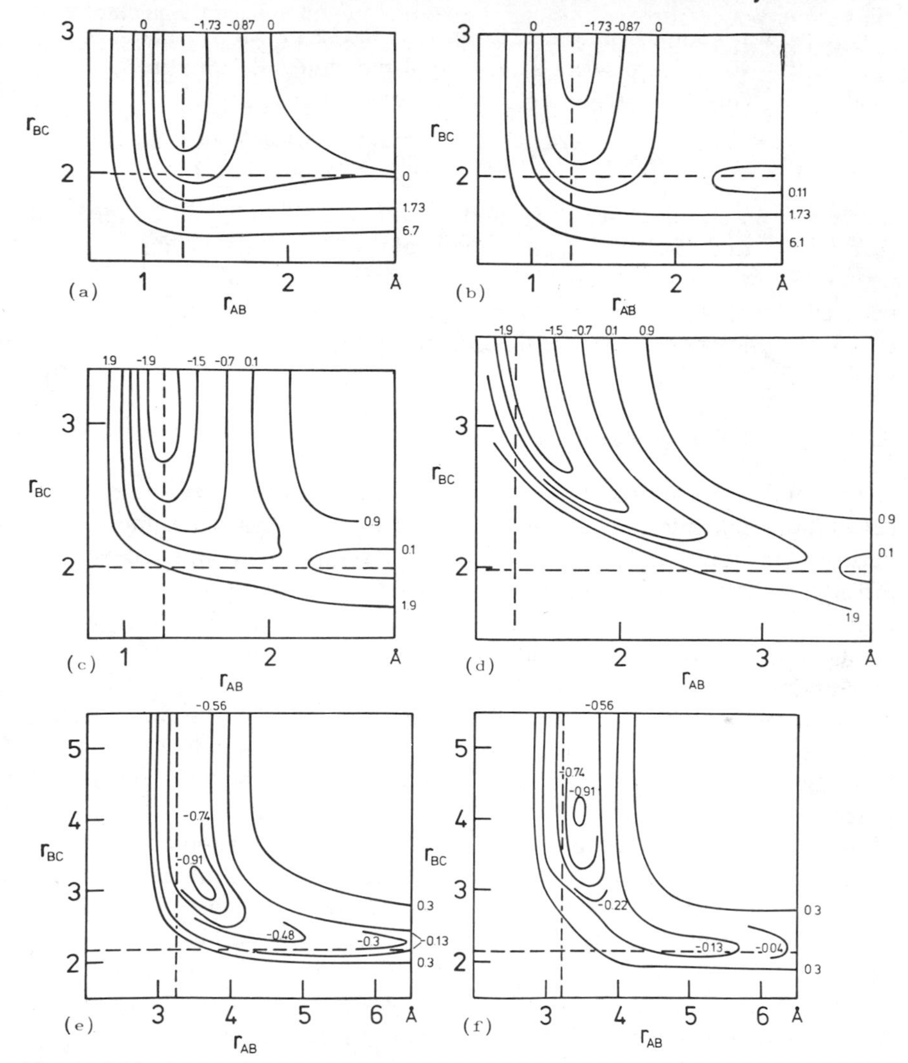

Figure 1.13 Examples of potential surfaces for collinear reactions of the type A + BC → AB + C. Energies are expressed in eV relative to separated A + BC at zero energy, and the dotted lines indicate the equilibrium internuclear distances of BC and AB.

(a) and (b) are LEPS surfaces [see equation (126)] for H + Cl_2 → HCl + Cl with $S^2 = 0.2$ and 0.0, respectively (313). The attractive character of the surface decreases as S^2 is reduced:

$$A_{\perp} = 72\% \text{ in (a) and } 33\% \text{ in (b).}$$

(c) and (d) are two of several surfaces used by Light and co-workers [298, 299] in classical and quantum-dynamical calculations on the H + Cl_2 reaction. The

A naive way to integrate the 12 equations in the transformed coordinates (Q_i and P_i) would be to select a very small time interval δt and to calculate small increments in the Q_i and P_i, using the equations

$$\delta Q_i = \frac{\partial H}{\partial P_i}\,\delta t \quad \text{and} \quad -\delta P_i = \frac{\partial H}{\partial Q_i}\,\delta t.$$

Starting from an initial set of Q_i and P_i, this procedure could be repeated successively, and with a sufficiently small value of δt it might yield reasonable results [3]. In practice, faster and more accurate numerical techniques have been employed with δt in the range $3\text{–}6 \times 10^{-16}$ sec. The accuracy can be tested by observing whether the final P_i values are significantly altered if δt is made smaller.

B. Potential-Energy Functions

The semiempirical potential functions which have been used in trajectory calculations fall into two categories: those based on pairwise interactions between the atoms, and more general interaction potentials which deal more directly with the three-atom reaction surface. In discussing their characteristics it is customary to concentrate on the surface for the collinear configuration since it is possible to represent this function by a contour diagram of the kind shown in Figure 1.13.

The commonest form of general potential function is one in which the Morse potential of BC, corresponding to V with A "infinitely" separated from BC, is converted smoothly along the reaction path to the Morse potential for AB [318, 305, 319, 298]. The inclusion of adjustable parameters allows systematic changes to be made in the position of the potential barrier (or col) on the surface, the curvature of the path passing from reactants with minimum energy, and the steepness of the potential along this reaction path.

For their calculations, Polanyi and co-workers and Karplus, Porter, and Sharma (20) have employed versions of the LEPS (London-Eyring-Polanyi-Sato) potential, which has some connection with formal theory since it is based on the London equation for a system of three atoms [320]:

$$V = Q_{\mathrm{AB}} + Q_{\mathrm{BC}} + Q_{\mathrm{CA}} - (J_{\mathrm{AB}}^2 + J_{\mathrm{BC}}^2 + J_{\mathrm{CA}}^2 - J_{\mathrm{AB}}J_{\mathrm{BC}} - J_{\mathrm{BC}}J_{\mathrm{CA}} - J_{\mathrm{CA}}J_{\mathrm{AB}})^{1/2}, \quad (126)$$

---

reaction path of minimum energy on these surfaces was less sharply curved than on the LEPS surfaces and in terms of the rectilinear classification, the surfaces were all markedly repulsive.

(e) and (f) are surfaces used by Blais and Bunker [48] in calculations on K + $CH_3I \rightarrow KI + CH_3$ (CH_3 was treated as a single particle). The long-range attraction evident in (e) is diminished in (f) and led to a reduction in the calculated vibrational excitation of KI.

where the Q and J are two-center Coulomb and exchange integrals. Sato [321, 322] proposed that the factor $1/(1 + S^2)$, where S is formally an overlap integral, should be retained as an adjustable parameter in this expression and that the individual Q and J for each diatomic pair could be evaluated by equating $(Q + J)/(1 + S^2)$ to the Morse potential, that is,

$$D_e \{\exp [-2\beta(r - r_e)] - 2 \exp [-\beta(r - r_e)]\},$$

and $(Q - J)/(1 - S^2)$ to an "anti-Morse" potential of the form

$$(D_e/2) \{\exp [-2\beta(r - r_e)] + 2 \exp [-\beta(r - r_e)]\}.$$

Although the resultant three-atom potential can only be justified empirically, since S^2 must be both a function of r and possess different values for different pairs of atoms, the Sato formulation yields a surface free from "basins" whatever value of S^2 is chosen.

Porter and Karplus [19] constructed a LEPS potential for $H + H_2$ including overlap and three-center terms in order to evaluate the energies of nonlinear configurations more realistically. Kuntz et al. [313] employed a modified LEPS function in a detailed investigation of metathetical reactions involving three atoms. Three adjustable parameters were included, instead of just S^2. This provided a more flexible potential, and it was possible to vary the nature of the potential surface quite considerably. Other potentials based on pairwise interactions have been used for calculations where AB is ionic [72–74, 306].

Mok and Polanyi's work [323] indicates a correlation between the potential barrier height and its position on the energy surface. For exothermic reactions, the barrier is generally in the "entry valley," that is, where $r_{BC} \approx r_{e,BC}$ and $r_{AB} > r_{e,AB}$, and it moves to progressively greater values of r_{AB} as its height is lowered. Conversely, for endothermic reactions, the barrier moves to successively later positions in the "exit valley" as its height increases.

C. Averaging Procedures

In order to obtain properly averaged results, either collision parameters for each trajectory must be selected by Monte Carlo methods or, when starting values are systematically chosen, the final results must be integrated over complete statistical distributions. The purpose of a Monte Carlo selection technique is to ensure that the distributions of each parameter within a sample of trajectories approach the true statistical distributions as the size of the sample grows. Some examples of how this can be done for different types of distribution function will be described below. Before starting the integration, it is generally necessary to transform the selected values of the collisional parameters, such as v and b, into the P_i and Q_i in which the total energy and motion equations are expressed.

The most straightforward application of the Monte Carlo technique arises where the probability of a parameter taking a particular value is constant over its entire range. For example, the initial angles between the BC internuclear axis, and a line joining A to the center of mass of BC and in the plane containing A, B, and C, are distributed in this way, and a value can be selected by multiplying π by a random number between -1 and 1.

A second case can be illustrated with reference to the selection of b, the impact parameter. The probability (P_b) of a collision having an impact parameter between b and $b + db$ is proportional to $2\pi b \cdot db$, so that a procedure is required that will produce a distribution appropriate to this weighting factor. To do this, $F(b)$ must be found such that b is uniformly distributed in F space, that is, so that $(dP_b/db) = (dF/db)(dP_b/dF)$, F must be chosen so that

$$F = \alpha \int_0^b dP_b$$

Therefore F is proportional to b^2, which means that b^2 must have a uniform distribution between zero and a value b^2_{max}, which is chosen so that reaction is very improbable for $b > b_{max}$. Accordingly the impact parameters are selected, using $b = S(Rb^2_{max})^{1/2}$, where R is a freshly generated random number between 0 and 1, and S is randomly $+1$ or -1.

For some variables, for example, the relative collision velocity, the cumulative distribution function does not have closed form, and then a third Monte Carlo method must be adopted. Here, another random number R is used to provide a value of v, but a decision on whether to accept this value is made on the outcome of a "game of chance" against a second random number. The probability that a value is accepted is proportional to the probability density in the statistical distribution at that value. The procedure is repeated until the game of chance is won, and the successful value of v is then incorporated into the set of starting parameters.

The amount of Monte Carlo selection that has been employed in different studies has varied. For example, Blais and Bunker [48, 305] used a complete Monte Carlo procedure in their studies of the $K + CH_3I$ reaction, although the distribution of one or more parameters could be suppressed, allowing them to observe how particular results depended on different features of the collisions. On the other hand, Karplus et al. [20] adopted a rather different approach in their investigation of the $H + H_2$ system. A batch of trajectories was calculated with particular values of v, b and vibrational and rotational energies of H_2. The remainder of the variables were chosen by Monte Carlo methods. The vibrational and rotational energies corresponded to individual rotational states in the zero-point vibrational level. By averaging the results over the b distribution, curves of $\sigma_R(v)$ were obtained for each vibration–rotation energy. The method was particularly appropriate to this reaction, because the H_2 quantum states are widely spaced and few are significantly populated.

D. Results

1. General

The effect on reactive collisions of changing the atomic masses and the nature of the potential function were examined some time ago in extensive two-dimensional calculations. Blais and Bunker [303] performed Monte Carlo calculations for various combination of light (2-amu), medium (16-amu), and heavy (128-amu) atoms. Kuntz et al. [313] employed modified LEPS hypersurfaces based on the $H + Cl_2$ system. The conditions at the start of each trajectory were chosen systematically, rather than by Monte Carlo procedures, and trajectories were computed for different combinations of light (1-amu) and heavy (80-amu) atoms.

The Toronto group [313] was concerned chiefly with finding useful definitions of the terms *attractive*, *repulsive*, and *mixed*, which can be used to describe either the character of the potential function or the way in which the energy of the reaction is liberated as the system passes from A + BC to AB + C. In terms of V, the *attractive* contribution could be defined as the energy released on following a rectilinear path across the collinear reaction surface from $r_{AB} = \infty$ and $r_{BC} = r_{e,BC}$ to the point where $r_{AB} = r_{e,AB}$ and $r_{BC} = r_{e,BC}(\mathscr{A}_\perp)$; the energy released on going from this point to products would then be classified as the *repulsive* character ($\mathscr{R}_\perp$). However, it seemed more useful to consider the actual path followed by a collision, and this depends on the relative atomic masses and the relative collision energy, as well as the energy hypersurface. For the purposes of the definition, collinear collisions were considered that possessed either the mean thermal energy or, in the case of a surface with an appreciable barrier, just sufficient energy to surmount this barrier. The *attractive energy release* ($\mathscr{A}$) was then defined as that released before r_{BC} exceeded the classical amplitude of the BC vibration, with mean energy equivalent to 300°K (BC was not given initial "zero-point energy"). The energy liberated after r_{AB} first reached $r_{e,AB}$ was termed *repulsive energy release* ($\mathscr{R}$) and the remainder was classified as *mixed energy release* ($\mathscr{M}$). These quantities were defined as percentages of the total energy available ($Q \approx -\Delta H + E_a$) and the sum of ($\mathscr{A} + \mathscr{M}$) showed a distinct correlation with the percentage of Q channeled into the AB vibrational mode.

Interesting features arose when one or more light atoms were present. For example, when B or C was light and A heavy, repulsion between B and C caused them to separate rapidly, leading to a large mixed energy release and vibrational excitation of AB, even on surfaces that, in terms of the rectilinear classification, were decidedly repulsive. The other extreme was where B and C were both heavy but A was light. Since B and C separated sluggishly but A approached rapidly, there was very little mixed energy release. When $\mathscr{A}_{\perp}$ was small, vibrational excitation of AB was reduced, because mixed energy release became repulsive release when A was light instead of heavy. This "*light-atom anomaly*" is sufficient to explain why HCl is more excited in the Cl + HI reaction than in H + Cl_2. On more attractive surfaces, there was less effect, since then $\mathscr{M}$ was converted to $\mathscr{A}$ rather than $\mathscr{R}$ when the attacking atom was light.

What seemed to be a similar "*energy anomaly*" had been reported by Blais and Bunker [303]. They also found that there was a reduction in the vibrational energy yield when A was light but their potential possessed considerable long-range attraction. However, this effect did not survive a modification of their potential and the extension of the calculations to three dimensions [305]. Nevertheless, the exact conditions under which the *light-atom anomaly* appears are not yet entirely clear. Calculations by Bunker and Parr [319] on a "wide-corner" potential, that is, one on which the reaction path on the collinear surface "cuts the corner" [210], showed no indication of the light-atom anomaly, and it seems that "still more study will be required before the interrelations between the way different surfaces display this phenomenon can be completely understood" [319].

One subsidiary characteristic that may influence the degree of vibrational excitation and reduce the correlation between $\langle W'_{vib} \rangle$ and $(\mathscr{A} + \mathscr{M})$ is the degree to which the reaction path on the collinear surface deviates from the rectilinear (sometimes described in terms of the curvature of the reaction path). That the degree of vibrational excitation was reduced—and the distribution broadened—was first observed by Blais and Bunker [305], and was noted also by Smith [210]. It seems that this factor is particularly important on surfaces which are, at least moderately, repulsive. Certainly, its most dramatic effect has been in the 10 quantum and classical calculations of Light and co-workers [298, 299], where the surfaces—in terms of the rectilinear classification—were all markedly repulsive. The degree of vibrational excitation was low, but rose as the curvature of the reaction path was increased, until the percentage of $\langle W'_{vib} \rangle \approx 10\%$ on the surface shown in Figure 1.13.

The observed yield of vibrational energy does appear to provide a useful indication of how much energy is released before AB and C start to separate. A further complication is that on highly attractive surfaces "secondary encounters" may occur. This arises because the atom C does not receive sufficient initial impetus to leave the vicinity of the vibrating AB molecule.

Consequently B (or possibly A) rebounds and interacts again with C. These secondary encounters may be of two types: either the interaction is of a repulsive kind, in which case C may abstract energy from AB broadening the vibrational energy distribution as it departs [267]; or A and C may interact attractively (for example, if B and C are the same), and AC may become the molecular product rather than AB. The probability of secondary encounters will be greatest when AB is ionic or has a small reduced mass, since then the vibrational amplitude will be large; the effect will be less likely where the reduced mass of the products is small.

The results of 2*D* and 3*D* calculations have been compared by Karplus and Raff [308] and by Blais and Bunker [305]. The yields of vibrational and translational energy were very similar, but the 2*D* calculations of the rotational energy and the angular scattering were less satisfactory. These conclusions imply that the overall rotational excitation is small, a result which has always been found experimentally and is only not found in the calculations when the potential is attractive at very long range, the total reaction cross section is very large, and the dynamics approach the extreme of "spectator stripping" (see below). Collisions where the energy is shared almost exclusively between product rotation and relative translation and where total angular momentum is conserved only because the final orbital and rotational momentum vectors have opposite directions are not common.

Conservation of angular momentum may be less restrictive in reactions like A + BCD → AB + CD, and it is conceivable that a large part of the reaction energy may cause rotational excitation. Product repulsion between AB and CD could cause the molecules to spin away from one another with their rotational momentum vectors approximately opposed and therefore little overall final angular momentum. This possibility has not been tested, as the only 3*D* calculation on a four-particle system [309] employed an attractive hypersurface with little energy released as repulsion. There appear to be three ways in which CD might be excited in reactions of this kind. First, excitation may occur in secondary encounters with highly excited AB [309]. This is most likely on attractive hypersurfaces, where AB and CD possess little relative translational energy. If the hypersurface became still more attractive, a complex would tend to form and in the limit energy would be shared statistically when this dissociated to AB and CD. This mode of exciting CD is not possible on hypersurfaces which are even moderately repulsive. CD would be excited vibrationally if its equilibrium distance were appreciably different from that in BCD; this is rarely so. A special case of excitation may be where CD is extended during the course of the reactive encounter and then must subsequently "relax" as the products separate. The final possibility is that CD is excited by the repulsive forces that operate as AB and CD separate. For appreciable excitation, the impulsiveness of this force may be more important than the total amount of energy released in this way [210, 324].

Recently, Monte Carlo trajectory studies have been performed [325, 326] on systems with appreciable energy barriers, with a view to discovering whether excitation of particular degrees of freedom in the reagents promotes reaction. For three-atom reactions, if the barrier lies in the exit valley, vibrational (rather than translational or rotational) excitation can be used most effectively for surmounting the barrier. Conversely, if the barrier is in the entry valley, it is most easily surmounted if energy is located in the relative translation of the products rather than in vibration. For appreciably endothermic reactions, the barrier is very likely to be in the exit valley [323], and the conclusion that vibrational excitation will considerably assist the occurrence of such reactions is supported by calculations based on the applications of microscopic reversibility to the detailed rate coefficients for several exothermic reactions [327, 227]. It appears that similar criteria apply to four-center reactions of the AB + CD → AC + BD type [317], and the effect of vibrational excitation on the rate of such reactions has been investigated [316, 317].

2. Reactions Forming Ionic Bond

In several Monte Carlo trajectory studies, attempts have been made to discover the nature of the potential that operates in reactions of alkali metal atoms which produce alkali halides. The results of the calculations are compared to the particularly detailed information provided by crossed-molecular-beam experiments. The reaction

$$K + CH_3I \rightarrow KI + CH_3, \qquad \Delta H_0^0 = -24 \text{ kcal/mole} \quad (-1.04 \text{ eV}) \qquad (50)$$

was chosen as the subject of the first extensive calculations [48]. Unfortunately, the experimental conclusion [43] that reaction (50) contained very nearly all of the released energy as internal excitation has had to be revised (45; see Section II.C.3), and this must be borne in mind when the results of the calculations are assessed.

Blais and Bunker [48] calculated the motion of three particles—the methyl group was regarded as a single species of 15 amu—in $2D$, using the potential expression

$$\begin{aligned} V = {} & D_{AB}\{1 - \exp[-\beta_{AB}(r_{AB} - r_{e,AB})]\}^2 \\ & + D_{BC}\{1 - \exp[-\beta_{BC}(r_{BC} - r_{e,BC})]\}^2 \\ & + D_{BC}[1 - \tanh(ar_{AB} + c)] \exp[-\beta_{BC}(r_{BC} - r_{e,BC})] \\ & + D \exp\{-\gamma(r_{AB}^2 + r_{BC}^2 - 2r_{AB}r_{BC}\cos\alpha)^{1/2} - r_0\}. \end{aligned} \qquad (127)$$

The first two terms are Morse potentials for K + I and CH_3 + I, the third term attenuates the attraction between CH_3 and I as K approaches I, and the final term introduces repulsion between CH_3 and K. The surface for the collinear system is shown in Figure 1.14. Most of the collisions were found to be simple and direct. The c.m. angular distribution is trapezoidal, with rather more KI scattered backward than forward. Blais and Bunker pointed out that when this distribution is transformed, there should be pronounced peaks at lab angles corresponding to $\theta = 180°$ and $\theta = 0°$ where $\sin \theta = 0$. They suggested that the weaker forward peak might be obscured by elastic scattering of K atoms close to $\Theta = 0°$. The energy released during the reaction was found largely in the KI vibration, which agreed with the experimental conclusions available at that time.

The attractive potential of Figure 1.13 could be changed to one in which the energy is released mainly as repulsion between the products by altering a and c in expression [127]. A much greater proportion of KI molecules was then scattered into the backward hemisphere and the average vibrational excitation of KI was reduced. In fact, these results are in much better agreement with the latest experimental data [45] than those obtained on the attractive surface, which Blais and Bunker chose to call the "normal" potential.

Blais and Bunker [303] extended their calculations to $3D$, using a more general form of potential, and Karplus and Raff [308, 310] have also performed $3D$ calculations on this system. They pointed out that the total reaction cross section, calculated using the attractive Blais and Bunker potential, exceeded the experimental value by an order of magnitude, and they proposed modifications to the function in order to reduce the long-range attraction. $3D$ trajectories on their three modified hypersurfaces gave similar results: the reaction cross section was brought into line with the experimental value; KI was scattered predominantly backward with maximum intensity at, or near, $\theta = 180°$, but most of the energy was still released into the KI vibration. Thus their results could be compared directly and precisely with the experimental data. Karplus and Raff transformed their results into the lab frame of reference. The correlation was encouraging. The agreement was much better than with lab distributions corresponding either to uniform c.m. scattering, on the one hand, or limiting backward scattering, that is, all the KI recoiling along $\theta = 180°$, on the other.

More recently, Blais [306], Godfrey and Karplus [72], and Polanyi and co-workers [73, 74] have concentrated on the alkali (M) + diatomic halogen (X_2) reactions. In contrast to the M + alkyl halide (RX) reactions, the experimental results show that MX scatters predominantly forward and absorbs a large proportion of the total energy as vibrational energy. Various versions of the "electron-jump" model have been tried in the attempts to simulate the dynamics. It is generally assumed that an electron leaves M while the M-X_2 distance is still large (~5–10 Å), and that thereafter the dynamics is dominated by Coulombic attraction between M^+ and an atomic halogen ion (X^-), with little or no interaction with the atomic product (X'). The results of all three investigations indicate that sharp forward scattering and high internal excitation of MX can be reproduced as the stripping limit is approached, that is, as the electron-jump distance is increased and the product repulsion is reduced to zero. However, in these collisions MX is excited rotationally [73, 74] much more than is found experimentally [34–36]. Polanyi's group found [73, 74] that the details of angular scattering, rotational distribution, and vibrational distribution could all be matched only if "*charge migration*" was permitted, that is, if the negative charge was located on X or X' depending on which particle was at that instant closer to M^+. With charge migration allowed, close agreement with the experimental result was found when the electron jump occurred at $r_{MX} = 5$ Å and the interaction between the halogen atom and ion corresponded to 0.5 kcal/mole (0.02 eV) being released as repulsion. It seems that the dynamics of the M + RX reactions differ on account of (a) the impossibility of charge migration, (b) the smaller distance at which the Coulombic $M^+ + X^-$ attraction becomes important, and (c) the greater amount of energy that is released as repulsion between R and X^-. It also appears likely that a number of unusual features combine to produce the sharp forward peaking observed in M + X_2 reactions and that this kind of behavior is probably unusual.

V. SUPPLEMENT

This article has suffered—even more than most of its kind—from delays in publication. It was first prepared in 1969, and included references up to the end of the previous year. In 1971 it was redrafted. The supplement that now follows was added early in 1974 in anticipation of impending publication and in an attempt to draw attention to some of the most significant developments in a burgeoning field of research. In this supplement no attempt has been made to mention all the work that has been reported during the last two or three years, and apologies are offered to those whose work has been overlooked.

In 1969 it was still possible to consider separately the results of experiments carried out with crossed molecular beams and spectroscopic measurements of the product states of "bulb" reactions. Not only had there been few attempts to fuse together these two techniques, but also the lists of reactions which, up to that time, had been studied by the two methods were almost mutually exclusive. One result of progress in the 1970s is that this clear distinction has now been removed. The operation of crossed-beam apparatus which utilizes mass spectrometric (and therefore universal) detection [31] has had much to do with this. Furthermore, *spectroscopic* measurements of various kinds are now being made on the products of chemical reactions in crossed-beam experiments: electronic chemiluminescence has been observed in several cases; as foreseen earlier (see ref. 4) internal state distributions have been measured with a molecular beam electric resonance spectrometer [330–332]; finally, and perhaps of greatest significance for the future, Zare and his co-workers have reported the use of a tunable dye laser to excite resonance fluorescence from reaction products in their ground electronic state and a range of vibrational (and rotational) levels [333, 334].

Progress with bulk spectroscopic measurements may have been less spectacular, but has been no less real. The "arrested relaxation" version of the infrared chemiluminescence technique (see Section III.B.3) has continued to provide vibrational–rotational state distributions of unparalleled excellence. Furthermore, Polanyi's group has begun to explore how the state distributions of excited products vary as the initial collision energy or vibrational states of the reactants are modified [335–337]. Also, Pimentel's belief that measurements on the output of chemical lasers could be used—under carefully controlled and selected experimental conditions—to determine the vibrational distributions of laser-active, chemically excited molecules has been amply justified by experiments in his own laboratory [226, 284, 338–341] and by Berry [342, 343]. The interest in lasers in general, and chemical lasers in particular, has served to stimulate work on bimolecular reactions that produce electronically excited species, in the hope that a chemical laser operating on an electronic transition in the visible or ultraviolet might be discovered. Quantitative determinations of relative cross sections or rate constants for reactions leading to excited and unexcited products have been reported from several laboratories, notably those of Zare [344, 345], Palmer [346–349], and Broida [350, 351].

Finally, in this catalogue of recent achievements, mention must be made of the measurements, by a variety of techniques, of the energy partitioning in reactions that involve long-lived, "association–dissociation" complexes. Examples are provided by the reactions of F atoms with unsaturated hydrocarbons, such as

$$F + C_4H_8 \rightarrow (C_4H_8F)^{\dagger} \rightarrow C_4H_7F + H \text{ or } C_3H_5F + CH_3, \qquad (128)$$

that have been studied in crossed-beam experiments by the Chicago group led by Lee [352–355]. These experiments, which constitute the microscopic equivalent of the chemical activation experiments [356] that have been used for some years to test theories of unimolecular reaction rates, are likely to have a significant impact on such theories over the next few years. Already a lively discussion is taking place as to what can be inferred from the observations of nonstatistical energy sharing in such systems [357].

These, then, are the topics that have been chosen for very brief review in the remainder of this supplement. Despite the overlap and merging of molecular beam studies with other types of measurement, it seems appropriate to arrange the material under headings similar to those used in the main part of this article.

A. Reactive Scattering

In attempting this hasty and brief review of recent reactive scattering measurements, the recent publication of the *Proceedings of the Faraday Discussion of the Chemical Society on Molecular Beam Scattering* (**55,** 1973) has been very helpful. This meeting attracted most of the leading experimentalists and theoreticians in this field, and several of them presented papers that summarized some of the most recent work in their laboratories. The stimulating lectures of Herschbach [358], reviewing reactive scattering experiments and their interpretation, and Polanyi [359], summarizing the proceedings of the meeting as a whole, are particularly valuable. Kinsey [360] has also recently reviewed molecular beam studies of chemical reactions.

The flowering of molecular beam chemistry has been delightfully illustrated by Herschbach [361]. He distinguishes 15 families of reaction of the atom + diatomic molecule type (103 reactions in all) that have been studied. Table 1.6 is reproduced from his paper and lists these classes of reaction; to complete this inventory it is necessary to add the reactions of (a) metal atoms with organic and inorganic polyatomic molecules, (b) methyl radicals

Table 1.6
Atom + Diatomic Molecule Reactions studied by Reactive Scattering [358]

$H + H_2 \rightarrow H_2 + H$ (1)[a]	$H + HX \rightarrow HX + H$ (3)	$H + X_2 \rightarrow HX + X$ (7)
	$X + H_2 \rightarrow HX + H$ (2)	$X + HX \rightarrow HX + X$ (2)
$H + M_2 \rightarrow MH + M$ (3, a)	$H + MX \rightarrow HX + M$ (7, a)	
	$M + HX \rightarrow MX + H$ (7, b)	
$M + M_2 \rightarrow M_2 + M$ (3, a)	$M + MX \rightarrow MX + M$ (26, c)	$M + X_2 \rightarrow MX + X$ (22, b)
	$X + M_2 \rightarrow MX + M$ (5, a)	$M + O_2 \rightarrow MO + O$ (2, d)
		$X + X_2 \rightarrow X_2 + X$ (8)
		$O + X_2 \rightarrow XO + X$ (5)

[a] M and X represent metal and halogen atoms, respectively; where more than one such atom appears in a reaction they are not distinguished, nor are H and D atoms. The figures in parentheses indicate the number of reactions of that type that have been investigated; (a) done only with M an alkali atom; (b) done with M either an alkali or alkaline earth atom; (c) done with M either an alkali or thallium atom; (d) done only with M an alkaline earth atom.

with halogens, (c) halogen atoms with alkyl halides, and (d) halogen atoms with unsaturated organic molecules, which are reactions of the association–dissociation type alluded to above. At the present time, however, there remains a fairly wide variation in the degree of sophistication in different experiments, and on the amount of detailed information that is available concerning the collisional dynamics of different reactions.

The expansion in the scope of reactive scattering measurements has largely resulted from rapid progress in the technology of these experiments. The most important advance—described in Section II.A—has been the improvement in mass spectrometric detection which has ended the old reliance on surface ionization and carried reactive scattering measurements firmly out of the "alkali-only" era. The acquired capability of universal detection has naturally stimulated research on beam sources of various atomic and polyatomic free radical species. As well as experiments using beams of alkali and alkaline earth metal atoms, beams of alkali dimer molecules, halogen atoms from F to I, H or D atoms, O atoms and methyl radicals have all now been used in studies of reactive scattering. This improvement in variety of the radical beam has occurred in parallel with developments in seeded supersonic nozzle molecular beams which provide [358] high intensity (typically $\sim 10^{18}$ molecules/sr-sec) narrow velocity spreads (often $<10\%$), high and adjustable translational energy (up to ~ 5 10 eV for heavy molecules), and very low internal temperatures (often $<100°K$). These improvements allow the reactive cross section to be probed as a function of collision energy [362], and they permit the investigation of reactions with appreciable threshold energies. A final development is the increasingly standard measurement of the products' velocity distribution, most usually by time-of-flight techniques. This means that there is no longer any need to rely on simple, and occasionally over-optimistic, arguments based on the most probable Newton diagram for collisions between reactants, to extract the most probable recoil energy for the separating products (see Section II.B.2). As a result, the conclusions from molecular beam experiments regarding the partitioning of energy between relative translational motion (i.e., E') and internal degrees of freedom (W') are much more reliable.

Molecular beam experiments perhaps remain most restricted in their ability (or inability) to yield the distributions of product molecules among the various available internal energy states. Where electronically excited species are produced, the low pressure will ensure that no relaxation will occur before emission and the chemiluminescent spectrum then directly reflects the vibrational–rotational distribution produced chemically within the emitting state. Several chemiluminescent reactions have now been studied in beams, but it is not easy to estimate the ratio of the cross sections for production of excited- and ground-state product. Furthermore, collisions leading to these different states must occur over different potential hypersurfaces or involve nonadiabatic transitions. Certainly, the vibrational–rotational distribution in the chemiluminescent state can provide no guide to the dynamics of what is probably the major process, that leading to electronically unexcited products. Indeed, for the reactions,

$$Ba(^1S) + Cl_2(X^1\Sigma_g^+) \rightarrow BaCl(X^2\Sigma) + Cl(^2P) \qquad (129a)$$

and

$$Ba(^1S) + Cl_2(X^1\Sigma_g^+) \rightarrow BaCl^*(C^2\Pi) + Cl(^2P), \qquad (129b)$$

drastically different angular distributions have been reported [363] and indicate quite different collision dynamics for the two processes.

Two particularly interesting chemiluminescent reactions have been discovered using beams of alkali dimer molecules (M_2). These species are formed in high yield in an alkali metal nozzle beam source and they can be separated from the remaining atoms by means of an inhomogeneous magnetic field. This technique has made it possible to study reactions of these diatomic species by molecular beam methods for the first time [364] and as a result it has been demonstrated conclusively that reactions of the type

$$M_2 + X \rightarrow MX + M^* \qquad (130)$$

occur *directly* and not via the two-step energy transfer process suggested by M. Polanyi from an analysis of his early diffusion flame data (see Sections I and II.C.1). Although emission from the ^2P resonant state of the atomic product is strongest, in the 9 reactions that have been studied so far atoms are produced in all the electronically excited states that are energetically accessible, including some that can only be reached by utilizing a portion of the initial collision energy. These conclusions are supported by the results of single beam + scattering gas experiments [365] and of recent diffusion flame measurements [366]. Struve et al. [364] go so far as to suggest that the cross sections for the channels leading to excited and unexcited atoms are comparable, which should be encouraging to those searching for reactions that might pump visible chemical lasers.

The reactions of alkali dimers with halogen molecules also exhibit a multiplicity of reaction pathways with production of electronically excited atoms, electronically excited molecules, and ions all apparently possible [364, 367–370]. In these reactions, however, the dominant pathway, with a reactive cross section $\geq$150 Å^2, leads to products in their electronic ground states:

$$M_2 + X_2 \rightarrow M + MX + X. \qquad (131)$$

Both M and MX are scattered strongly forward relative to the incident M_2 beam, and it appears that these reactions occur by an electron jump mechanism similar to that for the $M + X_2$ series of reactions (see Section II.C.1).

This electron transfer is likely to occur when M_2 and X_2 are still far apart (~7–8 Å) with subsequent dissociation of X_2^-, and this appears to be the reason that the most exoergic reaction, resulting in formation of two MX molecules, does not occur to any appreciable extent. The cross sections for formation of M* and MX* in this class of reaction are roughly two orders of magnitude below that for reaction (131). This appears to be also true of the process leading to chemionization which requires the occurrence of a second electron jump.

Where the products of a beams reaction are formed in their electronic ground states and their relative velocity is analyzed, then application of equation (11) yields an approximate distribution for the total energy going into vibration (V') plus rotation (R'). Many more reactions have been studied using velocity analysis since the main part of this article was written. Two spectroscopic techniques have been used to explore the final distributions of V' and R' directly. In two laboratories, partial V', R' distributions have been obtained by passing the reactively scattered molecular product through a molecular beam electric resonance spectrometer [330–332]. These beautiful experiments are unfortunately limited by the fact that the Stark splitting of rotational states decreases rapidly with rotational quantum number and measurements have consequently been limited to a few low rotational levels of several vibrational states. Since, in general, the rotational and vibrational product state distributions may be strongly coupled, there is no reason to suppose that $N_{v'J'}/\sum_{J'=0}^{\infty} N_{v'J'}$ is the same for all values of v', and therefore the observed distribution may not correspond to the overall (summed over J') vibrational distribution.

For the reaction,

$$Cs + SF_6 \rightarrow CsF + SF_5, \tag{132}$$

the observed $N_{v',J'=1-4}$ distributions are Boltzmann ($T_{v'} \sim 1200°K$) [330, 331] and suggest that only about 5% of the available energy enters the CsF vibration. For this reaction, "conventional" beam measurements of angular and velocity distributions [371] show that reaction proceeds via a long-lived collision complex, and in this case, the observed distribution may correspond closely to the overall distribution; certainly the observed yield of vibrational energy is in accord with statistical sharing of the available energy, most of which excites modes in the SF_5 fragment because of its large number of vibrational degrees of freedom. The Li + SF_6 reaction has also been studied by this technique [332] with intriguing results.

The LiF distribution among levels $v' = 0$–3, $J' = 1$ indicates that vibrational excitation of this product is only half of what would be expected on the basis of statistical sharing. A study of the angular scattering [372], but without velocity analysis, indicates a direct reaction in this case with little energy being released as relative translation. One possible explanation is that a major product channel is to LiF + SF_4 + F rather than Li + SF_5 but if this is not so then one appears to have an example of a direct reaction in which energy is preferentially channeled into the product containing the "old" bonds.

The second spectroscopic method is due to Zare and his co-workers [333–334]. In their experiments, a beam of Ba atoms from an effusion source ($1000°K < T < 1100°K$) was either passed into a scattering chamber containing hydrogen halide at low pressure or crossed a beam of hydrogen halide molecules from a nozzle. The wavelength emitted from a repetitively pulsed dye laser that was focused into the reaction zone was continuously varied through the range where the barium halide product absorbs, and the intensity of the resonance fluorescence was recorded as a function of the excitation wavelength. By measuring the relative intensity of bands within this spectrum, it was possible to derive the vibrational distributions of the barium halides in their ground electronic states, and from the width of the bands the degree of rotational excitation could also be estimated. The degree of vibrational excitation was in all cases rather low; along the sequence Ba + HF, HCl/DCl, HBr, HI the values of $\bar{f}_{v'}$ are 0.12, 0.28/0.29, 0.36, 0.18. In view of the position of the light H atom in this reaction, one might expect a rather high degree of mixed energy release and consequent vibrational excitation (see Section IV.D.1). The most likely explanation of the low yields actually observed would appear to be that very little energy is released until the BaX distance has almost reached its equilibrium value. Repulsive energy released subsequently would be channeled very inefficiently into BaX vibration because of the light mass of the departing H atom. The reaction hypersurfaces might well have this character if there are activation barriers on them, which seems at least possible in view of relatively small values of the potential energy release. The experimental method employed in these measurements is limited only by the requirement that the molecular product must have a discrete spectrum with absorption in the wavelength range where tunable lasers are available. We can, with confidence, expect to hear much more about experiments of this kind.

Next, we briefly consider some reactions where measurements of the energy partitioning have been made by both beam [373, 374] and infrared chemiluminescence [228, 265, 376] experiments, namely the family

$$(D) + XY \rightarrow HX\,(DX) + Y \qquad (133)$$

where X and Y represent the same or different halogen atoms (it is easier to detect D atoms than H atoms mass spectrometrically, hence the preference of experimental beamists for the deuterated versions of these reactions). With both beams (D and XY) at ~350°K, McDonald et al. [374] found that the reaction yield was undetectable with $XY = Cl_2$, just detectable with Br_2, and large with I_2, in agreement with normal kinetic data [375] concerning the activation energies for these reactions. However, with a beam of D atoms from a furnace source operated at ~2800°K, angular and product velocity distributions could be determined for all three of these reactions as well as for D + IBr and D + ICl. The results of these experiments are, in a strict sense, not directly comparable to those from measurements of infrared chemiluminescence, because of the difference in temperature, and because only $D + Cl_2$ has been studied by the latter technique, other results pertain to the H atom reactions. However, there is evidence that $\bar{f}_{T'}$ will not be altered much by either of these changes; consequently, the fairly close agreement shown in Table 1.7 is highly satisfactory.

Table 1.7

Comparison of Experimental Data on Energy Partitioning in the Reactions of H or D Atoms with Halogen Molecules

REACTION	$\bar{Q}^a$	$\bar{f}_{V'}$	$\bar{f}_{R'}$	$\bar{f}_{T'}$
(a) From IR chemiluminescence [228, 265, 376]				
$H + Cl_2 \rightarrow HCl + Cl$	48.4	0.40	(— 0.60 —)	
		0.39	0.07	(0.54)
$D + Cl_2 \rightarrow DCl + Cl$	49.6	0.39	0.10	(0.51)
$H + Br_2 \rightarrow HBr + Br$	43.6	0.55	0.04	(0.41)
(b) From reactive scattering [374]				
$D + Cl_2 \rightarrow DCl + Cl$	55	(— 0.56 —)		0.44
$D + Br_2 \rightarrow DBr + Br$	51	(— 0.69 —)		0.31
$D + I_2 \rightarrow DI + I$	44	(— 0.72 —)		0.28
DBr + I	54	(— 0.72 —)		0.28
D + IBr				
DI + Br	38	(— 0.68 —)		0.32
D + ICl → DI + Cl	30	(— 0.76 —)		0.24

[a] The mean energy appearing as product vibration (V'), rotation (R'), and relative translation (T') is expressed as a fraction of $\bar{Q}$, the average energy available for distribution. For the thermal, infrared chemiluminescence experiments, $\bar{Q}$ is put equal to $\Delta D_0 + E_{act} + \frac{5}{2}RT$, whereas for the beams experiments, $\bar{Q}$ is put equal to ΔD_0 plus the mean collision energy (the deuterium halide issued from supersonic nozzle and the D atoms from a source at ~2800°K). The numbers enclosed in brackets are not obtained directly but by subtracting the measured values of $\bar{f}$ from unity.

It is likely that the potential hypersurfaces for these reactions are in large part *repulsive* (see Section IV.D.1), and Herschbach [358, 374, 377] has compared their dynamics to those of photodissociation of the halogen molecules. This interesting approach is based on two main assumptions: First, the H/D atom is so light relative to X and Y that its influence on the reaction dynamics is rather small; second, when the H/D atom approaches XY, three-atom molecular orbitals form, and the singly occupied 2σ orbital, which is the highest populated orbital, is similar to the orbital to which an electron is promoted by absorption in the long-wavelength continuous spectrum of XY. The observed distribution of recoil velocities is, for the lighter halogens, remarkably close to that calculated for photodissociation from the known absorption spectrum allowing for Franck–Condon effects. For heavy halogen reactants the agreement becomes less valid but this is consistent with a change in character of the 2σ three-atom orbital, in which the antibonding node shifts its position from midway between the Cl atoms in H—Cl—Cl to a location close to the central atom in H—I—I.

These simple but illuminating ideas have also been applied [358] to the alkali atom (M) + alkyl halide (RX) reactions (see Section II.C.3), for which a wealth of experimental data are now available [378]. The molecular orbital picture is now more complicated but the electron passed to RX in the "electron jump" enters a strongly antibonding orbital with a node between the X atom and the adjacent C atom. A qualitative description can be obtained by "shrinking" the alkyl group to a united atom limit (i.e., $CH_3 \rightarrow F$) when the parallels between these reactions and those of the H/D + XY type become clear. However, the dynamics are somewhat more complicated than before, because the mass of the reactant atom is no longer much smaller than that of the transferred atom, and because R is polyatomic. In the impulsive limit it is straightforward to apply the conservation equations and calculate the proportions of the repulsive energy that will enter relative translational energy and vibrational energy of MX and of R (R is usually treated as a "pseudo-diatomic"). The agreement between the experimental recoil velocity distributions and those calculated, even with this very simple limiting approach, is encouraging, and it is improved by slightly "softening" the repulsive force between the separating products [379]. Other, less empirical, treatments of repulsive energy release have also appeared [380–382].

Finally, we consider some reactive scattering measurements where the molecular collision dynamics is dominated by the formation of strong collision complexes. Experiments using crossed molecular beams have now scanned the entire range from systems undergoing direct reaction, through osculating complexes [82] lasting for about a rotational period, to collision complexes, such as halogenated alkyl radicals, which are bound by some tens of kilocalories per mole relative either to separated reactants or to any product channel. In beams experiments such species can be formed by the association of an F or Cl atom with an unsaturated species. The internally energized complex in the absence of stabilizing collisions will subsequently dissociate, giving an angular distribution symmetric about $\theta = 90°$. Where more than one product channel is energetically accessible, one should expect the relative cross sections to be given by RRKM theory with the most exoergic channel being predominant.

These expectations are largely borne out by the experimental results for a large number of F-atom reactions with unsaturated species [352–355], and for the reactions of Cl atoms with bromo-olefins [383, 384]. What has caused some dispute [357], however, is the interpretation of the velocity distributions of the scattered products particularly from F-atom reactions. In these, either a CH_3 group or an H atom can be expelled from the energized radical, but both processes involve passing over a significant energy barrier which is ~1–5 kcal/mole high, relative to the final products. In their papers, Lee and co-workers show that the measured recoil energy distribution does not correspond to what they term RRKM calculations (although this title is disputed by Marcus [385]) but is matched by a phase-space calculation based on a model involving a reduced number of oscillators. On the other hand, Herschbach [358, 386] has shown that agreement with experiment is much improved, but not perfect, if one does a calculation of the RRKM-type but additionally requires that the separating products can overcome a centrifugal barrier in the exit valley. Calculations of this type have been carried out and have been shown to agree rather well with experiment for a number of systems that show no discernible exit potential barrier and, therefore, where the activated complex is of the "loose" type. However, both the beam studies and conventional [387] kinetic experiments on reactions of this type indicate that this is not the situation in the systems studied by Lee. The fact that potential energy is released after the system passes through the activated state implies a "tight" complex. One possibility is that the equilibrium configuration of the atoms that form the molecular product is different at this point from the configuration in the final state of this product. If this were the case, then not all of the potential energy liberated after passing through the activated complex would be channeled into relative translation. In general, the sharing of this energy would seem to depend on the nature of the potential energy hypersurface traversed by the products as they separate. An extreme example of a reaction of this type is provided by the reaction

$$CH_3 + CF_3 \rightarrow (CH_3\text{–}CF_3)^{\dagger} \rightarrow CH_2CF_2 + HF \qquad (120)$$

which has been studied by infrared chemiluminescence [286] and is considered below.

B. Spectroscopic Observations: Association Reactions

The experiments discussed at the end of the previous section provided information about the *translational* excitation of the products of unimolecular fragmentation of energized species formed in association reactions. The distributions of *vibrational* energy in the products of some reactions of this, and related, types have been determined by chemical laser measurements and by observations of infrared chemiluminescence. Some of these studies were referred to in Section III.C, other reactions have been studied more recently [388–392]. In all of these investigations, the product which has been observed is HF or HCl formed in what is frequently termed a "snap-out" reaction. These processes require that, almost simultaneously, two bonds break, the HX bond forms, and the order of a bond in the other product is increased. The reverse reaction, a "four-centre" (bimolecular) one, has a high activation barrier, so in the "snap-out" process a considerable proportion of the total energy is released after the system passes through the activated state. Thus reaction (120)

$$CH_3 + CF_3 \rightarrow (CH_3\text{–}CF_3)^\dagger \rightarrow CH_2CF_2 + HF, \tag{120}$$

studied by Clough, Polanyi, and Taguchi [286], involves a $(CH_3\text{–}CF_3)^\dagger$ species containing 99 kcal/mole internal energy of which 72 kcal/mole is available for redistribution among the degrees of freedom of the CH_2CF_2 and HF products. However, approximately 42 kcal/mole, or almost 60% of that available, is released after passage over the activation barrier. Clearly, this is an extreme case where the final distribution of energy cannot be expected to be "statistical" but must depend to a large extent on the form of the potential hypersurface between the transition state and the separated products. However, as yet, in no reaction of this type proceeding through a relatively long-lived collision complex have total vibrational population inversions (i.e., $N_{(v+1)'} > N_{v'}$) been observed; this does not preclude laser action, since vibrational–rotational lasers can operate on "partial" inversions [393] [$N_{(v+1)'} \leq N_{v'}$, but $N_{(v+1)',J'} > N_{v',(J+1)'}$]. Moreover, the energy yields in any single vibrational degree of freedom are less than those frequently found in the "new" bond product of a direct, atom-exchange reaction. On the other hand, the hydrogen halide is generally somewhat more excited than would be the case if the total available energy was simply shared statistically among all the available degrees of freedom. This excess presumably arises because of some attractive energy release after the system has passed through the transition state. This will occur when the H–X distance is larger in the transition state than in the separated diatomic, but the relatively low vibrational yields in the reactions that have been studied so far suggest that in these instances this difference may be quite small.

In both the molecular beams and spectroscopic experiments that have just been considered, partial energy distributions were measured among the break-up products of an energized association complex. It is much more difficult to discover how the energy is distributed within the complex itself. If it is a fairly simple species its lifetime will probably be extremely short, and while an increase in the number of atoms such a complex contains lengthens its lifetime, any spectrum that is taken to determine the internal state distribution is very complex. However, a very recent paper by Moehlmann and McDonald [394] gives a preliminary report of the first spectroscopic observations from a reaction complex of this type. $O(^3P)$ atoms were reacted with cyclooctonene in a low pressure ($P \sim 2 \times 10^{-4}$ torr) reactor; the biradicals formed in the initial step undergo very rapid isomerization to vibrationally excited cyclooctanone molecules that lasted sufficiently long before further unimolecular reaction for their spectrum to be recorded. This was done using a Fourier transform spectrometer with the entire optical system, as well as the reaction chamber, cooled to 77°K. The observed distribution was consistent with a statistical sharing of energy among the different types of vibrational mode within the molecule. These experiments represent a new step forward in measurements of infrared chemiluminescence.

Another preliminary communication on a related topic is worthy of note. Von Rosenberg and Trainor [395] have observed time-resolved, infrared emission from $O_3^\dagger$ during its formation following short-wavelength flash photolysis of O_2. A somewhat oversimplified reaction scheme is

$$O_2 + h\nu\ (\lambda \geq 1800\ \text{Å}) \rightarrow O(^3P) + O(^3P) \tag{134}$$

$$O(^3P) + O_2 + O_2 \rightarrow O_3^\dagger + O_2 \tag{135}$$

$$O_3^\dagger + O_2 \rightarrow O_3 + O_2. \tag{136}$$

A series of filters were used to show that the emission included contributions from vibrational bands higher than the lowest fundamentals.

Radiative recombination via electronically excited states continues to engage the interest of both experimentalists and theoreticians. There seem to have been less dramatic advances in this subject over the last two or three years than in some other areas covered by this review, but a number of very useful articles by leading experts in this field have appeared. Carrington has provided a general review of electronic chemiluminescence [396] and has also discussed the theoretical ground rules and how these principles should be applied and modified in considering processes of various molecular complexity [397–399]. Kaufman has "revisited" the air afterglow and provided an excellent summary of the current state of knowledge on the NO + O radiative recombination [400]. Clyne has placed particular emphasis on the

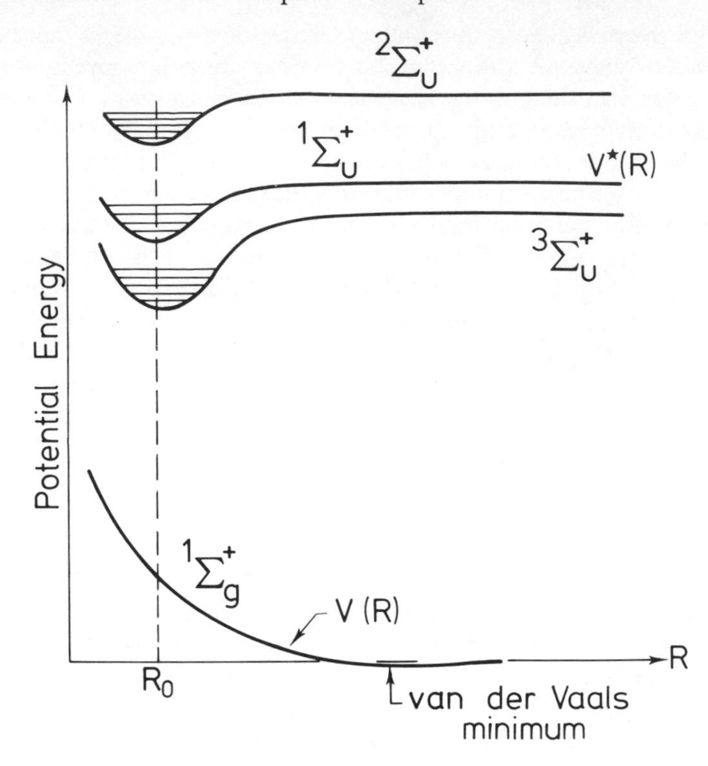

Figure 1.14 Typical potential energy curves for a homonuclear rare gas systems illustrating the non-bound $^1\Sigma_g^+$ state, the first excited $^3\Sigma_u^+$ and $^1\Sigma_u^+$ states, and the $^2\Sigma_u^+$ ionic states [410]. The bound states all have potential minima at an internuclear separation $\sim R_0$.

use of chemiluminescence as a means of detecting and measuring the concentrations of atoms in discharge-flow systems [402]. Finally, Golde and Thrush have written an authoritative and comprehensive review entitled *Afterglows* in which they critically discuss the proposed mechanisms of all the well-known and moderately well known radiative recombination processes involving simple species [402].

At the end of their article, Golde and Thrush comment that population inversions must occur in some afterglows, but that it is difficult to obtain a sufficient (absolute) overpopulation of the excited state to sustain laser action. The so-called "bound-free" molecular systems, which are characterized by a repulsive ground state and a bound excited state, have attracted a great deal of attention as potential lasers [403–410]. Examples of such systems are (noble gas + noble gas) or (noble gas + alkali metal atom);

typical potential curves for a noble gas pair are shown in Figure 1.14. Clearly, one excited- and one ground-state atom can undergo intermolecular recombination into the bound upper state whence they can radiate and fall to the lower nonbound state. Since they then immediately dissociate, the population of the ground state does not build up significantly. From an applications point of view such a laser system is extremely attractive since emission occurs at short wavelengths, and is in principle tunable across the rather wide spread of wavelengths present in the spontaneous emission. The practical problem is, of course, to obtain sufficiently fast rates of atomic excitation and recombination for a significant gain. However, the occurrence of stimulated emission has now been demonstrated in Xe_2, Kr_2, and XeAr, using relativistic electron beams and high gas pressures ($\geq$17 atm).

C. Spectroscopic Observations: Atom-Exchange Reactions

As already indicated, the desire to discover visible or ultraviolet chemical lasers is serving as a powerful stimulus to studies of electronic excitation in atom-exchange reactions. No such laser has yet been successfully operated, but in several laboratories the relative yields of excited- and ground-state products have been measured. The reactions of metal atoms with oxidants (O_3, N_2O, and NO_2) and halogens appear to be receiving greatest attention. The spin and orbital correlation rules [411, 412] that determine whether there is an electronically adiabatic pathway leading from the reactants to the desired products can help to explain or predict which electronic states of the products are most likely to be found. Thus it is interesting, but no surprise, to have confirmed [413] that the reaction,

$$N(^4S) + NO(X^2\Pi) \rightarrow N_2(^1\Sigma_g^+) + O(^3P), \tag{74}$$

occurs with no significant formation of the spin-disallowed products, $N_2(^1\Sigma_g^+)$ and $O(^1D)$.

There have been other recent papers on the reactions discussed in Section III.B.1 and III.B.2. Provencher and McKenney [414, 415] have further investigated the complex excitation of CN which occurs when small quantities of carbonaceous compound are added to active nitrogen. One of their main proposals is that $CN(B^2\Sigma^\dagger)$ is excited by transfer of energy from vibrationally excited N_2 rather than from a metastable, electronically excited molecule, as implied in reaction (79). However, it seems difficult to reconcile the large rate constant which they estimate for this process with the comparative inefficiency of N_2 in quenching $CN(B^2\Sigma^\dagger)$ [416, 417]. Recent experiments [418, 419] do appear to have confirmed that reaction (87) does excite CO to emit chemiluminescence in the vacuum ultraviolet. The overall rate constant for the reaction of $O(^3P)$ with C_2O has been determined in flash photolysis experiments [418] to be 9.5×10^{-11} cm^3/molecule sec with an estimated 1–10% of the reaction producing electronically excited CO.

Some interesting experiments have been carried out on reaction (92) between NO and O_3. Gauthier and Snelling [420] have shown that there is no significant production of either $O_2(^1\Delta_g)$ or $O_2(^1\Sigma_g^+)$ although both processes are allowed by the adiabatic correlation rules and are possible energetically if NO_2 is formed in its electronic ground state. Redpath and Menzinger [421], in beam and scattering gas experiments, have observed how the cross section for the chemiluminescent reaction producing $NO_2(\tilde{A}^2B_1)$ varies with collision energy. They determined a threshold energy of 3.0 kcal/mole, compared with the activation energy of 4.18 kcal/mole [192], clearly demonstrating the distinction between these two quantities (see Section I.A.2). Third, Gordon and Lin [422] have observed enhancement in the intensity of chemiluminescence when O_3 is excited vibrationally by absorbing laser radiation. They propose that effective excitation is not in the ν_3 mode which is excited initially but in the $\nu_2 = 1$ level which is populated as relaxation takes place.

Mention has already been made of beam experiments in which electronic chemiluminescence has been observed from reactions of metal atoms. Palmer's [423–426] and Broida's [427, 428] groups have studied reactions of this type in low-pressure diffusion flames. In most cases, the yield of excited species is disappointingly low. The most efficient of the reactions investigated so far (this excludes the reactions of the type $X + M_2 \rightarrow MX + M^*$ studied by Herschbach's group) appears to be

$$Ba(^1S) + N_2O(X^1\Sigma) \nearrow BaO^*(A^1\Sigma) + N_2(X^1\Sigma_g^+) \quad (137a)$$

$$Ba(^1S) + N_2O(X^1\Sigma) \searrow BaO(X^1\Sigma) + N_2(X^1\Sigma_g^+) \quad (137b)$$

The photon yield is sensitive to the total pressure reaching a maximum of 0.2 per molecular reaction with 10–15 torr of argon added. Jones and Broida [428] suggest that nonradiating states are formed initially and Ar induces transitions into $A^1\Sigma$ through levels of "mixed" character that are known from spectroscopic observations.

D. Spectroscopic Methods: Vibrational and Rotational Excitation

The hope that chemical reactions can be used to pump efficient and powerful lasers is a major stimulus to studies of the chemical production of excited species [429]. There have now been a very large number of papers [430] on lasers which oscillate on the vibration–rotation transitions of molecules produced in atom-transfer reactions, although laser oscillation following direct chemical excitation has only been achieved for the hydrogen and deuterium halides and for CO. Here we shall only refer to the relatively small number of recent studies in which quantitative information has been obtained concerning the vibrational distribution of the chemical product from measurements on the laser output.

The equal-gain temperature technique was described briefly in Section III.B.3. In its simplest form, it suffers from the drawback that information is only obtained about the relative populations in the pair of vibrational levels that are connected by the transition that exhibits highest gain. Other chemical laser techniques, which provide more information, have now been developed. These include the tandem chemical laser method, the grating equal gain method, and the zero gain temperature method, all of which have been compared by Molina and Pimentel [431]. Berry, in a well-documented paper [343] on the $F + H_2$, D_2, and HD reactions, has described experiments using a grating + mirror optical cavity so that in any single experiment oscillation is possible on only a single line. In this way the relative gains on many lines within several vibrational bands could be determined and the gain equations used to find accurate, relative vibrational populations. These experiments provided the first direct measurements of relative vibrational populations in $v' = 0$, a piece of information that cannot be obtained by chemiluminescence experiments.

In laser measurements on hydrogen halides, the system under investigation is itself all, or part, of the active laser medium. Similar experiments have not yet been made on other molecules. However, in related experiments on CO, in the author's laboratory [242, 432] and elsewhere [433], a CO cw laser has been used to probe the vibrational distributions in a reaction system outside the optical cavity that may act as an amplifier or absorber of lines from the laser. Time-resolved observations can be made for as long as one chooses after the reaction is initiated. If tunable infrared lasers become readily available this technique is likely to be applied more widely.

Observations of infrared chemiluminescence have still been made almost exclusively from hydrogen (and deuterium) halides produced in reactions such as

$$H + XR \rightarrow HX + R \tag{138}$$

and

$$X + HR \rightarrow HX + R \tag{139}$$

where X represents a halogen atom and R can be H [in the case of (139)], a halogen atom or another radical species. The recent literature includes papers from the laboratories of Polanyi [335–337, 434–437], Setser [438–444], and Jonathan [445, 446]. In nearly all of the latest experiments of the first two groups, gas-phase relaxation is "arrested" by working at low total pressures;

in Jonathan's work, relaxation is measured and allowed for in a fast-flow system.

Vibrational distributions are now rather well established for a number of reactions of the types represented by (138) and (139). The reactions between H atoms and diatomic halogens were considered earlier in this supplementary section since results from molecular beam experiments could be compared with spectroscopically determined data.

The kinematics of these collisions are rather different from those where the light H atom occupies the "central" position as in (139). One difference is that the collisional reduced mass in (138) is much less than in (139). In general, this will also mean that on average the initial orbital angular momentum in reactive collisions of type (138) is less than in reactive collisions not involving H atoms. The need to conserve total angular momentum restricts the extent of energy that can be released into the rotation of the molecular product, and infrared chemiluminescence experiments [434, 435] of the "arrested relaxation" type have shown $\bar{f}_{R'}$ to be significantly less in H-atom reactions, such as $H + Cl_2$, than in reactions (139) like Cl + HI.

In most spectroscopic studies of reactions such as (139), the atomic reaction has been fluorine. The reaction

$$F + H_2 \rightarrow HF + H \tag{102}$$

has attracted a great deal of attention. To show the level of agreement between chemical laser and chemiluminescence measurements, data from different studies of this reaction and its isotopic variants are presented in Table 1.8. Coombe and Pimentel's laser experiments [339, 340] have shown

Table 1.8
Product Vibrational State Ratios for the Reactions $F + H_2$, D_2, HD at Room Temperature [343]

REACTION	N_4/N_3	N_3/N_2	N_2/N_1	N_1/N_0	TECHNIQUE AND REF.[a]
$F + H_2 \rightarrow HF + H$		0.58 ± 0.12	3.6 ± 0.2		IRC(MR); 445
		0.53 ± 0.10	3.4 ± 0.7		IRC(AR); 436
		0.56 ± 0.06	3.4 ± 0.3		IRC(AR); 443
		0.48 ± 0.01	3.3		CL(ZGT); 339
		0.63 ± 0.04	3.4 ± 0.1	5.2 ± 0.4	CL(GS); 343
$F + D_2 \rightarrow DF + D$	0.66 ± 0.13	1.5 ± 0.3	2.3		IR(AR); 436
		1.5 ± 0.1			CL(ZGT); 339
	0.4 ± 0.2	1.8 ± 0.1	2.35 ± 0.1	2.3 ± 0.2	CL(GS); 343
$F + HD \rightarrow HF + D$		0.15 ± 0.05	3.1 ± 0.2	5.2 ± 0.4	CL(GS); 343
$F + DH \rightarrow DF + H$	0.3 ± 0.2	1.6 ± 0.1	2.25 ± 0.1	2.2 ± 0.2	CL(GS); 343

[a] Symbols used to denote the experimental techniques are IRC, infrared chemiluminescence with (MR) measured relaxation or (AR) arrested relaxation; CL, chemical laser using the (ZGT) zero gain temperature method or with (GS) grating selection.

that the relative values of the detailed rate coefficients, $k_{v'}$, are temperature dependent and are also altered when para-H_2 replaces normal-H_2 as the reactant, indicating that the excitation of the products is effected both by the collision energy and by the rotational state distribution of the reactants.

For all the exothermic reactions between halogen atoms and hydrogen halides, product vibrational distributions have been obtained by infrared chemiluminescence measurements. Values of $\bar{f}_{v'}$ are given in Table 1.9. In

Table 1.9
Average Yields of Vibrational Energy from X + HY Reactions expressed as $\bar{f}_{v'}$, a Fraction of the Total Available Energy

	HY = HCl	HBr	HI	REF.
X = F	0.58	0.54	0.56	445
Cl		~0.5	0.70	434
Br			~0.5–0.6	434

these reactions a light atom is transferred between two much heavier atoms. Trajectory calculations [447] reveal that quite a high degree of vibrational excitation may occur even if the hypersurface is quite strongly repulsive because of the high degree of mixed energy release (see Section IV.D) resulting from the low inertia of the light atom. However, these systems are particularly prone to "secondary encounters," because the products tend to separate slowly relative to the oscillation of the newly formed molecule. This may be the reason for the rather broad vibrational distributions which Jonathan et al. [230, 445] have measured for the F + HX (X ≡ Cl, Br, I) reactions.

Setser and co-workers [438–444] have studied the emission from HF produced in a large number of F-atom reactions. Plots of $k_{v'}$ against $f_{v'}$ for these reactions are nearly all rather similar, suggesting that the reactions proceed directly with little disturbance of the atoms or bonds not directly involved in the reaction.

There continue to be rather few quantitative measurements, by either laser or chemiluminescence techniques, on nonhydrides. No rotationally resolved data on such species have yet been obtained. There have, however, now been five different investigations [242, 243, 433, 448, 449] of the reaction

$$O + CS \rightarrow CO + S. \tag{111}$$

It seems generally agreed that the rates into different vibrational levels of CO reach a maximum at $v \approx 13$, but there remains some disagreement as to how large the relative rates are into levels below about $v = 6$. The closely related reaction of atomic oxygen with CSe has also been studied [450]; the variation of $k_{v'}$ with $f_{v'}$ appears to be very similar for these two reactions.

Three other spectroscopic studies of excitation of nonhydrides will be mentioned briefly. First, Ogryzlo, Reilly, and Thrush [451] have reported emission from $CO^{\dagger}(v \leq 17)$ on adding O_2 to active nitrogen containing free C atoms. Absence of emission from higher levels prompts them to suggest that CO is excited in the reaction

$$C(^3P) + O_2 \rightarrow CO(^1\Sigma^+) + O(^1D), \qquad \Delta H_0^0 = -29.5 \text{ kcal/mole } (4.02 \text{ eV}) \quad (140)$$

rather than the more exothermic channel yielding $O(^3P)$. Secondly [452], we call attention to measurements of the infrared emission from $NO^{\dagger}$ formed in the reaction

$$N + O_2 \rightarrow NO + O, \qquad \Delta H_0^0 = -32.1 \text{ kcal/mole } (1.39 \text{ eV}) \quad (141)$$

This reaction is distinguished from others studied by the infrared chemiluminescence method by its high activation energy (6.25 kcal/mole; 0.27 eV). Since any NO formed reacts rapidly with N atoms, the observed vibrational distribution should correspond quite closely to that actually formed in reaction (141). Unfortunately, the emission signal was very weak and a spectrum could only be obtained using interference filters. This was synthesized quite well by assuming an NO vibrational temperature of 5000°K corresponding to $\bar{f}_{v'}$ of between 0.2 and 0.25. Thirdly, we note the first use of Raman scattering to observe vibrationally excited species from chemical reactions [453]. The rapid reaction

$$N + NO \rightarrow N_2 + O, \qquad \Delta H_0^0 = -75 \text{ kcal/mole } (-3.25 \text{ eV}) \quad (74)$$

was allowed to occur in a discharge-flow apparatus and the Raman scattering from a point downstream of the reaction was recorded using Ar ion laser. Only $N_2(v = 1)$ was overpopulated, because of rapid vibrational energy exchange between more highly excited species and $N_2(v = 0)$. However, because such processes conserve vibrational quanta, and vibrational–translational energy transfer is very slow, Black et al. [453] concluded that the concentration of $N_2(v = 1)$ corresponds to the concentration of vibrational quanta from reaction (74). Hence, they determined $\bar{f}_{v'} \approx 0.25$, that is, $\langle W'_{\text{vib}} \rangle \approx 19$ kcal/mole ($\approx$0.8 eV), in good agreement with Phillips and Schiff's value [289, 290] (see Section III.D).

Infrared chemiluminescence experiments are now being used to investigate how the product energy distributions change if excess energy of a particular kind is supplied to the reactants. The reactions Cl + HI [335], F + HCl

[336, 337], $F + D_2$ [337], and $H + Cl_2$ [337] have been studied. The relative translational energy is enhanced by using a heated source or a supersonic nozzle for the atomic reagent keeping the molecular source, and hence the vibration–rotation state distribution, at room temperature. The effect of internal excitation is observed by heating the molecular source and determining how the product state distributions differ from the case where only the collision energy is enhanced. Rotational excitation of the reagent is shown to be relatively unimportant in trajectory calculations, so any changes arise from reactions of vibrationally excited species.

For these substantially exothermic reactions, translational energy is found to be more effective than vibrational excitation in promoting reaction, in agreement with the results of trajectory calculations on surfaces with potential barriers in the "entry valley" [325, 454]. On the average, "excess" vibrational energy is converted to vibrational excitation of the product, whereas "excess" translational energy becomes predominantly translational and rotational energy in the products. In addition to these experiments, Polanyi's group has begun to investigate experimentally how substantially endothermic reactions can be promoted, particularly by vibrational excitation of the molecular reactant [455].

E. Theory

Considerable use continues to be made of classical trajectory calculations in relating the experimentally determined attributes of electronically adiabatic reactions to the features in the potential energy surface that determine these properties. However, over the past 3 or 4 years, considerable progress has been made with semiclassical and quantum mechanical calculations with the result that it is now possible to predict with some degree of confidence the situations in which a purely classical approach to the collision dynamics will give acceptable results. Application of the semiclassical method, which utilises "classical dynamics plus the superposition of probability amplitudes" [456], has been pioneered by Marcus [457–466] and by Miller [456, 467–476].

These calculations are, appropriately enough, intermediate in difficulty between classical and totally quantum mechanical treatments, and since they employ classical methods to find the trajectory through the collision, there is no restriction on the form of potential energy hypersurface. In the results one can distinguish between collisions leading to classically allowed transitions and those with small transition probabilities and which are therefore classically forbidden. For classically allowed processes, if interference effects are neglected, semiclassical and classical approaches lead to essentially the same results and in practice the limitations on selection and resolution in any real experiment on a three-atom, reactive system will almost certainly lead to interference effects being washed out. On the other hand, semiclassical (or quantum mechanical) calculations are particularly needed for systems where the number of classical trajectories leading to a particular final result is zero or at least very small. This means that they are required in "threshold regions," where this term can mean close to threshold for overall reaction or close to threshold for formation of a product in a particular final state. Where collisions close to threshold are predominant, as in a thermal experiment where the threshold energy $E_0 \gg RT$, then classical trajectory calculations are likely to be of limited accuracy. Of course, this has generally not been the situation either in reactive scattering experiments or in most spectroscopic studies of energy partitioning since the cross sections or rate constants are large for most of the reactions that have been investigated. Consequently, the use of classical mechanics in calculations designed to model the results of these experiments has probably not led to significant errors. Conversely, the experiments themselves do not provide a sensitive probe for quantum dynamical effects [477].

A second interesting new theoretical development of a quite different kind has been the application of an information theory approach to the classification of experimental results [478–488]. The outcome of collisions, for example, the distribution of product translational or vibrational energies, is compared with the statistically predicted result for a closed system and the difference between the two is recorded as the *surprisal.* The surprisal is defined by an expression of the type found in information theory: for example, if the probability of a reaction producing a molecular product in a final vibrational state v', corresponding to a fractional energy yield $f_{v'}$, is $P(f_{v'})$ and the corresponding statistical expression for a closed system is $P^0(f_{v'})$, then the surprisal is given by

$$I(f_{v'}) = -\log \frac{P(f_{v'})}{P^0(f_{v'})} \, .$$

In the majority of cases which have been studied, the surprisal for each degree of freedom varies linearly with temperature. Although the reason for this is not apparently understood at the present time, the result itself is very striking and allows one to reduce a considerable body of experimental data into a small number of "temperature-like" parameters, such as $\lambda_{v'} = dI(f_{v'})/df_{v'}$, which are negative where population inversions occur.

In conclusion, we again draw attention to the recent *Proceedings of the Faraday Discussion of the Chemical Society on Molecular Beam Scattering* because this contains not only several interesting theoretical papers but also a very useful article by Marcus [466] reviewing the various theoretical approaches to reactive collisions that are being pursued at the present time.

ACKNOWLEDGMENTS

Professors J. C. Polanyi and H. B. Palmer were both kind enough to read the original draft of this article and I am grateful to them for several helpful suggestions. I should also like to thank the members of my own research group who have helped me with the preparation of this review.

REFERENCES

1. R. Wolfgang, *Prog. React. Kinet.*, **3,** 97 (1965).
2. D. R. Herschbach, *Advan. Chem. Phys.*, **10,** 319 (1966).
3. J. C. Polanyi, *J. Quant. Spectry. Radiative Transfer*, **3,** 481 (1963).
4. W. Klemperer, private communication.
5. M. Polanyi, *Atomic Reactions*, Williams and Norgate, London, 1932.
6. R. A. Ogg and M. Polanyi, *Trans. Faraday Soc.*, **31,** 604, 1375 (1935).
7. J. L. Magee, *J. Chem Phys.*, **8,** 687 (1940).
8. M. G. Evans and M. Polanyi, *Trans. Faraday Soc.*, **35,** 178 (1939).
9. E. F. Greene and A. Kuppermann, *J. Chem. Educ.* **45,** 361 (1968).
10. J. O. Hirschfelder, C. F. Curtiss, and R. B. Bird, *Molecular Theory of Gases and Liquids*, Wiley, New York, 1964.
11. E. A. Mason and L. Monchick, *Advan. Chem. Phys.*, **12,** 329 (1967).
12. I. Amdur and J. E. Jordan, *Advan. Chem. Phys.*, **10,** 29 (1966).
13. R. B. Bernstein, *Advan. Chem. Phys.*, **10,** 75 (1966).
14. R. B. Bernstein and J. T. Muckerman, *Advan. Chem. Phys.*, **12,** 389 (1967).
15. E. F. Greene, A. L. Moursund, and J. Ross, *Advan. Chem. Phys.*, **10,** 135 (1966).
16. R. B. Bernstein, *Science*, **144,** 141 (1964).
17. A. Kuppermann, *Nobel Symposium* 5, *Fast Reactions and Primary Processes in Chemical Kinetics*, Interscience, New York, 1967, 131.
18. A. Kuppermann and J. M. White, *J. Chem. Phys.*, **44,** 4352 (1966).
19. R. N. Porter and M. Karplus, *J. Chem. Phys.*, **40,** 1105 (1964).
20. M. Karplus, R. N. Porter, and R. D. Sharma, *J. Chem. Phys.*, **43,** 3259 (1965).
21. W. L. Fite and S. Datz, *Ann. Rev. Phys. Chem.*, **14,** 61 (1963).
22. H. Pauly and P. J. Toennies, *Advan. Atom. Mol. Phys.*, **1,** 195 (1965).
23. D. R. Herschbach, *Appl. Opt. Suppl.*, **2,** 128 (1965).
24. A. R. Blythe, M. A. D. Fluendy, and K. P. Lawley, *Quart. Rev. Chem. Soc.*, **20,** 465 (1966).
25. J. P. Toennies, *Chemische Elementarprozesse*, Springer-Verlag, Berlin, 1968, p. 157.
26. E. H. Taylor and S. Datz, *J. Chem. Phys.*, **23,** 1711 (1955).
27. R. R. Herm, R. Gordon, and D. R. Herschbach, *J. Chem. Phys.*, **41,** 2218 (1964).
28. R. J. Gordon, R. R. Herm, and D. R. Herschbach, *J. Chem. Phys*,, **49,** 2684 (1968).
29. D. D. Parrish and R. R. Herm, *J. Chem. Phys.*, **51,** 5467 (1970).

30. Y. T. Lee, J. D. McDonald, P. R. Le Breton, and D. R. Herschbach, *J Chem. Phys.*, **49,** 2447 (1968).
31. Y. T. Lee, J. D. McDonald, P. R. Le Breton, and D. R. Herschbach, *Rev. Sci. Instr.*, **40,** 1402 (1969).
32. A. E. Grosser and R. B. Bernstein, *J. Chem. Phys.*, **42,** 1268 (1965).
33. A. E. Grosser and R. B. Bernstein, *J. Chem. Phys.*, **43,** 1140 (1965).
34. R. R. Herm and D. R. Herschbach, *J. Chem. Phys.*, **43,** 2139 (1965).
35. C. Maltz and D. R. Herschbach, *Discussions Faraday Soc.*, **44,** 176 (1967).
36. R. Grice, J. E. Mosch, S. A. Safron, and J. P. Toennies, *J. Chem. Phys.*, **53,** 3376 (1970).
37. M. C. Moulton and D. R. Herschbach, *J. Chem. Phys.*, **44,** 3010 (1966).
38. K. T. Gillen and R. B. Bernstein, *Chem. Phys. Lett.*, **5,** 275 (1970).
39. Ch. Ottinger and R. N. Zare, *Chem. Phys. Lett.*, **5,** 243 (1970).
40. J. R. Airey, E. F. Greene, K. Kodera, G. P. Reck, and J. Ross, *J. Chem. Phys.*, **46,** 3287 (1967).
41. J. R. Airey, E. F. Greene, G. P. Reck, and J. Ross, *J. Chem. Phys.*, **46,** 3295 (1967).
42. E. F. Greene, L. F. Hoffmann, M. W. Lee, J. Ross, and C. E. Young, *J. Chem. Phys.*, **50,** 3450 (1969).
43. D. R. Herschbach, *Discussions Faraday Soc.*, **33,** 149 (1962).
44. W. B. Miller, S. A. Safron, and D. R. Herschbach, *Discussions Faraday Soc.*, **44,** 108 (1967).
45. E. A. Entemann and D. R. Herschbach, *Discussions Faraday Soc.*, **44,** 289 (1967).
46. L. R. Martin and J. L. Kinsey, *J. Chem. Phys.*, **46,** 4834 (1967).
47. R. Wolfgang and J. B. Cross, Jr., *J. Phys. Chem.*, **73,** 743 (1969).
48. N. C. Blais and D. L. Bunker, *J. Chem. Phys.*, **37,** 2713 (1962).
49. J. C. Polanyi, *Discussions Faraday Soc.*, **44,** 293 (1967).
50. W. B. Miller, S. A. Safron, and D. R. Herschbach, *Discussions Faraday Soc.*, **44,** 292 (1967).
51. S. B. Jaffe and J. B. Anderson, *J. Chem. Phys.*, **51,** 1057 (1969).
52. M. Krocsak and G. Schay, *Z. Phys. Chem.* (Leipzig), **B19,** 344 (1932).
53. E. Roth and G. Schay, *Z. Phys. Chem.* (Leipzig), **B28,** 323 (1935).
54. S. Datz and R. E. Minturn, *J. Chem. Phys.* **41,** 1153 (1964).
55. K. R. Wilson, G. H. Kwei, J. A. Norris, R. R. Herm, J. H. Birely, and D. R. Herschbach, *J. Chem. Phys.*, **42,** 1154 (1964).
56. R. E. Minturn, S. Datz, and R. L. Becker, *J. Chem. Phys.*, **44,** 1149 (1966).
57. J. H. Birely and D. R. Herschbach, *J. Chem. Phys.*, **44,** 1690 (1966).
58. T. Warnock, A. E. Grosser, and R. B. Bernstein, *J. Chem. Phys.*, **46,** 1685 (1967).
59. J. H. Birely, D. R. Herschbach, R. R. Herm, and K. R. Wilson, *J. Chem. Phys.*, **47,** 993 (1967).
60. R. Grice and P. Empedocles, *J. Chem. Phys.*, **48,** 5352 (1968).
61. G. H. Kwei and D. R. Herschbach, *J. Chem. Phys.*, **51,** 1742 (1969).
62. J. H. Birely, E. A. Entemann, R. R. Herm, and K. R. Wilson, *J. Chem. Phys.*, **51,** 5461 (1969).

63. D. D. Parrish and R. R. Herm, *J. Chem. Phys.*, **51,** 5467 (1969).
64. D. C. Brodhead, P. Davidovits, and S. A. Edelstein, *J. Chem. Phys.*, **51,** 3601 (1969).
65. D. R. Herschbach, G. H. Kwei, and J. A. Norris, *J. Chem. Phys.*, **34,** 1842 (1961).
66. C. Riley, K. T. Gillen, and R. B. Bernstein, *J. Chem. Phys.*, **47,** 3672 (1967).
67. K. T. Gillen, C. Riley, and R. B. Bernstein, *J. Chem. Phys.*, **50,** 4019 (1969).
68. A. C. Roach, *Chem. Phys. Lett.*, **6,** 389 (1970).
69. G. H. Kwei, J. A. Norris, and D. R. Herschbach, *J. Chem. Phys.*, **52,** 1317 (1970).
70. D. R. Herschbach, G. H. Kwei, and J. A. Norris, *J. Chem. Phys.*, **34,** 1842 (1961).
71. P. J. Kuntz, M. H. Mok, E. M. Nemeth, and J. C. Polanyi, *Discussions Faraday Soc.*, **44,** 229 (1967).
72. M. Godfrey and M. Karplus, *J. Chem. Phys.*, **49,** 3602 (1968).
73. P. J. Kuntz, E. M. Nemeth, and J. C. Polanyi, *J. Chem. Phys.*, **50,** 4607 (1969).
74. P. J. Kuntz, M. H. Mok, and J. C. Polanyi, *J. Chem. Phys.*, **50,** 4623 (1969).
75. W. B. Miller, S. A. Safron, and D. R. Herschbach, *Discussions Faraday Soc.*, **14,** 292 (1967).
76. G. M. Wieder and R. A. Marcus, *J. Chem. Phys.*, **37,** 1835 (1962).
77. J. C. Light, *Discussions Faraday Soc.*, **44,** 14 (1967).
78. D. O. Ham, J. L. Kinsey, and F. S. Klein, *Discussions Faraday Soc.*, **44,** 174 (1967).
79. D. O. Ham and J. L. Kinsey, *J. Chem. Phys.*, **48,** 939 (1968).
80. D. O. Ham and J. L. Kinsey, *J. Chem. Phys.*, **53,** 285 (1970).
81. R. R. Herm and D. R. Herschbach, *J. Chem. Phys.*, **52,** 5783 (1970).
82. Y. T. Lee, J. D. McDonald, P. R. Le Breton, and D. R. Herschbach, *J. Chem. Phys.*, **49,** 2447 (1968).
83. Y. T. Lee, P. R. Le Breton, J. D. McDonald, and D. R. Herschbach, *J. Chem. Phys.*, **51,** 455 (1969).
84. D. Beck, F. Engelke, and H. J. Loesch, *Ber. Bunsenges. Phys. Chem.*, **72,** 1105 (1968).
85. J. B. Cross and N. C. Blais, *J. Chem. Phys.*, **50,** 4108 (1969).
86. N. C. Blais and J. B. Cross, *J. Chem. Phys.*, **52,** 3580 (1970).
87. G. A. Fisk, J. D. McDonald, and D. R. Herschbach, *Discussions Faraday Soc.*, **44,** 228 (1967).
88. S. Datz and E. H. Taylor, *J. Chem. Phys.*, **39,** 1896 (1963).
89. J. Geddes, H. F. Krause, and W. L. Fite, *J. Chem. Phys.*, **52,** 3296 (1970).
90. J. M. White, *Chem. Phys. Lett.*, **4,** 441 (1969).
91. T. P. Schafer, P. E. Siska, J. M. Parson, F. P. Tully, Y. C. Wong, and Y. T. Lee, *J. Chem. Phys.*, **53,** 3385 (1970).
92. K. G. Anlauf, P. E. Charters, D. S. Horne, R. G. MacDonald, D. H. Maylotte, J. C. Polanyi, W. J. Sirlac, D. C. Tardy, and K. B. Woodall, *J. Chem. Phys.*, **53,** 4091 (1970).
93. J. C. Polanyi and D. C. Tardy, *J. Chem. Phys.*, **51,** 5717 (1969).

94. J. H. Parker and G. C. Pimentel, *J. Chem. Phys.*, **51,** 91 (1969).
95. S. B. Jaffe and J. B. Anderson, *J. Chem. Phys.*, **51,** 1057 (1969).
96. I. M. Campbell and B. A. Thrush, *Ann. Rept. Chem. Soc.*, **62,** 17 (1965).
97. B. A. Thrush, *Chem. in Britain*, **2,** 287 (1966).
98. B. A. Thrush, *Ann. Rev. Phys. Chem.*, **19,** 371 (1968).
99. T. Carrington and D. Garvin, *Comprehensive Chemical Kinetics*, Vol. 3, The Formation and Decay of Excited Species, Elsevier, Amsterdam, 1969, Chap. 3.
100. G. Porter, *Discussions Faraday Soc.*, **33,** 198 (1962).
101. J. K. Cashion and J. C. Polanyi, *Proc. Roy. Soc.* (London), **A258,** 529 (1960).
102. J. K. Cashion and J. C. Polanyi, *Proc. Roy. Soc.* (London), **A258,** 570 (1960).
103. J. K. Cashion and J. C. Polanyi, *J. Chem. Phys.*, **30,** 317 (1959).
104. J. R. Airey, P. D. Pacey, and J. C. Polanyi, Eleventh International Symposium on Combustion, 1967, p. 85.
105. R. A. Young and R. L. Sharpless, *J. Chem. Phys.*, **44,** 1071 (1963).
106. R. A. Young and G. Black, *J. Chem. Phys.*, **44,** 3741 (1966).
107. R. A. Young, *Can. J. Chem.*, **47,** 1927 (1969).
108. D. W. Setser and B. A. Thrush, *Proc. Roy. Soc.* (London), **A288,** 275 (1965).
109. G. Herzberg, *Molecular Spectra and Molecular Structure I: Spectra of Diatomic Molecules*, 2nd ed., Van Nostrand, New York, 1950.
110. J. G. Calvert and J. N. Pitts, *Photochemistry*, Wiley, New York, 1966.
111. F. R. Gilmore, *J. Quant. Spectry. Radiative Transfer*, **5,** 369 (1965).
112. D. L. Baulch, D. D. Drysdale, D. G. Horne, and A. C. Lloyd, *High Temperature Reaction Rate Data*, No. 4 (Department of Physical Chemistry, The University, Leeds, 1969).
113. R. A. Young and R. L. Sharpless, *Discussions Faraday Soc.*, **33,** 228 (1962).
114. R. A. Young and R. L. Sharpless, *J. Chem. Phys.*, **39,** 1071 (1963).
115. R. W. F. Gross and N. Cohen, *J. Chem. Phys.*, **48,** 2851 (1968).
116. D. W. Setser and B. A. Thrush, *Nature*, **200,** 864 (1963).
117. A. B. Callear and I. W. M. Smith, *Discussions Faraday Soc.*, **37,** 96 (1964).
118. A. B. Callear and I. W. M. Smith, *Trans. Faraday Soc.*, **61,** 2383 (1965).
119. C. H. Dugan, *J. Chem. Phys.*, **45,** 87 (1966).
120. K. H. Welge, *J. Chem. Phys.*, **45,** 166 (1966).
121. K. H. Becker and K. D. Bayes, *J. Phys. Chem.*, **71,** 37 (1967).
122. R. A. Young and G. A. St. John, *J. Chem. Phys.*, **48,** 898 (1968).
123. A. B. Callear, M. J. Pilling, and I. W. M. Smith, *Trans. Faraday Soc.*, **64,** 2296 (1968).
124. A. B. Callear and M. J. Pilling, *Trans. Faraday Soc.*, **66,** 1886 (1970).
125. A. B. Callear and M. J. Pilling, *Trans. Faraday Soc.*, **66,** 1618 (1970).
126. R. W. Fair and B. A. Thrush, *Trans. Faraday Soc.*, **65,** 1208 (1969).
127. D. L. Baulch, D. D. Drysdale, and D. G. Horne, *High Temperature Reaction Rate Data*, No. 5 (Department of Physical Chemistry, The University, Leeds, 1970).
128. G. H. Myers, D. M. Silver, and F. Kaufman, *J. Chem. Phys.*, **44,** 718 (1966).
129. M. A. A. Clyne and B. A. Thrush, *Proc. Roy. Soc.* (London), **A269,** 404 (1962).
130. R. R. Reeves, P. Harteck, and W. H. Chace, *J. Chem. Phys.*, **41,** 764 (1964).

131. G. Doherty and N. Jonathan, *Discussions Faraday Soc.*, **37,** 73 (1964).
132. K. H. Becker, W. Groth, and F. Joo, *Ber. Bunsenges. Phys. Chem.*, **72,** 157 (1968).
133. N. Jonathan and R. Petty, *Trans. Faraday Soc.*, **64,** 1240 (1968).
134. A. McKenzie and B. A. Thrush, *Chem. Phys. Lett.*, **1,** 681 (1968).
135. F. Kaufman and J. R. Kelso, *Symposium on Chemiluminescence*, Duke University, 1965.
136. D. B. Hartley and B. A. Thrush, *Proc. Roy. Soc.* (London), **A37,** 220 (1964).
137. K. H. Becker, W. Groth and D. Thran, *Chem. Phys. Lett.*, **6,** 583 (1970).
138. M. A. A. Clyne and B. A. Thrush, *Proc. Roy. Soc.* (London), **A261,** 259 (1961).
139. P. Cadman, J. C. Polanyi, and I. W. M. Smith, *J. Chem. Phys.*, **64,** 111 (1967).
140. P. Cadman and J. C. Polanyi, *J. Phys. Chem.*, **72,** 3715 (1968).
141. R. J. Donovan and D. Husain, *Trans. Faraday Soc.*, **62,** 1050 (1966).
142. A. B. Callear and J. F. Wilson, *Trans. Faraday Soc.*, **63,** 1983 (1967).
143. R. J. Fallon, J. T. Vanderslice, and R. D. Cloney, *J. Chem. Phys.*, **37,** 1097 (1962).
144. F. A. Jenkins, *Phys. Rev.*, **31,** 539 (1928).
145. N. H. Keiss and H. P. Broida, Seventh International Symposium on Combustion, 1959, p. 207.
146. K. D. Bayes, *Can. J. Chem.*, **39,** 1074 (1961).
147. D. W. Setser and B. A. Thrush, *Proc. Roy. Soc.* (London), **A288,** 256 (1965).
148. H. P. Broida and S. Golden, *Can. J. Chem.*, **38,** 1666 (1960).
149. H. E. Radford and H. P. Broida, *J. Chem. Phys.*, **38,** 644 (1963).
150. K. M. Evenson and H. P. Broida, *J. Chem. Phys.*, **44,** 1637 (1966).
151. T. Iwai, M. I. Savadatti, and H. P. Broida, *J. Chem. Phys.*, **47,** 3861 (1967).
152. K. Schofield and H. P. Broida, *Photochem. Photobiol.* **4,** 989 (1965).
153. S. M. Read and J. T. Vanderslice, *J. Chem. Phys.*, **36,** 2366 (1962).
154. E. A. Ballik and D. A. Ramsey, *J. Chem. Phys.*, **31,** 1128 (1959).
155. J. Drowart, R. P. Burns, G. De Maria, and M. G. Ingram, *J. Chem. Phys.*, **31,** 1131 (1959).
156. L. Brewer, W. T. Hicks, and O. H. Krikorian, *J. Chem. Phys.*, **36,** 182 (1962).
157. G. Herzberg, *Phys. Rev.*, **70,** 762 (1946).
158. C. Kunz, P. Harteck, and S. Dondes, *J. Chem. Phys.*, **46,** 4157 (1967).
159. M. I. Savadatti and H. P. Broida, *J. Chem. Phys.*, **45,** 2390 (1967).
160. W. J. Miller and H. B. Palmer, *J, Chem. Phys.*, **38,** 278 (1963).
161. W. J. Miller and H. B. Palmer, Ninth International Symposium on Combustion, 1963, p. 90.
162. W. J. Miller and H. B. Palmer, *J. Chem. Phys.*, **40,** 3701 (1964).
163. D. W. Naegeli and H. B. Palmer, *J. Chem. Phys.*, **48,** 2372 (1968).
164. D. W. Naegeli and H. B. Palmer, *J. Mol. Spectry*, **26,** 152 (1968).
165. A. Tewanson, D. W. Naegeli, and H. B. Palmer, Twelfth International Symposium on Combustion, 1969, p. 415.
166. D. W. Naegeli and H. B. Palmer, Eleventh International Symposium on Combustion, 1967, p. 1161.
167. A. Tewanson and H. B. Palmer, *J. Mol. Spectry.*, **22,** 117 (1967).
168. D. W. Naegeli and H. B. Palmer, *J. Mol. Spectry.*, **26,** 277 (1968).

169. C. W. Hand, *J. Chem. Phys.*, **36,** 2521 (1962).
170. E. M. Bulewicz, P. J. Padley, and R. E. Smith, *Proc. Roy. Soc.* (London), **A315,** 129 (1970).
171. J. B. Homer and G. B. Kistiakowsky, *J. Chem. Phys.*, **45,** 1359 (1966).
172. D. Gutman and S. Matsuda, *J. Chem. Phys.*, **52,** 4155 (1970).
173. S. L. N. G. Krishnamachari and H. P. Broida, *J. Chem. Phys.* **34,** 1709 (1961).
174. N. Jonathan, F. F. Marmo, and J. P. Padur, *J. Chem. Phys.*, **42,** 1463 (1965).
175. K. H. Becker and K. D. Bayes, *J. Chem. Phys.*, **45,** 3967 (1966).
176. K. H. Becker and K. D. Bayes, *J. Chem. Phys.*, **48,** 653 (1968).
177. K. D. Bayes, *J. Chem. Phys.*, **52,** 1093 (1970).
178. A. Williams and D. B. Smith, *Chem. Rev.*, **70,** 267 (1970).
179. H. B. Palmer and W. D. Cross, *Carbon*, **3,** 475 (1966).
180. H. B. Palmer, private communication.
181. J. H. Bleekrode and W. C. Nieuwport, *J. Chem. Phys.*, **43,** 3680 (1965).
182. W. Brennen and T. Carrington, *J. Chem. Phys.*, **46,** 7 (1967).
183. K. H. Becker, D. Kley, and R. J. Norstrom, Twelfth International Symposium on Combustion, 1969, p. 405.
184. H. B. Palmer, Twelfth International Symposium on Combustion, 1969, p. 412.
185. H. P. Broida, *J. Chem. Phys.*, **36,** 444 (1962).
186. T. Cawthorn and J. D. McKinley, *J. Chem. Phys.*, **25,** 585 (1956).
187. K. G. Anlauf, R. G. Macdonald, and J. C. Polanyi, *Chem. Phys. Lett.*, **1,** 619 (1968).
188. G. Herzberg, *Molecular Spectra and Structure, III: Electronic Spectra and Electronic Structure of Polyatomic Molecules*, Van Nostrand, New York, 1966.
189. Y. Tanaka and M. Shimaya, *J. Sci. Res. Instr.*, Tokyo, **43,** 241 (1949).
190. J. C. Greaves and D. Garvin, *J. Chem. Phys.*, **30,** 348 (1959).
191. M. A. A. Clyne, B. A. Thrush, and R. P. Wayne, *Trans. Faraday Soc.*, **60,** 359 (1964).
192. P. N. Clough and B. A. Thrush, *Trans. Faraday Soc.*, **63,** 915 (1967).
193. A. E. Douglas and K. P. Huber, *Can. J. Phys.*, **43,** 74 (1965).
194. C. J. Halstead and B. A. Thrush, *Proc. Roy. Soc.* (London), **A295,** 380 (1966).
195. F. J. Lipscomb, R. G. W. Norrish, and B. A. Thrush, *Proc. Roy. Soc.* (London), **A233,** 455 (1956).
196. N. Basco and R. G. W. Norrish, *Can. J. Chem.*, **38,** 1769 (1960).
197. W. D. McGrath and R. G. W. Norrish, *Proc. Roy. Soc.* (London), **A242,** 265 (1957).
198. O. R. Lundell, R. D. Ketcheson, and H. I. Schiff, Twelfth International Symposium on Combustion, 1969, p, 307.
199. D. Biedenkapp and E. J. Bair, *J. Chem. Phys.*, **52,** 6119 (1970).
200. R. J. Donovan and D. Husain, *Chem. Rev.*, **70,** 489 (1970).
201. J. H. Kiefer and R. W. Lutz, Eleventh International Symposium on Combustion, 1967, p. 67.
202. R. V. Fitzsimmons and E. J. Bair, *J. Chem. Phys.*, **40,** 451 (1964).
203. V. D. Baiamonte, D. R. Snelling, and E. J. Bair, *J. Chem. Phys.*, **44,** 673 (1966).

204. D. R. Snelling, V. D. Baiamonte, and E. J. Bair, *J. Chem. Phys.*, **44,** 4137 (1966).
205. G. R. Hebert and R. W. Nicholls, *Proc. Phys. Soc.* (London), **78,** 1024 (1961).
206. D. L. Baulch, D. D. Drysdale, and D. G. Horne, *High Temperature Reaction Rate Data*, No. 5 (Department of Physical Chemistry, The University, Leeds, 1968).
207. R. A. Kane, J. J. McGarvey, and W. D. McGrath, *J. Chem. Phys.*, **39,** 840 (1963).
208. A. M. Bass and D. Garvin, *J. Chem. Phys.*, **39,** 840 (1963).
209. M. A. A. Clyne and J. A. Coxon, *Trans. Faraday Soc.*, **62,** 1175 (1966).
210. I. W. M. Smith, *Discussions Faraday Soc.*, **44,** 194 (1967).
211. I. W. M. Smith, *Trans. Faraday Soc.*, **64,** 328 (1968).
212. A. A. Westenberg and N. de Haas, *J. Chem. Phys.*, **50,** 707 (1969).
213. K. H. Homann, G. Krome, and H. Gg. Wagner, *Ber. Bunsenges. Phys. Chem.*, **72,** 998 (1968).
214. I. W. M. Smith, *Trans. Faraday Soc.*, **64,** 3183 (1968).
215. C. Morley and I. W. M. Smith, *Trans. Faraday Soc.*, **67,** 2575 (1971).
216. M. A. A. Clyne, C. J. Halstead and B. A. Thrush, *Proc. Roy. Soc.* (London), **A295,** 355 (1966).
217. W. D. McGrath and R. G. W. Norrish, *Proc. Roy. Soc.* (London), **A260,** 293 (1961).
218. W. D. McGrath and R. G. W. Norrish, *Z. Physik. Chem.*, **15,** 245 (1958).
219. W. D. McGrath, *J. Chem. Phys.*, **33,** 297 (1968).
220. R. J. Donovan, D. Husain, and P. T. Jackson, *Trans. Faraday Soc.*, **64,** 1798 (1968).
221. F. P. Del Greco and F. Kaufman, *Discussions Faraday Soc.*, **33,** 128 (1962).
222. J. C. Polanyi, *J. Chem. Phys.*, **34,** 347 (1961).
223. J. C. Polanyi, *Appl. Opt. Suppl*, 2: Chemical Lasers, 1965.
224. C. B. Moore, *Ann. Rev. Phys. Chem.*, **22,** 387 (1971).
225. G. A. Kuipers, *J. Mol. Spectry.*, **2,** 75 (1958).
226. J. H. Parker and G. C. Pimentel, *J. Chem. Phys.*, **51,** 91 (1969).
227. J. C. Polanyi and D. C. Tardy, *J. Chem. Phys.*, **51,** 5717 (1969).
228. K. G. Anlauf, P. E. Charters, D. S. Horne, R. G. Macdonald, D. H. Maylotte, J. C. Polanyi, W. J. Skrlac, D. C. Tardy, and K. B. Woodall, *J. Chem. Phys.*, **53,** 409 (1970).
229. N. Jonathan, C. M. Melliar-Smith, and D. H. Slater, *Mol. Phys.*, **20,** 93 (1971).
230. N. Jonathan, C. M. Melliar-Smith, D. Timlin, and D. H. Slater, *Appl. Opt.*, **10,** 1821 (1971).
231. J. V. V. Kasper and G. C. Pimentel, *Phys. Rev. Lett.*, **5,** 231 (1965).
232. P. H. Corneil and G. C. Pimentel, *J. Chem. Phys.*, **49,** 1379 (1968).
233. J. R. Airey, *J. Chem. Phys.*, **52,** 156 (1970).
234. J. R. Airey, *IEEE J. Quantum Electron.*, **QE-3,** 208 (1967).
235. H. L. Chen, J. C. Stephenson, and C. B. Moore, *Chem. Phys. Lett.*, **2,** 593 (1968).
236. M. A. Pollack, *Appl. Phys. Lett.*, **8,** 237 (1966).
237. D. W. Gregg and S. J. Thomas, *J. Appl. Phys.*, **39,** 4399 (1968).

238. S. J. Arnold and G. H. Kimbell, *Appl. Phys. Lett.*, **15,** 351 (1969).
239. G. Hancock and I. W. M. Smith, *Chem. Phys. Lett.*, **3,** 573 (1969).
240. G. Hancock and I. W. M. Smith, *Trans. Faraday Soc.*, **67,** 2586 (1971).
241. A. B. Callear, *Proc. Roy. Soc.* (London), **A276,** 401 (1963).
242. G. Hancock, C. Morley, and I. W. M. Smith, *Chem. Phys. Lett.*, **12,** 193 (1971).
243. C. Morley, Ph.D. thesis, University of Cambridge, 1971.
244. R. M. Osgood, Jr., W. C. Eppers, Jr., and E. R. Nichols, *IEEE J. Quantum Electron.*, **QE-6,** 145 (1970).
245. G. Hancock, B. A. Ridley, and I. W. M. Smith, *J.C.S Faraday II*, **68,** 2117 (1972).
246. H-L. Chen, J. C. Stephenson, and C. B. Moore, *Chem. Phys. Lett.*, **2,** 593 (1968).
247. R. W. F. Gross, *J. Chem. Phys.*, **50,** 1889 (1969).
248. T. A. Cool, R. R. Stephens, and T. J. Falk, *Intl. J. Chem. Kinet.*, **1,** 495 (1969).
249. T. A. Cool, T. J. Falk, and J. J. Stephens, *Appl. Phys. Lett.*, **15,** 318 (1969).
250. T. A. Cool, J. A. Shirley, and R. R. Stephens, *Appl. Phys. Lett.*, **17,** 278 (1970).
251. T. A. Cool and R. R. Stephens, *J. Chem. Phys.*, **51,** 5175 (1970).
252. T. A. Cool and R. R. Stephens, *Appl. Phys. Lett.*, **16,** 55 (1970).
253. T. A. Cool and R. R. Stephens, *J. Chem. Phys.*, **52,** 3304 (1970).
254. T. A. Cool, R. R. Stephens, and J. A. Shirley, *J. Appl. Phys.*, **41,** 4038 (1970).
255. G. Hancock and I. W. M. Smith, *Chem. Phys. Lett.*, **8,** 41 (1971).
256. G. Hancock and I. W. M. Smith, *Appl. Opt.*, **10,** 1827 (1971).
257. W. Q. Jeffers and C. E. Wiswall, *Appl. Phys. Lett.*, **17,** 67 (1970).
258. R. D. Suart, S. J. Arnold, and G. H. Kimbell, *Chem. Phys. Lett.*, **7,** 337 (1970).
259. R. A. Meinzer, *Intl. J. Chem. Kinet.*, **2,** 335 (1970).
260. P. E. Charters, R. G. Macdonald, and J. C. Polanyi, *Appl. Opt.*, **10,** 1747 (1971).
261. G. Karl, P. Kruus, and J. C. Polanyi, *J. Chem. Phys.*, **46,** 224 (1967).
262. G. Karl, P. Kruus, J. C. Polanyi, and I. W. M. Smith, *J. Chem. Phys.*, **46,** 244 (1967).
263. P. E. Charters and J. C. Polanyi, *Discussions Faraday Soc.*, **33,** 107 (1962).
264. K. G. Anlauf, P. J. Kuntz, D. H. Maylotte, P. D. Pacey, and J. C. Polanyi, *Discussions Faraday Soc.*, **44,** 183 (1967).
265. P. D. Pacey and J. C. Polanyi, *Appl. Opt.*, **10,** 1725 (1971).
266. D. M. Creek, R. Petty, and N. Jonathan, *J. Sci. Instr.*, **1,** 582 (1968).
267. K. G. Anlauf, J. C. Polanyi, W. H. Wong, and K. B. Woodall, *J. Chem. Phys.*, **49,** 5189 (1968).
268. R. L. Johnson, M. J. Perona, and D. W. Setser, *J. Chem. Phys.*, **52,** 6372 (1970).
269. M. J. Perona, R. L. Johnson, and D. W. Setser, *J. Chem. Phys.*, **52,** 6384 (1970).
270. H. Heydtmann and J. C. Polanyi, *Appl. Opt.*, **10,** 1738 (1971).
271. J. D. McKinley, D. Garvin, and M. Boudart, *J. Chem. Phys.*, **23,** 784 (1955).
272. D. Garvin, H. P. Broida, and H. J. Kostkowski, *J. Chem. Phys.*, **32,** 880 (1960).
273. D. Garvin, H. P. Broida, and H. J. Kostkowski, *J. Chem. Phys.*, **37,** 193 (1962).

274. P. E. Charters and J. C. Polanyi, *Can. J. Chem.*, **38,** 1742 (1960).
275. J. K. Cashion and J. C. Polanyi, *J. Chem. Phys.*, **35,** 600 (1961).
276. P. E. Charters, B. N. Khare, and J. C. Polanyi, *Nature*, **193,** 367 (1962).
277. K. G. Anlauf, Ph.D. thesis, University of Toronto, 1968.
278. R. G. W. Norrish and G. A. Oldershaw, *Proc. Roy. Soc.* (London), **A262,** 1 (1961).
279. N. Basco, *Proc. Roy. Soc.* (London), **A283,** 302 (1965).
280. P. N. Clough and B. A. Thrush, *Discussions Faraday Soc.*, **44,** 205 (1967).
281. P. N. Clough and B. A. Thrush, *Proc. Roy. Soc.* (London), **A309,** 419 (1969).
282. M. J. Berry and G. C. Pimentel, *J. Chem. Phys.*, **49,** 5190 (1968).
283. M. J. Berry and G. C. Pimentel, *J. Chem. Phys.*, **51,** 2274 (1969).
284. M. J. Berry and G. C. Pimentel, *J. Chem. Phys.*, **53,** 3453 (1970).
285. T. D. Padrick and G. C. Pimentel, *J. Chem. Phys.*, **54,** 720 (1971).
286. P. N. Clough, J. C. Polanyi, and R. T. Taguchi, *Can. J. Chem.*, **48,** 2919 (1970).
287. P. N. Clough and B. A. Thrush, *Chem. Commun.*, 1351 (1968).
288. J. E. Morgan, L. F. Phillips, and H. I. Schiff, *Discussions Faraday Soc.*, **33,** 118 (1962).
289. L. F. Phillips and H. I. Schiff, *J. Chem. Phys.*, **36,** 1509 (1962).
290. L. F. Phillips and H. I. Schiff, *J. Chem. Phys.*, **36,** 3283 (1962).
291. L. F. Phillips and H. I. Schiff, *J. Chem. Phys.*, **37,** 1233 (1962).
292. A. Mathias and H. I. Schiff, *Discussions Faraday Soc.*, **37,** 38 (1964).
293. L. F. Phillips and H. I. Schiff, *J. Chem. Phys.*, **42,** 3171 (1965).
294. K. T. Tang, B. Kleinman, and M. Karplus, *J. Chem. Phys.*, **51,** 1119 (1969).
295. B. Kleiman and K. T. Tang, *J. Chem. Phys.*, **51,** 4587 (1969).
296. E. A. McCullogh and R. E. Wyatt, *J. Chem. Phys.*, **51,** 1253 (1969).
297. R. E. Wyatt, *J. Chem. Phys.*, **51,** 3489 (1969).
298. C. C. Rankin and J. C. Light, *J. Chem. Phys.*, **51,** 1701 (1969).
299. D. Russell and J. C. Light, *J. Chem. Phys.*, **51,** 1720 (1969).
300. F. T. Wall, L. A. Hiller, and J. Mazur, *J. Chem. Phys.*, **29,** 255 (1958).
301. F. T. Wall, L. A. Hiller, and J. Mazur, *J. Chem. Phys.*, **35,** 1284 (1961).
302. D. L. Bunker, *J. Chem. Phys.*, **37,** 393 (1962).
303. N. C. Blais and D. L. Bunker, *J. Chem. Phys.*, **39,** 315 (1963).
304. D. L. Bunker, *J. Chem. Phys.*, **40,** 1946 (1964).
305. D. L. Bunker and N. C. Blais, *J. Chem. Phys.*, **41,** 2377 (1964).
306. N. C. Blais, *J. Chem. Phys.*, **49,** 9 (1968); **51,** 856 (1969).
307. M. Karplus, R. N. Porter, and R. D. Sharma, *J. Chem. Phys.*, **40,** 2033 (1964).
308. M. Karplus and L. M. Raff, *J. Chem. Phys.*, **41,** 1267 (1964).
309. L. M. Raff, *J. Chem. Phys.*, **44,** 1202 (1966); **50,** 2276 (1969).
310. L. M. Raff and M. Karplus, *J. Chem. Phys.*, **44,** 1212 (1966).
311. M. Karplus, R. N. Porter, and R. D. Sharma, *J. Chem. Phys.*, **45,** 3871 (1966).
312. J. C. Polanyi and S. D. Rosner, *J. Chem. Phys.*, **38,** 1028 (1963).
313. P. J. Kuntz, E. M. Nemeth, J. C. Polanyi, S. D. Rosner, and C. E. Young, *J. Chem. Phys.*, **44,** 1168 (1966).
314. D. L. Bunker and M. D. Pattengill, *J. Chem. Phys.*, **53,** 3041 (1970).

315. L. M. Raff, L. B. Sims, D. L. Thompson, and R. N. Porter, *J. Chem. Phys.*, **53,** 1606 (1970).
316. L. L. Poulsen, *J. Chem. Phys.*, **53,** 1987 (1970).
317. M. H. Mok and J. C. Polanyi, *J. Chem. Phys.*, **53,** 4588 (1970).
318. F. T. Wall and R. N. Porter, *J. Chem. Phys.*, **36,** 3256 (1962).
319. D. L. Bunker and C. A. Parr, *J. Chem. Phys.*, **52,** 5700 (1970).
320. F. London, *Z. Electrochem.*, **35,** 552 (1929).
321. S. Sato, *J. Chem. Phys.*, **23,** 592 (1955).
322. S. Sato, *J. Chem. Phys.*, **23,** 2465 (1955).
323. M. H. Mok and J. C. Polanyi, *J. Chem. Phys.*, **51,** 1451 (1969).
324. D. D. Parrish and R. R. Herm, *J. Chem. Phys.*, **53,** 2431 (1970).
325. J. C. Polanyi and W. H. Wong, *J. Chem. Phys.*, **51,** 1439 (1969).
326. L. M. Raff, L. B. Sims, D. L. Thompson, and R. N. Porter, *J. Chem. Phys.*, **53,** 1606 (1970).
327. K. G. Anlauf, D. H. Maylotte, J. C. Polanyi, and R. B. Bernstein, *J. Chem. Phys.*, **51,** 5716 (1969).
328. D. D. Wagman, W. H. Evans, V. B. Parker, I. Halow, S. M. Bailey, and R. H. Schumm, N.B.S. Technical Note 270–3, Selected Values of Chemical Thermodynamic Properties, 1968.
329. V. I. Vedeneyer, L. V. Gurvick, V. N. Kondrat'yev, V. A. Medredev, and Y. L. Frankevich, *Bond Energies, Ionization Potentials and Electron Affinities*, published in English by Edward Arnold, London, 1966.
330. S. M. Freund, G. A. Fisk, D. R. Herschbach, and W. Klemperer, *J. Chem. Phys.*, **54,** 2510 (1971).
331. H. G. Bennewitz, R. Haerten, and G. Muller, *Chem. Phys. Lett.*, **12,** 335 (1971).
332. R. P. Mariella, D. R. Herschbach, and W. Klemperer, *J. Chem. Phys.*, **58,** 3785 (1973).
333. A. Schultz, H. W. Cruse, and R. N. Zare, *J. Chem. Phys.*, **57,** 1354 (1972).
334. H. W. Cruse, P. J. Dagdigian, and R. N. Zare, *Discussions Faraday Soc.*, **55,** 277 (1973).
335. L. T. Cowley, D. S. Horne, and J. C. Polanyi, *Chem. Phys. Lett.*, **12,** 144 (1971).
336. L. J. Kirsch and J. C. Polanyi, *J. Chem. Phys.*, **57,** 4498 (1972).
337. A. M. G. Ding, L. J. Kirsch, D. S. Perry, J. C. Polanyi, and J. L. Schreiber, *Discussions Faraday Soc.*, **55,** 252 (1973).
338. J. H. Parker and G. C. Pimentel, *J. Chem. Phys.*, **55,** 857 (1971).
339. R. D. Coombe and G. C. Pimentel, *J. Chem. Phys.*, **59,** 251 (1973).
340. R. D. Coombe and G. C. Pimentel, *J. Chem. Phys.*, **59,** 1535 (1973).
341. M. J. Molina and G. C. Pimentel, *J. Chem. Phys.*, **56,** 3988 (1972).
342. D. E. Klimek and M. J. Berry, *Chem. Phys. Lett.*, **20,** 141 (1973).
343. M. J. Berry, *J. Chem. Phys.*, **59,** 6229 (1973).
344. C. D. Jonah, R. N. Zare, and Ch. Ottinger, *J. Chem. Phys.*, **56,** 263 (1972).
345. J. L. Gole and R. N. Zare, *J. Chem. Phys.*, **57,** 5331 (1972).
346. R. H. Obenhauf, C. J. Hsu, and H. B. Palmer, *Chem. Phys. Lett.*, **17,** 455 (1972).

347. R. H. Obenhauf, C. J. Hsu, and H. B. Palmer, *J. Chem. Phys.*, **57,** 5607 (1972); **58,** 2674 (1973).
348. R. H. Obenhauf, C. J. Hsu, and H. B. Palmer, *J. Chem. Phys.*, **58,** 4693 (1973).
349. R. H. Obenhauf, C. J. Hsu and H. B. Palmer, *Proceedings of the European Combustion Symposium*, Academic Press, London and New York, 1973, p. 41.
350. G. S. Capelle, R. S. Bradford, and H. P. Broida, *Chem. Phys. Lett.*, **21,** 418 (1973).
351. C. R. Jones and H. P. Broida, *J. Chem. Phys.*, **59,** 6677 (1973).
352. J. M. Parson and Y. T. Lee, *J. Chem. Phys.*, **56,** 4658 (1972).
353. J. M. Parson, K. Shobatake, Y. T. Lee, and S. A. Rice, *J. Chem. Phys.*, **59,** 1402 (1973).
354. K. Shobatake, Y. T. Lee, and S. A. Rice, *J. Chem. Phys.*, **59,** 1416, 1427, 1435 (1973).
355. J. M. Parson, K. Shobatake, Y. T. Lee, and S. A. Rice, *Discussions Faraday Soc.*, **55,** 344 (1973).
356. D. W. Setser in *Chemical Kinetics*, ed. J. C. Polanyi, M.T.P. International Review of Science, Butterworths, Oxford, 1972 *Physical Chemistry*, Series 1, **9,** 1.
357. See comments by Marcus, Lee, Rice, and others in *Discussions Faraday Soc.*, **55** (1973).
358. D. R. Herschbach, *Discussions Faraday Soc.*, **55,** 233 (1973).
359. J. C. Polanyi, *Discussions Faraday Soc.*, **55,** 389 (1973).
360. J. L. Kinsey in *Chemical Kinetics*, ed. J. C. Polanyi, M.T.P. International Review of Science, Butterworths, Oxford, 1972, *Physical Chemistry*, Series 1, **9,** 213.
361. See Figure 2 of ref. 358.
362. Beautiful experiments of this kind have been carried out on the K + CH_3I reaction by M. E. Gersh and R. B. Bernstein, *J. Chem. Phys.*, **55,** 4661 (1971) and **56,** 6131 (1972).
363. M. Menzinger and D. J. Wren, *Chem. Phys. Lett.*, **18,** 431 (1973).
364. W. S. Struve, T. Kitagawa, and D. R. Herschbach, *J. Chem. Phys.*, **54,** 2759 (1971).
365. See comments by P. J. Dagdigian, *Discussions Faraday Soc.*, **55,** 311 (1973).
366. D. O. Ham, *Discussions Faraday Soc.*, **55,** 313 (1973).
367. W. S. Struve, J. R. Krenos, D. L. McFadden, and D. R. Herschbach, *Discussions Faraday Soc.*, **55,** 314 (1973).
368. D. R. King and D. R. Herschbach, *Discussions Faraday Soc.*, **55,** 331 (1973).
369. S. M. Lin and R. Grice, *Discussions Faraday Soc.*, **55,** 370 (1973).
370. S. M. Lin, J. C. Whitehead, and R. Grice, *Mol. Phys.*, **27,** 741 (1974).
371. S. J. Riley and D. R. Herschbach, *J. Chem. Phys.*, **58,** 27 (1973).
372. D. D. Parrish and R. R. Herm, *J. Chem. Phys.*, **54,** 2519 (1971).
373. J. Grosser and H. Haberland, *Chem. Phys. Lett.*, **7,** 442 (1970).
374. J. D. McDonald, P. R. Le Breton, Y. T. Lee, and D. R. Herschbach, *J. Chem. Phys.*, **56,** 769 (1972).
375. B. A. Thrush, *Prog. Reac. Kinetics*, **2,** 3, 65 (1965).

376. K. G. Anlauf, D. S. Horne, R. G. Macdonald, J. C. Polanyi, and K. B. Woodall, *J. Chem. Phys.*, **57,** 1561 (1972).
377. D. R. Herschbach in *Potential Energy Surfaces in Chemistry*, ed. W. A. Lester, IBM Research Lab., San Jose, Calif., 1970, p. 44.
378. A useful summary of the "state-of-the-art" for the K + CH_3I reaction is provided by R. B. Bernstein and A. M. Rulis, *Discussions Faraday Soc.*, **55,** 293 (1973).
379. D. D. Parrish and R. R. Herm, *J. Chem. Phys.*, **53,** 2431 (1970).
380. P. J. Kuntz, M. H. Mok, and J. C. Polanyi, *J. Chem. Phys.*, **50,** 4623 (1969).
381. P. J. Kuntz, *Trans. Faraday Soc.*, **66,** 2980 (1970).
382. P. J. Kuntz, *Mol. Phys.*, **23,** 1025 (1972).
383. J. T. Cheung, J. D. McDonald, and D. R. Herschbach, *J. Amer. Chem. Soc.*, **95,** 7889 (1973).
384. J. T. Cheung, J. D. McDonald, and D. R. Herschbach, *Discussions Faraday Soc.*, **55,** 377 (1973).
385. R. A. Marcus, *Discussions Faraday Soc.*, **55,** 381 (1973).
386. S. A. Safron, N. D. Weinstein, and D. R. Herschbach, *Chem. Phys. Lett.*, **12,** 564 (1972).
387. R. J. Cvetanovic and R. S. Irwin, *J. Chem. Phys.*, **46,** 1694 (1967).
388. T. D. Padrick and G. C. Pimentel, *J. Phys. Chem.*, **76,** 3125 (1972).
389. M. J. Molina and G. C. Pimentel, *J. Chem. Phys.*, **56,** 3988 (1972).
390. D. E. Klimek and M. J. Berry, *Chem. Phys. Lett.*, **20,** 141 (1973).
391. H. W. Chang, D. W. Setser, and M. J. Perona, *J. Phys. Chem.*, **75,** 2070 (1971).
392. K. C. Kim and D. W. Setser, *J. Phys. Chem.*, **72,** 283 (1972).
393. J. C. Polanyi, *Appl. Opt. Suppl.* **2,** 109 (1965).
394. J. G. Moehlmann and J. D. McDonald, *J. Chem. Phys.*, **59,** 6683 (1973).
395. C. W. Von Rosenberg and D. W. Trainor, *J. Chem. Phys.*, **59,** 2142 (1973).
396. T. Carrington and J. C. Polanyi in *Chemical Kinetics*, ed. J. C. Polanyi, M.T.P. International Review of Science, Butterworths, Oxford, 1972, *Physical Chemistry*, Series 1, **9,** 135.
397. T. Carrington, *J. Chem. Phys.*, **57,** 2033 (1972).
398. T. Carrington in *Chemiluminescence and Bioluminescence*, eds. M. J. Cormier, D. M. Hercules, and J. Lee, Plenum Press, New York, 1973, p. 43.
399. T. Carrington, *Acct. Chem. Res.*, **7,** 200 (1974).
400. F. Kaufman in *Chemiluminescence and Bioluminescence*, eds. M. J. Cormier, D. M. Hercules, and J. Lee, Plenum Press, New York, 1973.
401. M. A. A. Clyne in *Physical Chemistry of Fast Reactions*, ed. B. P. Levitt, Plenum Press, London and New York, 1973, **1,** 245.
402. M. F. Golde and B. A. Thrush, *Rep. Prog. Phys.*, **36,** 1285 (1973).
403. N. G. Basov, V. A. Danilychev, Yu. M. Popov, and D. D. Khodkevich, *Sov. Phys., J.E.T.P. Lett.*, **12,** 329 (1971).
404. N. G. Basov, V. A. Danilychev, and Yu. M. Popov, *Sov. J. Quant. Electron.*, **1,** 18 (1971).
405. H. A. Koehler, L. J. Ferderber, D. L. Redhead, and P. J. Ebert, *Appl. Phys. Lett.*, **21,** 198 (1971).

406. P. W. Hoff, J. C. Swingle, and C. K. Rhodes, *Opt. Commun.*, **8,** 128 (1973).
407. J. B. Gerarado and A. W. Johnson, *I.E.E.E. J. Quant. Electron.*, **QE–9,** 748 (1973).
408. P. W. Hoff, J. C. Swingle, and C. K. Rhodes, *Appl. Phys. Lett.*, **23,** 245 (1973).
409. F. H. Mies, *Mol. Phys.*, **26,** 1233 (1973).
410. C. K. Rhodes, *I.E.E.E. J. Quant. Electron.*, **QE–10,** 153 (1974).
411. K. E. Shuler, *J. Chem. Phys.*, **21,** 624 (1953).
412. R. J. Donovan and D. Husain, *Chem. Rev.*, **70,** 489 (1970).
413. W. Felder and R. A. Young, *J. Chem. Phys.*, **57,** 572 (1972).
414. G. M. Provencher and D. J. McKenney, *Chem. Phys. Lett.*, **10,** 365 (1971).
415. G. M. Provencher and D. J. McKenney, *Can. J. Chem.*, **50,** 2527 (1972).
416. W. M. Jackson and J. L. Faris, *J. Chem. Phys.*, **56,** 95 (1972).
417. C. K. Luk and R. Bersohn, *J. Chem. Phys.*, **58,** 2153 (1973).
418. W. L. Shackleford, F. N. Mastrup, and W. C. Kreye, *J. Chem. Phys.*, **57,** 3933 (1972).
419. A. Fontijn and S. E. Johnson, *J. Chem. Phys.*, **59,** 6193 (1973).
420. M. Gauthier and D. R. Snelling, *Chem. Phys. Lett.*, **20,** 178 (1973).
421. A. E. Redpath and M. Menzinger, *Can. J. Chem.*, **49,** 3063 (1971).
422. R. J. Gordon and M. C. Lin, *Chem. Phys. Lett.*, **22,** 262 (1973).
423. R. H. Obenhauf, C. J. Hsu, and H. B. Palmer, *Chem. Phys. Lett.*, **17,** 455 (1972).
424. R. H. Obenhauf, C. J. Hsu, and H. B. Palmer, *J. Chem. Phys.*, **57,** 5607 (1972); **58,** 2164 (1973).
425. R. H. Obenhauf, C. J. Hsu, and H. B. Palmer, *J. Chem. Phys.*, **58,** 4693 (1973).
426. R. H. Obenhauf, C. J. Hsu, and H. B. Palmer, *Proceedings of the European Combustion Symposium*, Academic Press, London and New York, 1973, p. 41.
427. G. A. Capelle, R. S. Bradford, and H. P. Broida, *Chem. Phys. Lett.*, **21,** 418 (1973).
428. C. R. Jones and H. P. Broida, *J. Chem. Phys.*, **59,** 6677 (1973).
429. C. B. Moore, *Ann. Rev. Phys. Chem.*, **22,** 387 (1971).
430. A bibliography of papers published before the end of 1971 is provided by S. J. Arnold and H. Rojeska, *Appl. Opt.*, **12,** 169 (1973).
431. M. J. Molina and G. C. Pimentel, *IEEE J. Quant. Electron.*, **QE–9,** 64 (1973).
432. I. W. M. Smith and C. Wittig, *J.C.S. Faraday II*, **69,** 939 (1973).
433. H. T. Powell and J. D. Kelley, *J. Chem. Phys.*, **60,** 561 (1974).
434. D. H. Maylotte, J. C. Polanyi, and K. B. Woodall, *J. Chem. Phys.*, **57,** 1547 (1972).
435. K. G. Anlauf, D. S. Horne, R. G. Macdonald, J. C. Polanyi, and K. B. Woodall, *J. Chem. Phys.*, **57,** 1561 (1972).
436. J. C. Polanyi and K. B. Woodall, *J. Chem. Phys.*, **57,** 1574 (1972).
437. J. C. Polanyi and J. J. Sloan, *J. Chem. Phys.*, **57,** 4988 (1972).
438. H. W. Chang, D. W. Setser, M. J. Perona, and R. L. Johnson, *Chem. Phys. Lett.*, **9,** 587 (1971).
439. M. J. Perona, *J. Chem. Phys.*, **54,** 4024 (1971).
440. H. W. Chang, D. W. Setser, and M. J. Perona, *J. Phys. Chem.*, **75,** 2070 (1971).

441. K. C. Kim and D. W. Setser, *J. Phys. Chem.*, **76,** 283 (1972).
442. K. C. Kim and D. W. Setser, *J. Phys. Chem.*, **77,** 2493 (1973).
443. H. W. Chang and D. W. Setser, *J. Chem. Phys.*, **58,** 2298 (1973).
444. W. H. Duewer and D. W. Setser, *J. Chem. Phys.*, **58,** 2310 (1973).
445. N. Jonathan, C. M. Melliar-Smith, S. Okuda, D. H. Slater, and D. Timlin, *Mol. Phys.*, **22,** 561 (1971).
446. N. Jonathan, S. Okuda, and D. Timlin, *Mol. Phys.*, **24,** 1153 (1972).
447. C. A. Parr, J. C. Polanyi, and W. H. Wong, *J. Chem. Phys.*, **58,** 5 (1973).
448. K. D. Foster, *J. Chem. Phys.*, **57,** 2451 (1972).
449. S. Tsuchiya, N. Nielsen, and S. H. Bauer, *J. Phys. Chem.*, **77,** 2455 (1973).
450. C. Morley, B. A. Ridley, and I. W. M. Smith, *J.C.S. Faraday II*, **68,** 2127 (1972).
451. E. A. Ogryzlo, J. P. Reilly, and B. A. Thrush, *Chem. Phys. Lett.*, **23,** 37 (1973).
452. F. Hushfar, J. W. Rogers, and A. T. Stair, Jr., *Appl. Opt.*, **10,** 1843 (1971).
453. G. Black, R. L. Sharpless, and T. G. Slanger, *J. Chem. Phys.*, **58,** 4792 (1973).
454. C. A. Parr, J. C. Polanyi, and W. H. Wong, *J. Chem. Phys.*, **58,** 5 (1973).
455. D. J. Douglas, J. C. Polanyi, and J. J. Sloan, *J. Chem. Phys.*, **59,** 6679 (1973).
456. W. H. Miller and A. W. Raczkowski, *Discussions Faraday Soc.*, **55,** 45 (1973).
457. R. A. Marcus, *Chem. Phys. Letters*, **7,** 525 (1970).
458. R. A. Marcus, *J. Chem. Phys.*, **54,** 3965 (1971).
459. J. N. L. Connor and R. A. Marcus, *J. Chem. Phys.*, **55,** 5636 (1971).
460. W. H. Wong and R. A. Marcus, *J. Chem. Phys.*, **55,** 5663 (1971).
461. W. H. Wong and R. A. Marcus, *J. Chem. Phys.*, **56,** 311 (1972).
462. W. H. Wong and R. A. Marcus, *J. Chem. Phys.*, **56,** 3548 (1972).
463. J. Stine and R. A. Marcus, *Chem. Phys. Lett.*, **15,** 536 (1972).
464. R. A. Marcus, *J. Chem. Phys.*, **57,** 4903 (1972).
465. R. A. Marcus, *Discussions Faraday Soc.*, **55,** 34 (1973).
466. R. A. Marcus, *Discussions Faraday Soc.*, **55,** 9 (1973).
467. W. H. Miller, *J. Chem. Phys.*, **53,** 1949 (1970).
468. W. H. Miller, *J. Chem. Phys.*, **53,** 3578 (1970).
469. W. H. Miller, *Chem. Phys. Lett.*, **7,** 431 (1970).
470. W. H. Miller, *J. Chem. Phys.*, **54,** 5386 (1971).
471. W. H. Miller, *Acc. Chem. Res.*, **4,** 161 (1971).
472. C. C. Rankin and W. H. Miller, *J. Chem. Phys.*, **55,** 3150 (1971).
473. W. H. Miller and T. F. George, *J. Chem. Phys.*, **56,** 5668 (1972).
474. T. F. George and W. H. Miller, *J. Chem. Phys.*, **57,** 5722 (1972).
475. J. D. Doll and W. H. Miller, *J. Chem. Phys.*, **57,** 5019 (1972).
476. J. D. Doll, T. F. George, and W. H. Miller, *J. Chem. Phys.*, **58,** 1343 (1973).
477. The author is grateful for a discussion with Professor W. H. Miller in which these points were clarified.
478. R. D. Levine, F. A. Wolf, and J. A. Maus, *Chem. Phys. Lett.*, **10,** 2 (1971).
479. R. D. Levine and R. B. Bernstein, *Chem. Phys. Lett.*, **11,** 552 (1971).
480. R. D. Levine and R. B. Bernstein, *J. Chem. Phys.*, **56,** 2281 (1972).
481. R. B. Bernstein and R. D. Levine, *J. Chem. Phys.*, **57,** 434 (1972).
482. A. Ben-Shaul, R. D. Levine, and R. B. Bernstein, *J. Chem. Phys.*, **57,** 5427 (1972).

483. A. Ben-Shaul, R. D. Levine, and R. B. Bernstein, *Chem. Phys. Lett.*, **15,** 160 (1972).
484. R. D. Levine, B. R. Johnson, and R. B. Bernstein, *Chem. Phys. Lett.*, **19,** 160 (1972).
485. R. D. Levine and R. B. Bernstein, *Discussions Faraday Soc.*, **55,** 100 (1973).
486. G. L. Hofacker and R. D. Levine, *Chem. Phys. Lett.*, **9,** 617 (1971).
487. G. L. Hofacker and R. D. Levine, *Chem. Phys. Lett.*, **15,** 165 (1972).
488. A. Ben-Shaul, *Chem. Phys.*, **1,** 244 (1973).

CHAPTER TWO

POTENTIAL-ENERGY SURFACE CONSIDERATIONS FOR EXCITED-STATE REACTIONS

Joyce J. Kaufman

Department of Chemistry, The Johns Hopkins University, Baltimore, Maryland 21218
and
Department of Anesthesiology, The Johns Hopkins University School of Medicine, Baltimore, Maryland 21205

Contents

I. INTRODUCTION

Many reactions, neutral–neutral, ion–neutral, ion–ion, electron–neutral, and so on, involve electronically excited states of reactants, products, or intermediate species. The relevant potential-energy surfaces for these systems are of great value in elucidating the detailed mechanisms of such reactions. The philosophy behind this chapter is to present to experimentalists an outline of the various methods used for constructing potential-energy surfaces. No attempt has been made to give complete detail; rather, careful consideration has been given to refer the reader to the most pertinent references.

It is first necessary to construct the potential-energy surfaces for the various electronic states of the intermediate complex and then to correlate these with the electronic states of the reactants and products. Consideration of these surfaces is vital whether the intermediate complex is a stable bound entity or a completely repulsive state, since the reactants must approach and the products must recede along the potential-energy curves of the intermediate quasimolecule. These intermediates are of importance because their symmetries and stabilities have a profound effect on the paths along which, and the rates at which, reactions proceed. The survival probability (or detachment of negative ions that could be formed in electron–neutral collisions) may be determined by their stability against predissociation (or autodetachment). In the case of ion–neutral or ion–ion and electron–neutral or electron–ion reactions, these curves are supplemented by the long-range forces, usually attractive.

The course of the reaction is determined by the shapes of the potential-energy curves and by the kinds of curve crossings which do or do not take place. In this regard there are apparently two types of behavior followed:

1. adiabatic, where the reactants approach each other slowly and, at least for states of the same symmetry and spin, the curves do not cross;
2. diabatic, where the reactants approach each other rapidly and diabatic transitions occur even between states of the same symmetry and spin.

For diatomic systems with states of different symmetry and spin, the probability of crossing was derived independently by Landau and Zener. Even for diatomics the Landau–Zener formula has been shown to be inapplicable in certain instances and has been revised. For polyatomic molecules the Landau–Zener formula is not applicable as such, and modifications have been made to permit its use for specified cases.

II. POTENTIAL-ENERGY SURFACES

A. Correlation Rules

1. Diatomics

There are three different schemes for building up the electronic states of diatomic molecules: (a) from separated atoms, (b) from the united atom, and (c) from the molecular orbitals of the diatomic molecule itself. It is the correlation between the electronic states of the diatomic molecule as built up from the separated atoms and as determined from the molecular orbitals of the diatomic which is most valuable for any general consideration of reactions and excited states. The correlation of molecular states obtained by these two methods is not limited solely to diatomic molecules but also forms a valid approach for polyatomic molecular systems. The correlation of separated atoms with the hypothetical united atom has value for diatomics and has been applied to simple polyatomic molecules, especially those with a heavy atom or two and a number of hydrogen atoms. However, it is conceptually less appealing even for simple polyatomic molecules and completely inapplicable for complex polyatomic molecules.

a. From Separated Atoms

The correlation rules for determining what types of molecular electronic states result from given electronic states of the separated atoms were derived quantum mechanically by Wigner and Witmer [1] and are discussed in great detail by Herzberg [2]. These correlation rules hold for the adiabatic potential curves of the electronic states[1] when Russell–Saunders coupling is valid for the separated atoms as well as the molecule.

Briefly, the derivation is as follows. In general, in an atom which contains several electrons, the orbital momenta $\mathbf{l}_1, \mathbf{l}_2, \mathbf{l}_3, \ldots$ and the spins $\mathbf{s}_1, \mathbf{s}_2, \mathbf{s}_3, \ldots$ of the individual electrons are strongly coupled among themselves. (This is called Russell–Saunders coupling.) By quantum-theoretical addition of angular momentum vectors the addition of the $\mathbf{l}_i$ gives the resultant angular momentum vector $\mathbf{L}$. For $\mathbf{l}_1 + \mathbf{l}_2$,

$$\mathbf{L} = (l_1 + l_2), (l_1 + l_2 - 1), (l_1 + l_2 - 2), \ldots, (l_1 - l_2).$$

If there are several angular momenta, the strongly coupled vectors are best

[1] In general, electronic motions are very fast compared to nuclear motions. For slowly varying changes in the internuclear distance the electronic motion adjusts itself to each change in nuclear configuration without undergoing electronic transitions. This behavior is called adiabatic.

added first to form partial resultants, and then to these are added the more weakly coupled partial resultants to give the total resultant **L**. For $L = 0, 1, 2, 3, \ldots$ the energy levels of an atom are designated as S, P, D, F, . . . terms, respectively. The spins $\mathbf{s}_i$ (where $\mathbf{s}_i$ is always equal to $\frac{1}{2}$) are added in the same manner to give a resultant spin **S**. The multiplicity of the atom is $2S + 1$. The terms of an atom (the parity of the atomic state) are even (subscript g) or odd (subscript u) according as $\sum l_i$, summed over all the electrons of the atom, is even or odd.

Unlike Atoms

The quantum number Λ of the molecule formed when an atom with L_1 is brought up to an atom with L_2 is

$$\Lambda = |M_{L_1} + M_{L_2}|,$$

where where $\mathbf{M}_{L_1}$ and $\mathbf{M}_{L_2}$ are the components of the space quantization of $\mathbf{L}_1$ and $\mathbf{L}_2$ with reference to the internuclear axis. (This quantization is caused by the inhomogeneous electric field that arises in the direction of the line joining the nuclei.) All possible Λ values arise that can be generated by combination of all possible M_{L_i} values.

Similarly to atoms, $\Lambda = 0, 1, 2, 3, \ldots$ correspond to molecular states Σ, Π, Δ, Φ, and so on. (Two coinciding Σ states split into two Σ states of different energy, Σ^+ and Σ^-, neither of which can be ascribed to a particular configuration.) A table of these correlations is given in Herzberg [2].

Like Atoms

If two identical atoms are in the same state, some of the resulting molecular states are even or some are odd. (These correlations are tabulated in Herzberg [2].) If the two like atoms are not exactly in the same state (in regard to both symmetry and energy), each of the molecular states occurring for unlike atoms occurs twice for like atoms, once as an odd and once as an even state.

For diatomics composed of either like or unlike atoms the resultant molecular spin vector **S** can take on the values

$$S = (s_1 + s_2), (s_1 + s_2 - 1), (s_1 + s_2 - 2), \ldots, (s_1 - s_2).$$

The multiplicity of the molecular state is $(2S + 1)$.

b. From the United Atom

It should be stressed that this is a purely hypothetical concept which involves splitting the nucleus of the atom into two nuclei [2]. As for the true case of separated atoms, these two hypothetical new nuclei produce an inhomogeneous electric field in the direction of the internuclear axis. **L** and **S** of the united atom in general are considered decoupled by this field (as in Russell–Saunders coupling):

$$\Lambda = |M_L| = L, L - 1, L - 2, \ldots, 0$$

The spin **S** of the diatomic remains the same as in the corresponding atomic state. The correlation diagram is constructed on the basis of the nonintersection of curves which have the same symmetry and spin.

Unlike Atoms

The molecular states are specified by Λ and **S** as described above for building up of states from separated atoms. If $L + \Sigma l_i$ for the united atom is even, a resulting Σ state is Σ^+; if it is odd, the resulting Σ state is Σ^-.

Like Atoms

The resulting molecular states are all even or all odd depending on whether the state of the united atom was even or odd.

c. *From the Molecular Orbitals of the Diatomic Molecule*

Theoretical Background

Separation of Electronic and Nuclear Motion. Because, in general, electrons move with much greater velocities than nuclei, to a first approximation electron and nuclear motions can be separated (Born–Oppenheimer theorem [3]). The validity of this separation of electronic and nuclear motions provides the only real justification for the idea of a potential-energy curve of a molecule. The eigenfunction Ψ for the entire system of nuclei and electrons can be expressed as a product of two functions Ψ_e and Ψ_n, where Ψ_e is an eigenfunction of the electronic coordinates found by solving Schrödinger's equation with the assumption that the nuclei are held fixed in space and Ψ_n involves only the coordinates of the nuclei [4].

The exact Hamiltonian operator may be written as

$$\mathscr{H} = -\sum_A \frac{\hbar^2}{2M_A} \nabla_A^2 - \sum_i \frac{\hbar^2}{2m} \nabla_i^2 + V_{nn} + V_{ne} + V_{ee},$$

where the first term represents the kinetic energy of the nuclei, the second represents the kinetic energy of the electrons, and V_{nn}, V_{ne}, and V_{ee} are the contributions to the potential energy arising from nuclear, nuclear-electronic, and electronic interactions, respectively. If the nuclei were assumed to be fixed in space, the Hamiltonian for the electrons would be

$$\mathscr{H}_e = -\sum_i \frac{\hbar^2}{2m} \nabla_i^2 + V_{ne} + V_{ee}.$$

The remaining terms are represented by $\mathscr{H}_n$:

$$\mathscr{H}_n = -\sum_A \frac{\hbar^2}{2M_A} \nabla_A^2 + V_{nn}$$

and

$$\mathscr{H} = \mathscr{H}_e + \mathscr{H}_n.$$

Ψ_e is defined as the function which satisfies the equation

$$\mathscr{H}_e \Psi_e = E_e \Psi_e,$$

where E_e is the electronic energy.

For the total system

$$\mathscr{H}\Psi_e\Psi_n = E\Psi_e\Psi_n.$$

(This holds only in a "loose" perturbation sense for $\mathscr{H}\Psi = E\Psi$. Ψ should be expanded in a complete set.) In the Born–Oppenheimer approximation

$$(\mathscr{H}_n + E_e)\Psi_n = E\Psi_n$$

since small terms coupling the electronic and nuclear motion have been neglected. (Often by convention V_{nn} is included in $\mathscr{H}_e$ rather than in $\mathscr{H}_n$, so that with $\mathscr{H}_e$ defined in this second manner $\mathscr{H}_e\Psi$ gives rise to the molecular energy.) Thus the molecular energy as a first approximation can be considered as the sum of two terms: the energy that the electrons would have if the nuclei were at rest plus the repulsive mutual energies of these fixed nuclei. The neglected terms may be treated as a perturbation and will give rise to energy terms representing the interaction of electronic and nuclear motions. These perturbations prove of importance later when discussing probabilities for crossing from one potential curve to another.

Group Theory and Quantum Mechanics. [4] If the Schrödinger equation

$$\mathscr{H}\Psi_i = E_i\Psi_i$$

for an atomic or molecular system is subjected to some transformation of coordinates R, which interchanges like particles in the system, then

$$R\mathscr{H}\Psi_i = RE_i\Psi_i.$$

Since R interchanges only like particles, it can have no effect on the Hamiltonian, so that $R\mathscr{H} = \mathscr{H}R$.

R commutes with the constant E_i; therefore

$$\mathscr{H}R\Psi_i = E_iR\Psi_i;$$

that is, the function $R\Psi_i$ is a solution of the Schrödinger equation with the eigenvalue E_i. If E_i is a nondegenerate eigenvalue, then Ψ_i or constant multiples of Ψ_i are the only eigenfunctions satisfying the above equation. If

E_i is k-fold degenerate, then any linear combination of the functions Ψ_{i1}, $\Psi_{i2}, \ldots, \Psi_{ik}$ will be a solution of the above equation; in this case

$$R\Psi_{il} = \sum_{j=1}^{k} \Psi_{ij}\alpha_{jl},$$

where the α_{jl} must satisfy the relation

$$\sum_{j=1}^{k} \alpha_{jl}^2 = 1.$$

The matrices obtained from the coefficients in the expansion of $R\Psi_{il}$ are unitary and form a representation of the group of operations that leave the Hamiltonian unchanged. The set of eigenfunctions $\Psi_{i1}, \ldots, \Psi_{ik}$ forms a basis for the representation of the group. The representations generated by the eigenfunctions corresponding to a single eigenvalue are irreducible representations.

If Γ_j is an irreducible representation of dimension k, and if $\Psi_1^j, \Psi_2^j, \ldots, \Psi_k^j$ is a set of degenerate eigenfunctions that form the basis for the jth irreducible representation of the group of symmetry operations, these eigenfunctions transform according to the relation

$$R\Psi_i^j = \sum_{l=1}^{k} \Gamma_j(k)_{li}\Psi_l^j.$$

For a symmetrical atomic or molecular system, these considerations place a severe restriction on the possible eigenfunctions of the system. All possible eigenfunctions must form bases for some irreducible representation of the group of symmetry operations. The form of the possible eigenfunctions is also determined to a large extent since they must transform in a quite definite way under the operations of the group.

Various LCAO–MO Methods. In order to render tractable the problem of determining the molecular electronic eigenfunction, Ψ_e, it is customary to assume the individual molecular orbitals to be functions of the atomic electron eigenfunctions, χ_r, centered on each atom. The molecular orbitals (MO's), ϕ_i, are taken to be linear combinations of the atomic orbitals (LCAO's), χ_r:

$$\phi_i = \sum_r \chi_r c_{ri}.$$

A configurational wave function Φ (here Φ denotes the molecular electronic eigenfunction Ψ_e) is represented by an antisymmetrized product wave function [5]:

$$\Phi = (N!)\, \phi_1{}^{[1}\, \phi_2 \cdots \phi_N{}^{N]},$$

since only those states can occur whose eigenfunctions are antisymmetric with respect to exchange of any two electrons.

The total wave function Φ is normalized:

$$(\Phi \mid \Phi) = \int \Phi^* \Phi \, d\tau = 1$$

and

$$E_{el} = \frac{\int \Phi^* \mathscr{H}_{el} \Phi \, d\tau}{\int \Phi^* \Phi \, d\tau}.$$

The Rayleigh–Ritz variational method is used to determine the coefficients c_{ri} corresponding to the best approximation to the minimum energy, E_{el}, for the system. It is convenient to make linear combinations of the atomic orbitals that form symmetry orbitals of the molecule in question. The allowed combinations of atomic eigenfunctions that form molecular eigenfunctions of the various symmetry types are determined (for a molecule of any given symmetry) directly from the character table of the irreducible representations of this group. (The various symmetry operations, the classification of molecular symmetry types according to the behavior of molecular electronic eigenfunctions with respect to the symmetry operations, and character tables for all of the groups of diatomic and polyatomic molecules are listed in Herzberg [2, 6].)

Since the discussion of molecular orbitals for the intermediate species involved in collisions will be used extensively in this chapter, it seems appropriate to describe now the various LCAO–MO methods in most common usage for calculating molecular orbitals. This presentation will be general and is equally applicable to diatomic and polyatomic molecules. This description of MO methods is not intended as a detailed comprehensive review of quantum chemical calculational techniques, but rather as an indication of the different methods available.

These molecular orbital computational methods fall into several rather well delineated categories: nonempirical (including electron-electron repulsion and solving all the interatomic integrals exactly), semirigorous (including electron–electron repulsion and solving some of the atomic integrals involved and estimating others), and semiempirical (neglecting electron–electron repulsion and estimating the atomic integrals). A schematic block diagram relating the various theoretical methods is outlined in Chart 2.A.

Non-empirical Methods

The methods under the category of nonempirical fall into two subclasses. The first consists of the well known Hartree–Fock–Roothaan [7, 8] LCAO–MO–SCF (self-consistent field) methods. The second is an even more rigorous

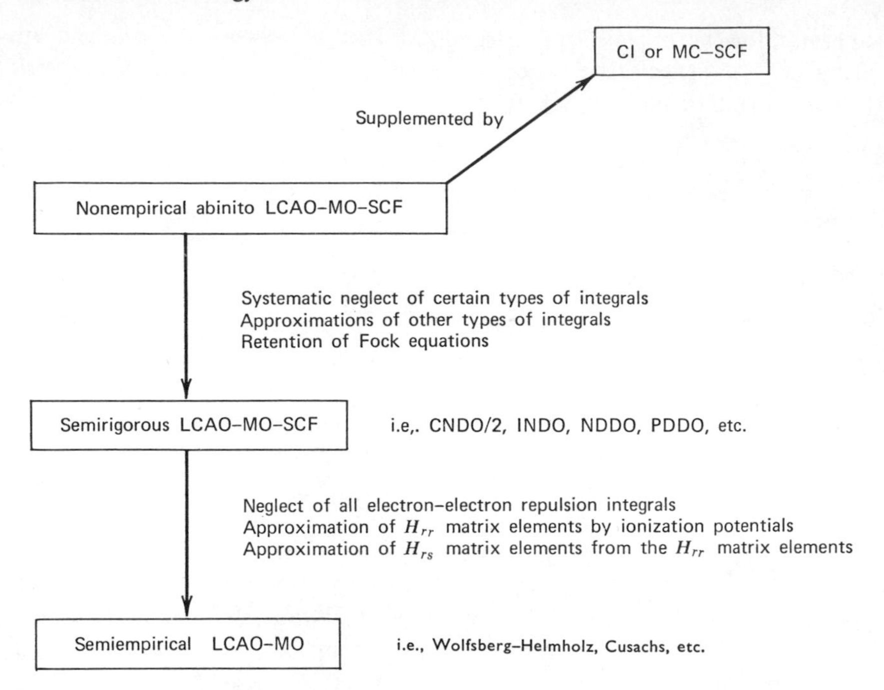

Chart 2.A. Schematic block diagram relating the various theoretical methods.

technique that includes either the effect of configuration interaction on the eigenfunction obtained in a nonempirical calculation or the multiconfiguration SCF method. Configuration interaction or the multiconfiguration SCF method correct for the deficiency in the Hartree–Fock method which is necessary for calculating correlation energy corrections to the molecular energies. For the problem of collisions, appropriate configuration interaction is vital because it leads to separated atoms (or atoms plus molecular fragments) in the proper electronic states, while the Hartree–Fock method, in general, does not lead to separated species in the proper states.

Hartree–Fock–Roothaan: *Closed-Shell Theory*. Here [7], the molecular spin-orbitals ϕ_i, where the subscript labels the different MOs, are functions of (x^μ, y^μ, z^μ) (where μ stands for the coordinate of the μth electron) and a spin function. The configurational wave function is represented by a single determinantal antisymmetrized product wave function. The total Hamiltonian operator $\mathscr{H}$ is defined by

$$\mathscr{H} = \sum_{\mu} H^\mu + \tfrac{1}{2}e^2 \sum_{\mu \neq v} \frac{1}{r^{\mu v}};$$

H^μ is the Hamiltonian operator for the nuclear field plus the kinetic-energy operator for the μth electron; $r^{\mu\nu}$ is the distance between the μth and νth electron. The expectation value of the energy becomes a sum of integrals over the orbitals, namely,

$$\langle\Phi|\,\mathscr{H}\,|\Phi\rangle = 2\sum_i H_i + \sum_{ij}(2J_{ij} - K_{ij}),$$

where

$$H_i = \langle\phi_i|\,H\,|\phi_i\rangle,$$

$$J_{ij} = \langle\phi_i|\,\mathscr{J}_j\,|\phi_i\rangle = \langle\phi_j|\,\mathscr{J}_i\,|\phi_j\rangle,$$

$$K_{ij} = \langle\phi_i|\,\mathscr{K}_j\,|\phi_i\rangle = \langle\phi_j|\,\mathscr{K}_i\,|\phi_j\rangle$$

and

$$\mathscr{J}_i^{\mu}\phi^{\mu} = \left(\int \phi_i^{\nu}\phi_i^{\nu}\frac{1}{r^{\mu\nu}}\,dV^{\nu}\right)\phi^{\mu},$$

$$\mathscr{K}_i^{\mu}\phi^{\mu} = \left(\int \phi_i^{\nu}\phi^{\nu}\frac{1}{r^{\mu\nu}}\,dV^{\nu}\right)\phi_i^{\mu}.$$

$\mathscr{J}_i$ and $\mathscr{K}_i$ are commonly called the Coulomb and exchange operators, respectively, associated with the orbital ϕ_i; they are defined by how they operate on an arbitrary one-electron function ϕ.

The Hartree–Fock equation is obtained by requiring that the orbitals minimize the expectation value of the energy. Those orbitals satisfy

$$\mathscr{F}\phi = \phi\varepsilon,$$

where the Hartree–Fock Hamiltonian operator $\mathscr{F}$ is given by

$$\mathscr{F} = H + \sum_i (2\mathscr{J}_i - \mathscr{K}_i),$$

and ε is a Hermitian matrix of Lagrangian multipliers that are introduced by the orthonormality constraints. We may single out that set of orbitals ϕ_i in which ε becomes diagonal, so that all the orbitals satisfy

$$\mathscr{F}\phi_i = \varepsilon_i\phi_i.$$

This is a pseudoeigenvalue problem, inasmuch as the operator $\mathscr{F}$ is defined in terms of the solutions. This is called the Hartree–Fock equation.

One guesses at an initial set of wave functions, ϕ_i, and constructs the Hartree–Fock Hamilton $\mathscr{F}$ which depends on the ϕ_i through the definitions of the Coulomb and exchange operators, $\mathscr{J}_i$ and $\mathscr{K}_i$. One then calculates the new set of ϕ_i and compares it (or the energy or the density matrix) to the input set (or to the energy or density matrix computed from the input set). This procedure is continued until the appropriate self-consistency is obtained.

The total energy can be expressed in terms of the orbital energies ε_i and the one-electron integrals, namely,

$$E = \sum_i (H_i + \varepsilon_i).$$

The term $-\varepsilon_i$ is approximately equal to the ionization potential for the removal of an electron occupying ϕ_i. (This is known as Koopmans' theorem [9], and it assumes that the molecular orbitals of the molecule and its positive ion are the same.)

Open Shell Theory

The total wave function for open-shell systems [8] is, in general, a sum of several antisymmetrized products, each of which contains a closed-shell core and a partially occupied open shell. The combinal set of orbitals ϕ is defined by

$$\phi = (\phi_c, \phi_o)$$

and is assumed to be orthonormal, so that the two sets ϕ_c and ϕ_o are orthonormal and mutually orthogonal. The indices k, l refer to closed-shell orbitals; m, n refer to open-shell orbitals; and i, j refer to orbitals of either set. The expectation value of the energy is given by

$$E = 2\sum_k H_k + \sum_{kl} (2J_{kl} - K_{kl})$$
$$+ f\left[2\sum_m H_m + f\sum_{mn} (2aJ_{mn} - bK_{mn}) + 2\sum_{km} (2J_{km} - K_{km})\right],$$

where a, b, and f (the fractional occupancy) are numerical constants depending on the specific case. The first two sums represent the closed-shell energy, the next two sums the open-shell energy, and the last sum the interaction energy of the closed and open shell. The variational method is applied to the above equation. Without going into the details of the derivation, the problem is reduced to a pseudoeigenvalue problem in which the closed- and open-shell orbitals are solutions of the same eigenvalue equation,

$$\mathscr{F}\phi_i = \varepsilon_i \phi_i,$$

instead of the equivalent set ϕ_l, ϕ_m satisfying the equations

$$\mathscr{F}_c\phi_k = \eta_k\phi_k, \qquad \mathscr{F}_o\phi_m = \eta_m\phi_m.$$

(For details the reader is referred to Roothaan's original paper [8].)

Configuration Interaction and Multiconfiguration SCF: Configuration Interaction. A single determinantal wave function does not represent an exact solution of Schrödinger's equation; by determining the spin-orbitals according to the Hartree–Fock equations, one gets the best approximate solution in the form of a single determinant. To obtain a better solution, one must set up many determinantal functions, formed from different spin-orbitals, and must use an approximate wave function that is a linear combination of these determinantal functions, with coefficients to be determined by minimizing the energy. This process is called configuration interaction (CI) or superposition of configurations.

The determinantal functions must be linearly independent and eigenfunctions of the spin operators S^2 and S_z, and preferably they belong to a specified row of a specified irreducible representation of the symmetry group of the molecule [10, 11]. Definite spin states can be obtained by applying a spin projection operator to the spin-orbital product defining a configuration [12]. Suppose Φ_0 to be the solution of the Hartree–Fock equation. From functions of the same symmetry as Φ_0 one can build a wave function Φ,

$$\Phi = a_0\Phi_0 + a_1\Phi_1 + a_2\Phi_2 + \cdots.$$

By optimizing the orbitals in each function and by variationally selecting the CI coefficients $a_0, a_1, a_2, \ldots$, one gets an eigenfunction as good as or better than Φ_0. If the series is sufficiently long, one can reach an exact solution. The trouble is that the necessary series is too long and usually converges slowly. In practice a truncated series is used.

Multiconfiguration SCF (MC–SCF–LCAO–MO). If $\Phi_1, \Phi_2, \ldots$ are constructed at the same time as Φ_0 from a common orthonormal set of orbitals, and one solves, not for the best possible Φ_0, but for the best Φ; then the variational principle, used simultaneously on both the a (the CI coefficients) and the χ (the atomic orbitals), will ensure that the Φ_i will overlap as much as possible.

Complete multiconfiguration–self consistent-field (CMC–SCF) technique designates the method where a given occupied molecular orbital of the set is excited to all unoccupied molecular orbitals. If an occupied orbital is excited to one or more, but not all, of the unoccupied orbitals, the technique is described as incomplete MC–SCF (IMC–SCF). The reader is referred to refs. 13 and 14 for details of the derivation. The CMC–SCF formalism differs from most many body techniques presented to date insofar as the Hartree–Fock energy is not assumed to be the zero order energy.

Semirigorous Methods

Semirigorous LCAO–MO–SCF methods start with the complete many-electron Hamiltonian and make certain approximations for the integrals and for the form of the matrices to be solved. Several years ago, such a method was derived starting with the correct many electron Hamiltonian (in which interelectronic interactions are included explicitly) and the LCAO–MO–SCF equations of Roothaan and then making a consistent series of systematic approximations for the integrals involved [15]. The molecular orbitals ϕ_i were still represented by a linear combination of atomic orbitals χ_r,

$$\phi_i = \sum_r \chi_r c_{ri},$$

and the notation for the integrals is defined as

$$(a'c''' \,|G|\, b''d^{\prime\mathrm{v}}) = \iint a'^*(1)c'''^*(2)(r_{12})^{-1}b''(1)d^{\prime\mathrm{v}}(2)\, dv_1\, dv_2.$$

There were several different levels of approximation that evolved, depending on how restrictive one made the conditions for neglecting $a'(1)a''(1)$, where a' and a'' are two orbitals on atom A, and $a'(1)b''(1)$, where a is an orbital on atom A and b is an orbital on atom B, $A \neq B$, and for neglecting $(a'c''' \,|G|\, b''d^{\prime\mathrm{v}})$, where a, b, c, and d are orbitals on atoms A, B, C, and D, respectively. The following set of successively less restrictive approximation methods was outlined:

A. $a'^*(2)a''(2) = 0$, $a' \neq a''$ for any atom A.
B. $a'^*(2)b''(2) = 0$, $B \neq A$, and $(a'c''' \,|G|\, b''d^{\prime\mathrm{v}}) = 0$ unless $B = A$ and $D = C$, but in general $b'^*(2)b''(2)$ need not be zero; all one-center integrals and two-center Coulomb integrals allowed.
C. $(a'c''' \,|G|\, b''d^{\prime\mathrm{v}}) \neq 0$ also in the case $A = D$ and $B = C$. In addition all two-center exchange integrals allowed.
D. $(a'c''' \,|G|\, b''d^{\prime\mathrm{v}}) \neq 0$ also in the case $A = B = C$. In addition all two-center hybrid integrals allowed.

In the article procedures were presented for estimating the various integrals involved: $H^{\mathrm{core}}_{a'a'}$, $H^{\mathrm{core}}_{a'a''}$, and $H^{\mathrm{core}}_{a'b''}$. Preliminary calculations were made on a test system and these results compared both with experiment and with those of rigorous non-empirical computations. These results indicated that method A seemed not to be a physically justifiable method and that one should go at least to method B. The two most restrictive approximations A and B of this semirigorous scheme are similar conceptually to the CNDO (complete neglect of differential overlap) method published by Pople at the same time and to his later revisions, the NDDO (neglect of diatomic differential overlap) and INDO (intermediate neglect of diatomic differential overlap) methods (where only valence orbitals and valence electrons are included) [16–23]. Method A resembles Pople's CNDO method. Method B resembles Pople's NDDO or INDO methods.[1]

[1] Subsequent to the original completion of this chapter, a book and several useful review articles on all valence electron semirigorous self-consistent-field calculations appeared [23–26]. For details of the CNDO and INDO methods the reader is referred to ref. 23. For a comparative and critical review of various semirigorous methods, the reader is referred to refs. 24–26.

The CNDO/2 method (a particular type of parametrization of the CNDO method) has been and is being used widely by various investigators to determine approximate electronic wave functions and energies.

Semi-Empirical Methods

The semi-empirical extended Hückel method, which takes into account all valence electrons in a molecule, was introduced by Wolfsberg and Helmholz [27]; it has been used over the years extensively by Lipscomb and co-workers, especially Hoffmann, [28–31] and has been applied by a large number of investigators.

From a molecular orbital ϕ_i built up as a linear combination of atomic orbitals χ_r,

$$\phi_i = \sum_r \chi_r c_{ri},$$

and by application of the variation principle for the variation of energy, the following set of equations for the expansion coefficients is obtained:

$$(\alpha_r + ES_{rr})c_r + \sum_{r \neq s} (\beta_{rs} - ES_{rs})C_s = 0,$$

$$S_{rs} = \int \chi_r^* \chi_s \, dv = \text{overlap integral},$$

$$H_{rr} = \alpha_r = \int \chi_r^* \mathscr{H} \chi_r \, dv = \text{Coulomb integral},$$

$$H_{rs} = \beta_{rs} = \int \chi_r^* \mathscr{H} \chi_s \, dv = \text{resonance integral} \quad (r \neq s).$$

$\mathscr{H}$ is an effective one-electron Hamiltonian representing the kinetic energy, the field of the nuclei, and the smoothed out distribution of the other electrons. Electron–electron repulsion is neglected in this method.

The diagonal elements are set equal to the effective valence-state ionization potentials of the orbitals in question. The off-diagonal elements, H_{rs}, can be evaluated in several ways. The two expressions in most common usage are the original Wolfsberg–Helmholz expression,

$$H_{rs} = 0.5k(H_{rr} + H_{ss})S_{rs}$$

with $k = 1.75 - 2.00$ (Hoffmann set $k = 1.75$) and the more recent Cusachs' expression [32]

$$H_{rs} = \frac{(H_{rr} + H_{ss})}{2} S_{rs}(2 - |S_{rs}|).$$

The total electronic energy of a particular system is taken to be the sum of the orbital energies, ε_i, times their occupation numbers. For a closed-shell system $E_{\text{elect}} = 2\sum \varepsilon_i$. The total molecular energy can be written as

$$E = 2\sum \varepsilon_i + \sum_{n,n'} E_{nn'} - \sum_{e,e'} E_{ee'},$$

where $E_{nn'}$ and $E_{ee'}$ are nuclear–nuclear and electron–electron repulsion energies. The success of these extended Hückel calculations in predicting preferred geometrical conformations from calculated minimum energies lies in the fact that the method of selecting the H_{rs} values must simulate, within the calculated electronic energies, the contribution of nuclear repulsions to the total energy [30]. The nuclear–nuclear and electron–electron repulsion energies cancel approximately [5], and thus the simple sum of one-electron energies behaves similarly to the true molecular energy.

Because of neglect of electron–electron repulsion, the calculated energy values are, unfortunately, equal for the identical molecular states with different multiplicities.

Electronic Configuration

The electrons are fed into the molecular orbitals according to the *aufbau* principle. First the allowed orbitals are determined and then the electrons are fed one at a time into these levels, beginning with the lowest, and satisfying the Pauli exclusion principle by allowing only two electrons to each of the orbitals. For the ground state of the molecule all of the lowest molecular orbitals are filled, and in the case of non-totally filled degenerate orbitals, the electrons are fed in according to Hund's rule where the state of lowest energy is almost invariably the state of highest multiplicity. For excited states the electrons are fed into all of the low-lying molecular orbitals with now one or more of the electrons from higher-lying filled orbitals being promoted into upper levels.

To derive the term type (species) from the electronic configuration, Russell–Saunders coupling is assumed as defined eaılier,

$$\mathbf{\Lambda} = \sum \boldsymbol{\lambda}_i,$$

$$\mathbf{S} = \sum \mathbf{S}_i,$$

where, for nonequivalent electrons, the quantities $\boldsymbol{\lambda}_i$ and $\mathbf{S}_i$ are added vectorially in all possible combinations. If the electrons are equivalent, the Pauli principle must be taken into account when adding the $\boldsymbol{\lambda}_i$ and the $\mathbf{S}_i$; the electrons must differ in m_l or m_s. The energy difference between corresponding states of different multiplicity is due to the electrostatic interaction of the electrons. The state with the greatest multiplicity almost invariably lies lower in energy.

If equivalent as well as nonequivalent electrons are present, the resulting states are found by first forming the resulting states of each group of equivalent electrons and then forming the direct product of the species so obtained. Since closed shells always give a single totally symmetric singlet state, they can be entirely neglected in the determination of the resulting states.

Herzberg gives tables of the terms arising from nonequivalent electrons (diatomics [2]; polyatomics [6]), equivalent electrons (diatomics [2]; polyatomics [6]), and equivalent as well as nonequivalent electrons (diatomics [2]).

In addition, for diatomics consisting of like atoms (or certain symmetrical linear polyatomic molecules), the resulting states are also either even (g) or odd (u). They are even if the number of "odd" electrons (σu, πu, . . .) is even, whereas they are odd if the number of "odd" electrons is odd.

These terms of electron configurations arising from different partially filled orbitals are very valuable in deciphering what excited electronic states of molecules and intermediates are possible and lie in an energy range accessible to the particular experiment.

The ordering of the molecular orbitals is dependent on the relative positions of the nuclei. In analyzing a collisional problem, care must be taken to feed the electrons into the appropriate molecular orbitals correlating the reactants or products with the proper intermediate state as permitted by the symmetry and spin restrictions outlined above.

The above discussion of the LCAO–MO method and the terms of the electronic configurations is not restricted to diatomic molecules. It is general and completely applicable to polyatomic molecules; hence, the emphasis in this chapter on the correlation between the reaction intermediates arising from states of the separated atoms (or in the next section on polyatomics from the separated molecular fragments and atoms) and arising from the molecular orbitals of the intermediates.

Some references to particularly pertinent molecular-orbital or configuration-interaction calculations of potential curves of diatomic molecules will be mentioned below in this section. These references are not intended to be exhaustive but should serve to indicate some of the calculations which have been carried out and the type of information which is available. (The most comprehensive compilation of nonempirical molecular-orbital calculations is a 1967 survey by Krauss [33], which includes both diatomic and polyatomic molecules.)

Note Added in Proof

In order to update the material in this chapter references 185 through 299 have been added. In some instances substantial additions to the text have been included as well. They can be identified by the associated reference numbers.

With the advent of great interest in possible electronic excitation molecular chemical lasers and selective preparation of excited atomic lasers, this entire area of potential surface considerations and, especially, for excited state reactions takes on an even greater significance. The basic concepts and references remain the same.

During the period since the latest revision of this chapter in 1972, a book has appeared by Schaefer on a survey of rigorous quantum mechanical results of the electronic structure of atoms and molecules [185].

A promising semi-rigorous molecular orbital method, PRDDO, partial retention of diatomic differential overlap has been reported in the interim [186]. Also, the method of diatomics-in-molecules has been revived and the derivations extended to include p orbitals appropriately [187].

Unlike Atoms—Heteronuclear Diatomic Molecules

Heteronuclear diatomic molecules belong to the symmetry group $C_{\infty v}$. States which are invariant under rotation about the symmetry axis are called Σ states. These are the states for which $\Lambda = 0$. Π states are those for which $\Lambda = 1$, Δ states those for which $\Lambda = 2$, and so on. In addition, the Σ states are further characterized by the designation $+$ and $-$ according to whether they remain invariant or change sign when subjected to the operation σ_v (reflection in a plane in which the symmetry axis lies).

Diatomic Hydrides—AH. A listing of electronic configurations and term types for the ground and first excited states of the higher diatomic hydrides as well as the first- and second-row diatomic hydrides is given in Herzberg [2].

Very accurate Hartree–Fock calculations of the electronic structure of first and second row diatomic hydrides have been performed by the University of Chicago group [34] and are now being intensified by the IBM research group at San Jose [35, 36]. For diatomic molecules, these calculations are literally the criterion of excellence for such Hartree–Fock computations. The ordering of occupied molecular orbital energy levels for hydrides and the term of the ground states of these hydrides to which these configurations correspond are listed at top of facing page [2].

Cade's article [34] also contains tables of computed total molecular energies as a function of distance from which potential curves of the ground states can be constructed. It should be reemphasized that the Hartree–Fock calculations do not, in general, separate to the correct states of the dissociated atoms. One case which does separate to the correct states of the dissociated atoms is $Li^+ + He$. A very accurate Hartree–Fock potential curve (which picked up all of the long-range ion-induced dipole attraction) was calculated using Gaussian basis functions, and with this curve the quantum scattering

AH	Configuration	Term
LiH	$1\sigma^2 2\sigma^2$	$X\,^1\Sigma^+$
BeH	$1\sigma^2 2\sigma^2 3\sigma$	$X\,^2\Sigma^+$
BH	$1\sigma^2 2\sigma^2 3\sigma^2$	$X\,^1\Sigma^+$
CH	$1\sigma^2 2\sigma^2 3\sigma^2 1\pi$	$X\,^2\Pi$
NH	$1\sigma^2 2\sigma^2 3\sigma^2 1\pi^2$	$X\,^3\Sigma^-$
OH	$1\sigma^2 2\sigma^2 3\sigma^2 1\pi^3$	$X\,^2\Pi_i$
HF	$1\sigma^2 2\sigma^2 3\sigma^2 1\pi^4$	$X\,^1\Sigma^+$
NaH	$KL4\sigma^2$	$X\,^1\Sigma^+$
MgH	$KL4\sigma^2 5\sigma$	$X\,^2\Sigma^+$
AlH	$KL4\sigma^2 5\sigma^2$	$X\,^1\Sigma^+$
SiH	$KL4\sigma^2 5\sigma^2 2\pi$	$X\,^2\Pi_r$
PH	$KL4\sigma^2 5\sigma^2 2\pi^2$	$X\,^3\Sigma^-$
SH	$KL4\sigma^2 5\sigma^2 2\pi^3$	$X\,^2\Pi_i$
KCl	$KL4\sigma^2 5\sigma^2 2\pi^4$	$X\,^1\Sigma^+$

where $K \equiv 1\sigma^2$ and $L \equiv 2\sigma^2 1\pi^4 3\sigma^2$ or $2\sigma^2 3\pi^2 1\pi^4$. The orbital configuration is written in order of decreasing orbital energies.

calculations were performed reproducing experiment [37]. (A CI calculation on this same system [38] gave too shallow a well since the atoms were over-correlated relative to the intermediate.)

Hartree-Fock calculations have been carried out for the $^7\Sigma^+$ and $^7\Pi$ states of MnH [188].

Some potential curves of ground or excited states of first-row diatomic hydrides have been calculated by the valence-bond method or other types of configuration-interaction approach.

An accurate CI study of all first-row diatomic hydrides used large normalized STO basis sets [39]. An SCF calculation was performed, then the CI was solved with only selected single excitations and natural orbitals were constructed from this function. The natural orbitals with the largest occupation numbers were defined to be the occupied orbitals for a single configuration CI wave function. The space configurations were selected using an energy-contribution criterion. Other recent pertinent examples are OH [40] (which included extensive CI for the ground and a number of excited states) and HF^- (dissociating to $H^- + F$ and $H + F^-$) [41].

For the particular case of H_2, very accurate calculations including correlation effects have been carried out for the potential curves of ground and some excited states of H_2 by Kolos and Wolniewicz [42–44] and by Davidson [45–47]. Extended Hartree–Fock calculations for the ground state [48] and double configuration SCF calculations for some excited states of H_2 [49] have been carried out by Wahl and co-workers. For H_2^+ a very rigorous calculation has been carried out by Bates and co-workers [50].

Configuration-interaction curves for the ground state and a number of excited states of HeH^+ were calculated by Michels [51]. Similar studies of HeH have been carried out by Harris and co-workers [52, 53], by Davidson and Bender [54], and by Schaefer [55].

CI calculations have been carried out for the lowest six Σ states of HeH^+ [189] and for HeH [190]. MC-SCF calculations have been carried out for LiH $\tilde{X}\,^1\Sigma^+$ and the excited states $A\,^1\Sigma^+$, $B\,^1\Pi$, $^3\Sigma^+$, $^3\Pi$ [191] and generalized valence bond (GVB) calculations have been carried out for the first five $^1\Sigma^+$ states, the first four $^3\Sigma^+$ states and the first three $^1\Pi$ and $^3\Pi$ states [192]. CI calculations have been carried out for BeH $X\,^2\Sigma^+$, $A\,^2\Pi$ [193]. CI calculations have been carried out for BH $\tilde{X}\,^1\Sigma^+$, $B\,^1\Sigma^+$, $b^1\,^3\Sigma^+$, $C\,^3\Sigma^+$ [194] and for $B\,^1\Sigma^+$ and lowest $^3\Sigma^+$ [195]. GVB (GI and SØGI calculations have been carried out for BH $X\,^1\Sigma^+$, $a\,^3\Pi$, $A\,^1\Pi$, and $^3\Sigma^+$ [196, 197]. Extensive CI calculations have been carried out for the $X\,^1\Sigma^+$, $A\,^1\Pi$, $^3\Pi$ and $^3\Sigma^+$ states of CH^+ [198] and the $a\,^4\Sigma^-$, $A\,^2\Delta$, and $B\,^2\Sigma^-$ state of CH [199] by the IBM group at San Jose. Other CI calculations have been performed for CH^+ excited states [200, 201]. CI calculations have been carried out for HF^+ [202] and for OH, HF, HF^-, NeH^+, HeH [203], NH [204] and HCl [205]. Among the most interesting recent developments is ab-initio calculations of potential energy surfaces in the complex plane. It was earlier pointed out that the main features of CI treatment of avoided crossings of potential energy curves of identical spin and symmetry could be reproduced by single determinant wave functions provided that the molecular orbitals are permitted to be complex [206]. The motivation for this recent development [207, 208] stemmed from the interest of the collision theorists in the analytic continuation of potential energy surfaces for complex values of nuclear coordinates. The theory was applied to a simple model system, HeH^{2+} of two potential curves with an avoided crossing for real R. Their crossings and electron distributions were investigated in the complex R plane.

Diatomics—AB. The ordering of the molecular orbitals for diatomics composed of two first-row atoms near their equilibrium internuclear distance is

$$(1\sigma)^2\ (2\sigma)^2\ (3\sigma)^2\ (4\sigma)^2\ (5\sigma)^2\ (1\pi)^4\ (2\pi)^4,$$

where (5σ) and (1π) are sometimes interchanged. The scheme continues similarly for diatomics of a first- and second-row atom or two second-row atoms.

Nonempirical potential-energy curves have been calculated for a number of heteronuclear diatomic molecules. Hartree–Fock curves include a number of unpublished ones from IBM [56–58], which are listed in the NBS report [33]: among them LiF, LiCl, BeO, BF, CO, NaF, NaCl, MgO, AlF, SiO, PN, CaO, SrO, and others, which were also mentioned at the recent Sanibel

Conference [36]; some rare-gas–rare-gas curves [59]: HeNe, HeAr, NeAr; a series of papers by Wahl and co-workers: OF, SF, SeF [60], and CaO [61]; and so on.

Among the interesting Hartree-Fock calculations for potential curves of AB diatomics are included $Be^{2+}\cdot He$, $Be^{2+}\cdot Ne$ [209], NaO and its ions [210], pnicogen monofluorides [211], CF, SiF and their positive and negative ions [212], LiHe and NaHe($\tilde{X}\,^2\Sigma^+$, $A\,^2\Pi$, $B\,^2\Sigma^+$) and their $X\,^1\Sigma^+$ ions [213] and PbO [214].

Curves that go beyond the Hartree–Fock method have been calculated for certain systems, and the list grows monthly. Of special interest is the series on ground [62–64] and excited states [63, 64] of CO, on NaLi and $NaLi^+$ using an extended Hartree–Fock method with optimized double-valence configurations [65], HeLi by a valence-bond method [66], and so on. A quite complete listing of all nonempirical potential-energy curves calculated through 1967 is included in the NBS report [33].

CI calculations have been carried out for a number of species including LiO [215], Li + Na [216], BeO [217, 218], CN [219], CO [220], SiO [221], FeO (the lowest $^5\Sigma^+$ state and related states) [222]. At the very recent conference on electronic transition lasers, CI calculations were reported for the ground and excited states of BeO and a few selected points for MgO, BaO, and AlF [223].

Like Atoms—Homonuclear Diatomics—AA

The ordering of molecular orbitals for first-row homonuclear diatomics follows the same pattern as for first-row heteronuclear diatomics, with the addition of the *g* and *u* subscripts for even or odd behavior of the wave functions:

$$(1\sigma_g)^2\ (1\sigma_u)^2\ (2\sigma_g)^2\ (2\sigma_u)^2\ (3\sigma_g)^2\ (1\pi_u)^4\ (1\pi_g)^4.$$

Again the $(3\sigma_g)$ and the $(1\pi_u)$ orbitals may be interchanged.

Among potential-energy curves of particular interest (some including various excited states) are those going beyond the Hartree–Fock approximation and thus having the property of asymptotically connecting to the correct atomic states for He_2^{++} [67], He_2^+ [67–69], He_2 [70–75], Li_2 [76, 77], Be_2 [78], B_2 [79], C_2 [80], N_2 [63, 81, 82] (ref. 81 is a very extensive Hartree–Fock SCF calculation), O_2 [83, 84], F_2 [76, 85–87], and Cl_2 [87, 88]. There are some Hartree–Fock calculations of pertinence on other systems that have not yet had molecular orbitals calculated by a more rigorous technique. These include F_2^- [89, 90], Ne_2 [91], and Ar_2 [91].

The He–He system has generated considerable interest. MC–SCF calculations have been performed for He–He [224], CI calculations were performed for the He–He Van der Waal's interaction [225], large CI calculations have been performed for the $^1\Sigma_g^+$, $^3\Sigma_u^+$, $^5\Sigma_g^+$ states arising from triplet metastable He atoms [226], and for a number of He_2 states [227]. He_2^+ excited states have been investigated by GVB [228]. MC–SCF calculations have been performed for F_2 [229] and Hartree-Fock calculations have been performed for the $^2\Sigma_g^+$ and $^2\Pi_u$ states of Li_2^+ [230]. Generalized VB calculations have been carried out for the ground and excited states of N_2 [231, 232]. CI calculations have been carried out for O_2 [233, 234] and for the three lowest states of O_2 [235]. MC–SCF calculations have been carried out for O_2^- [236]. GVB calculations have been carried out for the low lying Rydberg states of O_2 [237].

There is great interest in the potential energy surfaces arising from ground and excited rare gas atoms since these form an electron lasing system by transition from the excited state dimer to the repulsive ground state dimer dissociates into two ground state atoms. Calculations have just been reported for the ground and excited states of Ne_2 [238].

2. Polyatomics

There are four different schemes for building up the electronic states of polyatomic molecules: (a) from the united atom or molecule, (b) by starting from a geometrical conformation of different symmetry (either higher or lower), (c) from separated atoms or molecular fragments, and (d) from the molecular orbitals of the polyatomic molecule. The most useful correlations for elucidation of the mechanism of collisions involving polyatomic intermediates are those between the states of the molecule formed from separated atoms or molecular fragments and those built up from the molecular orbitals of the polyatomic intermediate. In all of the correlations, as long as spin-orbit coupling is small, there must be spin correlation; singlets correlate with singlets, doublets with doublets, triplets with triplets, and so on.

a. From the United Atom or Molecule

To obtain the molecular electronic states corresponding to a given state of the united atom, the species of the united atom must be resolved into the species of the point group to which the molecule belongs. (This is not a simple one-to-one transformation, since the united atom must first be transformed into a hypothetical diatomic which is then further transformed into a hypothetical polyatomic.) This resolution is performed by finding the characters for the symmetry elements of the point group of the molecule. Herzberg's *Polyatomics* [6] gives tables with the resolution of the first ten species of the spherical point group of free atoms into those of point groups O_h, T_d, $D_{\infty h}$, D_{6h}, D_{4h}, C_{3v}, D_{2d}, D_{2h}, C_{2v}, and C_s and with the resolution of the first 12 species of linear molecules (point groups $D_{\infty h}$ and $C_{\infty v}$) into those of molecules of lower symmetry: D_{6h}, C_{6v}, D_{4h}, D_{3h}, C_{3v}, D_{2d}, D_{2h}, C_{2v}, C_{2h}, and C_s. The table includes the correlation for several possible orientations of the symmetry elements $D_{\infty h}$ with respect to those of the point group under consideration. The question of the orientation of the fragments to one another is vital in considering the question of collisions, and the table following in Herzberg [6] gives the resolution of point groups C_{2v}, D_{2h}, D_{3h}, D_{4h}, and T_d into those of point groups of lower symmetry. These correlations were first derived by Bethe [92] and Mulliken [93].

b. From a Geometrical Conformation of Different Symmetry

For the question of collisions this correlation is of importance because during the collisional process the symmetry of the intermediate often changes. For example, even if a completely symmetrical intermediate complex results during the reaction of, say, an atom plus a diatomic, the complex is only completely symmetrical when it is in its equilibrium conformation. During the formation and break-up, the intermediate has had a conformation of lower symmetry. A table in Herzberg's *Polyatomics* gives the correlation of species of different point groups corresponding to different conformations of a given molecule: $D_{2h} \rightarrow D_2 \rightarrow D_{2d}$, $D_{3d} \rightarrow D_3 \rightarrow D_{3h}$, $C_{2v} \rightarrow C_2 \rightarrow C_{2h}$ [6]. Herzberg points out that while the correlations of infinite species of a point group of higher symmetry into those of a point group of lower symmetry are always unambiguous, the reverse correlations are not always unambiguous. The actual correlation depends on the path chosen. It is straightforward to decipher which states of a distorted molecule arise from certain states of a more symmetric molecule; the reverse correlation is not always unambiguous.

c. From Separated Atoms or Molecular Fragments

The correlation between the electronic states of a polyatomic molecule and those of the separated atoms or molecular fragments can be derived from a generalization of the Wigner–Witmer correlation rules for diatomic molecules.

Linear Molecules

Unsymmetrical Molecules ($C_{\infty v}$). The possible values for the quantum number Λ of the molecule are given by algebraic addition of the M_{L_i} values of all atoms (or fragments):

$$\Lambda = |\sum M_{L_i}|$$

where $M_{L_i} = L_i, L_i - 1, L_i - 2, \ldots, -L_i$. When an atom of angular momentum $\mathbf{L}$ approaches a linear diatomic of angular momentum Λ_d with the formation of a linear complex, the resultant Λ is given by

$$\Lambda = |M_{L_i} + M_{L_2}|,$$

where $M_{L_i} = L, L - 1, \ldots, -L$ and $M_{L_2} = \pm\Lambda_d$.

When two diatomics or two linear polyatomics approach to form a linear intermediate, the resultant Λ is given by the sum of the M_{L_i} $(= \pm\Lambda_i)$. A most useful table of both of these correlations is given in Herzberg [6]. The resultant permitted spins $\mathbf{S}$ are obtained by vector addition of the individual spins $\mathbf{S}_i$:

$$\mathbf{S} = \sum \mathbf{S}_i.$$

If the states are built up from separated atoms then partial resultants have to be formed according to

$$S_{ik} = S_i + S_k,\ S_i + S_k - 1, \ldots, |S_i - S_k|,$$

which are then added, also using partial resultants. The permitted multiplicities of the states are $|2S + 1|$. Each of the resultant states occurs with each of the possible multiplicities.

A table of molecular electronic states of linear molecules resulting from certain states of the separated atoms is given in Herzberg [6].

Symmetrical Molecules ($D_{\infty h}$). If two like linear fragments (but in different electronic states) are brought together, the correlations given in the previous section are the ones used. Each state of that table now occurs twice since resolution of the resonance degeneracy leads to splitting into a *g* and a *u* state. If two identical linear fragments (in identical electronic states) are brought together, there is no resonance degeneracy and only the same total of states arises as for unequal groups. The symmetry character of the states alternates for different possible spin values, and therefore some of the states are *g* and some are *u*. Table 23 in Herzberg's *Polyatomics* [6] lists the electronic states of symmetrical linear molecules ($D_{\infty h}$) resulting from identical states of the separated equal groups.

Nonlinear Molecules

The building up of nonlinear molecules from individual atoms leads to a very great number of molecular states. The procedure is outlined in Herzberg's *Polyatomics* [6] but will not be discussed here since it is not easily usable for the purposes of studying possible collisional intermediate complexes. Much more useful for this purpose is the building up of the intermediate molecules from separated fragments.

Unlike Groups. If neither of the two parts that are to be combined has a symmetry lower than that of the final molecule and if the full symmetry is retained during the approach of the two parts, then the correlation is the same as if the molecule were built up from separated atoms.

If the symmetry of at least one of the parts is lower than that of the complete molecule and if during the approach of the parts the molecule has the geometrical conformation of the part with the lower symmetry, the resulting states are obtained in terms of the lower symmetry. If in the final molecule the symmetry is now higher, the correlation rules between deformed conformations are used to find the relation of the states of the intermediate to the states of the final molecule.

A third case is that in which the symmetry of both parts is higher than that of the final molecule being built up but where the symmetry of the intermediate during the formation process is lower than that of the final molecule. The states of both parts must be resolved into the species of the final molecule.

The resulting molecular states of the final molecule are obtained by multiplication of the states of the two parts. For symmetric final molecules this resolution is not completely unambiguous.

Like Groups. If two like fragments are brought together (even if the symmetry of the parts is lower than that of the final molecule), it is usually possible to determine unambiguously the resulting molecular states. This is possible because even at large separation of the two fragments the full symmetry of the molecule may exist.

If the two like fragments are in different electronic states, there is a resonance degeneracy that leads to a splitting into a symmetric and an antisymmetric state. If the two fragments are identical (in the identical electronic states), there is no resonance degeneracy. The symmetries of the resulting states are either g or u or else $'$ or $''$. These symmetries are determined by using the correlation with the corresponding diatomic or linear polyatomic molecules. (A paper by Shuler [94] in which adiabatic orbital and spin correlation rules applicable to study of elementary chemical reactions involving nonlinear polyatomic intermediate complexes were formulated is often quoted. The same paper also presents some pertinent correlation tables. Two other useful references discussing the same general topic are a paper by Laidler and Shuler [95] and Laidler's *Excited States* [96].)

d. From the Molecular Orbitals of the Polyatomic Molecule

In a previous section of this chapter a general discussion of the theoretical background of the molecular orbital method was presented. There are correlations of orbitals between large and small internuclear distances and also between different geometrical conformations of the final molecule.

Triatomic Molecules

Calculations on triatomic molecules, because of their theoretical and experimental significance, will be mentioned below in some detail. Herzberg's *Polyatomics* [6] gives diagrams of the correlation of orbitals between large and small internuclear distances in linear AH_2 molecules and in linear AB_2 molecules [6]. Walsh, in a classic series of papers [97], gave correlations of molecular orbitals between linear and bent AH_2 molecules and linear and

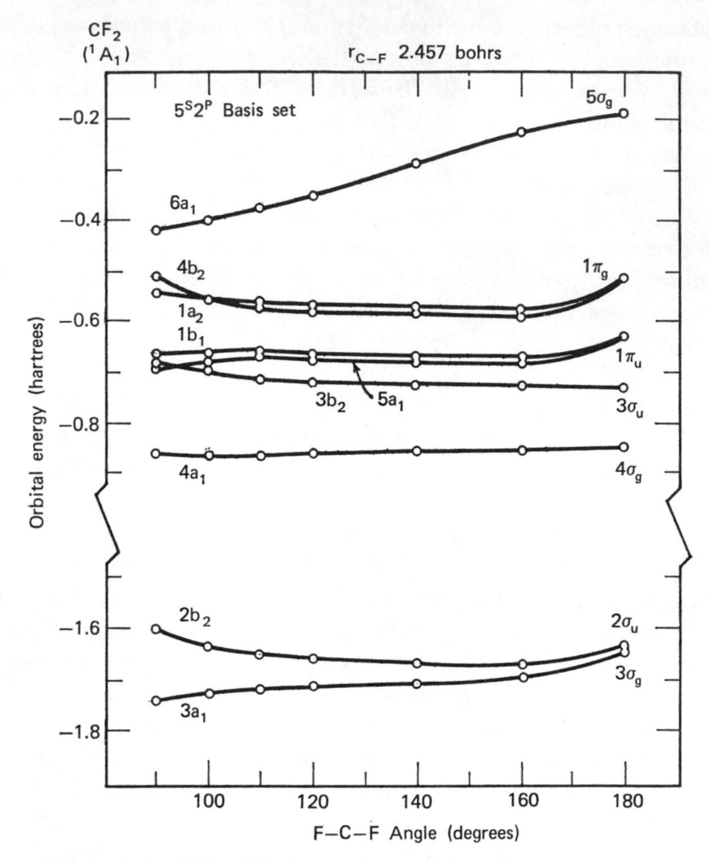

Figure 2.1 CF_2 orbital energies versus angle.

bent AB_2 molecules, and linear and bent HAB molecules. Walsh diagrams (as they are now called) for XH_2 and XY_2 are reproduced in Herzberg's *Polyatomics* [6]. As an example, Figure 2.1 shows a Walsh diagram for CF_2 calculated by a nonempirical LCAO–MO–SCF procedure using Gaussian basis orbitals [98].

During the past several years[1] there has been an intensive emphasis on calculating such diagrams by nonempirical LCAO–MO–SCF techniques as well as by semiempirical LCAO–MO methods. (Most of these results postdate

[1] So many Hartree–Fock calculations on triatomics have appeared in these past several years, and more appear each month, that to attempt to list them here would be overwhelming. Thus the only new references added to this section will be selected CI or MC–SCF results.

publication of Herzberg's *Polyatomics* [6].) A representative listing of Walsh diagrams (in most cases the results are given for internuclear angles from 90° to 180°; however, in certain instances results are given at only several different internuclear angles) calculated nonempirically using Gaussian or Gaussian lobe functions includes the following: BeH_2 [99, 100] (ref. 99 also includes the behavior of orbital energies as a function of Be-H distance), BH_2^+ [100], BH_2^- [100–102], NH_2^+ [100, 101] (ref. 100 gives the angular dependence of the orbitals and ref. 103 gives the internuclear dependence), CH_2 [103–105] (ref. 105, while it does not present the Walsh diagram, is especially pertinent since it is a CI calculation to determine if the ground state of CH_2 is linear or bent), NH_2 [103], H_2O, H_2O^+ [103, 106], HCN [107], NF_2^+ [108], F_2O, Li_2O, FOH, LiOH [109], CO_2, BeF_2 [101], O_3, N_3^- [110], FNO [111], and NNO and NON [112].

Many potential surfaces have been calculated for H + H_2. Two of the most recent, an accurate CI calculation [113] and a superposition of configuration calculation [114], review also the prior work on this problem. Ab initio potential surfaces have been calculated for a number of systems of interest. Some of the more pertinent are He + H_2 [115], He^+ + H_2 [116], Li^+ + H_2 [117], Li + H_2 [118], and Li + HF [119].

An exceedingly accurate CI calculation on the surface of H_3 recently has appeared [239]. CI calculations on the avoided intersection of potential energy surfaces for H^+ + H_2 and H + H_2^+ [240] compared surprisingly well to the earlier diatomics-in-molecules surface of Preston and Tully [241]. CI calculations were reported for He + H_2 $B\,^1\Sigma_u^+$, and compared to SCF calculations for He + $\tilde{X}\,^1\Sigma_g^+$ H_2 and $b\,^3\Sigma_u^+$ H_2 [242]. CI calculations have been carried out for He_3^+ [243]. CI calculations have been reported on BeH_2 [244] the ground ($^2A_1$) and first excited $\{^2B_1(\Pi)\}$ states of CH_2^+ and BH_2 [245] as well as CH_2 [246, 247] including its singlet-triplet separation and Walsh and Mulliken diagrams [248]. CI calculations have been reported for the lowest and triplet states of NH_2^+ [249]. CI calculations had also been reported earlier for the low lying excited states of NH_2^+ + CH_2 [250]. H_2O continues to attract attention. A number of CI calculations have been reported [247, 251, 252]. MC–SCF calculations have been reported on the lowest triplet ($^3B_1$) state of H_2O [253, 254] as well as GVB calculations on thirty-two excited states of H_2O [255]. A particularly interesting discussion of the potential energy surfaces for ground and first excited singlet states of H_2O covering insertion of O^1D into H_2 and abstraction of H to yield OH + H is given by Gangi and Bader [256].

CI calculations were reported on Li^+ + H_2 [257]. GVB functions have been calculated for H_2 + D and for LiH + H [258].

The advent of production of vibrationally excited HX lasers by the exothermic reactions of $X + H_2 \rightarrow HX + H$ or $H + X_2 \rightarrow HX + X$ (where

X is a halogen) have led to extensive SCF and CI calculations on the $F + H_2$ surface [259] and $H + F_2$ surface [260]. The $F + Li_2$ system was also investigated [261]. There has also been interest in the HO_2 radical; CI calculations (first order wave functions and iterative natural orbitals) have been reported for the ground state [262] and for the ground and low lying $^2A_1$ state [263] and GVB and CI calculations have been reported for the low lying excited states [264]. A few additional calculations or surfaces pertinent to collisional, laser or reactive problems are LiO_2 [265], O_3 and O_3^- [266], O_3 [267, 268], GVB calculations on up to 24 excited states of O_3 [269, 270], CI calculations on low lying electronic states of CO_2 [271], MC–SCF calculations on CO_2^- [272], CI calculations on HCO^+ [273], GVB calculations on CH_n and CF_n (including $^2\Pi$ and $^4\Sigma^-$ states of CH and CF) [274], GVB calculations on $TiCO^+$ and TiCO [275], and the alkaline earth halides; BeF_2, MgF_2, CaF_2 [276, 277].

For heavier triatomic systems, Hartree–Fock potential surfaces had earlier been calculated for $H + NO \rightarrow HNO$ [120] and $O + NO$ [121]. Recently Pipano and Kaufman [122] performed an ab initio large-scale configuration-interaction calculation for the reaction $O^+ + N_2 \rightarrow NO^+ + N$ that included all single and double excitations relative to the lowest configurations of the $^2\Pi_i$, $^2\Sigma^+$, $^4\Pi$, $^4\Sigma^-$ states of the intermediate NNO^+ as a function of internuclear distance. These results, as an example of the melding together of the way the theoretical techniques outlined earlier complement each other and experimental observations, will be discussed in more detail in Section IV of this chapter.

Mention should be made here of a series of recent papers by Pople and coworkers on self-consistent molecular-orbital methods in which they use mainly small optimized sets of Gaussian orbitals as fits to STOs or energy-optimized atomic orbitals to study a wide variety of molecules, including triatomics [123–130]. (One of the papers [124] also discusses a variant of a semirigorous technique, PDDO (projection of diatomic differential overlap).)[1]

Semirigorous LCAO–MO–SCF computations of the CNDO, INDO, and NDDO types have been performed on numerous AH_2 and AB_2 molecules as a function of internuclear angle [16–23]. (In these references the calculated equilibrium geometries are given and are in reasonably good agreement with experiment; however, the individual orbital correlations are not given.) The semiempirical Wolfsberg–Helmholz method has proven to give incorrect results for calculated equilibrium geometries of diatomic and triatomic molecules. The Cusachs modification of this method has been more successful in this respect.

[1] So many papers have appeared in these intervening years using these small optimized sets of Gaussian orbitals that it is outside the scope of this present paper to review this area.

Larger than Triatomic Molecules

Among the ab initio non-empirical LCAO–MO–SCF calculational results of interest to potential surface considerations are those on BeH_3 [100], BH_3 [11, 100, 131, 132], CH_3^+ [100, 133], CH_3 [134], CH_3^- [135], NH_3 [136–139], H_3O^+ [106, 140], and BH_4^- and CH_4 [103].

A Walsh diagram for the correlation of orbitals between nonplanar and planar XH_3 is depicted in Herzberg [6].

Any further listing of ab initio calculations for potential surfaces could only be out of date by the time this book is published. Availability of computer programs to perform ab initio LCAO–MO–SCF computations including various types of configuration interaction and MC–SCF computations coupled with the increase in speed, size, and availability of large computers makes the calculation of good potential surfaces for collisions a reality (albeit an expensive one). The ever-growing importance of accurate potential-energy surfaces was emphasized at a recent excellent meeting cosponsored by IBM and the University of California [141].

There has been such a multitude of ab initio results reported in these intervening years on barriers to internal rotation, large molecule calculations for ground and excited states, stretching and bending potential surfaces, that it is again outside the scope or desire of the present paper to attempt to list them specifically. A few of these calculations will be mentioned because of their CI results and/or relevance to collisional problems. H_4 remains a popular subject: linear symmetric [278], as a function of various geometries [279–283], and long range. Calculations have been reported for BH_3 with correlation included in the independent electron pair approximation using natural orbitals [284], excited coupled many pair electron theory [285], GVB calculations on BH_3 (also B, BH and BH_2) [286]. Calculations have been reported for CH_3, CH_3^+ and CH_3^- (CI) [287], MC–SCF for NH_3 [288], CI for H_3O^+ [289]. There are several SCF calculations on dimers or polymers of interest in collisional problems: dimers of LiF and of NaH [290], H_2O [291]. There is an impressive series of papers by Clementi and coworkers on large scale SCF calculations of hydration of ions [292–294]. There have been CI calculations on reactive collisions: H + CH_4, H abstraction and exchange, [295], the reaction pathway for triplet methylene ($^3B_1$) abstraction from H_2 [296]; UHF calculations on H + CH_4 [297]. Several papers have dealt with reactions of the excited state of H_2CO: the H bond energy between lower excited states of H_2CO and H_2O [298] and photodissociation to form H + HCO [299].

Potential surfaces have also been calculated by less rigorous LCAO–MO–SCF or LCAO–MO computations. Publications in this area are mushrooming rapidly, and for this reason, as well as because of their inaccuracy noted below, no detailed listing of these will be given.

However, it should be emphasized that use of these less rigorous methods for the calculation of potential-energy surfaces should be viewed with great caution. Even for a system which dissociates properly in the Hartree–Fock approximation, recent research by Kaufman and co-workers [142] has shown that even the INDO method is not capable of giving an accurate or even realistic surface for $Li^+ + H_2$ when compared point by point to Lester's accurate Hartree–Fock surface [117].

B. Correlations for Negative Ions

This subject will not be discussed in the present chapter since an excellent review by Chen of the entire field of the theory of transient negative ions and their role in collisional processes recently appeared [143]. Since most molecular negative ions of interest are not stable, Chen's review presents a dynamic description of the lifetime of transient molecular negative ions in terms of the S matrix. In the Born–Oppenheimer separation approximation, the potentials of compound electronic negative-ion states that appear as intermediates in the collisions are in general complex. The physical significance of curve crossings and crossings with the electronic continua are discussed. The role of these transient negative ions in molecular reactions is discussed for a few reactions, elastic and inelastic scattering by molecules, dissociative attachment, atomic collisions with negative ions, and internal energy transfer.

The reader is referred to the references in the above-mentioned review, especially to those of Chen, for any detailed description of phenomena involving negative ions. A recent paper by Chen and Mittleman [144] will bring the reader up to date as of the writing of the present chapter.

Other recent papers on the role of negative ions in collisional processes that supplement the treatment of Chen are those by Bardsley [145], which discuss the theory of configuration interaction in relation to molecular states that have sufficient energy to decay by electron emission (or predissociation) and the theory of dissociative recombination.

A diagram of potential curves illustrating various types of electron capture that the reader might find graphically illustrative for the behavior of negative ions is shown in Figure 2.2. (The diagram and the following description of the various curves are reproduced by permission of Dr. Christophorou [146].)

Figure 2.2*a* illustrates the most general type of electron capture by a diatomic molecule. The asymptote of the dissociative AB^- curve lies below the asymptote of the AB curve by an amount equal to A(B), the electron affinity of B. Since the speed of the incident electron will be large, the nuclei may be considered at rest during the time in which attachment takes place. The Franck–Condon principle then states that dissociative attachment

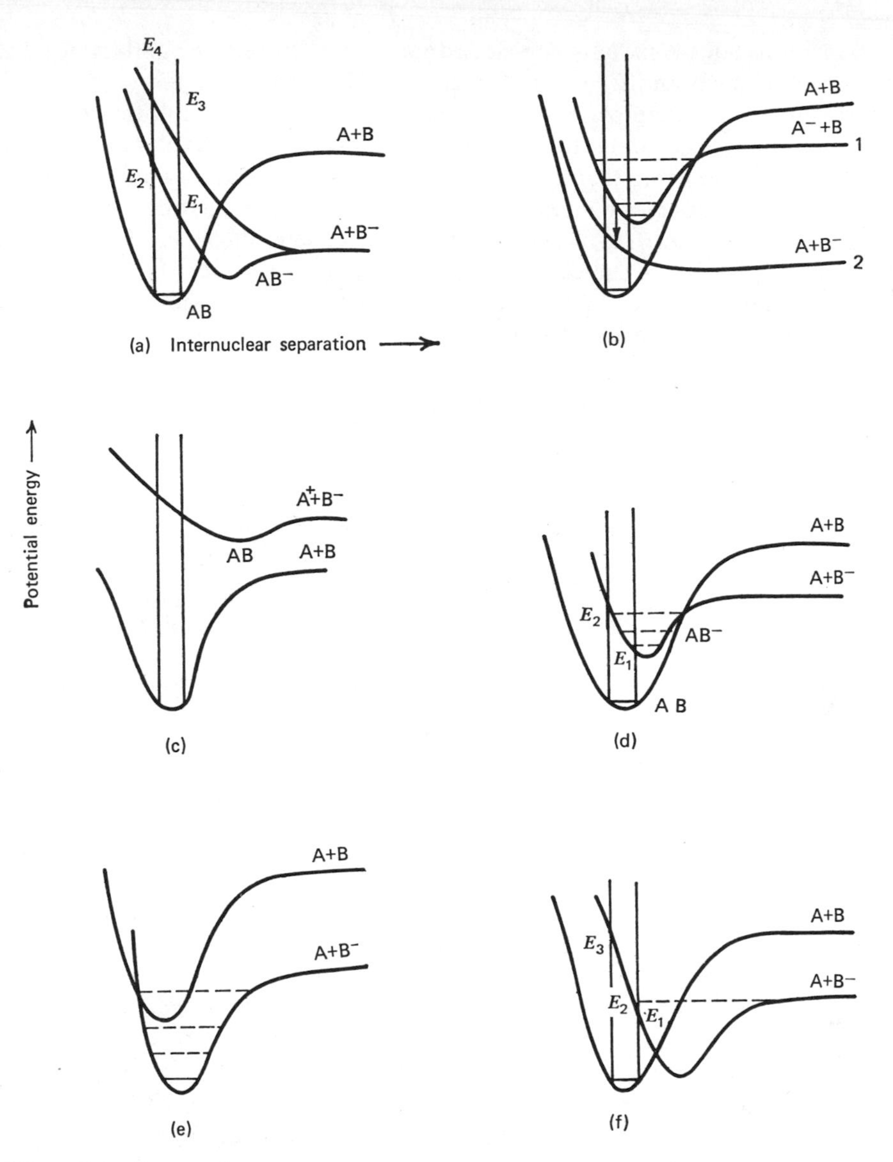

Figure 2.2 Potential curves illustrating various types of electron capture.

transition can occur only for electron energies between E_1 and E_2 for the attractive states and E_3 and E_4 for the repulsive states.

Capture may take place to bound states of the negative ion that undergo radiationless transitions to repulsive states of the negative ion resulting in dissociative electron capture. These radiationless intramolecular transitions (Auger transitions) are a result of overlap of the discrete states with a continuum of states of AB^-. These types of processes are illustrated for diatomic molecules in Figure 2.2*b*. Electrons are first captured into the bound state represented by curve 1, and before autodetachment can take place an intramolecular radiationless transition occurs to curve 2 resulting in dissociation into A and B^-.

In ion-pair production the incident electron excites the molecule to an electronic level which leads to A^+ and B^- (Figure 2.2*c*).

Figure 2.2*d* shows the simplest type of nondissociative electron capture into discrete states of AB^- that will occur between energies E_1 and E_2, resulting in a vibrationally excited AB^- molecular ion. If the capture process remains an isolated event, the electron will be ejected by autodetachment (Auger process) within a time comparable with a vibration time.

Figure 2.2*e* presents a capture process leading to temporary negative-ion formation that is distinctly different from that in Figure 2.2*d* since the potential curve of AB^- does not cross the Franck–Condon region. The mechanism can be described as vibrational excitation of the neutral molecule and subsequent capture of the incident electron.

The negative-ion curve shown in Figure 2.2*f* crosses the Franck–Condon region in a manner which will allow both dissociative and nondissociative electron attachment.

C. Supplementary Long-Range Forces

In general there are three different types of long-range interactions: the direct electrostatic, the polarized or induced, and the dispersion or fluctuation [147]. Direct electrostatic interaction covers interaction between two charged species (ions with either like or unlike charges). It also includes interaction between two permanent molecular dipoles, the average interaction energy of which is proportional to

$$\bar{E}_{AB} = -\frac{2}{3kT}\frac{\mu_A^2\mu_B^2}{R^6},$$

where μ_A and μ_B are the dipole moments of molecules A and B. This expression only holds when the separation R between two molecules A and B is larger than the sum of the greatest radii of the two molecules.

Polarization or induced interaction covers both charge-induced dipole and dipole-induced dipole interactions. An atom or a molecule A which has a charge C_A produces an electric field $+C_A/R^2$ at the position of the atom or molecule B. This induces a dipole moment $\mu^{(ind)} = \alpha_B \varepsilon_B$ at the position of B, where α_B is the polarizability of B and ε_B is the electric field at the position of B. The dominant term in the energy of interaction is proportional to $-C_A \alpha_B/2R^4$. In ion-neutral reactions this is an important term and can give rise to an attractive potential of 1 eV or more. Langevin gave the orbits of such a potential in connection with the calculation of mobilities of ions in gases as far back as 1905 [148]. In 1958 Gioumousis and Stevenson [149] applied the Langevin formulation to a theoretical treatment of ion-molecule reactions. Their conclusions were derived under the special assumptions: "i) that formation of the activated complex depends only on the long range forces, ii) that their method of calculation of the probability of formation is valid if certain conditions are met, to wit: no dipole moment, $r_c \leq b_0/\sqrt{2}$ (the critical radius is less than or equal to the impact parameter/$\sqrt{2}$), etc., and iii) that while in some cases the activated complex always goes on to react, so that their expression for the rate constant is very close to experimental, this need not always be so." Numerous succeeding investigators of ion–molecule reactions have invoked the Gioumousis and Stevenson conclusions in cases where the special conditions under which they were derived are *not* fulfilled. In particular, in certain cases the formation of the activated complex can be influenced also by the short-range so-called chemical forces [150] and by the spin and symmetry restrictions on the electronic states of the intermediate that are permitted to be formed from the spins and symmetries of the reactants or products. Also, the Gioumousis and Stevenson conclusions have been invoked where there are dipole moments present, although in this case there should be terms supplementary to the $-C_A^2 \alpha_B/2R^4$ [151]. Even when there are no dipole moments present, the term which is proportional to $1/R^4$ is only the leading term of a more exact expression for the polarization potential [152]. The question arises as to whether, for an example where the interacting species separate appropriately to closed-shell ground-state species, a single-determinant Hartree–Fock LCAO–MO–SCF calculation was capable of giving the proper long-range potential behavior $\sim 1/R^4$ when there were no permanent dipole moments present. Recent ab initio LCAO–MO–SCF calculations by Catlow et al. [37] of the potential curve of He + Li^+ have shown that the attractive potential $\sim 1/R^4$ is given by the single determinant Hartree–Fock wave function. The same was shown by Lester in his Hartree–Fock calculation of $Li^+ + H_2$ [117].

Similarly for polarization interactions a molecule that possesses a permanent dipole moment in a direction θ_A with respect to the internuclear axis can induce a moment in B that interacts with the original dipole moment to produce the energy

$$E_{AB}^{(polar)} = \frac{-\mu_A^2 \alpha_B (3 \cos^2 \Theta_A + 1)}{2R^6}.$$

Dispersion interactions arise because there are rapid fluctuations in electronic charge distributions. Even a nonpolar molecule A has an instantaneous dipole moment $\mu_A^{(instant)}$ that induces a dipole moment in molecule B that interacts with the instantaneous dipole to produce the instantaneous energy:

$$E_{AB}^{(instant)} = \frac{-[\mu_{AB}^{(instant)}]^2 \alpha_B [3 \cos^2 \Theta_A + 1]}{2R^6}.$$

By time averaging $E_{AB}^{instant)}$, the London dispersion energy is obtained.

Recently a book on the theory of intermolecular forces that treats the subject in a comprehensive manner [153] has been published.

III. CROSSING OR PSEUDOCROSSING OF MOLECULAR POTENTIAL-ENERGY SURFACES

The course of behavior followed in a collisional process is greatly influenced by the crossings or pseudocrossings of relevant molecular potential-energy surfaces. For diatomic molecules these surfaces are simply curves, crossing in a point. But for a general molecule with N atoms, the surfaces are in a configuration space of dimension $3N - 6$ (for a linear molecule $3N - 5$) [154]. For convenience of visualization, these energy surfaces are often referred to as curves, and that practice will be followed in the present section.

The well-known Landau–Zener [155–158] formula relating to the probability of an electronic jump near the crossing point of two potential-energy curves or surfaces has been seriously critiqued [4, 154]. New treatments of greater validity have been formulated [154, 159, 160].

As a necessary background for the subsequent discussion of curve crossings, it seems germane here to present the physical equations which are used to describe this phenomena. This presentation recaps the treatment as given by Coulson [154] using his notation.

An exact molecular wave function is denoted by $\phi(\mathbf{r}, \mathbf{R})$, where $\mathbf{r}$ represents all the electronic coordinates and $\mathbf{R}$ all the nuclear coordinates. Since $\phi(\mathbf{r}, \mathbf{R})$ is very difficult to obtain, it is usual to separate the nuclear and electronic motions by the zero-order approximation

$$\phi(\mathbf{r}, \mathbf{R}) = \chi_{jn}(\mathbf{R}) \psi_j(\mathbf{r} \mid \mathbf{R}), \tag{1}$$

where the electronic and nuclear wave functions satisfy the equations

$$\left\{H + \sum_{\kappa} \frac{\hbar^2}{2M_\kappa} v_\kappa^2\right\} \psi_j(\mathbf{r} \mid \mathbf{R}) = E_j(\mathbf{R}) \psi_j(\mathbf{r} \mid \mathbf{R}), \tag{2}$$

$$\left\{-\sum_{\kappa} \frac{\hbar^2}{2M} \nabla_\kappa^2 + E_j(R)\right\} \chi_{jn}(\mathbf{R}) = E_{jn} \chi_{jn}(\mathbf{R}). \tag{3}$$

The sum over κ extends over all the nuclei in the molecule, and H is the exact molecular Hamiltonian.

Since ψ_j involves $\mathbf{R}$ as well as $\mathbf{r}$, the functions $\chi_{jn}\psi_j$ are not true eigenfunctions of the exact Hamiltonian, but only of an approximate H_0. The perturbation operator for this problem is defined by the equation

$$H = H_0 + V. \tag{4}$$

If V were zero, there would be complete separation of nuclear and electronic motions (Born–Oppenheimer approximation). It is the presence of V that mixes the motions. It is assumed that

$$\varepsilon_{kj}(\mathbf{R}) \equiv \int \psi_k^*(\mathbf{r} \mid \mathbf{R}) v(\mathbf{r}, \mathbf{R}) \psi_j(\mathbf{r} \mid \mathbf{R})\, d\mathbf{r}. \tag{5}$$

Equation (3) describes the motion of the nuclei in an average potential field $E_j(\mathbf{R})$; for different electronic eigenfunctions $\psi_j(\mathbf{r} \mid \mathbf{R})$, different potential functions will be obtained. The nuclear coordinates $\mathbf{R}$ enter into (2) only as a parameter. This causes the following difficulty, whose elucidation is essential to any discussion of the "crossing" of potential-energy curves. Supposing that one knows all the functions $\psi(\mathbf{r} \mid \mathbf{R})$ for each nuclear configuration and wants to know which of the functions $\psi(\mathbf{r} \mid \mathbf{R} + d\mathbf{R})$ is the continuation of any chosen $\psi(\mathbf{r} \mid \mathbf{R})$, say, $\psi_1(\mathbf{r} \mid \mathbf{R})$. This is simple if the energy levels $E_j(\mathbf{R})$ are well separated in the vicinity of $\mathbf{R}$, since one can impose the condition that $E_1(\mathbf{R})$ should vary continuously with R, and thus define $\psi_1(\mathbf{r} \mid \mathbf{R})$ for all $\mathbf{R}$. However, if at some $\mathbf{R} = \mathbf{R}_0$, two electronic configurations $E_j(\mathbf{R})$ and $E_k(\mathbf{R})$ are degenerate and have the same energy, then the curves come very near to each other (or "cross"). (The two different electronic configurations have the same energy; this introduces a resonance energy which separates the surface slightly, so that they never actually intersect but only approach each other closely [4].) Where two curves apparently "cross" it is convenient to match together functions $\psi_1(\mathbf{r} \mid \mathbf{R} + d\mathbf{R})$ to fit $\psi_1(\mathbf{R})$ that describe similar physical situations. For cases where "crossing" actually occurs, this corresponds to assuming the continuity of the derivative of $E_j(\mathbf{R})$ along the chosen path.

A formula for the probability $P(t)$ that for diatomic molecules or atom–atom scattering the system is in state 1 at time t, if it is known that it is in state 1 at time $t = 0$, was derived by Landau [155, 156], Zener [157], and Stueckelberg [158]:

$$P = \exp\left\{\frac{-2\pi}{\hbar v}\frac{|\varepsilon_{12}(R_0)^2|}{|F_1 - F_2|}\right\}, \tag{6}$$

where v is the relative classical speed of the nuclei, $F_j = \partial E_j/\partial R$, ε_{12} is defined by

$$\varepsilon_{kj}(\mathbf{R}) = \int \psi_k^*(\mathbf{r} \mid \mathbf{R}) V(\mathbf{r}, \mathbf{R}) \psi_j(\mathbf{r} \mid \mathbf{R})\, d\mathbf{r},$$

and all functions are evaluated at $\mathbf{R} = \mathbf{R}_0$. The Landau–Zener formula (6), as it is often called, was derived under very strong assumptions, some of which are unlikely to be fulfilled for real systems. In particular, it was assumed that the interaction $\varepsilon_{12}(\mathbf{R})$ does not depend on the internuclear distance R. The error thus introduced was partly compensated for by assuming that the difference $E_1(R) - E_2(R)$ varies linearly with R; this makes the perturbation relatively small and unimportant for large internuclear distances. (Bates has severely criticized this assumption [159] and shown that transitions may readily also occur well away from the crossing. In a subsequent paper Bates [160] considered the coupled equations of the semiclassical two-state approximation in terms of the solutions to these equations at the point where the colliding atoms are closest. These equations are of the same form as the equations of a two-state approximation involving the eigenfunctions of the quasi-molecule consisting of the colliding atomic systems [159]. He derived the probability of a transition occurring for given impact parameter which was similar to, but had revealing differences from, the Landau–Zener approximation [160].)

Coulson derived a new method for computing $P(t)$ that gives the Landau–Zener formula as a special case. He considered three cases: (1) transition between two discrete states [approximate internal conversion when the molecule passes from one discrete (bound) state to another], (2) transition from a bound state to the continuum (predissociation), and (3) transitions between two states of the continuum (corresponding to scattering problems). Coulson omitted in his article the computational details for $P(t)$ for case (1). For case (2) he gave

$$P = e^{-2\nu T}, \qquad \text{where} \quad \nu = \frac{\Pi\rho}{\hbar^2}|V|^2; \tag{7}$$

ρ is the density of states in the continuum and T is the half-period. This formula reduces to the Landau–Zener formula if

$$|V|^2 = \frac{\hbar\,|\varepsilon_{12}(R_0)|^2}{|F_1 - F_2|\, v\rho T}. \tag{8}$$

For case (3) Coulson gave

$$P = \exp\left\{\frac{2\pi}{\hbar^2}\frac{|V|^2\,\rho_2^T\hbar^2}{1+|V|^2\,\rho_1\rho_2\pi^2}\right\}. \tag{9}$$

Coulson's analyses yielded different results for cases (2) and (3), in conflict with all previous work on this problem. Prior to this work it had been usual to apply the Landau–Zener formula both to predissociation and to scattering.

For numerical evaluation of the matrix elements the result for transitions of case (2),

$$P = \exp\left\{-\frac{2}{\hbar^2 v^2}\left|\int_{-\infty}^{\infty}\varepsilon_{12}(R)\,dR\right|^2\right\}, \tag{10}$$

arises when the product v_1v_2 has been replaced by v^2, some average of v_1 and all the possible v_2. The length ΔR characterizes the range of the potential $\varepsilon_{12}(R)$. Under assumptions to specify the effective range ΔR, Coulson derived the following expression for P, which is good for high energies (the low-energy limit being about 1 eV), and not excessively large ranges:

$$P = \exp\left\{-\frac{1}{\hbar^2 v^2}\left|\int \varepsilon_{12}(R)\,dR\right|^2\right\}. \tag{11}$$

He then went on to show that the conventional Landau–Zener formula is scarcely ever justified by its derivation and that the above formula seems to be more frequently applicable. However, its numerical evaluation is very difficult, because of the difficulty of computing $\int \varepsilon_{12}(R)\,dr$.

A fairly recent paper that discusses avoided crossings in bound potential-energy curves of diatomic molecules [161] presents a mathematically well-defined procedure for going from potential curves exhibiting an avoided crossing (such as would be obtained from solving exactly the eigenvalue problem associated with the complete electronic Hamiltonian for fixed nuclei) to two crossing potential curves and an interaction function.

Theoretical justification for the noncrossing rule for diatomic molecules is a special case of a more general treatment which was presented by von Neumann and Wigner [162]. The rule states that vibrational potential-energy curves for electronic states of like symmetry do not cross and, as a result, situations of avoided crossings occur where two potential-energy curves approach each other closely and then move apart. In Figure 2.3*a* is presented an avoided crossing, and in Figure 2.3*b* two normal potential curves are constructed from two potential curves exhibiting an avoided crossing. A time-dependent treatment is particularly appropriate when considering scattering or predissociation problems, but less appropriate when considering the discrete rotational and vibrational levels of two bound electronic states exhibiting an avoided crossing. The derivation of a time-independent

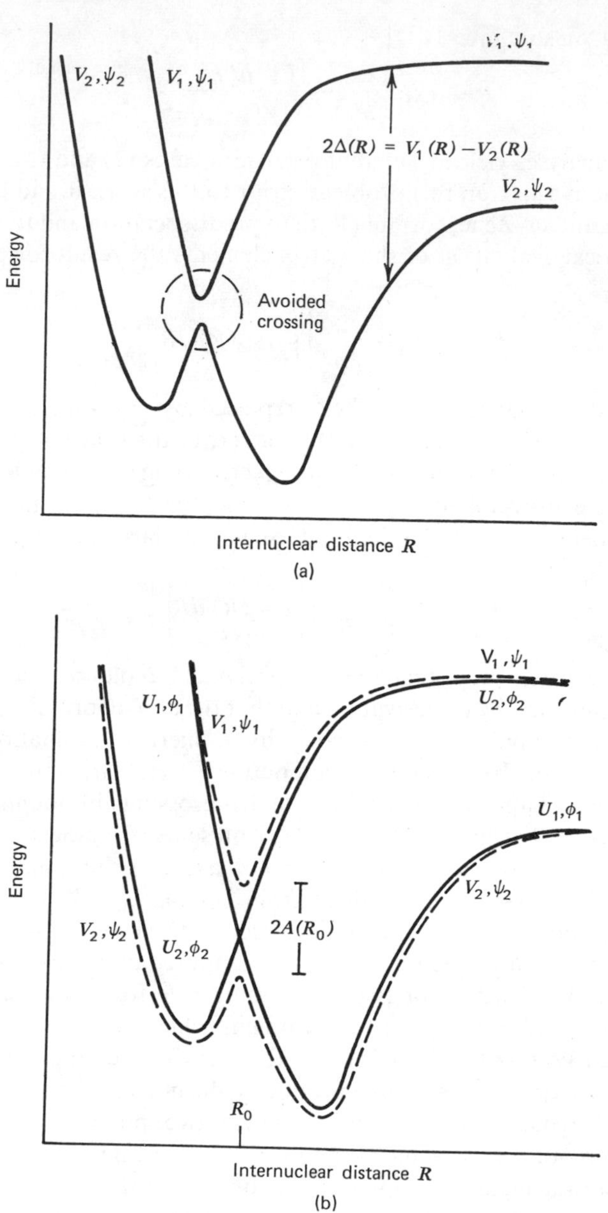

Figure 2.3 (a) Avoided crossing of potential-energy curves (energy versus internuclear distance). The curves are labeled by the potential functions they represent and by the corresponding electronic wavefunctions. (b) Two normal potential curves (solid lines) constructed from two potential curves exhibiting an avoided crossing (dotted lines).

vibrational Hamiltonian for a molecule with a pair of potential-energy curves exhibiting an avoided crossing is presented in ref. 161. The functions $V_1(R)$, $V_2(R)$, $\psi_1(R)$, and $\psi_2(R)$ shown in Figure 2.3*a* are those obtained from an exact calculation of the electronic energies and wave functions for fixed nuclei, using the complete electronic Hamiltonian plus the internuclear repulsion (total Hamiltonian with the nuclear kinetic energy removed), and thus have an exact physical meaning. It is sometimes convenient to use an approximate Hamiltonian that leads to potential curves like those shown by the solid lines in Figure 2.3*b* for which the crossing is not avoided. At some stage the approximate Hamiltonian must be supplemented by a perturbation operator that reintroduces the required crossing. (Often, unavoided crossings and their associated wave functions arise naturally in zeroth-order molecular-orbital calculations [163].)

A series of papers by Lichten [164–167] on the general topic of molecular wave functions and inelastic atomic collisions introduced a new descriptive terminology that has since been used extensively. In Figure 2.3*b* if the atoms approach each other slowly in state ϕ_1, an adiabatic transition from ϕ_1 to ϕ_2 will occur. Lichten coined the expression "diabatic" to describe the behavior if the two atoms approached each other rapidly and a transition from ϕ_1 to ϕ_1 occurred. This type of a transition is called a "diabatic" transition. Recent results in low-energy ion–atom and ion–molecule collisions indicated that these processes can be highly effective in producing optical excitation in contradiction to the well-known "adiabatic criterion" [168]. The concept of diabatic transitions has proven particularly useful in delineating the processes in resonant and quasiresonant charge-exchange processes [165, 166, 108]. In ref. 167 Lichten discusses some of the effects which cause transitions among diabatic MOs at crossings: electron penetration, electron correlation, and electronic interaction with nuclear motion.

The role of potential curve crossing in subexcitation molecular collisions has been investigated numerically by exact (two-state) computation versus decoupling approximations [169]. In this work, attention was restricted to elastic scattering, in low energy (subexcitation) collisions, when the (approximate) potential-energy curves cross. The results showed that even in the subexcitation region a considerable range of behavior is to be expected, from a fully adiabatic description (noncrossing) to the pure distortion one. (In this work are a number of references to related theoretical investigations by those authors.)

The role of curve crossing in atomic and molecular collisions is a subject of considerable interest at the present time. Fairly recently there was an international conference entirely devoted to this subject [170]. A fine review article on radiationless molecular electronic transitions has also appeared rather recently [171].

IV. AN EXAMPLE OF AN APPLICATION OF THESE CONCEPTS

The utility of the concepts described earlier in this chapter as an aid in deciphering and predicting experimental results in molecular collisional processes may be illustrated by the following theoretical justification derived for the experimentally observed apparently anomalous low-energy behavior of the $O^+ + N_2$ ion–molecule reaction.

It is generally observed that cross sections for exothermic ion–molecule reactions decrease with increasing ion kinetic energies, indicating a negligible activation energy. An important exception to this behavior is exhibited by the exothermic reaction $O^+ + N_2 \rightarrow NO^+ + N$, the cross section of which rises with energy, goes through a broad maximum at 10 eV, and then decreases [172]. This is contrary to the expected behavior of an exothermic ion–molecule reaction since on the basis of the simple considerations of the Gioumousis and Stevenson [149] picture, the cross section is expected to decrease with increasing energy. A recent review has mentioned that there was no satisfactory explanation for this apparent anomalous behavior. More recently Schmeltkopf et al. [173] studied this reaction in the thermal region and found that it proceeded with no activation energy and furthermore that the rate of reaction increased rapidly with vibrational excitation of the N_2. It was very desirable to have detailed knowledge of the potential-energy surfaces of this system, since this knowledge would permit a more definitive interpretation of the experimental results. When the theoretical investigation was initiated, no accurate quantum-chemical calculations had yet been made on these surfaces. (These computations are being performed at the present time, and the preliminary computational results will be mentioned later in this chapter.) However, application merely of the concepts described earlier in this chapter combined with the available theoretical, optical spectroscopic, and mass spectroscopic data that existed at that time for this system permitted one to outline schematically some of the pertinent potential-energy curves [150]. From these curves it was possible to explain qualitatively the apparently anomalous features of this reaction.

The intermediate of the $O^+ + N_2$ reaction, N_2O^+, has 15 valence electrons, and thus it is assumed to have a linear configuration (N–N–O$^+$) in accordance with Walsh's rules [97] for a 15 valence electron triatomic. Experimentally the ground state of N_2O^+ is known to be $^2\Pi_i$ and its first excited state, lying at 3.45 eV, to be $^2\Sigma^+$ [174]. A second excited state with no symmetry assigned had been reported at 7.19 eV [6]. McLean and Yoshimine at IBM had done accurate LCAO–MO–SCF calculations on the parent system N_2O itself [57]. The bond lengths of ground state N_2O^+ and N_2O are quite similar [174].

Since N_2O is a closed-shell system, Koopmans' theorem [11] is considered valid for the ionization potential and thus also for the total energy of the remaining N_2O^+ system. The ordering of the top-most filled levels for $N_2O^+(\tilde{X}\,^2\Pi_i)$ is $(1\pi)^4\,(7\sigma)^2\,(2\pi)^3$ [57], the first excited state $^2\Sigma^+$ must have the configuration $(1\pi)^4\,(7\sigma)^1\,(2\pi)^4$, and the second excited state must be a $^2\Pi_i$ state arising from $(1\pi)^3\,(7\sigma)^2\,(2\pi)^4$. From very recent photoelectron spectroscopic measurements on N_2O [175, 176] it is substantiated that the ground and first excited states of N_2O^+ and $\tilde{X}\,^2\Pi_i$ and $^2\Sigma^+$ (3.49 eV above ground state N_2O^+) and that the second excited state, $^2\Pi_i$, lies only about 1.3 eV above the $^2\Sigma^+$ state. This $^2\Pi_i$ state is attractive since transitions to at least 14 different vibrational levels were reported. The third excited state, $^2\Sigma^+$, corresponds to the earlier state reported at ~7.19 eV above the ground state of N_2O^+ and is also attractive since transitions are reported to five different vibrational levels. From these levels for N_2O^+ (Table 2.1), the experimental data for the reactants $O^+ + N_2$ (or $O + N_2^+$, which must also be taken into account when drawing a full potential curve) (Table 2.2), the products $NO^+ + N$ (or $NO + N^+$) (Table 2.3), the experimental dissociation energy of N_2O^+ to $NO^+ + N$ (or $NO + N^+$) [177] and *most importantly* from the

Table 2.1
Energy Levels N_2O and N_2O^+

N_2O	Occupied	$(1\sigma)^2\,(2\sigma)^2\,(3\sigma)^2\,(4\sigma)^2\,(5\sigma)^2\,(6\sigma)^2\,(1\pi)^4\,(7\sigma)^2\,(2\pi)^4$	
	Unoccupied	$(3\pi)\,(8\sigma)$	
		LCAO–MO–SCF (MCLEAN AND YOSHIMINE) (eV)	PHOTOELECTRON IP (EXPT) (eV)
	$\varepsilon_{6\sigma}$	−22.6314	20.11
	$\varepsilon_{1\pi}$	−20.7289	18.23 (vertical) 17.65 (adiabatic)
	$\varepsilon_{7\sigma}$	−19.0112	16.38
	$\varepsilon_{2\pi}$	−13.3717	12.89
N_2O^+			ENERGY (eV)
		$\tilde{X}\,^2\Pi_i\;(6\sigma)^2\,(1\pi)^4\,(7\sigma)^2\,(2\pi)^3$	0
		$A\;^2\Sigma^+\;(6\sigma)^2\,(1\pi)^4\,(7\sigma)\,(2\pi)^4$	3.49
		$B\;^2\Pi_i\;(6\sigma)^2\,(1\pi)^3\,(7\sigma)^2\,(2\pi)^4$	4.76 (adiabatic) 5.34 (vertical)
		$C\;^2\Sigma^+\;(6\sigma)\,(1\pi)^4\,(7\sigma)^2\,(2\pi)^4$	7.22

Table 2.2
NNO⁺ Reactant Energy Levels and Symmetry– and Spin–Permitted Intermediates

REACTANTS	ENERGY SCALE (RELATIVE TO SEPARATED NEUTRAL ATOMS) (eV)
$N_2 + O^+ \rightarrow N{-}N{-}O^+$	3.859
$\tilde{X}\,^1\Sigma_g^+$ ^4S_u $^4\Sigma^-$	
$N_2^+ + O \rightarrow N{-}N{-}O^+$	5.821
$\tilde{X}\,^2\Sigma_g^+$ ^3P_g $^{2,4}\Sigma^-, ^{2,4}\Pi$	
$N_2^+ + O \rightarrow N{-}N{-}O^+$	6.941
$A\,^2\Pi_u$ ^3P_g $^{2,4}\Sigma^+, ^{2,4}\Sigma^-, ^{2,4}\Pi, ^{2,4}\Delta$	
$N_2 + O^+ \rightarrow N{-}N{-}O^+$	7.184
$\tilde{X}\,^1\Sigma_g^+$ ^2D_u $^2\Sigma^-, ^2\Pi, ^2\Delta$	

rules for the possible symmetry- and spin-permitted combinations of these reactants and products into the intermediate states of N_2O^+, the following schematic potential diagram was drawn for the system (Figure 2.4). The energies indicated are those relative to the separated neutral atoms. (The N—N distance in N_2O^+ ($\tilde{X}\,^2\Pi_i$ 1.115 Å; $A\,^2\Sigma^+$ 1.140 Å) is not very different from the N—N distance in N_2 (1.094 Å) [2] or N_2^+ (1.1162 Å) [2], and the N—O distance in N_2O^+ ($\tilde{X}\,^2\Pi_i$ 1.185 Å; $A\,^2\Sigma^+$ 1.141 Å) is similar to those in NO

Table 2.3
NNO⁺ Product Energy Levels and Symmetry- and Spin-Permitted Intermediates

PRODUCTS	ENERGY SCALE (RELATIVE TO SEPARATED NEUTRAL ATOMS) (eV)
$NO^+ + N \rightarrow N{-}N{-}O^+$	2.760
$\tilde{X}\,^1\Sigma^+$ ^4S_u $^4\Sigma^-$	
$NO^+ + N \rightarrow N{-}N{-}O^+$	5.144
$\tilde{X}\,^1\Sigma^+$ ^2D_u $^2\Sigma^-, ^2\Pi, ^2\Delta$	
$NO^+ + N \rightarrow N{-}N{-}O^+$	6.336
$\tilde{X}\,^1\Sigma^+$ ^2P_u $^2\Sigma^+, ^2\Pi$	
$NO + N^+ \rightarrow N{-}N{-}O^+$	8.025
$\tilde{X}\,^2\Pi_i$ ^3P_g $^{2,4}\Sigma^+, ^{2,4}\Sigma^-, ^{2,4}\Pi, ^{2,4}\Delta$	

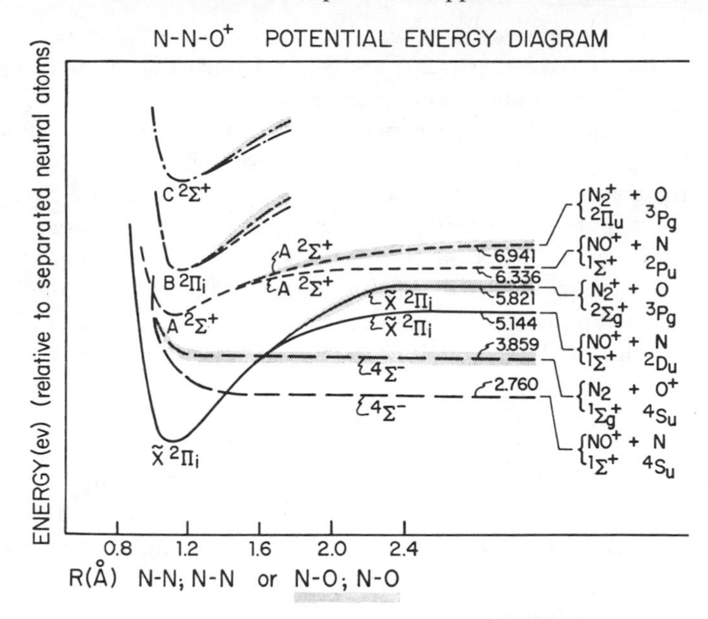

Figure 2.4 N–N–O$^+$ Potential-energy diagram (schematic). The stippled curve represents reactant paths. The unstippled curve represents product paths. — — —, $^4\Sigma^-$; ———, $\tilde{X}\,^2\Pi_i$; - - - - -, $A\,^2\Sigma^+$; — — — —, $B\,^2\Pi_i$; • — • — $C\,^2\Sigma^+$.

(1.1508 Å) [2] or NO$^+$ (~1.08 Å) [177]. Figure 2.4 depicts two-potential energy profiles through a section of a cube [96], the one for the reactants corresponding to the approximate N—N distance in N_2 and the one for the products corresponding to the approximate N—O distance in NO. Since there is very little change in the bond distances involved, these two profiles present an adequate representation for a qualitative understanding of the potential-energy surfaces involved.)

The stippled curves designate the reactants O$^+$ + N_2 or O + N_2^+; in these cases, the abscissa represents the N–O coordinate. The unstippled curves designate the products NO$^+$ + N or NO + N$^+$; here the abscissa represents the N–N distance. It becomes very obvious from examination of this potential diagram (and Table 2.2) that ground state O$^+$ + N_2 can combine uniquely only along the repulsive $^4\Sigma^-$ N_2O^+ curve (represented by a — — — curve)[1] that is then cut by the attractive ground state $\tilde{X}\,^2\Pi_i$ N_2O^+ curve [the solid curve arising from $N_2^+(^2\Sigma_g^+)$ + O(^3P_g)]. The $\tilde{X}\,^2\Pi_i$ N_2O^+ curve indicates that predissociation from the $\tilde{X}\,^2\Pi_i$ N_2O^+ to ground state NO$^+$ + N can

[1] The curves leading to each individual intermediate state either from reactants or products have different graphic representations.

take place since the repulsive $^4\Sigma^-$ lower curve (also a — — — curve), which separates to ground state $NO^+(^1\Sigma^+)$ and $N(^4S_u)$, cuts the attractive $\tilde{X}\,^2\Sigma_i$ ground-state solid curve that would separate to $NO^+(^1\Sigma^+)$ and $N(^2D_u)$. This behavior is substantiated by the experimental observation that (although spin-forbidden) predissociation to ground-state products $NO^+ + N$ is noted to occur in the mass-spectroscopic decomposition of ground state N_2O^+ [6].

Three experimental facts were in need of explanation: (1) the absence of an activation energy in the reaction $O^+ + N_2 \rightarrow NO^+ + N$, when studied at thermal energies and its very low reaction efficiency under these conditions; (2) the increase in rate constant with increasing vibrational energy of N_2; and (3) the higher-energy data, which indicate an activation energy for the reaction.

Referring to Figure 2.4, since the ground states of O^+ and N_2 can combine uniquely only along the indicated $^4\Sigma^-$ curve, the lack of an activation energy for the production of NO^+ at thermal energies implies that this $^4\Sigma^-$ curve, even though repulsive, must come in with an almost flat slope until the O–N distance is in the region of the NNO^+ equilibrium distance. The very low reaction efficiency under these conditions when N_2 is in its ground vibrational state indicates that the reactants in the $^4\Sigma^-$ state must make a forbidden curve crossing into another state (in this case into the $\tilde{X}\,^2\Pi_i$ state of NNO^+). This is in agreement with experiment since the appearance potential for the process $N_2O \rightarrow N_2(\tilde{X}\,^1\Sigma_g^+) + O^+(^4S_u)$ is reported as 15.3 eV [178, 179] (which coincides with the energy level of $N_2 + O^+$ on the same relative energy scale). The appearance potential for the process $N_2O \rightarrow NO^+(\tilde{X}\,^1\Sigma^+) + N(^4S_u)$ is 15.01 eV [180]. This scheme permits the curve of the ground-state reactants to cross into the attractive ground state of N_2O^+ with no activation energy and then to cross over into the $^4\Sigma^-$ state of the ground-state products. The spin forbiddenness of the crossovers then accounts for the observed low probability of reaction, as suggested by Schmeltekopf et al. [173].

The increase in the rate constant at low energies with increasing vibrational energy of the N_2 indicates that there is probably a barrier in the exit channel [181]. Since the ground-state products $NO^+ + N$ also must arise uniquely from a $^4\Sigma^-$ state of NNO^+, the schematic curve Figure 2.4 would have indicated that the thermal-energy reaction when N_2 is vibrationally excited either went through the $\tilde{X}\,^2\Pi_i$ NNO^+ state with another curve crossing back into the dissociative $^4\Sigma^-$ state or possibly through the $^4\Sigma^-$ state for the entire reaction. The accessibility of the portion of the $^4\Sigma^-$ state leading to the products would be enhanced with increasing vibrational excitation of the N_2. The appearance potential for $N_2O \rightarrow NO^+ + N$ of 15.01 eV indicates the presence of some type of barrier in this exit channel since the energy of the separated $NO^+ + N$ is lower than 15.01 eV on the same relative energy scale. The detailed quantum-chemical calculations for $O^+ + N_2 \rightarrow NO^+ + N$

[122], which will be described below, indicated an unsuspected channel, a low-lying $^4\Pi$ state that cut through both the $\tilde{X}\,^2\Pi_i$ and the $^4\Sigma^-$ states in the exit channel. To exit from the $^2\Pi_i$ state via the $^4\Pi$ state to the $^4\Sigma^-$ state would also be facilitated by an increase in the vibrational energy of the N_2.

The activation energy indicated by the data for higher kinetic energies of the O^+ is indicative of the fact that higher states of NNO^+ become accessible and thus more channels open up through which the reaction can proceed.

In addition to the curves shown on Figure 2.4, there are 27 more possible intermediate N–N–O^+ states resulting from reactants with energies from 3.859 to 10.01 eV (relative to separated neutral atoms) and 39 more possible intermediate states resulting from the products with energies from 2.760 to 11.316 eV (relative to separated neutral atoms). A great many of these will undoubtedly be repulsive states. It should be noted that the potential energy curves are compatible with the appearance potential of 16.4 eV [182] for the process $NO^+(\tilde{X}\,^1\Sigma^+) + N(^2D)$, of 17.4 eV (182) for the process $N_2^+(\tilde{X}\,^2\Sigma_g^+) + O(^3P_g)$ and 20 eV [179, 182] for the process $NO(\tilde{X}\,^2\Pi) + N^+(^3P)$.

A more comprehensive interpretation of this reaction depends on the availability of more detailed and more accurate potential curves. Pipano and Kaufman have performed ab initio large scale configuration interaction calculations of some of the pertinent states for the above reaction. The results are preliminary since within the limit of the computer time so far available we have used a minimal Slater orbital basis set and computed only the same two cuts along the reaction surface as our schematic curves (Figure 2.4). (The effect of bending the molecule was tested for the $^2\Pi$ and $^4\Pi$ states. These calculations were done for the experimental equilibrium internuclear distance and an angle of bending of 5° relative to the axis of the linear molecule. The results showed that the energy of both the A' and A'' components of both Π states increase with bending. It was already well known that energies of Σ states increase when the molecule is distorted to a lower symmetry. Thus the linear configuration is the preferred minimum energy path for the reaction.) A configuration interaction calculation (including all single and double excitations) was carried out at each point and for each state. For each state the following number of configurations were included:

$^2\Pi_i$	123
$^2\Sigma^+$	177
$^4\Pi$	1379
$^4\Sigma^-$	1310

The orbital composition of each configuration of each state was identified with a serial number. The configurations which have the largest weight in the resulting wave functions are listed below according to their serial numbers (here all the σ orbitals are listed first, followed by the π orbitals; they are not listed in order of increasing energies since this order changes as a function of internuclear distance):

$^1\Sigma^+$	(1)	$(1\sigma)^2 (2\sigma)^2 (3\sigma)^2 (4\sigma)^2 (5\sigma)^2 (6\sigma)^2 (7\sigma)^2 (1\pi)^4 (2\pi)^4$
$^2\Pi$	(1)	$(1\sigma)^2 (2\sigma)^2 (3\sigma)^2 (4\sigma)^2 (5\sigma)^2 (6\sigma)^2 (7\sigma)^2 (1\pi)^4 (2\pi)^3$
	(7)	$(1\sigma)^2 (2\sigma)^2 (3\sigma)^2 (4\sigma)^2 (5\sigma)^2 (6\sigma)^2 (7\sigma)^1 (8\sigma)^1 (1\pi)^4 (2\pi)^3$
$^2\Sigma^+$	(1)	$(1\sigma)^2 (2\sigma)^2 (3\sigma)^2 (4\sigma)^2 (5\sigma)^2 (6\sigma)^2 (7\sigma)^1 (1\pi)^4 (2\pi)^4$
	(8)	$(1\sigma)^2 (2\sigma)^2 (3\sigma)^2 (4\sigma)^2 (5\sigma)^2 (6\sigma)^2 (7\sigma)^2 (1\pi)^4 (2\pi)^2 (8\sigma)^1$
$^4\Pi$	(1)	$(1\sigma)^2 (2\sigma)^2 (3\sigma)^2 (4\sigma)^2 (5\sigma)^2 (6\sigma)^2 (7\sigma)^2 (1\pi)^4 (2\pi)^2 (3\pi)^1$
	(2)	$(1\sigma)^2 (2\sigma)^2 (3\sigma)^2 (4\sigma)^2 (5\sigma)^2 (6\sigma)^2 (7\sigma)^1 (1\pi)^4 (2\pi)^3 (8\sigma)^1$
	(4)	$(1\sigma)^2 (2\sigma)^2 (3\sigma)^2 (4\sigma)^2 (5\sigma)^2 (6\sigma)^2 (7\sigma)^2 (1\pi)^3 (2\pi)^3 (3\pi)^1$
$^4\Sigma^-$	(1)	$(1\sigma)^2 (2\sigma)^2 (3\sigma)^2 (4\sigma)^2 (5\sigma)^2 (6\sigma)^2 (7\sigma)^2 (1\pi)^4 (2\pi)^4 (8\sigma)^1$

The calculated curves are presented in Figures 2.5 and 2.6. An even greater wealth of structure than would have been anticipated prior to such detailed calculation is evident from these figures. The energy distances between curves

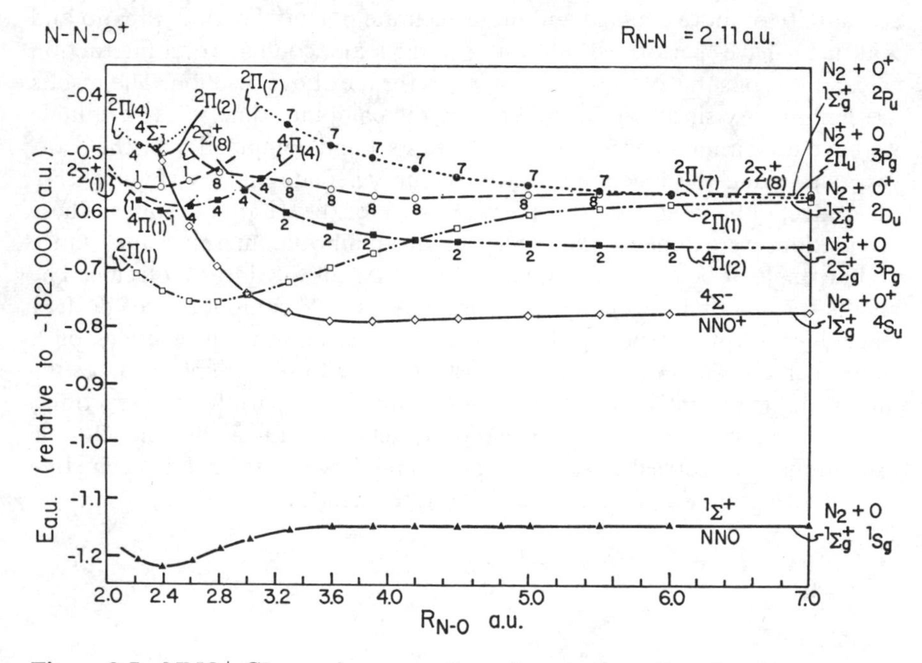

Figure 2.5 NNO$^+$ CI energies versus R_{N-O} for reactants (keeping r_{N-N} constant at 2.11 a.u.). —▲—▲—, $^1\Sigma^+$ (NNO); —◇—◇—, $^4\Sigma^-$; —■—•—•—, $^4\Pi$; —□—••—□—, $^2\Pi$; —○ — —○—, $^2\Sigma^+$; - - - ● - - ● - - , $^2\Pi$. The subscript numbers in parentheses or directly under the points represent the serial number of the particular configuration that is dominant at that point.

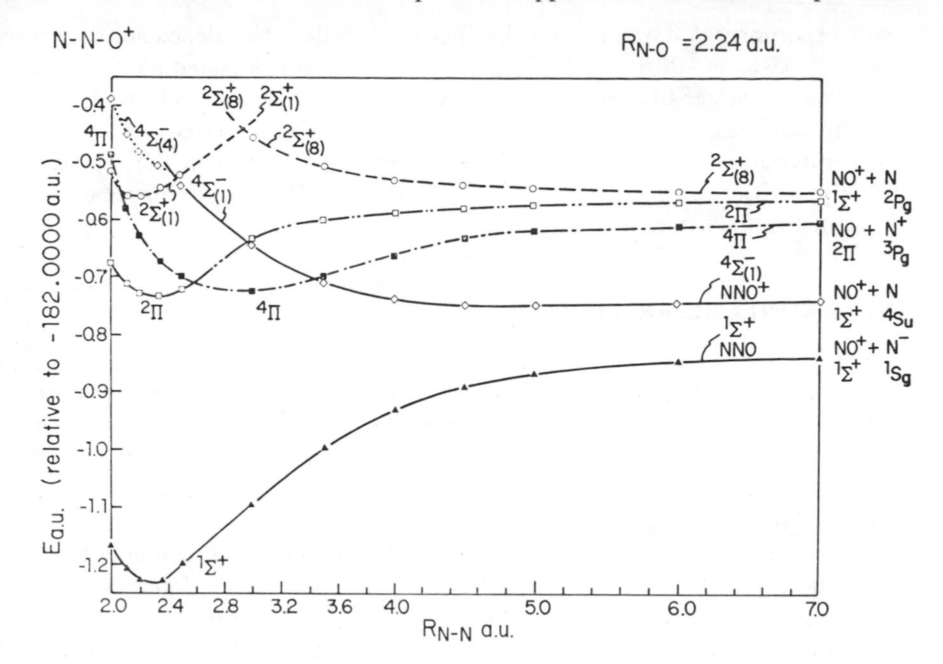

Figure 2.6 NNO$^+$ CI energies versus $R_{\text{N-N}}$ for products. —▲—▲—, $^1\Sigma^+$ (NNO); —◇—◇—, $^4\Sigma^-$; —□—••—, $^2\Pi$; —■—•—, $^4\Pi$; - - - ○ - - - ○, $^2\Sigma^+$. The subscript numbers in parentheses or directly under the points represent the serial number of the particular configuration that is dominant at that point.

near the separation limit are in reasonable agreement with experiment. The two apparent discrepancies in case of the two $^2\Pi$ states of reactants [$^2\Pi(1)$ and $^2\Pi(7)$] and the $^2\Sigma^+(8)$ and $^2\Pi(1)$ states of products can be accounted for by looking at the corresponding curves in Figures 2.5 and 2.6. In the case of reactants, the energy of the $^2\Pi(7)$ state decreases and the energy of the $^2\Pi(1)$ state increases while increasing the N–O distance, so that it is quite possible that the corresponding curves cross at larger N–O separations. The same argument holds for the $^2\Sigma^+(8)$ and the $^2\Pi(1)$ states, which separate to the same products. As no attempt was made simultaneously to optimize the N–O distance, while changing the N–N distance or vice versa, the resulting lower ground state energy of reactants, compared to the ground state energy of products, near the dissociation limits has no significance. Moreover, due to a steeper energy drop along the $^4\Sigma^-$ curve, when the N–O distance is increased, (compared to the effect of increasing the N–N distance) it is anticipated that such a simultaneous optimization would yield the correct results. Although the energy separations between potential curves are in reasonable agreement with experimental data, the results inevitably reflect the deficiencies of a minimal basis set. Both equilibrium and dissociation calculated equilibrium distances are larger than one should expect, due to an insufficient number of polarization functions (*p*-, *d*-, and *f*-type functions) in the basis.

A situation similar to that of $O^+ + N_2$ appears to exist in the reaction $C^+(^2P_u) + O_2(\tilde{X}\,^3\Sigma_g^-) \rightarrow CO^+(\tilde{X}\,^2\Sigma^+) + O(^3P_g)$ [184], and it also can be explained in the same manner.

ACKNOWLEDGMENTS

Special thanks are due to the Air Force Office of Scientific Research, Propulsion Division, and to its Chief, Dr. Joseph Masi, for their long-time support of the author's research in theoretical and quantum chemistry, which led to the understanding necessary to write this chapter.

The author should also like to thank Professor Walter S. Koski, who suggested the $O^+ + N_2$ problem to her and who collaborated in its theoretical treatment, and Dr. Aaron Pipano, who collaborated in the configuration-interaction computation of the $O^+ + N_2$ reaction. The configuration-interaction calculations on the $O^+ + N_2$ potential-energy surface were supported in part by BRL under Contract No. DAAD05-70-0027 and in part by the Atomic Energy Commission.

REFERENCES

1. E. Wigner and E. E. Witmer, *Z. Physik*, **51,** 859 (1928).
2. G. Herzberg, *Molecular Spectra and Molecular Structure. I. Spectra of Diatomic Molecules*, 2nd ed., Van Nostrand, New York, 1950, p. 315.
3. M. Born and R. Oppenheimer, *Ann. Physik*, **84,** 457 (1927).
4. H. Eyring, J. Walter, and G. E. Kimball, *Quantum Chemistry*, Wiley, New York, 1944.
5. J. C. Slater, *Quantum Theory of Molecules and Solids*, Vol. 1, McGraw-Hill, 1963.
6. G. Herzberg, *Molecular Spectra and Molecular Structure. III. Electronic Spectra and Electronic Structure of Polyatomic Molecules*, Van Nostrand, New York, 1966. See Subject Index for tables for individual symmetry groups. (In the text of the present article this book is often referred to as Herzberg's *Polyatomics.*)
7. C. C. J. Roothaan, *Rev. Mod. Phys.*, **23,** 69 (1951).
8. C. C. J. Roothaan, *Rev. Mod. Phys.*, **32,** 179 (1960).
9. T. Koopmans, *Physica*, **1,** 104 (1933).
10. Z. Gershgorn and I. Shavitt, *Int. J. Quantum Chem.*, **1S,** 403 (1967).
11. A. Pipano and I. Shavitt, *Int. J. Quantum Chem.*, **II,** 741 (1968).
12. F. E. Harris, *J. Chem. Phys.*, **46,** 2769 (1967).

13. E. Clementi, *Chem. Rev.*, **68,** 341 (1968).
14. J. Hinze and C. C. J. Roothaan, *Prog. Theor. Phys. Osaka*, Suppl., **40,** 37 (1967).
15. Joyce J. Kaufman, *J. Chem. Phys.*, **43,** S152 (1965).
16. J. A. Pople, D. P. Santry, and G. A. Segal, *J. Chem. Phys.*, **43,** S129 (1965).
17. J. A. Pople and G. A. Segal, *J. Chem. Phys.*, **43,** S136 (1965); **44,** 3289 (1966).
18. D. P. Santry and G. A. Segal, *J. Chem. Phys.*, **47,** 158 (1967).
19. G. A. Segal, *J. Chem. Phys.*, **47,** 1876 (1967).
20. J. A. Pople, D. L. Beveridge, and P. A. Dobosh, *J. Chem. Phys.*, **47,** 2026 (1967).
21. J. A. Pople and M. Gordon, *J. Am. Chem. Soc.*, **89,** 4253 (1967).
22. G. A. Segal and M. L. Klein, *J. Chem. Phys.*, **47,** 4236 (1967).
23. J. A. Pople and D. A. Beveridge, *Approximate Molecular Orbital Theory* McGraw-Hill, New York, 1970.
24. H. H. Jaffe, *Acct. Chem. Res.*, **2,** 136 (1969).
25. B. J. Nicholson, *Advan. Chem. Phys.*, **18,** 249 (1970).
26. K. Jug, *Theor. Chim. Acta*, **14,** 91 (1969).
27. M. Wolfsberg and L. Helmholz, *J. Chem. Phys.*, **29,** 837 (1952).
28. W. H. Eberhardt, B. Crawford, Jr., and W. N. Lipscomb, *J. Chem. Phys.*, **22,** 989 (1954).
29. R. Hoffmann and W. N. Lipscomb, *J. Chem. Phys.*, **36,** 2179 (1962); **37,** 2872 (1962).
30. R. Hoffmann, *J. Chem. Phys.*, **39,** 1397 (1963); **40,** 2474, 2480 (1964).
31. R. Hoffmann, *Tetrahedron*, **22,** 521 (1966).
32. L. C. Cusachs, *J. Chem. Phys.*, **43,** S157 (1965); **45,** 2717 (1966).
33. M. Krauss, "Compendium of ab initio Calculation of Molecular Energies and Properties." NBS Technical Note 438, 1967 (unpublished).
34. P. E. Cade and W. M. Huo, *J. Chem. Phys.*, **47,** 614, 649 (1967).
35. The major contributors to this effort at IBM are A. D. McLean and M. Yoshimine.
36. Presented at the International Symposium on Atomic, Molecular and Solid State Theory and Quantum Biology, Sanibel Island, Florida, January 1971 (unpublished).
37. G. W. Catlow, M. R. C. McDowell, Joyce J. Kaufman, L. M. Sachs, and E. Chang, *J. Phys. B: Atom. Molec. Phys.*, **3,** 833 (1970).
38. B. F. Junker and J. C. Browne, *Abstracts of the Sixth International Conference on the Physics of Electronic and Atomic Collisions*, MIT Press, Cambridge, Mass., 1969, p. 220.
39. C. F. Bender and E. R. Davidson, *Phys. Rev.*, **183,** 23 (1969).
40. F. E. Harris and H. H. Michels, *Chem. Phys. Lett.*, **3,** 441 (1969).
41. F. E. Harris, H. H. Michels, and J. C. Browne, *J. Chem. Phys.*, **48,** 2821 (1968).
42. W. Kolos and C. C. J. Roothaan, *Rev. Mod. Phys.*, **32,** 219 (1960).
43. W. Kolos and L. Wolniewicz, *Rev. Mod. Phys.*, **35,** 473 (1963).
44. W. Kolos and L. Wolniewicz, *J. Chem. Phys.*, **41,** 3663 (1964); **43,** 2429 (1965).
45. C. B. Wakefield and E. R. Davidson, *J. Chem. Phys.*, **43,** 834 (1965).
46. W. M. Wright and E. R. Davidson, *J. Chem. Phys.*, **43,** 840 (1965).

47. S. Rothenberg and E. R. Davidson, *J. Chem. Phys.*, **44,** 730 (1966); **45,** 2560 (1966).
48. J. D. Bowman, Jr., J. O. Hirschfelder, and A. C. Wahl, *J. Chem. Phys.*, **53,** 2743 (1970).
49. W. T. Zemke, P. G. Lykos, and A. C. Wahl, *J. Chem. Phys.*, **51,** 5635 (1969).
50. D. R. Bates, K. Ledsham, and A. L. Stewart, *Phil. Trans. Roy. Soc.* (London), Ser. A, **246,** 215 (1953).
51. H. H. Michels, *J. Chem. Phys.*, **44,** 3834 (1966).
52. H. H. Michels and F. E. Harris, *J. Chem. Phys.*, **39,** 1464 (1963).
53. H. S. Taylor and F. E. Harris, *Mol. Phys.*, **7,** 287 (1964).
54. C. F. Bender and E. R. Davidson, *J. Phys. Chem.*, **70,** 2675 (1966).
55. H. M. Miller and H. F. Schaefer, III, *J. Chem. Phys.*, **53,** 1421 (1970).
56. A. D. McLean and M. Yoshimine, *J. Chem. Phys.*, **46,** 3682 (1967).
57. A. D. McLean and M. Yoshimine, *Supplement to IBM J. Res. Dev.* 11/67.
58. A. D. McLean and M. Yoshimine, to be published.
59. R. L. Matcha and R. K. Nesbet, *Phys. Rev.*, **160,** 72 (1967).
60. P. A. G. O'Hare and A. C. Wahl, *J. Chem. Phys.*, **53,** 2469, 2834 (1970).
61. K. D. Carlson, K. Kaiser, C. Moser, and A. C. Wahl, *J. Chem. Phys.*, **52,** 4678 (1970).
62. E. R. Davidson and A. K. Q. Siu, *Int. J. Quantum Chem.*, **IV,** 223 (1970).
63. R. K. Nesbet, *J. Chem. Phys.*, **43,** 4403 (1965).
64. S. V. O'Neil and H. F. Schaefer, III, *J. Chem. Phys.*, **53,** 3994 (1970).
65. P. J. Bertoncini, G. Das, and A. C. Wahl, *J. Chem. Phys.*, **52,** 5112 (1970).
66. S. B. Schneiderman and H. H. Michels, *J. Chem. Phys.*, **42,** 3706 (1965).
67. H. Conroy and B. L. Bruner, *J. Chem. Phys.*, **47,** 921 (1967).
68. C. Edmiston and M. Krauss, *J. Chem. Phys.*, **45,** 1833 (1966).
68. J. C. Browne, *J. Chem. Phys.*, **45,** 2707 (1966).
70. G. H. Matsumoto, C. F. Bender, and E. R. Davidson, *J. Chem. Phys.*, **46,** 402 (1967).
71. R. D. Poshusta and F. A. Matsen, *Phys. Rev.*, **132,** 307 (1963).
72. D. R. Scott, E. M. Greenwalt, J. C. Browne, and F. A. Matsen, *J. Chem. Phys.*, **44,** 2981 (1966).
73. H. J. Kolker and H. Michels, *J. Chem. Phys.*, **50,** 1762 (1969).
74. H. F. Schaefer, III, D. R. McLaughlin, F. E. Harris, and B. J. Alder, submitted to *Phys. Rev. Lett.* **25,** 988 (1970).
75. A. C. Wahl, submitted for publication.
76. G. Das and A. C. Wahl, *J. Chem. Phys.*, **44,** 87 (1966).
77. G. Das, *J. Chem. Phys.*, **46,** 1568 (1967).
78. C. F. Bender and E. R. Davidson, *J. Chem. Phys.*, **47,** 4972 (1967).
79. C. F. Bender and E. R. Davidson, *J. Chem. Phys.*, **46,** 3313 (1967).
80. P. F. Fougere and R. K. Nesbet, *J. Chem. Phys.*, **44,** 285 (1966).
81. P. E. Cade, K. D. Sales, and A. C. Wahl, *J. Chem. Phys.*, **44,** 1973 (1966).
82. H. H. Michels, *J. Chem. Phys.*, **53,** 841 (1970).
83. H. F. Schaeffer, III and F. E. Harris, *J. Chem. Phys.*, **48,** 4946 (1968).
84. R. H. Pritchard, C. F. Bender, and C. W. Kern, *Chem. Phys. Lett.*, **5,** 529 (1970).

85. F. E. Harris and H. H. Michels, *Int. J. Quantum Chem.*, **3S,** 461 (1970).
86. G. Das and A. C. Wahl, *Phys. Rev. Lett.*, **24,** 440 (1970).
87. H. F. Schaefer, III, *J. Chem. Phys.*, **52,** 6241 (1970).
88. T. G. Heil, S. V. O'Neil, and H. F. Schaefer, III, *Chem. Phys. Lett.*, **5,** 253 (1970).
89. A. C. Wahl, P. J. Bertoncini, G. Das, and T. L. Gilbert, *Int. J. Quantum Chem.*, **1S,** 123 (1967).
90. T. L. Gilbert and A. C. Wahl, *J. Chem. Phys.* **55,** 5247 (1971).
91. T. L. Gilbert and A. C. Wahl, *J. Chem. Phys.*, **47,** 3425 (1967).
92. H. Bethe, *Ann. Physik.*, **3,** 133 (1929).
93. R. S. Mulliken, *Phys. Rev.*, **43,** 279 (1933).
94. K. E. Shuler, *J. Chem. Phys.*, **21,** 624 (1953).
95. K. J. Laidler, and K. E. Shuler, *Chem. Rev.*, **48,** 157 (1951).
96. K. J. Laidler, *The Chemical Kinetics of Excited States*, Clarendon, Oxford, England, 1955.
97. A. D. Walsh, *J. Chem. Soc.*, 2260 (1953).
98. L. M. Sachs, M. Geller, and J. J. Kaufman, *J. Chem. Phys.*, **51,** 2771 (1969).
99. Joyce J. Kaufman, L. M. Sachs, and M. Geller, *J. Chem. Phys.*, **49,** 4369 (1968).
100. S. D. Peyerimhoff, R. J. Buenker, and L. C. Allen, *J. Chem. Phys.*, **45,** 734 (1966).
101. S. D. Peyerimhoff, R. J. Buenker, and J. L. Whitten, *J. Chem. Phys.*, **46,** 1707 (1967).
102. L. M. Sachs, M. Geller, and Joyce J. Kaufman, *J. Chem. Phys.*, **52,** 974 (1970).
103. M. Krauss, *J. Res. Nat. Bur. Std. U.S.*, **68A,** 635 (1964).
104. J. F. Harrison and L. C. Allen, *J. Am. Chem. Soc.*, **91,** 807 (1969).
105. C. F. Bender and H. F. Schaefer, III, *J. Am. Chem. Soc.*, **92,** 4984 (1970).
106. J. W. Moskowitz and M. C. Harrison, *J. Chem. Phys.*, **43,** 3550 (1965).
107. D. C. Pan and L. C. Allen, *J. Chem. Phys.*, **46,** 1797 (1967).
108. L. M. Sachs, M. Geller, and Joyce J. Kaufman, presented before the Division of Physical Chemistry, 153rd American Chemical Society National Meeting, Miami Beach, Florida, April 1967, Abstract 153.
109. R. J. Buenker and S. D. Peyerimhoff, *J. Chem. Phys.*, **45,** 3682 (1966).
110. S. D. Peyerimhoff, *J. Chem. Phys.*, **47,** 1953 (1967).
111. S. D. Peyerimhoff and R. J. Buenker, *Theor. Chim. Acta* (Berl.), **9,** 103 (1967).
112. S. D. Peyerimhoff and R. J. Buenker, *J. Chem. Phys.*, **49,** 2473 (1968).
113. I. Shavitt, R. M. Stevens, F. L. Minn, and M. Karplus, *J. Chem. Phys.*, **48,** 2700 (1968).
114. C. Edmiston and M. Krauss, *J. Chem. Phys.*, **49,** 192 (1968).
115. M. Krauss and F. H. Mies, *J. Chem. Phys.*, **42,** 2703 (1965).
116. E. Hayes, presented at the Symposium on Chemical Dynamics in honor of Professor H. Eyring's 70th Birthday, Salt Lake City, Utah, February 1971; P. J. Brown and E. F. Hayes, *J. Chem. Phys.* **55,** 5132 (1971).
117. W. A. Lester, Jr., *J. Chem. Phys.*, **53,** 1511 (1970).
118. M. Krauss, *J. Res. Nat. Bur. Std. U.S.*, **72A,** 553 (1968).
119. W. A. Lester, Jr. and M. Krauss, *J. Chem. Phys.*, **52,** 4775 (1970).

120. M. Krauss, private communication, 1968.
121. L. Burnelle, unpublished results.
122. A. Pipano and Joyce J. Kaufman, presented at the International Symposium on Atomic, Molecular and Solid State Theory and Quantum Biology, Sanibel Island, Florida, January 1971, and at the Symposium on Chemical Dynamics in honor of Professor H. Eyring, Salt Lake City, Utah, February 1971. *J. Chem. Phys.* **56,** 5258 (1972).
123. W. J. Hehre, R. F. Stewart, and J. A. Pople, *J. Chem. Phys.*, **51,** 2657 (1969).
124. M. D. Newton, *J. Chem. Phys.*, **51,** 3917 (1969).
125. M. D. Newton, W. A. Lathan, W. J. Hehre, and J. A. Pople, *J. Chem. Phys.*, **51,** 3927 (1969).
126. W. J. Hehre, R. Ditchfield, R. F. Stewart, and J. A. Pople, *J. Chem. Phys.*, **52,** 2769 (1970).
127. M. D. Newton, W. A. Lathan, W. J. Hehre, and J. A. Pople, *J. Chem. Phys.*, **52,** 4064 (1970).
128. R. Ditchfield, W. J. Hehre, and J. A. Pople, *J. Chem. Phys.*, **52,** 5001 (1970).
129. R. Ditchfield, D. P. Miller, and J. A. Pople, *J. Chem. Phys.*, **53,** 613 (1970).
130. W. J. Hehre, R. Ditchfield, and J. A. Pople, *J. Chem. Phys.*, **53,** 932 (1970).
131. B. D. Joshi, *J. Chem. Phys.*, **46,** 875 (1967).
132. Z. Gershgorn and I. Shavitt, *Int. J. Quantum Chem.*, **II,** 751 (1968).
133. R. E. Kari and I. G. Csizmadia, *J. Chem. Phys.*, **46,** 1817 (1967).
134. K. Morokuma, L. Pedersen, and M. Karplus, *J. Chem. Phys.*, **48,** 4801 (1968).
135. R. E. Kari and I. G. Csizmadia, *J. Chem. Phys.*, **46,** 4585 (1967).
136. U. Kaldor and I. Shavitt, *J. Chem. Phys.*, **65,** 888 (1966).
137. A. Veillard, J. M. Lehn and B. Munsch, *Theor. Chim. Acta* (Berl.), **9,** 275 (1965).
138. A. Pipano, R. R. Gilman, C. F. Bender, and I. Shavitt, *Chem. Phys. Lett.*, **9,** 583 (1970).
139. C. F. Bender, *Theor. Chim. Acta*, **16,** 401 (1970).
140. M. D. Newton and S. Ehrenson, *J. Am. Chem. Soc.* **93,** 4971 (1971).
141. W. A. Lester, Jr., Editor, *Conference on Potential Energy Surfaces in Chemistry*, held at University of California, Santa Cruz, California, August, 1970, published by IBM Research Laboratory, San Jose, Calif., 1971.
142. Joyce J. Kaufman and R. Predney, paper presented at the International Symposium on Atomic, Molecular and Solid State Theory and Quantum Biology, Sanibel Island, Florida, 1971. *Int. J. Quantum Chem.*, **5,** 235 (1971).
143. J. C. Y. Chen, "Theory of Transient Negative Ions of Simple Molecules," in *Advances in Radiation Chemistry*, edited by M. Burton and J. L. Magee, Wiley, New York, 1968, Volume I.
144. J. C. Y. Chen and M. H. Mittleman, *Phys. Rev.*, **174,** 185 (1968).
145. J. N. Bardsley, *J. Phys. B.* (*Proc. Phys. Soc.* (London)) Ser. 2, **1,** 349, 365 (1968).
146. R. N. Compton, G. S. Hurst, L. G. Christophorou, and P. W. Reinhardt, "Electron Capture Cross Sections and Negative Ion Lifetimes." Submitted by R. N. Compton to the University of Tennessee as a Ph.D. dissertation, March 1966.

147. J. O. Hirschfelder and W. J. Meath, "The Nature of Intermolecular Forces," in *Intermolecular Forces*, Advances in Chemical Physics, Volume 12, Interscience, New York, 1967, Chapter 1.
148. P. Langevin, *Ann. Chem. Phys.*, **5,** 245 (1905).
149. G. Gioumousis and D. P. Stevenson, *J. Chem. Phys.*, **29,** 294 (1958).
150. Joyce J. Kaufman and W. S. Koski, *J. Chem. Phys.*, **50,** 1942 (1969).
151. A. D. Buckingham, *Advan. Chem. Phys.*, **12,** 107 (1967).
152. H. Reeh, *Z. Naturforsch.*, **15a,** 377 (1960).
153. H. Margenau and N. R. Kestner, *Theory of Intermolecular Forces*, Pergamon, New York, 1969.
154. C. A. Coulson and K. Zalewski, *Proc. Roy. Soc.*, (London), **A268,** 437 (1962).
155. L. Landau, *Soviet Phys.*, **1,** 89 (1932).
156. L. Landau, *Z. Phys. Sowj.*, **2,** 46 (1932).
157. C. Zener, *Proc. Roy. Soc.* (London), **A137,** 696 (1932).
158. E. G. C. Stueckelberg, *Helv. Phys. Acta*, **5,** 369 (1932).
159. D. R. Bates, *Proc. Roy. Soc.* (London), **A257,** 22 (1960).
160. D. R. Bates, *Proc. Phys. Soc.* (London), **84,** 517 (1964).
161. J, K. Lewis and J. T. Hougen, *J. Chem. Phys.*, **48,** 5329 (1968).
162. J. von Neumann and E. Wigner, *Physik. Z.*, **30,** 467 (1929).
163. E. R. Davidson, *J. Chem. Phys.*, **35,** 1189 (1961).
164. W. Lichten, *Phys. Rev.*, **131,** 229 (1963); **139A,** 27 (1965).
165. U. Fano and W. Lichten, *Phys. Rev. Lett.*, **4,** 627 (1965).
166. W. Lichten, *Advances in Chemical Physics*, Volume XIII, edited by I. Prigogine, Interscience, New York, 1968, p. 41.
167. W. Lichten, *Phys. Rev.*, **164,** 131 (1967).
168. S. Dworetsky, R. Novick, W. W. Smith, and N. Tolk, *Phys. Rev. Lett.*, **18,** 939 (1967).
169. R. D. Levine, B. R. Johnson, and R. Bernstein, WIS-TCI-305, Theoretical Chemistry Institute, University of Wisconsin, July 12, 1968.
170. "Transitions Non Radiative Dan Les Molecules," Société de Chimie Physique, Paris, France, May 27–30, 1969.
171. M. Kasha and B. R. Henry, *Ann. Rev. Phys. Chem.*, **19,** 161 (1968).
172. C. F. Giese, *Advan. Chem. Ser.*, **58,** 20 (1966).
173. A. L. Schmeltekopf, E. E. Ferguson, and F. C. Fehsenfeld, *J. Chem. Phys.*, **48,** 2966 (1968).
174. J. H. Callomon, *Proc. Chem. Soc.*, **2,** 313 (1959).
175. C. Brundle and D. W. Turner, *J. Mass Spectry. Ion Phys.*, **2,** 195 (1966).
176. P. Natalis and J. E. Collin, *J. Mass Spectry. Ion Phys.*, **2,** 221 (1969).
177. F. Gilmore, *DASA Reaction Rate Handbook*, 1967, Chap. 4. DASA Information and Analysis center.
178. Y. Tanaka, A. S. Jursa, and F. J. LeBlanc, *J. Chem. Phys.*, **32,** 1205 (1960).
179. R. K. Curran and R. E. Fox, *J. Chem. Phys.*, **34,** 1590 (1961).
180. V. H. Dibeler, J. A. Walker, and S. K. Liston, *J. Res. Nat. Bur. Std. U.S.*, **71A,** 371 (1967).
181. J. C. Polanyi, *Proceedings of the Conference on Potential Energy Surfaces in*

Chemistry, Edited by W. A. Lester, Jr., IBM Research Lab., San Jose, Calif. 1971, p. 10.
182. G. L. Weissler, J. A. R. Samson, M. Ogawa, and G. R. Cook, *J. Opt. Soc. Am.*, **49,** 338 (1959).
183. Joyce J. Kaufman, Ellen Kerman, and W. S. Koski, *Int. J. Quantum Chem.*, **4S,** 205 (1970).
184. R. C. C. Lao, R. W. Rozett, and W. S. Koski, *J. Chem. Phys.*, **49,** 4202 (1968).

Additional References

185. H. F. Schaefer, III, "The Electronic Structure of Atoms and Molecules. A Survey of Rigorous Quantum Mechanical Results," Addison-Wesley Publishing Co., Reading, Mass. 1972.
186. W. N. Lipscomb and T. A. Halgren, *Proc. Natl. Acad. Sci.*, **69,** 652 (1972); *J. Chem. Phys.*, **58,** 1569 (1973).
187. J. C. Tully, *J. Chem. Phys.*, **58,** 1090 (1973); *ibid*, **58,** 1396 (1973).
188. P. S. Bagus and H. F. Schaefer, III, *J. Chem. Phys.*, **58,** 1844 (1973).
189. H. H. Michels, J. C. Browne, T. A. Green, and M. M. Masden, Abstracts of papers 8th Int. Conference on Physics of Atomic and Electronic Collisions, Belgrade, Yugoslavia, (1923), p. 185.
190. B. Ulrich, A. L. Ford, and J. C. Browne, Abstracts of papers of 7th Int. Conference on Physics of Atomic and Electronic Collisions, Amsterdam, Netherlands, (1971), p. 208.
191. K. Docker and J. Hinze, *J. Chem. Phys.*, **57,** 4928 (1972).
192. C. F. Melius and W. A. Goddard, III, *J. Chem. Phys.*, **56,** 3348 (1972); C. F. Melius, W. A. Goddard, III, and L. R. Kahn, *J. Chem. Phys.*, **56,** 3342 (1972).
193. P. S. Bagus, C. M. Moser, P. Goethals, and G. Verhagen, *J. Chem. Phys.*, **58,** 1886 (1973).
194. K. Hsu, R. C. Raffenetti, and I. Shavitt, 21st Symp. on Molecular Structure and Spectroscopy, Columbus, Ohio, (1973) p. 174.
195. P. K. Pearson, C. F. Bender, and H. F. Schaefer, III, *J. Chem. Phys.*, **55,** 5235 (1971).
196. R. J. Blint and W. A. Goddard, III, *J. Chem. Phys.*, **57,** 5296 (1972).
197. R. J. Blint and W. A. Goddard, III, *Chem. Phys.*, **3,** 297 (1974).
198. S. Green, P. S. Bagus, B. Liu, A. D. McLean, and M. Yoshimine, *Phys. Rev.*, **A5,** 1614 (1972).
199. B. Liu, G. C. Lie and J. Hinze, *J. Chem. Phys.*, **57,** 625 (1972).
200. J. N. Bardsley and B. R. Junker, *Ap. J.* (*Letters*), **183,** L135 (1973).
201. M. Krauss and P. S. Julienne, *Ap. J.* (*Letters*), **183,** L139 (1973).
202. P. S. Julienne, M. Krauss, and A. C. Wahl, *Chem. Phys. Letts.*, **11,** 16 (1971).
203. V. Bondbey, P. K. Pearson, and H. F. Schaefer, III, *J. Chem. Phys.*, **57,** 1123 (1972).
204. S. V. O'Neil and H. F. Schaefer, III, *J. Chem. Phys.*, **55,** 394 (1971).
205. J. L. Whitten and J. D. Petke, *J. Chem. Phys.*, **56,** 380 (1972).
206. J. A. Pople, *Int. J. Quantum Chem. Symp. Issue*, **5,** 175 (1971).
207. K. Morokuma and T. F. George, *J. Chem. Phys.*, **59,** 1959 (1973).

208. T. F. George and K. Morokuma, *Chem. Phys.*, **2,** 129 (1973).
209. E. F. Hayes and J. L. Gobe, *J. Chem. Phys.*, **55,** 5182 (1971).
210. P. A. G. O'Hare and A. C. Wahl, *J. Chem. Phys.*, **56,** 4516 (1972).
211. A. C. Wahl and P. A. G. O'Hare, *J. Chem. Phys.*, **54,** 4563 (1971).
212. A. C. Wahl and P. A. G. O'Hare, *J. Chem. Phys.*, **55,** 666 (1971).
213. A. C. Wahl, M. Krauss, and P. Maldonado, *J. Chem. Phys.*, **54,** 4944 (1971).
214. G. M. Schwenzer, D. H. Liskow, H. F. Schaefer, III, P. S. Bagus, B. Liu, A. D. McLean, and M. Yoshimine, *J. Chem. Phys.*, **58,** 3181 (1973).
215. M. Yoshimine, *J. Chem. Phys.*, **57,** 1108 (1972).
216. C. F. Melius and W. A. Goddard, III, *Chem. Phys. Letts.*, **15,** 524 (1972).
217. S. V. O'Neil, P. K. Pearson, and H. F. Schaefer, III, *Chem. Phys. Letts.*, **10,** 404 (1971).
218. P. K. Pearson, S. V. O'Neil, and H. F. Schaefer, III, *J. Chem. Phys.*, **56,** 3938 (1972).
219. H. F. Schaefer, III, and T. G. Heil, *J. Chem. Phys.*, **54,** 2573 (1971).
220. S. Green, *J. Chem. Phys.*, **56,** 739 (1972).
221. T. G. Heil and H. F. Schaefer, III, *J. Chem. Phys.*, **56,** 958 (1972).
222. P. S. Bagus and H. J. T. Preston, *J. Chem. Phys.*, **59,** 2986 (1973).
223. H. H. Michels, Presented at the First Summer Colloquium on Electronic Transition Lasers, Univ. of Cal., Santa Barbara, June 17–19, (1974).
224. A. C. Wahl and P. J. Bertoncini, *J. Chem. Phys.*, **58,** 1259 (1973).
225. D. R. McLaughlin and H. F. Schaefer, III, *Chem. Phys. Letts.*, **12,** 244 (1971).
226. B. J. Garrison, W. H. Miller, and H. F. Schaefer, III, *J. Chem. Phys.*, **59,** 3193 (1973).
227. L. Lenamon, J. C. Browne, and R. E. Olson, *Phys. Rev.*, **A8,** 2380 (1973).
228. S. L. Gubermann and W. A. Goddard, III, *Chem. Phys. Letts.*, **14,** 460 (1972).
229. A. C. Wahl and G. Das, *J. Chem. Phys.*, **56,** 3532 (1972).
230. G. A. Henderson, W. T. Zemke, and A. C. Wahl, *J. Chem. Phys.*, **58,** 2654 (1973).
231. T. H. Dunning, Jr. and D. C. Cartwright, 28th Symposium on Molecular Structure and Spectroscopy, Columbus, Ohio, June 1973, p. 93.
232. T. H. Dunning, Jr. and D. C. Cartwright 28th Symposium on Molecular Structure and Spectroscopy, Columbus, Ohio, June 1973, p. 43.
233. H. F. Schaefer, III, *J. Chem. Phys.*, **54,** 2207 (1971).
234. H. F. Schaefer, III and W. H. Miller, *J. Chem. Phys.*, **55,** 4107 (1971).
235. S. D. Peyerimoff and R. J. Buenker, *Chem. Phys. Letts.*, **16,** 235 (1972).
236. W. T. Zemke, G. Das and A. C. Wahl, *Chem. Phys. Letts.*, **14,** 310 (1972).
237. D. C. Cartwright, W. J. Hunt, W. Williams, S. Trajmar, and W. A. Goddard, III, *Phys. Rev.*, **A8,** 2436 (1973).
238. B. I. Schnieder and J. S. Cohen, "Ground and Excited States of Ne_2 and Ne_2^+. I. Potential Curves With and Without Spin-Orbit Coupling" and "Ground and Excited States of Ne_2 and Ne_2^+. II. Spectroscopic Properties and Radiative Lifetimes." Submitted to Journal of Chem. Phys.
239. B. Liu, *J. Chem. Phys.*, **58,** 1925 (1973).

240. C. W. Bauschlicher, Jr., S. V. O'Neil, R. K. Preston, H. F. Schaefer, III, and C. F. Bender, *J. Chem. Phys.*, **59,** 1286 (1973).
241. H. J. T. Preston and J. C. Tully, *J. Chem. Phys.*, **54,** 4297 (1971).
242. H. F. Schaefer, III, D. Wallach, and C F. Bender, *J. Chem. Phys.*, **56,** 1219 (1972).
243. C. Vauge and J. L. Whitten, *Chem. Phys. Letts.*, **13,** 541 (1972).
244. M. Gelus and W. Kutzelnigg, *Theoret. Chim. Acta*, **28,** 103 (1973).
245. C. F. Bender and H. F. Schaefer, III, *J. Mol. Spectro.*, **37,** 423 (1971).
246. S. V. O'Neil, H. F. Schaefer, III, and C. F. Bender, *J. Chem. Phys.*, **55,** 162 (1971).
247. D. R. McLaughlin, C. F. Bender, and H. F. Schaefer, III, *Theoret. Chim. Acta*, **25,** 352 (1972).
248. C. F. Bender, H. F. Schaefer, III, D. R. Franceschetta, and L. C. Allen, *J. Am. Chem. Soc.* **94,** 6888 (1972).
249. S. T. Lee and K. Morokuma, *J. Am. Chem. Soc.*, **93,** 6863 (1974).
250. S. Y. Chu, A. K. Q. Siu, and E. F. Hayes, *J. Am. Chem. Soc.*, **94,** 2969 (1972).
251. H. F. Schaefer, III and C. F. Bender, *J. Chem. Phys.*, **55,** 1720 (1971).
252. R. P. Hosteny, A. C. Wahl, and M. Krauss, Symposium on Molecular Structure and Spectroscopy, Columbus, Ohio, (1973), p. 180.
253. B. J. Rosenberg, W. C. Ermler, and I. Shavitt, Symposium on Molecular Structure and Spectroscopy, Columbus, Ohio, (1973), p. 110.
254. R. P. Hosteny, A. R. Hinds, A. C. Wahl and M. Krauss, *Chem. Phys. Letts.*, **23,** 9 (1973).
255. W. A. Goddard, III, and W. J. Hunt, *Chem. Phys. Letts.*, **24,** 464 (1974).
256. R. A. Gangi and R. F. W. Bader, *J. Chem. Phys.*, **55,** 5369 (1971).
257. W. Kutzelnigg, W. Staemmler, and C. Hoheisel, *Chem. Phys.*, **1,** 27 (1973).
258. W. A. Goddard, III, *J. Am. Chem. Soc.*, **93,** 6750 (1971).
259. C. F. Bender, P. K. Pearson, S. V. O'Neil, and H. F. Schaefer, III, *J. Chem. Phys.*, **56,** 4626 (1972).
260. S. V. O'Neil, P. K. Pearson, H. F. Schaefer, III, and C. F. Bender, *J. Chem. Phys.*, **58,** 1126 (1973).
261. P. K. Pearson, W. J. Hunt, C. F. Bender, and H. F. Schaefer, III, *J. Chem. Phys.*, **58,** 5358 (1973).
262. D. H. Liskow, H. F. Schaefer, III, and C. F. Bender, *J. Am. Chem. Soc.*, **93,** 6734 (1971).
263. J. L. Gole and E. F. Hayes, *J. Chem. Phys.*, **57,** 360 (1972).
264. T. H. Dunning, Jr. and W. A. Goddard, III, 28th Symposium on Molecular Structure and Spectroscopy, Columbus, Ohio, (1973), p. 159.
265. S. V. O'Neil, H. F. Schaefer, III, and C. F. Bender, *J. Chem. Phys.*, **59,** 3608 (1973).
266. M. M. Heaton, A. Pipano, and J. J. Kaufman, *Int. J. Quantum Chem.*, **6S,** 181 (1972).
267. A. C. Wahl, private communication, 1973.
268. A. K. Q. Siu and E. F. Hayes, *Chem. Phys. Letts.*, **21,** 573 (1973).
269. P. J. Hay and W. A. Goddard, III, *Chem. Phys. Letts.*, **14,** 46 (1972).

270. P. J. Hay, T. H. Dunning, Jr., and W. A. Goddard, III, *Chem. Phys. Letts.*, **23,** 4576 (1973).
271. N. H. Winter, C. F. Bender, and W. A. Goddard, III, *Chem. Phys. Letts.*, **20,** 484 (1973).
272. M. Krauss and D. Neumann, *Chem. Phys. Letts.*, **14,** 26 (1972).
273. W. Wahlgren, B. Liu, P. K. Pearson, and H. F. Schaefer, III, *Nature*, **246,** 149 (1973).
274. W. A. Goddard, III, T. H. Dunning, Jr., W. J. Hunt, and P. J. Hay, *Accts. Chem. Res.*, **6,** 368 (1973).
275. A. P. Mortola and W. A. Goddard, III, *J. Am. Chem. Soc.*, **96,** 1 (1974).
276. J. L. Gole, A. K. Siu, and E. Hayes, *J. Chem. Phys.*, **58,** 857 (1973).
277. D. R. Yarkony, W. J. Hunt, and H. F. Schaefer, III, *Mol. Phys.*, **26,** 941 (1972).
278. C. F. Bender, H. F. Schaefer, III, and P. A. Kollman, *Mol. Phys.*, **24,** 235 (1972).
279. M. Rubinstein and I. Shavitt, *J. Chem. Phys.*, **51,** 2014 (1969).
280. M. Gelus and W. Kutzelnigg, unpublished.
281. C. W. Wilson, Jr. and W. A. Goddard, III, *J. Chem. Phys.*, **56,** 5913 (1972).
282. E. Korhanski, B. Ross, P. Siegbahn, and M. H. Wood, *Theoret. Chim. Acta*, **32,** 151 (1973).
283. H. F. Schaefer, III, and C. F. Bender, *J. Chem. Phys.*, **57,** 217 (1972).
284. W. Kutzelnigg and M. Gelus, *Theor. Chim. Acta*, **28,** 103 (1973).
285. J. Paldus, J. Cizek, and I. Shavitt, *Phys. Rev.*, **A5,** 50 (1972).
286. W. A. Goddard, III and R. J. Blint, *Chem. Phys. Letts.*, **14,** 616 (1972).
287. R. Alrichs, *Theoret. Chim. Acta.*, **30,** 315 (1973).
288. P. Dejarden, E. Kochanski, A. Veillard, B. Roos, and P. Siegbahn, *J. Chem. Phys.*, **59,** 5546 (1973).
289. J. Almlof and U. Wahlgren, *Theoret. Chim. Acta*, **28,** 161 (1973).
290. C. P. Baskin, C. F. Bender, and P. A. Kollman, *J. Am. Chem. Soc.*, **95,** 5868 (1973).
291. J. E. Del Bene and J. A. Pople, *J. Chem. Phys.*, **58,** 3605 (1973).
292. E. Clementi, *J. Chem. Phys.*, **46,** 3851 (1967).
293. E. Clementi and H. Popkie, *J. Chem. Phys.*, **57,** 1077 (1972).
294. H. Popkie, H. Kistenmacher, and E. Clementi, *J. Chem. Phys.*, **59,** 1325 (1973).
295. K. Morokuma and R. E. Davis, *J. Am. Chem. Soc.*, **94,** 1060 (1972).
296. C. P. Baskin, C. F. Bender, C. W. Bauschlicher, Jr., and H. F. Schaefer, III, *J. Am. Chem. Soc.*, **76,** 2709 (1974).
297. S. Ehrenson and M. Newton, *Chem. Phys. Letts.*, **13,** 24 (1972).
298. S. Iwata and K. Morokuma, *J. Am. Chem. Soc.*, **95,** 7563 (1973).
299. D. McHayes and K. Morokuma, *Chem. Phys. Letts.*, **12,** 539 (1972).

CHAPTER THREE

VIBRATIONAL AND ROTATIONAL EXCITATION IN GASEOUS COLLISIONS

Robert C. Amme

Department of Physics and Astronomy, University of Denver, Denver, Colorado 80210

Contents

I. INTRODUCTION

The exchange of energy between translational and internal degrees of freedom plays an essential role in a vast number of molecular processes. Vibrational and rotational excitation by collision are the major mechanisms by which molecular reactants acquire the activation energy necessary to produce a chemical change. Once this occurs, the new products will be formed and stabilized at a rate that is determined by the efficiency of collisions in transferring their internal energy back into translational energy. Shock-tube studies [1, 2] of the homogeneous exchange reaction

$$H_2 + D_2 \rightarrow 2HD,$$

for example, have shown that the exchange rate is controlled by the excitation rate of the reactants to a critical vibrational level, from which the exchange proceeds rapidly. The same conclusion has been drawn from similar studies [3] with isotopes of N_2.

Since in general the probability that vibrational and rotational energy will be lost by spontaneous infrared radiation is low [4], deactivation usually proceeds via collisions. Thus, it is important to have detailed information on intermolecular force laws and to have the necessary theoretical tools for the prediction of energy transfer in collisions.

Probably the first experiment to indicate that vibrational-to-translational energy transfer probabilities might be measured in a fairly direct way was that by Pierce [5] in 1925. He observed that for CO_2 at room temperature the equilibration between translation and vibration proceeded at a rate sufficiently slowly that high-frequency acoustic waves exhibited velocity dispersion. This increase in ultrasonic velocity with increasing frequency occurs for CO_2 in the megacycle-per-second region, and results from the phase lag of the vibrational specific heat. Since collision rates are on the order of 10^{10} per second, this observation suggests that on an average only about one in every 10^4 collisions between CO_2 molecules results in vibrational excitation at this temperature.

In the years between 1925 and 1955, ultrasonics became a well-developed tool for the study of molecular energy transfer in gases, with temperatures ranging from 78°K to over 1200°K. Later in that period, the application of shock tubes to the problem greatly extended the range of temperature [6], and this technique is now being employed for studies up to nearly 10,000°K [7].

Recently, there has been expanding interest in molecular energy transfer as a result of gas laser development. This is true not only because lasers have proven to be an important tool in energy-transfer research, but also because such processes are relevant to laser operation. The ultimate performance of a molecular laser depends upon the achievable population inversion of the active vibrational levels. This, in turn, is determined by the various exchange processes, which can be quite numerous in gaseous mixtures. In general, a detailed analysis requires information about energy transfer between various degrees of freedom:

a. Translation↔ vibration (T–V).
b. Vibration↔ vibration (V–V).
c. Translation↔ rotation (T–R).
d. Vibration↔ rotation (V–R)

In V–V processes, one is concerned with those collisions in which a vibrational quantum of one mode within a polyatomic molecule is combined with translational energy to excite a second mode of different energy within that molecule (intramolecular V–V exchange), and also those cases in which the second mode is that of another molecule (intermolecular V–V exchange). The latter process may arise in mixtures of diatomic gases as well as in pure polyatomic gases.

In addition to the four processes given above, we may also list rotation–rotation energy transfer (R–R); however, very little study has been done of this process at the present time. In V–V, V–R, and R–R energy transfer processes, the energy difference between the initial and final energy states, $\Delta E = E_i - E_f$, is imparted to or removed from another degree of freedom, such as translation. When ΔE is negative, energy is taken up from the other degree of freedom. When it is zero or nearly so, a resonant condition arises, wherein the probability for the specified transfer process may become large.

In this chapter, we shall describe the basic theories of molecular energy transfer in nonreactive collisions, up to their present state of development. We shall then discuss the various experimental techniques of measuring collisional excitation or deexcitation probabilities. Finally, we will list some experimental results in both diatomic and polyatomic systems.

II. TRANSLATIONAL-TO-VIBRATIONAL ENERGY TRANSFER THEORY

A. Preliminary Remarks

Several reviews of the theory of vibrational energy transfer may be found in the literature. A book published in 1959 by Herzfeld and Litovitz [8] is one of the standard references to the subject, as well as that of Cottrell and McCoubrey [9], published in 1961. More recently, there are reviews by

Takayanagi [10, 11], Herzfeld [12], and Rapp and Kassal [13]. The latter, most recent, review presents critical comparisons of the classical, semiclassical, and quantum methods.

The model which has received the greatest theoretical attention is that of a colinear collision between an atom A and a diatomic molecule BC (Figure 3.1). The molecule is usually assumed to be a harmonic oscillator, and the interaction potential V_{AB} an exponential repulsion. The model was first

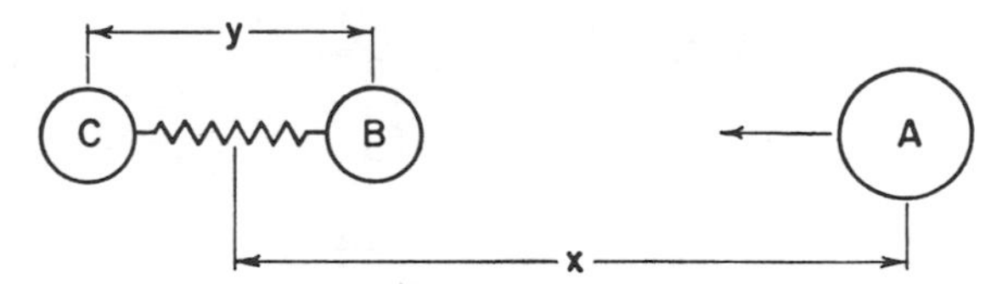

Figure 3.1 Colinear collision of an atom A with a diatomic molecule BC. Impact is presumed to be between atoms A and B.

investigated by the distorted-wave approximation, and then semiclassically, by Zener [14, 15]. It was first treated classically by Landau and Teller [16] in 1936. This early work has been thoroughly reviewed by Takayanagi [10]. In recent years, the approximations have been examined in greater detail, and several important discoveries have been made. These discoveries, with their attendant modifications to the older theories, are described in this section. We turn now to a brief description of the basic theories.

B. Classical Theory of V–T Transfer

Consider first the one dimensional, classical problem of a diatomic molecule BC colliding with an atom A (Figure 3.1) to yield either vibrational activation or deactivation of BC. Suppose that the molecule is a harmonic oscillator with frequency ν. If the interaction potential is

$$V(x) \propto \exp - \frac{x}{l} \tag{1}$$

and the initial relative velocity of approach is v_i, then the probability of energy transfer between translational and vibrational degrees of freedom may be deduced from the adiabatic principle [17]:

$$p(v_i) = C \exp - \frac{\tau_c}{\tau_0} = C \exp - \frac{4\pi^2 \nu l}{v_i}, \tag{2}$$

where $\tau_c = 2\pi l/v_i$ is the time required for the collision, $\tau_0 = 1/(2\pi\nu)$ relates in this case to the period of oscillation, and C is a constant. For a gas at temperature T, the average value for the transition probability is obtained by averaging over the appropriate Maxwellian distribution:

$$P(T) = C\left(\frac{\mu}{kT}\right)\int_0^\infty v \exp\left[-\frac{4\pi^2 \nu l}{v} - \frac{\mu v^2}{2kT}\right] dv. \tag{3}$$

This integral cannot be evaluated in closed form, but Landau and Teller noted that the temperature dependence may be estimated by observing that the exponent is a minimum when

$$\frac{d}{dv}\left[\frac{4\pi^2 \nu l}{v} + \frac{\mu v^2}{2kT}\right] = 0, \tag{4}$$

which gives

$$v_m = \left[\frac{4\pi^2 \nu k T l}{\mu}\right]^{1/3}. \tag{5}$$

The temperature dependence of the transition probability is therefore of the approximate exponential form

$$P(T) \sim \exp\left[-3\left(\frac{2\pi^4 \mu \nu^2 l^2}{kT}\right)^{1/3}\right]. \tag{6}$$

This is the famous result of Landau and Teller. In many of the more detailed theories, to be discussed below, this result appears explicitly as the principal contribution to the dependence of transition probability on temperature. In practical units, equation (6) is frequently written in the form

$$P(T) \sim \exp\left[-\frac{3}{2}\left(\frac{\theta'}{T}\right)^{1/3}\right], \tag{7}$$

where

$$\theta' = 0.8153\tilde{\mu}\theta^2 l^2, \tag{8}$$

in which θ is the characteristic temperature of the vibration:

$$\theta = \frac{h\nu}{k} = 1.4388\tilde{\nu}, \tag{9}$$

$\tilde{\nu}$ being the wave number, $\tilde{\mu}$ being the reduced molecular weight for the encounter, and l measured in angstroms.

More rigorously, it is necessary to distinguish between collisions giving rise to activation and those giving rise to deactivation. For a gas at or very near thermal equilibrium, one may write

$$\frac{n_1}{n_0} = \frac{g_1}{g_0} \exp - \frac{h\nu}{kT}, \tag{10}$$

where n_0 and n_1 are the numbers of molecules in the ground and first vibrational state, respectively, and g denotes the degeneracy of the level. To maintain equilibrium it is necessary that the average probability for activation per collision, P_{01}, and that for deactivation per collision, P_{10}, be similarly related:

$$\frac{P_{01}}{P_{10}} = \frac{g_1}{g_0} \exp - \frac{h\nu}{kT}, \tag{11}$$

where g_1 and g_0 are the respective statistical weights of the upper and lower levels.

In most of the more recent classical approaches [18], no allusion to Ehrenfest's (adiabatic) principle is employed, but rather the differential equations of motion from classical mechanics are solved, either exactly or approximately, subject to a set of initial conditions (masses, force constants, interaction potential, phase, and initial energies). The amount of energy, ΔE, transferred to the oscillator is obtained for these conditions. This quantity may then be averaged over all phases of the oscillating molecule. In approximate classical and semiclassical treatments, the interaction potential is expanded in a Taylor's series and only the first two terms are retained.

Exact calculations of ΔE have been carried out by Kelley and Wolfsberg [19] for colinear collisions between an atom and a diatomic molecule. The oscillator potential was considered to be both harmonic and Morse-type, and the interaction between the colliding pair was taken both as an exponential repulsion and as a Lennard–Jones 6:12 potential. Two important conclusions were reached: First, when the initial energy of the oscillator increases, the total energy transferred from translation to vibration, ΔE, decreases. Second, the effect of using a Morse-oscillator potential in place of the harmonic oscillator was generally to decrease ΔE, often by more than a factor of 10.

The harmonic-oscillator results from the exact calculation have been compared with an approximate formula for ΔE obtained by Rapp [20]:

$$\Delta E_{\mathrm{ap}} = 2\pi^2\omega^2 l^2 \left[\frac{m_{\mathrm{C}} m_{\mathrm{A}}^2 (m_{\mathrm{C}} + m_{\mathrm{B}})}{(m_{\mathrm{A}} + m_{\mathrm{B}} + m_{\mathrm{C}})^2 m_{\mathrm{B}}}\right] \operatorname{csch}^2 \frac{\pi\omega l}{v_0}, \tag{12}$$

where ω is the oscillator frequency. This equation can be shown [19] to violate energy conservation at large v_0 whenever the bracketed quantity exceeds 0.25. Rapp and Kassal [13] have found an approximate empirical relation for the ratio R of $\langle \Delta E \rangle$ and $\langle \Delta E_{\mathrm{ap}} \rangle$, the phase averaged energy transfer (exact and apparent):

$$R = \exp (1.685) m, \tag{13}$$

which applies when the oscillator has no initial excitation, m being a dimensionless mass-dependent quantity:

$$m = \frac{m_A m_C}{m_B} \frac{1}{m_A + m_B + m_C}. \tag{14}$$

Thus, for very light atoms colliding with heavy homonuclear molecules, the approximate case is in good agreement with the exact numerical calculation.

Unfortunately, the exact calculations have not proceeded to the point where they may be compared with experiment. The difficulty lies chiefly in obtaining the averaged result for all molecular orientations (the three-dimensional treatment). Introduction of a steric factor P_0, less than unity, has been shown [21] to be generally invalid. The concept has been pursued successfully, however, under limited conditions [22].

C. Semiclassical V–T Theory

A review of the semiclassical method is given by Cottrell and McCoubrey [9] and by Rapp and Kassal [13]. In this method, the translational motion is treated classically, while the molecule BC is assumed to have quantized vibrational levels. By converting the force $V'(x)$ on the oscillator due to the incident atom to $V'(t)$ by utilization of the classical trajectory $x(t)$, one may apply time dependent perturbation theory. The wave function for the perturbed system is written as a sum of the stationary-state wave functions: $\Psi_l(y) \exp(-i\omega_l t)$, with coefficients c_k given by

$$\frac{dc_k}{dt} = \frac{1}{i\hbar} \sum_l c_l(t) V'_{kl}(t) \exp(i\omega_{kl} t), \tag{15}$$

in which

$$V'_{kl}(t) = \int_{-\infty}^{\infty} \Psi_k V' \Psi_l \, dy, \tag{16}$$

where y is the vibrational coordinate, and $\omega_{kl} = (E_k - E_l)/\hbar$. The first order perturbation approximation leads to

$$p_{m,k} \equiv |c_k(\infty)|^2 \approx \frac{1}{\hbar^2} \left| \int_{-\infty}^{\infty} V'_{mk}(t) \exp(i\omega_{mk} t) \, dt \right|^2. \tag{17}$$

If it is assumed that the interaction is an exponential repulsion [equation (1)], where x is the distance between the impinging atom and the molecule, and that the amplitude of the vibrational displacement is much less than l, then one may show that [13]

$$p_{m\to m+1} - p_{m\to m-1} = \frac{\Delta E_{ap}}{\hbar\omega}, \tag{18}$$

where ΔE_{ap} is just that energy transfer calculated in the approximate classical treatment [equation (12)].

The influence of attractive forces can be allowed for in a rather direct way in the semiclassical theory. An analytic solution can be found for the case of a Morse-type interaction potential:

$$V(x) = \lambda \exp - \frac{x}{l} - \eta \exp - \frac{x}{2l}. \tag{19}$$

For the case of the $v = 1 \rightarrow v = 0$ transition, equation (17) is written [9] as

$$p_{10} = \frac{4}{\hbar^2} |X_{10}|^2 \left[\int_{-\infty}^{\infty} F(t) \cos \omega t \, dt \right]^2, \tag{20}$$

in which the zero of time has been taken at the distance of closest approach so that $F(t)$, the time-dependent force, is an even function. The quantity X_{10} is the matrix element of the vibrational coordinate. The energy to be transferred is simply $\hbar\omega$, where ω is the angular frequency of the oscillator. The time t is then given for the unperturbed trajectory by the integral

$$t = \int_{r_0}^{r} \left[v^2 - \frac{2V(r)}{\mu} \right]^{-1/2} dr, \tag{21}$$

where μ is the reduced mass and v is the velocity at $t = -\infty$. One finds that

$$F(t) = \frac{\mu v^2}{2l} \left[\frac{1 - \cosh (vt/2l) \sin \phi}{\cosh (vt/2l) - \sin \phi} \right], \tag{22}$$

where

$$\phi = \tan^{-1} \left(\frac{\eta^2}{2\mu v^2 \lambda} \right)^{1/2}. \tag{23}$$

The integral in equation (20) can be shown to be given approximately by $2\pi\mu\omega l \exp [-\omega l(\pi - 2\phi)/v]$ whenever the adiabatic condition $2\pi\omega l \gg v$ is fulfilled. The probability of a $1 \rightarrow 0$ transition is then found from equation (20) to be

$$p_{10} = \frac{16\pi^3\mu^2\omega l^2}{M\hbar} \exp - \frac{2\omega l(\pi - 2\phi)}{v}, \tag{24}$$

where it is assumed that the struck molecule is a harmonic oscillator, having $X_{10} = (\hbar/2M\omega)^{1/2}$ for the required matrix element. The reduced mass M for the oscillator is simply $m_B m_C/(m_B + m_C)$. One notes that if the attractive forces are negligible, $\phi = 0$, and equation (24) is of the Landau–Teller form [equation (2)] with the pre-exponential factor determined.

An approximate three dimensional semiclassical treatment of the harmonic oscillator interacting with an impinging atom according to a Morse potential has been presented by Calvert and Amme [23].

Shin [24] has utilized a semiclassical approach to examine orientational effects on vibrational excitation. He treats the problems of (XX–A) and (XY–A) collisions, where A is an atom interacting with atoms of a diatomic molecule XX or XY through a Morse potential, and also the problems of (XX–XX) collisions and (XY–XY) collisions. The orientation dependence of the transition probabilities for O_2–Ar, O_2–O_2, HBr–Ar, and HBr–HBr are plotted in polar diagrams. It is found, for example, that in the latter case only a very small range of angles gives the greatest contribution, namely, around that orientation in which the two H atoms lie between the Br atoms.

This same approach also was employed by Shin to calculate steric factors by averaging over orientation angles for XX–XX and XX–A collisions [25]. P_0 was found to vary from $\frac{1}{3}$ for H_2–Ar collisions to $\frac{1}{9}$ for I_2–Ar collisions. For XX–XX collisions, P_0 was found to be of the order of $\frac{1}{20}$ for H_2, and $\frac{1}{55}$ for I_2.

In another series of papers [26] Shin has used the WKB (Wentzel–Kramers–Brillouin) method for evaluating the vibrational transition matrix element, employing various forms of interaction potential. Comparisons are made with quantum-mechanical solutions.

Hansen and Pearson [27] have recently employed a linear superposition of exponential repulsive interaction potentials between an inert atom and the atoms of a homonuclear diatomic molecule. They then performed a three-dimensional semiclassical calculation of the vibrational-transition probability including simultaneous rotational transitions. They conclude that the effect of coupled rotational transitions leading predominantly to $\Delta J = \pm 2$ affect the vibrational transition rate by 50% or more.

D. Quantum-Mechanical V–T Theory

The colinear collision problem of atom A colliding with a molecule BC was first attempted quantum mechanically by Zener [14, 15] and then by Jackson and Mott [28] for the purpose of investigating thermal accommodation coefficients for atoms impinging on solid surfaces. An exponential repulsion was utilized, along with the harmonic-oscillator approximation. The distorted-wave (DW) method was employed to obtain a $1 \rightarrow 0$ transition probability of the form

$$p_{10} = \frac{16\pi^3\mu^2\nu l^2}{M\hbar} \frac{\sinh q_1 \sinh q_0}{(\cosh q_1 - \cosh q_0)^2}, \tag{25}$$

in which all the symbols have their prior meaning, and the quantities q_i are given by

$$q_i = \frac{2\pi m v_i l}{\hbar}. \tag{26}$$

It is this result which is the basis of the Schwartz–Slawsky–Herzfeld (SSH) calculation of vibrational relaxation times in gases [29]. In subsequent work, Schwartz and Herzfeld [30] extended Zener's approach to a collision in three dimensions, and developed a scheme for evaluating the potential parameter l in terms of the popular 6:12 Lennard–Jones potential function

$$V(r) = 4\varepsilon[(r_0/r)^{12} - (r_0/r)^6].$$

It was shown that r_0/l may be treated as a slowly varying function of temperature, and that, approximately, one may expect

$$l \approx \frac{r_0}{18}. \tag{27}$$

Their final formulation, as quoted in a practical form by Herzfeld and Litovitz, is obtained by averaging the DW transition probability, suitably expressed in three dimensions, over a normal distribution of velocities. The result is written

$$P_{10}(T) = [Z_o Z_{\text{osc}} Z'_{\text{tr}}\, \delta]^{-1} \exp \frac{\varepsilon}{kT}, \tag{28}$$

where

$$\delta = 1.017\left(\frac{r_0}{r_c}\right)^2 \left[0.76\left(1 + \frac{1.1\varepsilon}{kT}\right)\right] \tag{29}$$

is on the order of unity, and r_c is the distance of closest approach. The quantity Z_0^{-1} is the "steric factor," representing, qualitatively, the probability that the molecule be favorably oriented. As in the semiclassical arguments, Z_0 is usually taken to be 3. The other quantities are as follows:

$$Z_{\text{osc}} = \frac{M_B M_C (M_A + M_B + M_C)}{(M_B^2 + M_C^2) M_A} \frac{1}{\pi^2} \frac{\theta'}{\theta}, \tag{30}$$

in which θ' and θ are defined in equations (8) and (9),

$$Z'_{\text{tr}} = \pi^2 \left(\frac{\theta}{\theta'}\right)^2 \sqrt{\frac{3}{2\pi}} \left(\frac{T}{\theta'}\right)^{1/6} \exp\left[\frac{3}{2}\left(\frac{\theta'}{T}\right)^{1/3} - \frac{\theta}{2T}\right], \tag{31}$$

and ε is the Lennard–Jones well depth. The term $\exp(\varepsilon/kT)$ is introduced to allow for the effect of attractive forces in an approximate way. Introduction of the δ factor, equation (29), is done largely for the purpose of discounting those soft collisions (due to long-range forces) that contribute to the transport properties and thus to the calculated collision frequency, but not to the inelastic cross section.

Takayanagi and Kaneko [31] treated the case of He + O_2 collisions and concluded that for the vibrational transition $1 \to 0$ and for $J = 10$, collisions in which the rotational energy was also altered by $\Delta J = \pm 2$ were about $\frac{1}{4}$ as probable as those deactivating collisions for which $\Delta J = 0$.

In 1966, Secrest and Johnson [32] performed an exact quantum mechanical calculation of the one dimensional problem of A colliding with a harmonic oscillator BC, using a method referred to as that of "amplitude density functions" and yielding results reportedly correct to three significant figures. It was found that the DW result of Jackson and Mott [equation (25)] gives consistently excessive transition probabilities, by amounts depending upon the repulsive strength of the potential, the incident energy, and the mass parameter m [equation (14)]. In some cases it was found that double, or even triple, quantum jumps can be more important than single quantum jumps. Jackson and Mott's DW treatment involves an expansion of the transition amplitude. Keeping only the first-order term gives no accurate information about second- and higher-order transition probabilities. Higher-order terms have been considered by Thiele and Weare [33], who find that the DW expansion, including second- and third-order terms, yields reasonably good results when the incident atom is sufficiently light and has low kinetic energy. They find that the double-quantum transition probability in the second order, $P^{(2)}_{l+2\leftarrow l}$, is related to that in the first order, $P^{(1)}_{l+2\leftarrow l}$, through the approximate result

$$P^{(2)}_{l+2\leftarrow l} \approx P^{(1)}_{l+2\leftarrow l} + \tfrac{1}{4} P^{(1)}_{l+2\leftarrow l+1} P^{(1)}_{l+1\leftarrow l}. \tag{32}$$

Here, l is the initial oscillator state. The second term is generally found to be much greater than the first. Comparisons of the $0 \rightarrow 1$ transition probabilities calculated from first- through third-order DW results are compared, in their paper, to the exact values obtained numerically by Secrest and Johnson. Comparison is made as a function of energy and as a function of m. In Figure 3.2, $0 \rightarrow 1$ transition probabilities calculated by Thiele and Weare are plotted as a function of total energy in units of $h\nu$. First-, second-, and third-order DW results are shown for various values of $\alpha = 1/l$ and of m, and comparisons are presented with the exact results of Secrest and Johnson. One sees that for small m, the DW expansion can yield reasonably good agreement. Thus, one expects that for such processes as H_2 colliding with Cl_2 ($m = 0.028$), first-order DW theory will be valid. Caution is required, however, in comparing experiment with theories utilizing the saddle-point integration method in obtaining $P_{10}(T)$, since this method does not yield a good approximation when μ is small. Low-energy collisions may not be negligible, and the method outlined by Schwartz, Slawsky, and Herzfeld [29] for obtaining the repulsive range l from the Lennard–Jones potential may not be applicable. These authors discuss this problem for the Cl_2–H_2 mixture as an example.

Other quantum-mechanical treatments of V–T transfer include those of Takayanagi [34] and of Thompson [35] using the method of distorted waves

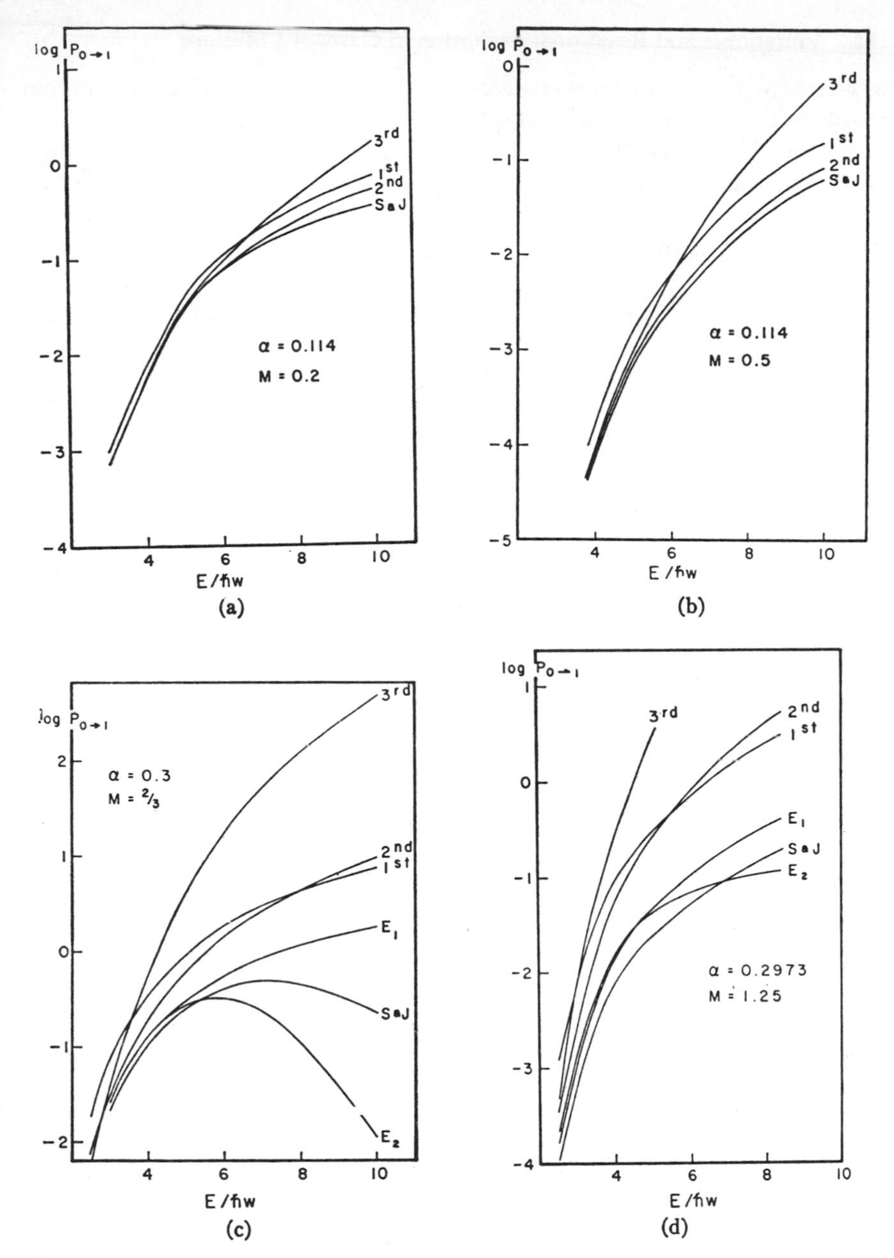

Figure 3.2 Theoretical transition probability $v = 0 \to v = 1$, according to Thiele and Weare (equation 33), as a function of total energy E and mass factor m [equation (14)]. α is a quantity depending upon the potential range. Curves labeled 1st, 2nd, and 3rd are the first-, second-, and third-order distorted-wave (DW) results. Curves labeled S and J are Secrest and Johnson's exact values [32]. Curves labeled E_1 and E_2 in (c) and (d) are additional approximations useful in the one-quantum-exchange case.

with a Morse-type interaction potential. This type of interaction was first considered quantum mechanically by Devonshire [36] in his work relating to accommodation coefficients. Allen and Feuer [37] discuss the application of this work to gaseous collisions and show that, in the limit, the resultant transition probability goes over to the semiclassical result for the Morse potential in one dimension [9].

E. Elementary Considerations of the Relaxation Process

With few exceptions, the experimental techniques that have been devised to study molecular vibrational and rotational energy transfer depend on the measurement of a relaxation time for a gas that is disturbed from its equilibrium condition. In the simplest case, one need consider only a single excitation mechanism in a pure two-state gas:

$$\mathrm{A} + \mathrm{A} \underset{k_{10}}{\overset{k_{01}}{\rightleftarrows}} \mathrm{A}^{\dagger} + \mathrm{A},$$

in which the energy transfer is simple T–V or T–R. The reaction rate coefficient k_{01} is thought of as the number of transitions per second per ground state molecule to the excited state. If it is assumed that the probability of excitation is independent of the state of the impinging molecule, then the rate coefficient is characterized by only one microscopic inelastic cross section. Solution of the rate equation in this case gives a single (isochoric–isothermal) relaxation time [8] (see Section VI.A):

$$\tau_{\mathrm{AA}} = (k_{10} + k_{01})^{-1}.$$

If the collision rate per molecule is denoted by $\mathscr{R}$, and the average excitation probability per collision is P_{01}, then we may write

$$\tau_{\mathrm{AA}}^{-1} = \mathscr{R}(P_{01} + P_{10}). \tag{33}$$

Detailed balancing requires that P_{01} and P_{10} be related according to equation (11), so that one has

$$\tau_{\mathrm{AA}}^{-1} = \mathscr{R}P_{10}\left[1 + \frac{g_1}{g_0}\exp\left(-\frac{\Delta E}{kT}\right)\right], \tag{34}$$

in which ΔE is the energy separation of the two levels. For a harmonic oscillator, the plus sign in this equation must be changed to a minus [16].

It is conventional to utilize the collision frequency at 1 atm, and thus the relaxation time is also referred to 1 atm. The collision frequency $\mathscr{R}$ is generally obtained from the viscosity η by use of the kinetic-theory relationship [8]

$$\mathscr{R}\eta = 1.27 \times 10^{6}\ (P/\mathrm{sec}). \tag{35}$$

The viscosity may be obtained from the Lennard–Jones parameters by aid of the formula

$$\eta \times 10^7 = \frac{266.9(MT)^{1/2}}{\sigma^2 \Omega^{(2,2)*}(kT/\varepsilon)}, \tag{36}$$

where the Ω functions of the reduced temperature kT/ε are tabulated quantities, but which can be represented approximately by the Sutherland-type expression (8)

$$\Omega^{(2,2)*} \frac{kT}{(\varepsilon)} \approx 0.76\left(1 + \frac{1.1\varepsilon}{kT}\right). \tag{37}$$

This is the quantity found in equation (29). Often, the reciprocal of the velocity-averaged transition probability, $P_{10}^{-1} = Z_{10}$, is reported in the literature. It is referred to as the average number of collisions required to deactivate a molecule from the first excited state to the ground state. Other authors have simply used Z, the so called collision number, which is the product $\mathscr{R}\tau$. In some instances, these two quantities, related for a harmonic oscillator by

$$Z_{10} = Z\left(1 - \frac{g_1}{g_0} \exp - \frac{h\nu}{kT}\right), \tag{38}$$

are considerably different in value.

The simplest gas mixture to analyze is one in which a monatomic gas B is added to a singly-relaxing gas. One observes again a single relaxation time τ, but this has been shifted from τ_{AA} according to the relationship

$$\frac{1}{\tau} = \frac{1 - X}{\tau_{AA}} + \frac{X}{\tau_{AB}}, \tag{39}$$

in which X is the mole fraction of the inert gas B. The relaxation time τ_{AB} characterizes a second V–T transition probability, P_{10}^{AB}, for binary collisions between molecules of types A and B. In reality, however, there need not be a one-to-one correspondence between observed relaxation times and transition probabilities, because several processes may be operative simultaneously. In many instances, the experimental observation may yield one or more relaxation times, each of which represents a number of microscopic processes. The general theory of relaxation phenomena is quite involved and will not be pursued further here. The reader is referred to the work by Herzfeld and Litovitz [8] and to the recent articles by Bauer [38], Kneser [39], and others [40, 41].

III. TRANSLATION–ROTATION ENERGY-TRANSFER THEORY

A. Classical and Semiclassical Theories

Much of the work done prior to 1959 on the theory of energy exchange between translation and rotation has been reviewed by Herzfeld and Litovitz

(8). In general, a high efficiency of exchange is anticipated at normal temperatures. For the rigid rotator, we may write

$$E_{\text{rot}} = \frac{\hbar^2}{2I} J(J+1) = k\theta_{\text{rot}} J(J+1), \tag{40}$$

where I is the moment of inertia and θ_{rot} is the characteristic rotational temperature. For homonuclear molecules obeying the optical selection rule $\Delta J = 2$, one has

$$\Delta E_{\text{rot}} = k\theta_{\text{rot}}(4J+6), \tag{41}$$

where J is the quantum number of the lower level.

From considerations of the rotational distribution function for a gas at temperature T, the most probable value of J is $(T/2\theta_{\text{rot}})^{1/2}$. One then has, approximately,

$$\frac{\Delta E_{\text{rot}}}{kT} \approx \left(\frac{8\theta_{\text{rot}}}{T}\right)^{1/2}, \tag{42}$$

which at 300°K ranges from greater than unity for H_2 to about 0.1 for Cl_2. Thus, for most diatomic species not containing hydrogen, ΔE may be small compared to kT, and first order perturbation theory is inadequate. Furthermore, optical selection rules are not necessarily obeyed, so that frequently a single collision may lead to larger energy transfer than that associated with $\Delta J = \pm 1$ or $\Delta J = \pm 2$. In their observations on rotational deactivation of $NO(A\ ^2\Sigma^+)$ using fluorescence techniques, Broida and Carrington [42] concluded that ΔJ values up to 5 were significant. With so many levels participating in the relaxation process, a large number of transition rates are involved. As a rule, experimental observations are insufficient to establish all of these rates unless some model is adopted regarding the relative probabilities of the various ΔJ. Carrington [43] has discussed, in some detail, transition probabilities in multilevel systems and the experimental limitations in calculating transition probabilities from optically observed population distributions.

A number of classical calculations have been attempted for T–R transfer. Parker [44], in 1959, employed a classical model of two homonuclear molecules colliding with coincident planes of rotation. Each molecule contained two centers of repulsion, separated by a distance d^* and symmetrically placed from the center of mass, and one center of attraction located at the center of mass. The repulsive force was taken to be of the form $e^{-\alpha r}$, where r is the distance between repulsive centers and α is the steepness parameter. The attractive force was taken to be $e^{-\alpha R/2}$, where R is the distance between centers of mass. This leads to an approximate potential of the form

$$V = Ce^{-\alpha R}(1 + \varepsilon \cos 2\theta_1)(1 + \varepsilon \cos 2\theta_2) - Be^{-\alpha R/2}, \tag{43}$$

where θ_1 and θ_2 represent the orientations of the respective molecular axes with respect to the line joining the molecular centers of mass, and ε is given by

$$\varepsilon = \frac{2I_2(\alpha d^*/2)}{I_0(\alpha d^*/2)}. \tag{44}$$

The integrals I_0 and I_2 arise from a Fourier expansion for hyperbolic cosines of trigonometric argument [45]. With this potential, Parker calculated both the average number of collisions to establish rotational equilibrium, Z_r, and that required to establish vibrational equilibrium, Z_v, as functions of temperature. One finds

$$Z_r = \frac{Z_r^\infty}{1 + \frac{1}{2}\pi^{3/2}(T^*/T)^{1/2} + (\frac{1}{4}\pi^2 + 2)T^*/T}, \tag{45}$$

in which kT^* is the well depth of the intermolecular potential and Z_r^∞ is the high temperature limit, given by

$$Z_r^\infty = \frac{1}{16}\left(\frac{\alpha d}{\varepsilon}\right)^2. \tag{46}$$

Here, d denotes the internuclear separation of the diatomic molecule. For the vibrational collision number, Parker writes

$$Z_v = \frac{1}{192}\left(\frac{\alpha d^*}{\varepsilon}\right)^2 \frac{h\nu/kT}{e^{h\nu/kT} - 1} \frac{1}{K(\beta)}, \tag{47}$$

where $K(\beta)$ is given by

$$K(\beta) = \int_0^{\pi/2} F \frac{\beta}{\cos\Theta} \cos^3\Theta \sin\Theta \, d\Theta, \tag{48}$$

and β is given by

$$\beta = \frac{\omega_0}{\alpha}\left(\frac{m}{2kT}\right)^{1/2}. \tag{49}$$

Θ is the scattering angle and ω_0 is the angular frequency of the oscillator. The function $F(\beta/\cos\Theta)$ is an integral of hyperbolic functions of the argument, which depends parametrically upon the value of β for $T = T^*$. Plots of this quantity, presented by Parker, show that, as the attractive forces decrease toward zero ($\beta^* \to \infty$), $F(\beta)$ varies with $-\beta$ in an approximately exponential manner. Thus, in Parker's theory, there are two quantities, Z_v and Z_r, to be determined, with two adjustable parameters, d^* and α. The observed temperature dependence of Z_v may be used to fix these quantities, and the value of Z_r may then be calculated directly and compared with experiment. For both N_2 and O_2, Parker concludes from equation (46)

$$Z_r^\infty \approx 15 \text{ collisions},$$

and consequently, from equation (45), that

$$Z_r(T = 300°\text{K}) \approx 4 \text{ collisions}.$$

One notes that in this theory the energy transfer is treated classically and thus no allowance is made for the effect of a quantum jump on the final velocity of the impinging molecule.

More recent classical calculations of T–R transfer include the work of Raff [46], Brau and Jonkman [47], and that of Benson and Berend [48, 49]. Raff examined specifically the cases of (H_2, He) and (D_2, He) collisions. He employed an accurate interaction potential due to Krauss and Mies [50] and a three body Monte Carlo calculation. Order-of-magnitude agreement with experiment was obtained.

When the duration of a collision is short compared with the rotational period, the collision is considered to be impulsive, while if the duration is comparatively long, the collision is considered to be adiabatic. Under the latter condition the probability of R–T transfer becomes very small. Brau and Jonkman [47] have shown that, in the case of nitrogen, molecules with rotational energies $E \geqslant kT$ typically experience adiabatic collisions, while those with $E \leqslant 0.2kT$ typically experience impulsive collisions. Molecules with high rotational energy therefore relax more slowly than those with low rotational energy, resulting in a rotational distribution function which becomes distorted from the Boltzmann distribution during the relaxation process. In Brau and Jonkman's theory, the rotational distribution function is described by a diffusion-equation approximation to the master equation for rotational relaxation. A rotational "diffusion coefficient" is calculated by considering classical head-on two-dimensional collisions in the impulsive and adiabatic limits. The results show that the absorption and dispersion of acoustic waves cannot be described strictly by a single rotational relaxation time. In the limits of very low and very high frequencies, "apparent relaxation times" can be used that have a temperature dependence similar to those predicted by Parker's repulsive collision model, but whose magnitudes are about twice as large.

Benson and Berend [48] have investigated classically the effect of initial rotation upon vibrational excitation, and found that at higher rotational energies the efficiency of collisions in exciting vibrations is reduced. Using the same two dimensional collision model, these investigators have also studied T–R transfer for two colliding diatomic molecules. Each of the four atoms was assumed to interact with the other three according to a Morse potential, and an energy cutoff was employed in determining the average collision numbers, so as not to count those collisions for which the relative energy is below excitation threshold. Application was made to the cases of H_2 colliding with H_2 and He, D_2 with He, and pure N_2 and O_2. In all but the latter two cases, the collision numbers were found to decrease with increasing temperature over the interval 100 to 700°K. For N_2 and O_2, higher-order transitions were accounted for simply by assuming that

$$\frac{P_{iJ}}{P_{iJ'}} = \frac{J'(J'+1)}{J(J+1)}, \tag{50}$$

which gives progressively lower probabilities to successively higher-order transitions. Comparisons of their theoretical result with experiment are given in Section VII.

An early paper by Takayanagi [51] discusses application of the semiclassical method to the calculation of rotational transitions in collisions between two diatomic molecules. A more recent semiclassical calculation of T–R energy transfer in collisions between a diatomic molecule and an atom has been performed by Lawley and Ross [52]. They considered the case in which (1) the energy exchange ΔE is about 20% of the relative kinetic energy, (2) the matrix elements are large, and (3) the period of interaction is on the order of $\hbar/\Delta E$. The calculation was applied to systems such as K + HBr. The spherically symmetric part of the interaction potential was of the Lennard–Jones [12, 6] form, and the asymmetric part was of the form $R^{-6}P_2(\cos\theta)$, where θ is the angle between the vector $\mathbf{R}$, joining the molecule's center of mass to that of the atom, and the internuclear axis. Probabilities of inelastic collisions and total inelastic cross sections were obtained for the $0 \rightarrow 2$, $2 \rightarrow 4$, $0 \rightarrow 4$, and $1 \rightarrow 3$ transitions, neglecting any possible contributions from potential terms of odd symmetry, which would give rise to $\Delta J = \pm 1$ transitions. They obtained cross sections equal to 5.74, 5.87, and 0.34 Å^2 for the $0 \rightarrow 2$, $2 \rightarrow 0$, and $0 \rightarrow 4$ transitions, respectively, using (symmetrized) velocities of 5×10^4 cm/sec for the $0 \rightarrow 2$ case and 4.65×10^4 cm/sec for the $2 \rightarrow 4$ case. The total average cross sections for the $2 \rightarrow 4$ transition and for the $1 \rightarrow 3$ transition were found to be 5.9 and 1.09 Å^2, respectively. These results neglect intermultiplet transitions, which, the authors estimated, could increase the calculated cross section by as much as 50%.

A semiclassical calculation of T–R energy transfer has also been performed by Raff [53] and the results compared with his previous classical theory as applied to p-H_2, He and o-D_2, He collisions. He concluded that the semiclassical cross sections are smaller than the classical values by a factor of about 2.

Zeleznik [54] has derived rotational collision numbers for pure polar gases from a classical perturbation theory in two dimensions, in which the polar molecule is taken to be a point dipole imbedded in a hard core. The calculation correct through third order gave the result, for chemical species 1 and 2 having n rotational degrees of freedom,

$$\frac{Z_r(1, T)}{Z_r(2, T_0)} = \frac{n_1}{n_2}\left(\frac{\mu_2}{\mu_1}\right)^4\left(\frac{M_1}{M_2}\right)\left(\frac{T}{T_0}\right)^3\left[\frac{\eta_2(T_0)}{\eta_1(T)}\right]^2\frac{\Theta_{11}(2, T_0)}{\Theta_{11}(1, T)}, \tag{51}$$

where T_0 is a reference temperature, μ is the dipole moment, η is the viscosity, and Θ_{11} is a function with argument $(16\pi/5)(I/M)\eta(\pi MkT)^{-1/2}$, in which I is an average moment of inertia. Zeleznik's derivation assumes, as does Parker's, that the system is initially unexcited, that is, that the rotational energy $E_r(t)$ equals zero at $t = 0$. The rotational relaxation equation

$$\frac{dE_r}{dt} = \frac{1}{\tau_r}\left[E_r(T) - E_r(t)\right]$$

thus gives

$$Z_r = \mathscr{R}\tau_r = \frac{\mathscr{R}E_r(T)}{(dE_r/dt)_{t=0}}, \tag{52}$$

where $\mathscr{R}$ is the collision rate.

Results show that the rotational collision number increases with increasing temperature and decreasing dipole moments and moments of inertia (see Section VII.A.2).

B. Quantum Theories of T–R Transfer

The early works of Brout [55], Takayanagi [56, 57], and a number of other investigators were summarized by Herzfeld and Litovitz [8] and later by Takayanagi [11]. Quantum-mechanical investigations in the past few years, however, have been extensive, especially in regard to simple collision systems. The basic problem of the scattering of an atom by a rigid rotator was treated in 1960 by Arthurs and Dalgarno [58]. They employed the total-angular-momentum representation, obtaining as an approximation a finite set of coupled radial differential equations. Simplification occurs from not distinguishing between initial and final states of the magnetic quantum number m_j in the collisions $(j, m_j) \rightarrow (j', m_j)$. Davison [59], in 1962, expanded the Arthurs–Dalgarno treatment to the inelastic collision of two rigid rotators and applied the theory to the $(0, 0) \rightarrow (2, 0)$ and $(2, 0) \rightarrow (4, 0)$ transitions in parahydrogen. The distorted-wave approximation was employed. A series of papers by Curtiss and co-workers (see references listed by Curtiss and Bernstein [60]) describes other treatments of the rotational excitation problem, and papers by Lester and Bernstein [61] and Allison and Dalgarno [62] describe the computational procedures employed in the "close-coupling" rotational excitation problem.

In the quantum-mechanical treatment of inelastic molecular collisions, the transition probability may fluctuate rapidly over short energy intervals, due to the temporary excitation of quasibound states during the collision [63]. These states arise because of the attractive term in the interaction potential. In a low-energy collision, some of the kinetic energy can be converted to internal excitation energy, forming a temporarily bound state. When the collision is over, this internal energy is reconverted to translational energy. Quasibound states may be formed when the relative kinetic energy at large separations is less than the threshold energy for excitation of the particular internal state. The influence of closed (i.e., energetically inaccessible) collision channels on the energy dependence of the excitation cross section has been discussed by several investigators [64–68]. Considerable effort has been directed at establishing the width of the subexcitation resonances to determine whether they are sufficient to be observed experimentally using crossed molecular beams. At the present time, there is no conclusive experimental evidence for these resonances [67].

Comparisons of quantum-mechanical calculations of T–R transfer with experiment have been limited for the most part to H_2, D_2, and collisions of these molecules with themselves or with noble-gas atoms. Davison [59] compared his theoretical results on H_2 with measurements by Rhodes [69] and Sluijter, Jonkman, Knaap, and Beenakker [70–73] compare their experimental values with the calculations of Takayanagi [55], Davison [57], and Roberts [74]. Valley and Amme's experimental work [75] on the parahydrogen–noble-gas systems at room temperature was also compared with these theories.

Consider as an example the application of Takayanagi's result for the inelastic cross section, $Q_{02}(k')$, where k' is the incident wave number of the relative motion, to the calculation of the rate constant, k_{02}, for the $0 \rightarrow 2$ rotational excitation of p-H_2 by He. Following Jonkman et al. [72, 73], we may write, for a mixture of He and p-H_2, in which the distribution is Maxwellian,

$$k_{02} = \sum_{i=1}^{2} x_i n (2\pi\mu_i kT)^{-3/2} \int_0^\infty Q_{02}^i(k') \frac{k'}{\mu_i} \exp\left(-\frac{k'^2}{2\mu_i kT}\right) 4\pi k'^2 \, dk', \qquad (53)$$

where $i = 1$ refers to H_2–H_2 and $i = 2$ refers to H_2–He collisions, x_1 to the mole fraction of H_2, and x_2 to that of He. The respective reduced masses are μ_1 and μ_2, and n is the total number density. Takayanagi's expression for $Q_{02}^i(k')$ was derived using the modified-wave approximation in the distorted-wave method, and an angle-dependent Morse potential of the form

$$V(R, \theta) = \varepsilon\{\exp[-2\alpha(R - R_0)][1 + \beta P_2(\cos\theta)] - 2\exp[-\alpha(R - R_0)]\}. \qquad (54)$$

In his notation, β is the nonsphericity parameter, similar to Parker's ε [equation (44)], and θ is the angle between the H_2 molecular axis and the line joining the centers of mass of the colliding particles. In the modified wave-number approximation, the quantity $k'^2 - J(J+1)/R^2$ in the Hamiltonian is replaced by $\tilde{k}'^2 = k'^2 - J(J+1)/R_c^2$, where R_c is assumed to be a constant approximately equal to the distance of closest approach. Defining $F_{02}(\tilde{k}') = Q_{02}(\tilde{k}')/\pi R_c^2$, we write the distorted-wave expression (see Jonkman et al. [73]) as

$$F_{02}(\tilde{k}') = \frac{\pi^2}{20} \frac{\sinh 2\pi q \sinh 2\pi q'}{(\cosh 2\pi q - \cosh 2\pi q')^2} \beta^2(\Delta q^2)^2 \left[(1+\delta)\Phi + \frac{1-\delta}{\Phi}\right], \quad (55)$$

where

$$q = \frac{\tilde{k}}{\alpha}, \qquad q' = \frac{\tilde{k}'}{\alpha}, \qquad \Delta q^2 = q'^2 - q^2 = \frac{2\mu\,\Delta E}{\alpha^2}.$$

The quantities d, δ, and Φ are given by

$$d = \frac{(2\mu\varepsilon)^{1/2}}{\alpha}, \qquad \delta = \frac{2d}{\Delta q^2},$$

and

$$\Phi = \frac{|\Gamma(\frac{1}{2} - d + iq)|}{|\Gamma(\frac{1}{2} - d + iq')|}.$$

Note that in this approximation, the transition probability is proportional to β^2.

The inelastic cross sections are given as a function of the temperature by

$$Q_{02}^i(T) = \frac{\pi R_c^2}{2\mu_i kT} \int_0^\infty F_{02}^i(\tilde{k}') \exp - \frac{\tilde{k}'^2}{2\mu_i kT}\, d(\tilde{k}')^2. \quad (56)$$

According to Jonkman et al., one can approximate adequately the theoretical cross section by simple quadratic expressions, unless the temperatures are very low, so that equation (56) can be integrated directly. Results from this procedure have been compared with measured experimental values by Sluijter et al. [70] and by Valley and Amme [75]. A numerical integration procedure was followed by Jonkman et al. [73] (see Section VII). The theoretical cross sections of Davison [59] for H_2–H_2 collisions, and of Roberts [74] for H_2–He and H_2–H_2 collisions, have been employed in a similar fashion by these two groups of investigators. These cross sections, which are also based on a Morse interaction potential, again give results proportional to β^2. Thus, for any of the three given theoretical developments, the ratio $Q_{20}(T)/Q_{42}(T)$ should be independent of β. Takayanagi's theory, for example, gives a ratio of 3.6 at room temperature for p-H_2–p-H_2 collisions, while Davison's theory gives 2.9. The 300°K measurements of Valley and Amme appear to show double relaxation in p-H_2. By assuming that the first relaxation step involves those transitions associated with the $J = 4$ level, they found a ratio $Q_{20}/Q_{42} \approx 1.6$. As they point out, their interpretation of sound-dispersion data in p-H_2 at room temperature neglects possible R–R processes, namely $(2, 2) \leftrightarrows (0, 4)$. They assume, further, that the process $(0, 0) \rightleftarrows (4, 0)$ is negligible at room temperature.

IV. VIBRATION–VIBRATION TRANSFER THEORY

In collisions involving polyatomic molecules, it is possible that the vibrational energy in one mode of frequency ν_i may be transformed to another mode of frequency ν_j, either within the same molecule (intramolecular transfer) or into another molecule (intermolecular transfer). Collisions between two dissimilar diatomic species may produce transfer of the latter type. Except in those cases where ν_i and ν_j are equal, translational and/or other internal degrees of freedom must participate to achieve energy balance. For a mixture of two diatomic gases, A and B, one then has the following processes:

$$\mathrm{A}^\dagger + \mathrm{A} \rightleftarrows \mathrm{A} + \mathrm{A} \qquad (\text{V–T}) \qquad (a)$$
$$\mathrm{B}^\dagger + \mathrm{B} \rightleftarrows \mathrm{B} + \mathrm{B} \qquad (\text{V–T}) \qquad (b)$$
$$\mathrm{A} + \mathrm{B}^\dagger \rightleftarrows \mathrm{A} + \mathrm{B} \qquad (\text{V–T}) \qquad (c)$$
$$\mathrm{A}^\dagger + \mathrm{B} \rightleftarrows \mathrm{A} + \mathrm{B} \qquad (\text{V–T}) \qquad (d)$$
$$\mathrm{A}^\dagger + \mathrm{B} \rightleftarrows \mathrm{A} + \mathrm{B}^\dagger \qquad (\text{V–V}) \qquad (e)$$

where the dagger denotes vibrational excitation. In many instances of V–V transfer, only one quantum of excitation is considered, although it may happen that higher levels are important, for example, when $\nu_i \approx 2\nu_j$. The average probability per collision for such a V–V process may be denoted $P(\mathrm{A}i_{10}, \mathrm{B}j_{02})$, in which mode i of molecule A goes from its first excited state to the ground state, and mode j of molecule B becomes excited from the ground state to its second excited state. If, rather, the near-resonance is between two vibrational modes i and j of the same molecule, A, then the probability for intramolecular transfer for this case would be written $P(\mathrm{A}i_{10}, \mathrm{A}j_{02}; \mathrm{B})$, where the vibrational states of B do not participate. The notation may readily be extended to include other such types of "complex" collisions. That vibrational energy may often be passed readily upon collision from one mode in a polyatomic molecule to another is evidenced by the experimental observation that ultrasonic velocity dispersion in gases of polyatomic molecules exhibits, with few exceptions, only one relaxation time. In many such cases, the vibrational specific-heat contribution from each of several modes is significant, and multiple sound-velocity dispersion would otherwise be anticipated. The effects of V–V (or, more generally, Vib → Vib ± Trans) transfer on ultrasonic dispersion behavior has been discussed by Bauer [38] and by Lambert and co-workers [76, 77]. For a gas consisting of molecules with two active vibrational modes, there are three distinct transitions, as illustrated in Figure 3.3. Relaxation times τ_1 and τ_2 correspond to separate V–T processes, and τ_{12} corresponds to the complex V–V process.

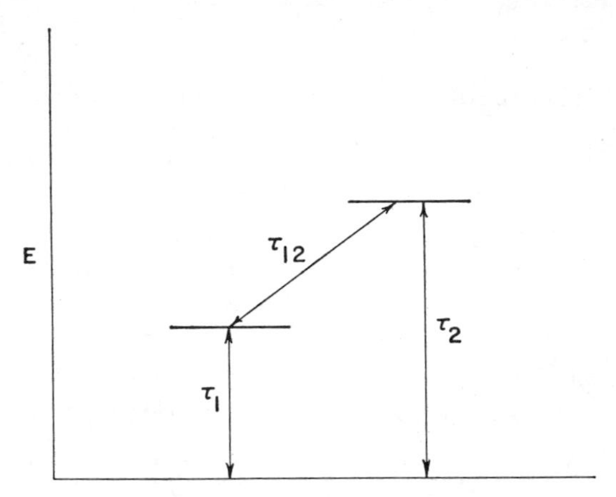

Figure 3.3 Complex relaxation process in which internal energy transfer occurs with relaxation time τ_{12}.

The common occurrence is that $\tau_{12} \ll \tau_1 \ll \tau_2$, in which the lowest-energy mode provides the channel through which energy is taken up from, and returned to, translational degıees of freedom. In this case, the single observed relaxation time τ is given by

$$\tau = \frac{C_{\text{vib}}}{C_1} \tau_1, \tag{57}$$

in which C_{vib} represents the total vibrational specific heat and C_1 is that due to the lowest mode.

In their 1952 formulation of V–T energy exchange, Schwartz, Slawsky, and Herzfeld [29] also examined the possibility of V–V exchange, employing the model of a purely repulsive interaction potential. This was followed by the work of Tanczos [78], who sought to describe the relaxation process in chlorinated methane vapors, and by Rapp and Englander-Golden [79], who studied collisions between a pair of diatomic molecules. More recently, Mahan [80] has reported a semiclassical treatment for resonant V–V transfer in the presence of long-range polar forces, and Berend and Benson [81] have employed a classical approach for the calculation of resonant and near-resonant transfer with a Morse potential. Dubrow and Wilson [82] have presented a classical calculation of V–T and V–V transfer for colinear collisions of two diatomic molecules and of an atom with a triatomic molecule, for example, C + H–C–C and C + C–C–H. The interaction was assumed to be that of a Lennard–Jones 6:12 potential.

According to Schwartz, Slawksy, and Herzfeld, when ν_i approaches ν_j and the energy transferred from translation approaches zero, the transition probability of Jackson and Mott [see equation (25)], for an exponential repulsive potential goes over to the form

$$p_{if} = V^2(i_1 \rightarrow f_1) \ldots \left(\frac{2\pi\mu v_0}{\alpha h}\right)^2, \tag{58}$$

where the V^2 coefficients represent a product of matrix elements for the vibrational states involved. This result may be integrated without difficulty over a Maxwellian distribution. Tanczos [78] writes, for the temperature-averaged probability of resonant energy exchange,

$$P(\mathrm{A, B}) = P_0(\mathrm{A})P_0(\mathrm{B})V^2(\mathrm{A})V^2(\mathrm{B})\frac{64\pi^2\mu kT}{\alpha^2h^2}\exp -\frac{\varepsilon}{kT}, \tag{59}$$

in which the P_0 coefficients are steric or orientation factors. Tanczos used equation (59) and a modified form of equation (31), in which θ and θ' of equations (8) and (9) are replaced by $\theta_1 - \theta_2$ and by

$$\theta'_{12} = 0.8153\tilde{\mu}l^2(\theta_1 - \theta_2)^2, \tag{60}$$

to represent the cases of exact resonance and nonresonance, respectively. For small values of $\theta_1 - \theta_2$, an interpolation was made between these two limiting cases. For CH_4 and the chlorinated methanes, he considered complex collisions involving total changes up to three quanta, and developed the relaxation equation to include the vibrational specific heats of all the normal modes.

Rapp and Englander-Golden [79] examined both the near-resonance and exact-resonance V–V conditions. A purely repulsive potential of the form

$$V(X, Y_1, Y_2) = E_0 \operatorname{csch}^2(kX + \xi)\exp\left[\left(\frac{m_\mathrm{A}}{m_\mathrm{A} + m_\mathrm{B}}\right)\frac{Y_1}{l} + \left(\frac{m_D}{m_\mathrm{C} + m_D}\right)\frac{Y_2}{l}\right] \tag{61}$$

was employed, where Y_1 and Y_2 are the vibrational amplitudes of molecules AB and CD, and $X = x - x_t$, where x is the separation of the centers of mass and x_t is the turning point. The parameter l determines the steepness of the molecular-interaction potential; $k = \pi/8l$; and $\operatorname{csch}^2 \xi = 1$. Atoms B and C are the innermost ones in the collision. An evaluation for the case of exact resonance in N_2: $(1, 0) \to (0, 1)$, gave

$$p_{10}^{01} = \sin^2 (1.5 \times 10^{-5} v_0), \tag{62}$$

in which v_0 is the initial speed of approach in centimeters per second. Thus, at low v_0, the probability for resonant exchange increases as v_0^2. For the near-resonance case, they find that p_{10}^{01} is reduced by an additional factor, $\operatorname{sech}^2 (\pi^2 \gamma / k v_0)$, where γ is the difference in vibrational frequencies. For the collision of CO with N_2, the frequencies are 2143 and 2345, so that $\gamma =$ 202 cm^{-1} (2.50×10^{-2} eV), and thus

$$p_{10}^{01} \approx \sin^2 (1.5 \times 10^{-5} v_0) \operatorname{sech}^2 (1.4 \times 10^{-5} / v_0). \tag{63}$$

In the vicinity of 1000°K, $v_0 \approx 10^5$ cm/sec, the second factor is about 0.2. Rapp [83] performed an approximate integration of equation (63) over a Maxwellian velocity distribution to obtain

$$P_{10}^{01} \approx 3.7 \times 10^{-6} T \operatorname{sech}^2 \frac{0.174\gamma}{T^{1/2}}, \tag{64}$$

applicable to the N_2–N_2 system ($\gamma = 0$), as well as to N_2–CO, N_2–O_2, CO–NO, and so on with similar values of μ. This result is plotted in Figure 3.4, along with Callear's experimental values [84].

Fisher and Kummler [85] have applied the model of Rapp and Englander-Golden for V–V exchange to the relaxation of pure gases and gas mixtures of anharmonic oscillators.

Berend and Benson's classical treatment of V-V transfer [81] employs a two-dimensional collision model of a pair of diatomics with identical Morse interaction potentials between each pair of atoms. The Morse range parameter α was determined from experimental data for the N_2–N_2 T–V process. In-all, six functions are employed, one between each pair of atoms (Figure 3.5). Molecule CD, oriented at an angle β relative to its velocity vector, collides with molecule AB, with impact parameter b. Molecule AB is taken to be oriented parallel to the velocity vector of CD. The instantaneous angle between the molecular axis of AB and the line joining the centers of mass is denoted η. Cross sections for the reactions

$$N_2 (v = 1) + CO (v = 0) \to N_2 (v = 0) + CO (v = 1) \tag{65}$$

and

$$N_2 (v = 2) + N_2 (v = 0) \to 2\, N_2 (v = 1) \tag{66}$$

were evaluated. Initial rotational excitation may be assumed in the model, and the phase angle ϕ of the initially vibrating AB molecule may be varied. After separation the change of vibrational energy ΔE_v is calculated, and the

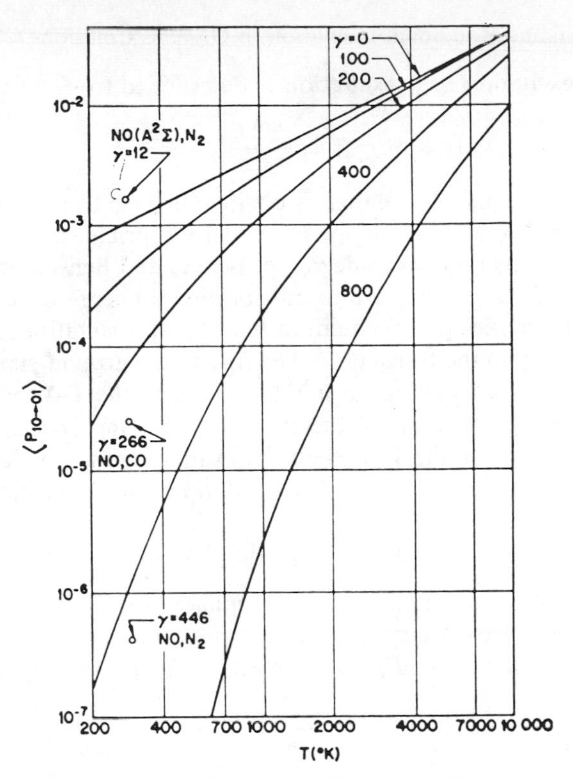

Figure 3.4 Temperature dependence of V–V transfer for systems such as N_2–N_2, N_2–CO, N_2–O_2, and CO–NO, according to Rapp [83], for a purely repulsive potential.

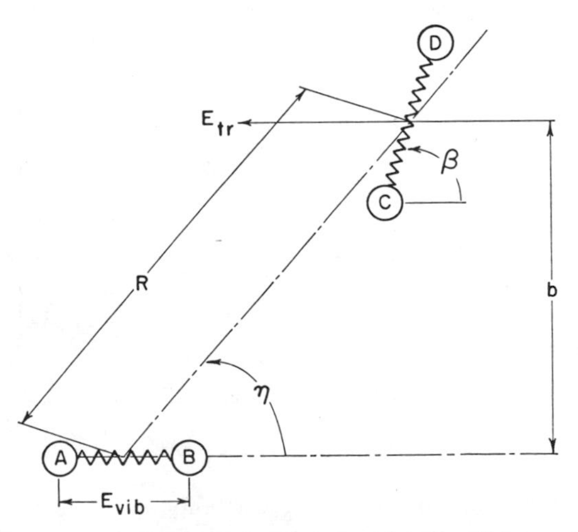

Figure 3.5 The two dimensional collision model for V–V transfer utilized by Berend and Benson [81]; an atomic Morse-type interaction is employed between each pair of atoms.

probability of excitation or deexcitation is determined from

$$p_{i \to i \pm 1} = \frac{\overline{\pm \Delta E_v}}{h\nu}, \tag{67}$$

depending upon whether the change is positive or negative. Here $\overline{\Delta E_v}$ is the value of ΔE_v obtained after averaging over all appropriate values of ϕ, β, and b leading to energy increase or to decrease. Berend and Benson observed that at low velocities of approach, a loss of vibrational energy occurred for the initially vibrating species [N_2 in equation (65)] at all orientations and phases, and a gain occurred in the translational energy of the ground-state molecule. At higher velocities, depending on initial geometry, the T–V process yields further vibrational excitation of the initially vibrating species. The V–V events are separated from the T–V events by examining the smaller of the two probabilities P_{10}(AB) and P_{01}(CD). The excitation transfer cross sections,

$$Q = 2\pi \int_0^\infty p(b) b \, \mathrm{d}b,$$

are computed from the orientation- and phase-averaged probabilities as functions of the relative velocity. The transfer probabilities, as functions of the relative translational energy, E_R, are shown in Figure 3.6. These quantities

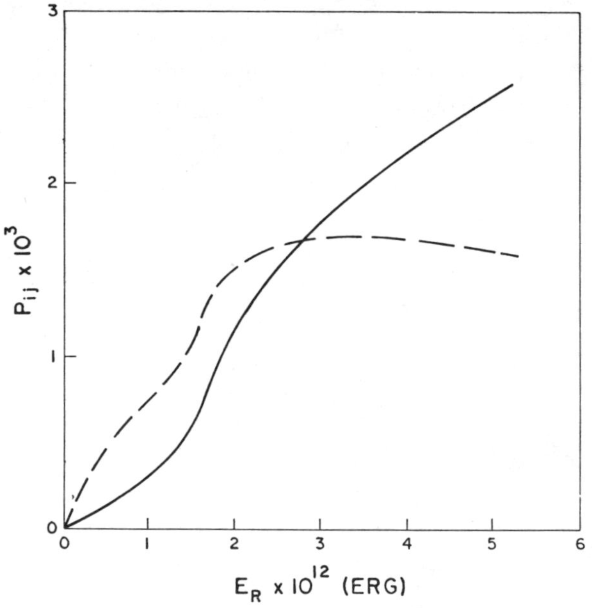

Figure 3.6 Vibration–vibration transfer probability as a function of the relative translational energy, according to Berend and Benson's classical model [81]. Solid line refers to P_{10}^{01} for $N_2^\dagger$ + CO, dashed line to P_{20}^{11} for $N_2^\dagger + N_2$.

may be averaged over a (two-dimensional) Boltzmann distribution to find the transfer probabilities as functions of the temperature. The results of these authors for the N_2–CO and N_2–N_2 processes given above are reproduced in Figure 3.7. The T–V process for N_2–N_2 $1 \rightarrow 0$ relaxation are shown for comparison. One sees that for this calculation the temperature dependence of both V–V processes appears relatively slight.

A semiclassical impact-parameter treatment of V–V transfer has been

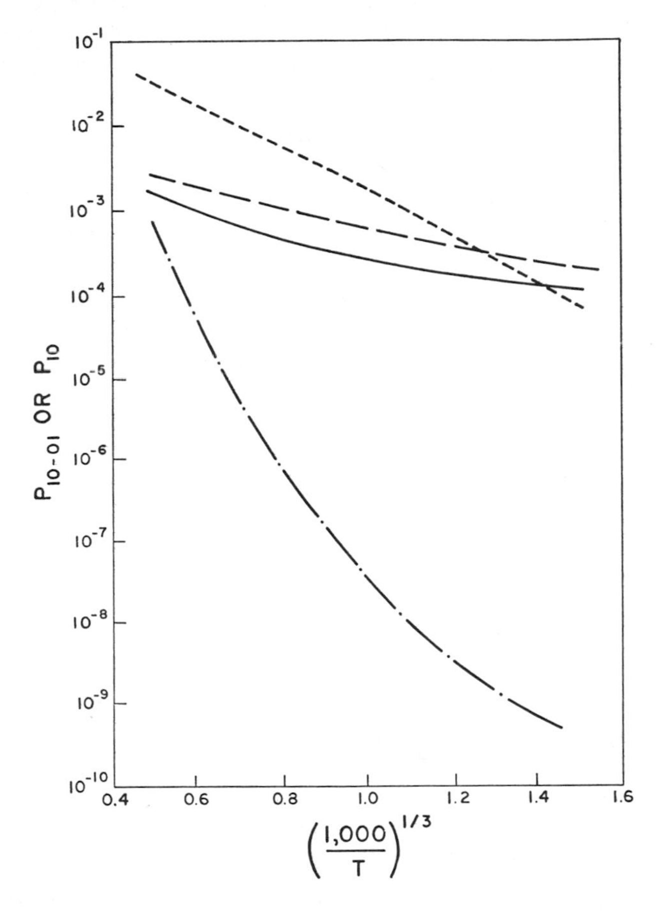

Figure 3.7 Vibration–vibration transfer probability as a function of temperature, according to Berend and Benson. Results for P_{10}^{01} $N_2^{\dagger}$ + CO are given by solid line, P_{20}^{11} for $N_2^{\dagger} + N_2$ by dashed line just above solid line; the higher dotted line represents the $\gamma = 0$ curve of Rapp from Figure 3.4, for comparison. The lower dash–dot line is the calculation for V–T transfer in N_2, which is made to fit experimental values in order to establish the Morse-potential range parameter.

developed by Sharma and Brau [86, 87] for the reaction

$$CO_2(001) + N_2(0) \rightarrow CO_2(000) + N_2(1) + \Delta E, \qquad (68)$$

in which $\Delta E = 18\ \text{cm}^{-1}$ (0.22×10^{-2} eV) and $97\ \text{cm}^{-1}$ (1.20×10^{-2} eV) for the nitrogen isotopes $^{14}N_2$ and $^{15}N_2$, respectively, and (v_1, v_2, v_3) is the familiar notation for the quantum number associated with the normal modes ν_1, ν_2, and ν_3 of carbon dioxide. In their preliminary paper, simultaneous rotational transitions were ignored, and the V–V transition probability was calculated using the orientation-averaged interaction potential $V = \mu Q/2r^4$, where μ is the long-range CO_2 instantaneous dipole moment and Q is the quadrupole moment of nitrogen. The vibrational energy transfer probability in a collision is then given by

$$p = \left|\frac{\mu_{10}Q_{01}}{2\hbar}\right|^2 \left|\int_{-\infty}^{\infty} \frac{\exp(i\omega t/\hbar)}{r^4(t)}\right|^2, \qquad (69)$$

in which μ_{10} is the dipole matrix element between the (000) and (001) states of CO_2, Q_{01} is the quadrupole matrix element between the $v = 0$ and $v = 1$ states of nitrogen, and $r^2(t) = b^2 + v^2t^2$. The angular frequency ω is that corresponding to the energy difference $\Delta E = \hbar\omega$. Sharma and Brau found a cross section of 0.23 Å^2 at 300°K, varying inversely with the temperature. They estimate that at temperatures above about 1000°K, the short-range repulsive forces will begin to dominate the V–V process for this case. In their later work [87], account is taken of rotational level changes. It was found that for $CO_2 + {}^{14}N_2$ collisions, only the low rotational levels of the two molecules contribute to reaction (68). For collisions with $^{15}N_2$, the contributing rotational levels are those which can undergo transitions cancelling most of the 97 cm^{-1} (1.20×10^{-2} eV) defect. The theory in this case is within ~30% of experiment.

Reaction (68) is important in the operation of the CO_2–N_2 laser [88], in which vibrationally excited nitrogen molecules collide with ground state CO_2 to produce excitation by the inverse process. The (001) level subsequently decays by stimulated emission of 10.6-micron radiation to the symmetric stretch (100) level (see Figure 3.8).

V. VIBRATION–ROTATION ENERGY–TRANSFER THEORY

During the late 1950s and early 1960s, there was growing evidence for the existence of very rapid relaxation rates in molecules containing one or more hydrogen atoms. Following a suggestion of Cottrell and co-workers [89, 90], Moore [91] developed, in 1965, a simple two-parameter model for vibration-to-rotation energy transfer for those cases in which the rotational velocity of

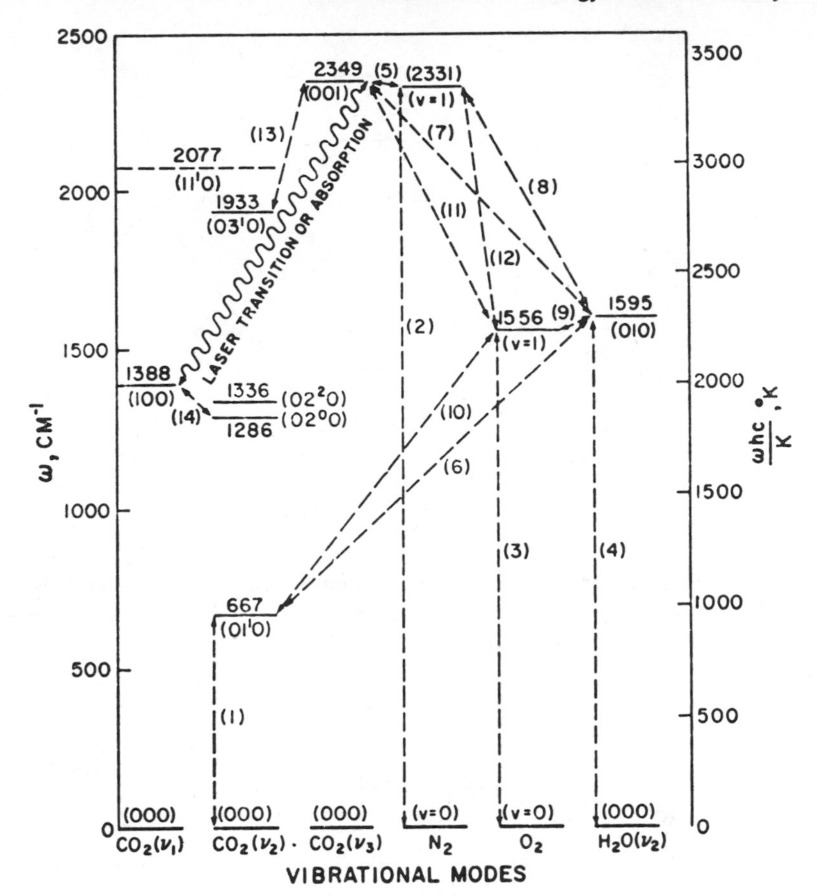

Figure 3.8 Vibrational energy-level diagram relevant to the CO_2–N_2 gas laser (from Taylor and Bitterman [88]). Dashed arrows numbered (1)–(14) represent the major collisional energy-transfer processes to be considered in the presence of O_2 and H_2O.

the hydrogen atoms far exceeds the translational velocity. He applied the semiclassical time-dependent perturbation treatment by Cottrell and McCoubrey [9] for V–T energy transfer to the vibrator–classical-rotator collision model depicted in Figure 3.9. A repulsive potential of the form

$$V = V_0 \exp(-\alpha^r)(1 + \alpha\,\Delta R) \tag{70}$$

is assumed, where ΔR is the vibrational displacement and r is the distance between the two colliding atoms when $\Delta R = 0$.

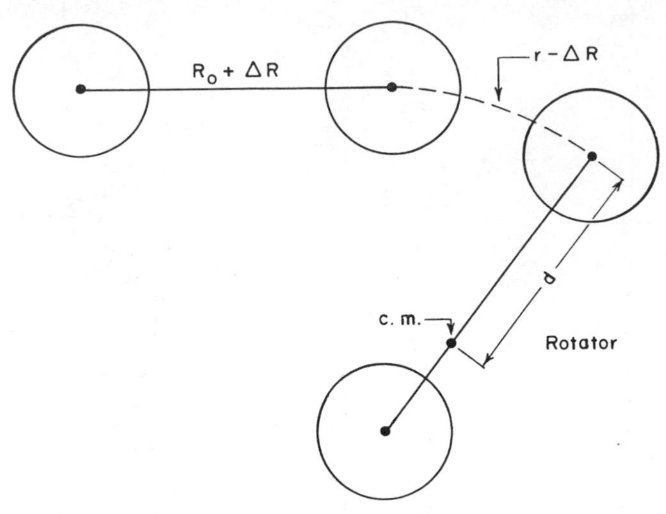

Figure 3.9 Moore's vibrator–rotator collision model, used to adapt semi-classical V–T theory to the explanation of V–R processes. ΔR is the vibrational displacement. The velocity of impact neglects translational contributions.

Moore obtained, in practical units,

$$P_{10} = \frac{17.1 I^{13/6} \nu^{4/3}}{Z_0 d^{13/3} T^{1/6} M \alpha^{7/3}} \exp\left[-1.78\left(\frac{I\nu^2}{d^2\alpha^2 T}\right)^{1/3}\right] \exp\left(\frac{0.7194\nu}{T}\right). \quad (71)$$

Here, I is in amu Å^2, M is in amu, d is in Å, α is in Å^{-1}, and ν is in cm^{-1}. The second exponential term results from a correction for the change in velocity due to the increase in rotational energy. The steric factor, Z_0, is introduced as in V–T theory to account for noncolinear collision orientations. Moore tested this equation, along with two other more approximate forms, on a series of molecules having one or more H (or D) atoms. He also examined V–R transfer for collisions of dissimilar molecules, such as CO_2–CH_4, Cl_2–HCl, and CH_4–Ar. Twenty-five different molecules having small moments of inertia were fitted with a single curve represented by equation (71) using $\alpha = 2.94$ Å^{-1} and $Z_0 = 5.0$, with at least qualitative success.

For those cases in which one of the colliding partners is molecular hydrogen or deuterium, the classical rotator approximation is particularly poor. One then anticipates efficient vibration–rotation exchange only when the rotational level spacings are such that the energy defects are relatively small. For such a gas at temperature T, it is also necessary to consider the population of the particular initial rotational state that can participate efficiently in the energy transfer. In mixtures of carbon dioxide with normal hydrogen, for example, we may consider the two hydrogen processes for the deactivation of the $CO_2(01^10)$ bending mode [667 cm^{-1} (8.27×10^{-2} eV)]:

$$CO_2(01^10) + H_2(v = 0, J = 1) \rightarrow CO_2(000) + H_2(v = 0, J = 3) + 80.5\ \text{cm}^{-1} \quad (1.00 \times 10^{-2}\ \text{eV}), \quad (72)$$

$$CO_2(01^10) + H_2(v = 0, J = 2) \rightarrow CO_2(000) + H_2(v = 0, J = 4) - 146.3\ \text{cm}^{-1} \quad (1.81 \times 10^{-2}\ \text{eV}). \quad (73)$$

These reactions have been studied in detail by Sharma [92], as well as V–R transfer involving D_2:

$$CO_2(01^10) + D_2(J = 4) \rightarrow CO_2(000) + D_2(J = 6) - 25.2\ \text{cm}^{-1} \quad (0.31 \times 10^{-2}\ \text{eV}). \quad (74)$$

Using a theory developed earlier [87] for the near-resonant V–V transfer, Sharma takes the interaction to be that between the instantaneous dipole moment of CO_2 and the quadrupole moment of H_2. The results show that reaction (72) involving orthohydrogen is an order of magnitude more effective than parahydrogen at room temperature in deactivating the CO_2 bending mode. Ultrasonic relaxation measurements in 95% CO_2–5% p-H_2 and 5% n-H_2 mixtures at 300°K performed in the author's laboratory (93) show very clearly that p-H_2 is indeed less effective than is n-H_2. The V–R deexcitation probability per collision resulting from Sharma's calculation for reaction (72) is about 4×10^{-4} at room temperature. For reaction (74) involving D_2, he obtains $P \approx 1.4 \times 10^{-3}$, in good agreement with the observed values of Cottrell and Day [94] at 30°C.

Sharma and Kern [95] have also performed a theoretical analysis, based on the Born–Bethe approximation, of V–R transfer between CO and para-H_2. A difference between p-H_2 and o-H_2 in the quenching rate of vibrationally excited CO was first observed in 1964 by Millikan and Osburg [96], and more recently over a larger temperature range by Millikan and Switkes [97]. The reaction

$$CO(v = 1) + H_2(v = 0, J) \rightarrow CO(v = 0) + H_2(v = 0, J + 4) + \Delta E \quad (75)$$

is exothermic by 88 cm^{-1} (1.09×10^{-2} eV) for $J = 2$ and endothermic by 327 cm^{-1} (4.05×10^{-2} eV) for $J = 3$; the $v = 1 \rightarrow 0$ transition in CO corresponds to 2143 cm^{-1} (26.57×10^{-2} eV). Equation (71), which is based on a classical rotator model and a short-range repulsive interaction, is not expected to reproduce the experimental temperature dependence nor the difference observed between the two hydrogen rotational species. For processes involving near-resonant energy transfer, the long-range forces should be more important than the short-range forces. Sharma and Kern assume that the cross sections due to short-range and long-range forces are additive, and that the former are the same for the two H_2 species. They utilize, in their model for $CO + H_2$ collisions, the first nonvanishing term in the multipole expansion connecting the initial and final states that arises from the coupling of the transition dipole moment of CO ($0.01D^2$) with the permanent hexadecapole moment of H_2 ($0.13D$ Å^3), an interaction varying as R^{-6}. (See Sections VII.A.4 and VII.B.1 for further discussion of these V–R processes.)

Kelley [98] has investigated the role of intramolecular V–R transfer in the deexcitation of a diatomic molecule, utilizing a two dimensional classical model. An atom A impinges on atom B of an oscillator bound by either a Morse or a quadratic potential, oriented at an angle θ_0 from the line joining A with the center of mass of BC. The molecule has initial vibrational energy but is not rotating. Several different mass combinations ($M_A + M_B - M_C$), Figure 3.1, were considered: $4 + 1 - 80$, $130 + 1 - 1$, $2 + 1 - 1$, and $4 + 80 - 80$. These combinations correspond to values of m [equation (14)]

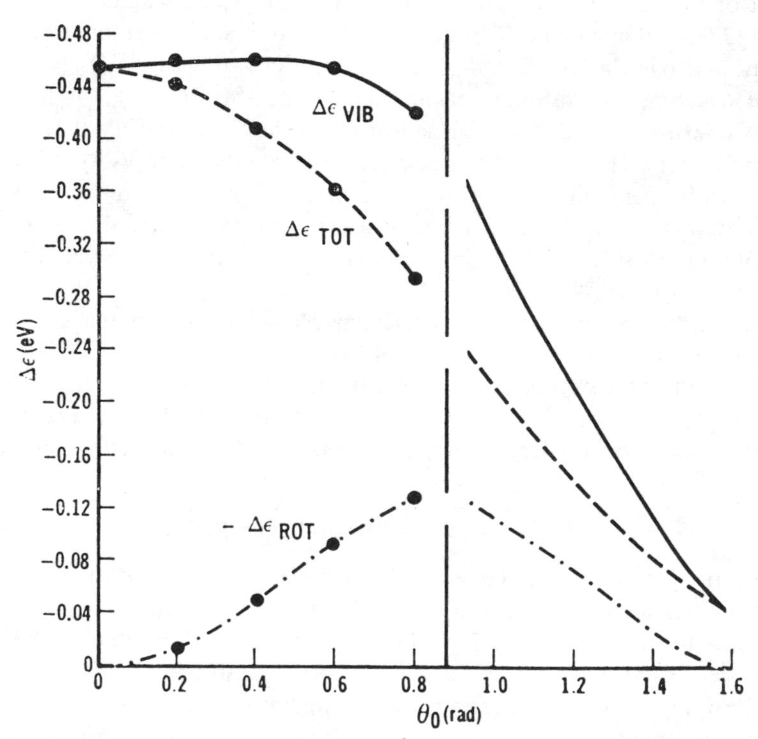

Figure 3.10 Vibrational deexcitation of a classical Morse oscillator as a function of the orientation angle θ_0 (see text), according to Kelley [98], for the case $m_A + m_B - m_C = 2 + 1 - 1$. Rotational energy is acquired via intramolecular V–R transfer. ΔE_{TOT} is the net internal energy lost by the molecule BC.

equal to 3.76, 0.986, 0.50, and 0.024, respectively. In each case, the dissociation energy was taken to be 5 eV; the initial oscillator energy, 2 eV; and the collision energy, 0.02 eV. An increasing influence of intramolecular V–R transfer was found as m increases. In the 4 + 80 — 80 case (e.g., He on Br_2), $m = 0.024$, the total energy transfer is essentially all vibrational. For the case in which $m = 0.5$ (a value corresponding also to collisions between identical homonuclear diatomic molecules, one of which replaces atom A), rotational energy gain of the Morse oscillator is found to account for about one third of the vibrational energy lost when the initial orientation angle θ_0 is about 55° (see Figure 3.10).

The case 4 + 1 — 80, representing, for example, He + HBr, was shown by Kelley to illustrate the difficulty with which vibrational energy is transferred when the Br end is struck ($m = 0.0008$) as opposed to the values obtained when the H atom is struck ($m = 3.8$). It was also demonstrated that, considering all values of θ_0 for this case, the predominant deexcitation mechanism is intramolecular V–R transfer. Since this V–R process depends upon particle masses and collision energy in a manner different from the V–T process, a colinear collision model will not always lead to a proper description of vibrational deexcitation. This conclusion probably applies as well to more complicated systems, such as noble gas collisions with CH_4.

VI. EXPERIMENTAL TECHNIQUES

A. Absorption and Dispersion of Sound

1. Acoustic Relaxation Processes

A number of excellent reviews and monographs have been written pertaining to the application of acoustics to the study of molecular energy transfer [8, 9, 38–41, 99–101]. An abbreviated description of the relaxation process is presented here.

At temperatures for which electronic excitation is negligible, the energy of a gas may be expressed as a sum of contributions from the remaining degrees of freedom:

$$E = E_{\mathrm{trans}} + E_{\mathrm{rot}} + E_{\mathrm{vib}}. \tag{76}$$

If the gas is suddenly disturbed thermally, a nonequilibrium distribution of internal states will result, and each degree of freedom is considered to relax to the new equilibrium distribution with a characteristic relaxation time τ. Now, if the period of an acoustic wave is long compared to the largest τ for the system, and if C_{vib} is the vibrational specific heat, then the total specific heat at constant volume,

$$C_v^0 = \frac{dE}{dT} = C_{\text{trans}} + C_{\text{rot}} + C_{\text{vib}}, \tag{77}$$

is the (static) specific heat which enters into the low frequency sound velocity for an ideal gas:

$$V^2 = \frac{RT}{M}\gamma = \frac{RT}{M}\left(1 + \frac{R}{C_v^0}\right). \tag{78}$$

Consider, for example, a single vibrational relaxation process, as in a gas of diatomic molecules exhibiting harmonic oscillations. Then for acoustic frequencies ω comparable to τ^{-1}, the sound velocity becomes frequency dependent. To find $V^2(\omega)$, the specific heat may be written in the complex form

$$C(\omega) = C_v^\infty + \frac{C_{\text{vib}}}{1 + i\omega\tau}.$$

Substituting $C(\omega)$ into equation (78) in place of C_v^0 and taking the real part, one obtains an approximation (9) for the sound-velocity dispersion in an ideal gas:

$$V^2(\omega) = \frac{RT}{M}\left[1 + R\frac{C_v^0 + \omega^2\tau^2 C_v^\infty}{C_v^{02} + \omega^2\tau^2 C_v^{\infty 2}}\right], \tag{79}$$

in which 7 is the true (energy) relaxation time [102], sometimes referred to as τ_{VT} [see equation (33)]. This relation may also be written in the form

$$\frac{V^2 - V_0^2}{V_\infty^2 - V^2} = \left(\frac{\omega\tau C_v^\infty}{C_v^0}\right)^2, \tag{80}$$

where V_0 and V_∞ are the lower and upper limits of $V(\omega)$, respectively. Generally, the dispersion curve equation [equation (79)] is plotted logarithmically, V^2 versus $\log \omega/2\pi$. In this case, the curve is symmetrical, with the midpoint, $V^2 = \frac{1}{2}(V_0^2 + V_\infty^2)$, occurring at the inflection-point frequency ω_i. Then equation (80) becomes

$$\frac{\omega_i \tau C_v^\infty}{C_v^0} = 1,$$

or

$$\omega_i = \frac{C_v^0}{C_v^\infty \tau}.$$

A slightly different approximation for the sound velocity has been derived by Herzfeld (8):

$$\frac{V_0^2}{V^2} = 1 - \frac{V_\infty^2 - V_0^2}{V_\infty^2}\frac{\omega^2\tau^2}{1 + \omega^2\tau^2}. \tag{81}$$

The quantity τ differs from $\tau' = \tau_{pS}$, the "isobaric, adiabatic" relaxation time, by the factor $\tau'/\tau = (C_p - C_{\mathrm{vib}})/C_p$. Both the sound-velocity dispersion and excess sound absorption α' can be written in terms of τ':

$$\frac{V_0^2}{V^2} = 1 - \frac{RC_{\mathrm{vib}}}{C_v^0(C_p^0 - C_{\mathrm{vib}})} \frac{\omega^2\tau'^2}{1 + \omega^2\tau'^2}, \tag{82}$$

$$\alpha'\lambda\left(\frac{V_0}{V}\right)^2 = \pi \frac{RC_{\mathrm{vib}}}{C_v^0(C_p^0 - C_{\mathrm{vib}})} \frac{\omega^2\tau'}{1 + \omega^2\tau'^2}. \tag{83}$$

Here, α' is the total (amplitude) sound-absorption coefficient, in units of cm^{-1}, less the classical (viscothermal) absorption [equation (84)]. The frequency at which equation (83) has its maximum value is given by $\omega_{\max} = \tau'^{-1}$. This relation is frequently utilized in reducing experimental data from sound-absorption measurements.

For gases which deviate significantly from ideality, one may either derive a dispersion formula for a given equation of state, or else correct the observed sound velocity or absorption to ideality before fitting data to equations (79), (81), (82), or (83).

From equation (83) it can be seen that the amplitude absorption coefficient α', due to relaxation, varies as ω^2 when $\omega\tau \ll 1$. This is the dependence upon frequency that is found for the classical (viscothermal) absorption coefficient, α_{cl}, for a pure gas having viscosity η:

$$\alpha_{cl} = \frac{\omega^2}{2\rho_0 V^3}\left(\frac{4}{3\eta} + \frac{\gamma - 1}{C_p^0} K\right), \tag{84}$$

in which γ is the ratio of specific heats, K the thermal conductivity, and ρ_0 the density. Thus, at low frequencies, the ratio $(\alpha' + \alpha_{cl})/\alpha_{cl}$ is found to be independent of ω. Herzfeld [8] writes, for linear molecules,

$$\frac{\alpha' + \alpha_{cl}}{\alpha_{cl}} \equiv \frac{\alpha}{\alpha_{cl}} = 1 + 0.0671 \frac{Z_{\mathrm{rot}} + c_{\mathrm{vib}}Z_{\mathrm{vib}}}{R} \tag{85}$$

for the sound absorption at low frequencies, which is valid when $C_{\mathrm{vib}}/R \ll 1$. Thus, low-frequency sound-absorption measurements lead to a collision number for rotational relaxation, which, of course, is based on the assumption that a single rotational relaxation time exists.

A review of the common methods of measuring acoustic absorption and dispersion is presented by Cottrell and McCoubrey [9]. The ultrasonic interferometer, the absorption tube, the condenser transducer, and the reverberation chamber are the standard types of apparatus.

2. The Ultrasonic Interferometer

The quartz-crystal interferometer has been used mainly for obtaining sound velocities with high accuracy (0.1% or better) over the range of frequency-to-pressure ratios from about 50 kHz/atm to 100 MHz/atm. The phase velocity $V = f\lambda$ is obtained through measurements of the wavelength λ at accurately known frequencies. The common design employs a precision-cut quartz crystal opposed by a movable reflector, between which standing waves

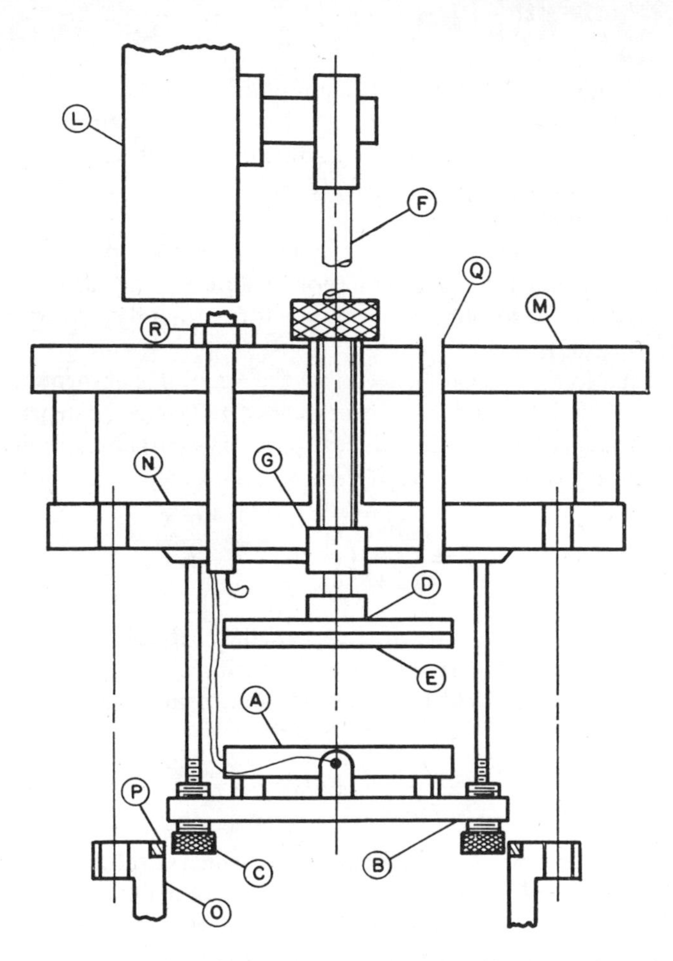

Figure 3.11 An acoustic interferometer of the type used in the author's laboratory (from Nethery [104]). A:X-cut quartz crystal, 100–600 KHz; B: crystal support mount and aligning screws. Optical flat E is attached to a movable reflector D for generation of ultrasonic standing waves. Invar rod F position is read from precision micrometer slide L.

can be established for separation distances equal to an integral number of half-wavelengths. The instrument currently in use at the University of Denver is a larger version of the design of Walker, Rossing, and Legvold [103], and employs precision stainless steel ball bushings for the (Invar) reflector drive rod. Crystals as large as 10 cm in diameter and 2.82 cm in thickness (100 kHz) may be accommodated. Figure 3.11, from Nethery [104], shows the details of construction. A Gaertner Model M-342 micrometer slide

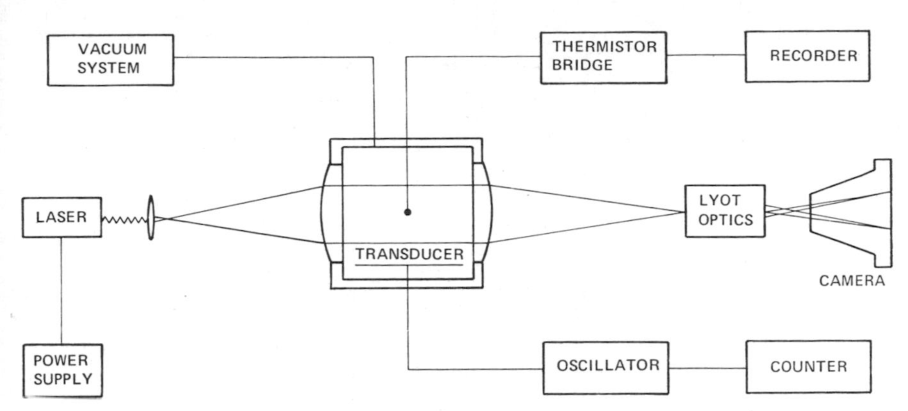

Figure 3.12 Optical diffraction apparatus for measurement of sound velocity (from Hancock and Decius [109]).

(direct reading to ± 0.001 mm) is used to move the reflector. At frequencies above about 50 MHz/atm, sound-velocity data as a rule must be corrected for the effects of translational dispersion [105] before they can be compared with equations (86)–(89). These corrections are generally less than 1% up to 100 MHz/atm [75, 106].

3. Optical Diffraction

A relatively new technique for the measurement of sound velocity has been described by Decius and co-workers [107–109] employing laser light diffraction and coronagraph optics. A He–Ne gas laser is used as the light source which, through a pinhole aperture, illuminates the sound field generated by 0.56–1.07-MHz transducers. The apparatus is sketched in Figure 3.12. The arrangement should find future application in high-temperature measurements. The f/p range that has been covered is from about 0.3 to 40 MHz/atm.

4. Condenser Transducers

The condenser microphone has been shown by Bauer and others [70, 110, 111] to be a very useful device for obtaining sound absorption coefficients at extremely high values of f/p. First described by Sell [112] in

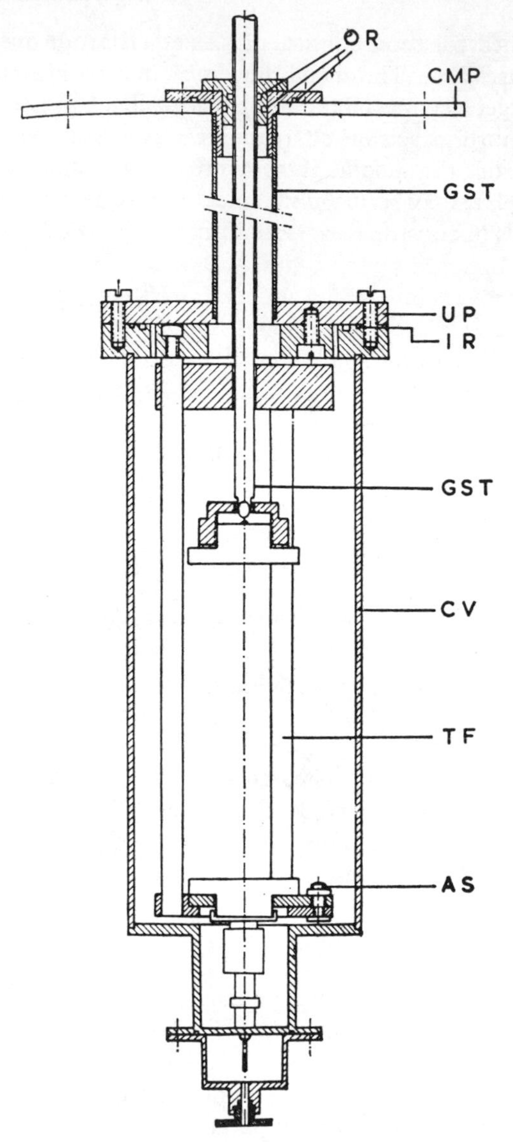

Figure 3.13 Condenser transducer arrangement utilized by Sluijter, Knaap, and Beenakker [70]. OR represent O-ring seals; CMP, cryostat mounting plate; GST, copper vessel; TE, transmitter frame; and AS, adjustment screws.

1937, this type of transducer consists of a metal electrode over which a plastic membrane is stretched. The membrane is coated with a thin metallic film to form the second electrode. It has the advantage relative to crystals of being useful over a continuous range of frequencies. A schematic diagram of a pair of condenser transducers, suitably mounted in an enclosing metal tube, is presented in Figure 3.13, taken from a paper by Sluijter, Knaap, and Beenakker [70]. The condenser transmitter is located at the bottom of the apparatus. It is opposed by the condenser microphone mounted on a sliding tube that enters the chamber from above. The transmitter is driven at voltages up to 50 V, with a frequency range from 0.1 to 0.5 MHz. A 100-V dc polarizing voltage is applied to both condensers. The signal from the microphone is preamplified and thence fed into a narrow-band amplifier. At sufficiently large separations, disturbances due to reflections from the receiver surface become negligible, and a simple exponential decrease of the amplitude with distance is observed.

5. Absorption Tubes

Over the past decade, a number of excellent measurements have been reported on sound absorption using tube methods. One such design, based on the "progressive-wave" technique of Angona [113], has been employed extensively by Shields and co-workers (see, for example, refs. 114–120). In their work, sound absorption is measured by observing directly the reduction in the output from a microphone as it is drawn away from a sound generator placed near one end of the tube. Sound-absorbing material (e.g., glass wool) is used at the tube end to eliminate standing waves. The observed absorption coefficient is given by

$$\alpha = \alpha' + \alpha_{\text{class}} + \alpha_w, \tag{86}$$

where α_w is a correction term due to the tube walls [121]. In Shields's apparatus, the sound source is a thin Inconel foil placed in a strong external magnetic field and driven at frequencies on the order of 10 KHz. The detector is also a thin foil, operating on the same principle as the source. The tube is of precision-bore Pyrex, 1.73 cm in diameter and 120 cm in length. Temperature range is from 20°C to 300°C.

The absorption technique employed by Parker [122] is a modified version of the resonance tube described by Knötzel and Knötzel [123]. It consists of a Pyrex glass cylinder, 5 cm in diameter and about 75 cm in length. A cylindrical brass piston is used to generate standing waves in the tube. The sound pressure P_n is observed for frequencies near any of the n resonant frequencies f_n for the tube, and the resonance half-width δ, in Hz, is determined:

$$\frac{|P_n|}{|P_n|_{\max}} = \left\{\left[\left(\frac{2\pi}{\alpha c}\right)(f - f_n)\right]^2 + 1\right\}^{-1/2}, \tag{87}$$

in which $f_n = nc/2l$, where c is the sound velocity and l is the tube length. The resonance half-width is given by $\delta = \alpha c/\pi$, in which α is the absorption coefficient. An instrumental correction, α_w, arising from wall effects, must again be subtracted from the observed absorption.

The resonance tube utilized by Henderson and co-workers [124] for measurements of the vibrational relaxation time in N_2, O_2, and other gases, is encased in lead to suppress wall movements and to permit operation at high pressures. The f/P range is from less than 5 to more than 20,000 Hz/atm.

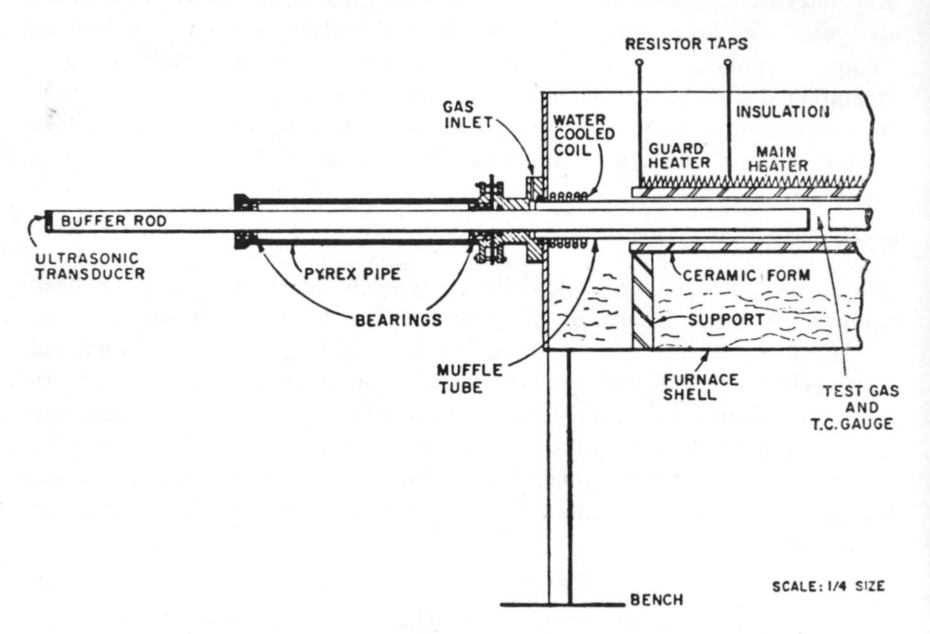

Figure 3.14 Muffle tube and externally mounted piezoelectric transducer for the measurement of ultrasonic absorption and dispersion at temperatures to 1300°K (from ref. 125).

Ultrasonic measurements of relaxation times at temperatures up to 1300°K have been described by Carnevale, Carey, and Larson [125]. Their method involves a muffle tube with an externally mounted piezoelectric transducer (1–3 MHz) driven by a pulsed oscillator (Figure 3.14). The acoustic energy is fed into the high-temperature test volume through a fused-silica buffer rod, $\frac{1}{2}$ in. in diameter, and 2 to 6 in. long. A receiving crystal, mounted externally on a second buffer rod that extends into the test volume from the opposite end, detects the transmitted signal. This technique has also been applied successfully by Winter and Hill [126] to the study of high-temperature rotational relaxation.

B. Aerodynamic Methods

The finite thickness of a shock front arises from the time interval required to reestablish thermal equilibrium. The measurement of the thickness of a shock front due to vibrational relaxation was first described by Smiley and Winkler [6]. An optical (Mach–Zehnder) interferometer was employed to observe the density profile in the relaxation region. This work showed that shock tubes could be employed for the measurement of T–V transfer at temperatures much above those which could be attained by existing acoustical methods. This technique, which has been reviewed by several authors [8, 9, 127], has become highly developed in recent years. Millikan and White [128], in 1963, observed the vibrational relaxation time of N_2 and of N_2 + CO mixtures using both optical interferometry and infrared emission from the CO. Temperatures ranged from 1900°K to 5400°K. Results of this work showed that V → V exchange from $N_2(v = 1)$ to $CO(v = 1)$ occurred with high efficiency (see Figure 3.7 and Section III).

The 1966 measurements of vibrational relaxation in D_2 by Kiefer and Lutz [129] required a greatly improved method because of rapidity of the relaxation and the low vibrational heat capacity. This was accomplished by employing a helium–neon 6328 Å gas laser to produce a narrow, intense light beam (Figure 3.15) that is deflected by the passing shock wave. The beam is reflected from mirror M_1, collimated through telescope T_1 and window W_1 into the shock tube. The emerging beam is focused onto a knife edge K, which forms a *schlieren* pattern of the density gradient. (Since the density falls off exponentially with time, its derivative exhibits the same decay rate.) A photomultiplier tube is used to detect the intensity change at the knife edge resulting from the deflection of the laser beam due to the density gradient. The second mirror, M_2, provides a baseline correction of the schlieren record for slight variations in the beam intensity. Relaxation times as short as 0.2 μsec can be resolved with this technique.

If we define $|\Delta V/V_0|$ to be the absolute fractional change in photomultiplier signal, h, then a plot of $\log_{10} h$ versus laboratory time produces a straight line whose slope S corresponds to an apparent relaxation time: $\tau_a = 1/(2.303\,|S|)$. The apparent, or laboratory, relaxation time differs from the true relaxation time τ because of the motion of the shock front. According to Blackman [130],

$$\tau P = \tau_a P \frac{\rho}{\rho_0} \frac{C_p}{C_p - C_{\text{vib}}}, \tag{88}$$

where P is the pressure, ρ is taken to be the postshock equilibrium density, ρ_0 the preshock density, and C_p is the total specific heat at constant pressure.

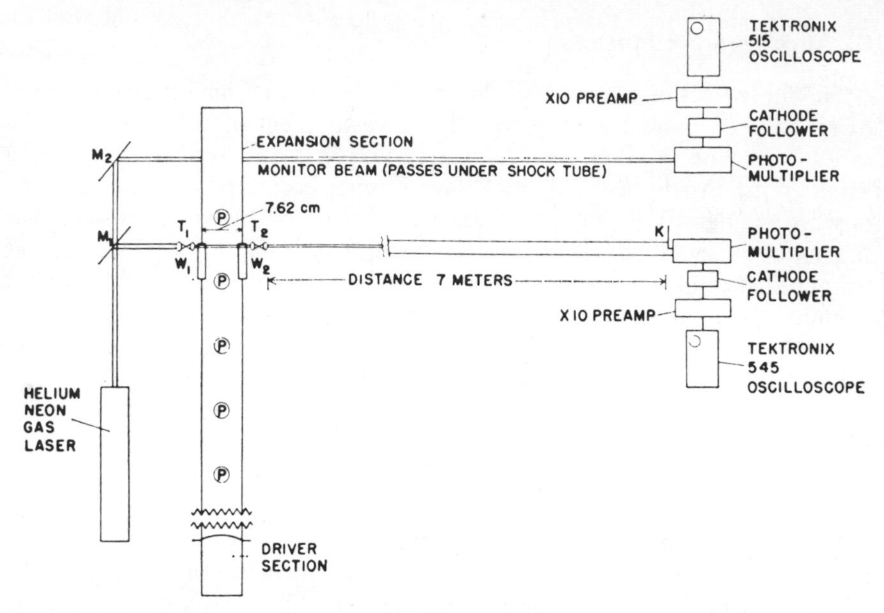

Figure 3.15 Shock-tube arrangement of Kiefer and Lutz [129], utilizing laser-beam deflection, for study of vibrational relaxation in D_2. T_1 and T_2 are collimating telescopes, W_1 and W_2 are windows; Ⓟ designates pressure transducers for measurement of shock velocity.

(An expression for the relaxation time in terms of the observed fringe shift appearing in an interferogram is given by Blackman; see also the discussion by Cottrell and McCoubrey [9].) Kiefer and Lutz [129] used an iterative technique to determine the relaxation times in D_2 and H_2, defined by the energy equation

$$\frac{dE_v}{dt} = \frac{1}{\tau}(E_v^{\text{eq}} - E_v), \tag{89}$$

and employing thermodynamic data for the gases. In this equation, E_v is the instantaneous vibrational energy and E_v^{eq} is the instantaneous equilibrium vibrational energy, which is dependent upon the translational temperature.

Appleton [131] has measured vibrational relaxation times in nitrogen up to 9000°K by utilizing vacuum ultraviolet light absorption in the shock wave. This method is based on an earlier observation that, for a given wavelength, the absorption coefficient varies with temperature in the same manner as the relative population density of a particular upper vibrational level v'' of the ground electronic state. Appleton used a wavelength of 1176 Å, which is absorbed principally by molecules populating the level $v'' = 10$. To achieve higher pressures and temperatures, he made his measurements behind the reflected shocks, just 3 mm from the end of the shock tube. In this case, the observed relaxation is not foreshortened due to the motion of the gas, as in measurements on the incident shock. At temperatures below 5500°K, the method of monitoring the $v'' = 0$ population rate was found to be in good agreement with the Mach–Zehnder interferometer results of Millikan and White [128], a result which supports the hitherto unconfirmed assumption that the vibrational relaxation in high-temperature shocks proceeds via a continuous evolution of Boltzmann distributions of vibrational state populations.

Two other aerodynamic techniques that have been employed for measuring vibrational relaxation rates are the impact (Pitot) tube method and the supersonic nozzle flow method. The former has been described by Huber and Kantrowitz [132] and others. A review of their work is presented by Cottrell and McCoubrey [9]. The latter method, described in detail by Hurle, Russo, and Hall [133], is of particular interest because the experimental conditions are the converse of those for shock-wave investigations. In a shock tube, the gas relaxes to a higher temperature, while in the supersonic nozzle flow, the rapid expansion causes large departures from equilibrium, with the gas relaxing to a lower temperature. The apparatus utilized by Hurle et al. includes a conventional 12-ft shock tube terminated by a 3 ft, 15° conical nozzle. The nozzle, which has a $\frac{3}{8}$-in.-diameter throat, is joined to a 6-ft discharge tank. A second diaphragm, separating the end of the shock tube from the nozzle, bursts when the pressure from the reflected shock rises to the desired value, which ranges between 24 and 82 atmospheres. A steady flow is thus established through the nozzle for a period of a few hundred microseconds, with reservoir temperatures ranging from 2800 to 4600°K. In their measurements of vibrational relaxation times in N_2, these investigators employed the sodium-spectrum-line reversal method for time-resolved temperature measurements in the nozzle [127]. Their conclusions were that the relaxation times, τ_e, in expansion flows differed from those observed at the same temperature in normal shocks, τ_s, by rather large factors, for example, $\tau_s/\tau_e \approx 15$ for N_2. Even larger ratios have been reported for CO. This result was thought to be related to anharmonicity: Normal shock measurements are made principally during the final stages of equilibration, involving the lower-lying states, whereas in expansion-flow measurements the final equilibrium is at a much higher temperature, where higher vibrational levels with less harmonic character are populated. While anharmonicity is expected to affect the results to some degree, it appears that the control of impurities is of great consideration. Von Rosenberg, Taylor, and Teare [134], in the examination of vibrational relaxation of CO in nonequilibrium nozzle flow, have pointed out that the high reservoir temperatures lead to outgassing from the chamber walls, as well as to dissociation and other processes that can introduce impurities, especially atomic hydrogen. The high efficiency of atomic hydrogen for vibrational energy transfer could account for most, if not all, of the observed difference between τ_e and τ_s. Future work may well establish the expansion nozzle as a useful device for studying the efficiency of atoms and radicals in the relaxation process.

C. Optic–Acoustic Effect

If a gas that is infrared active is subjected to a source of chopped radiation, the energy absorbed into the internal degrees of freedom will be transferred by collision to translational energy. The result is then a pressure increase, giving rise to acoustical energy. The time lag between the absorption of the radiation and the appearance of a sound pulse is a measure of the relaxation time. The effect is described in considerable detail by Cottrell and McCoubrey [9], together with some experimental results of various investigators. A 1967 publication by Cottrell, Macfarlane, and Reed [135] describes the use of such an apparatus ("spectrophone") in the study of V → V transfer in CO_2.

D. Vibrational Fluorescence

The 1963 measurements by Millikan [136] showed that for high-purity carbon monoxide at room temperature, the collisional lifetime of the first excited vibrational state was substantially greater than the radiative lifetime. By exposing carbon monoxide to infrared radiation under optically thin conditions, the decay constant of the vibrational fluorescence was found to be about 0.03 sec, in agreement with the value expected from consideration of the absorption intensity data. Introduction of impurities, such as H_2, He, and H_2O, gave rise to quenching of the CO fluorescence. For example, 4 ppm H_2O reduced the fluorescence signal to one-half its intrinsic value. Molecular oxygen showed similarly high efficiency for quenching, which was attributed to chemical affinity. This work demonstrated the usefulness of vibrational fluorescence techniques, and preceded by several years the laser-induced vibrational fluorescence techniques described below.

Three important papers, published at about the same time in 1966, demonstrated very dramatically the usefulness of lasers in the measurement of molecular energy transfer. The first of these, by DeMartini and Ducuing [137], reports a study of vibrational relaxation in normal H_2 using stimulated Raman scattering. The experimental arrangement is shown in Figure 3.16. Radiation from a Q-switched ruby laser was focused onto a pressure cell of H_2 gas at room temperature to produce about 10^{16} vibrationally excited H_2 molecules in a period of about 20 nsec. This excess population distribution

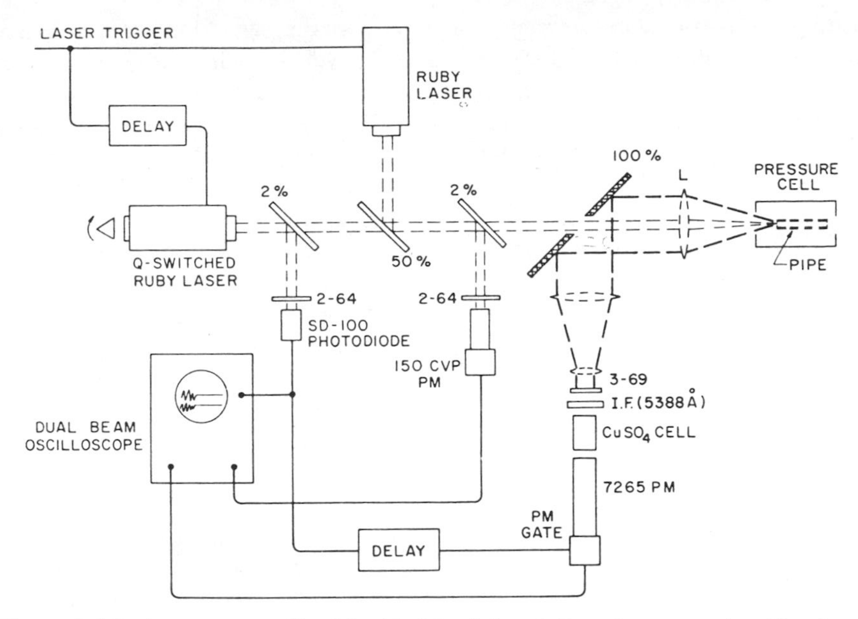

Figure 3.16 Apparatus utilized by DeMartini and Ducuing to study vibrational relaxation in n-H_2 by the use of stimulated Raman scattering. A Q-switched ruby laser produces stimulated Raman scattering in the pressure cell. The non-Q-switched laser produces a 500-μsec train of spikes, with peak power below threshold for stimulated scattering, for the probing beam (from ref. 137).

relaxes by collision to its thermal equilibrium value with a characteristic relaxation time of the $v = 1$ level. A second ruby laser, shown at the top of the figure, transmits a train of spikes that produce spontaneous Raman anti-Stokes scattering, and thus serves as a monitor of the number of molecules in the $v = 1$ state. Radiation at the anti-Stokes frequency is selected by appropriate filters and detected by a fast photomultiplier. The nearly perfect exponential decay of the excited-state population gives a measure of the vibrational relaxation time. The resulting relaxation time was observed over a range of pressures from about 26 to 60 atm and was found to vary linearly, indicating that binary collisions are responsible for the deactivation. The relaxation time was found to be 3×10^{-4} atm sec [138], which is in excellent agreement with the distorted wave calculation by Calvert [139] (see Section VII.A.1) and is close to the value extrapolated from high-temperature shock-tube measurements [129].

Since the vibrational specific heat of H_2 at this temperature is negligible, there are no ultrasonics measurements with which these results can be compared. Application of the method to gases such as N_2, for which some sound-absorption measurements have been made, would be useful. Since Raman excitation selects the $v = 1$, $J = 1$ level of hydrogen, the study of various mixtures of n-H_2 and p-H_2 would also be of interest, as would measurements at still lower temperatures (see Section VII.A.4).

Shortly after DeMartini and Ducuing's publication, there appeared an article by Hocker, Kovacs, Rhodes, Flynn, and Javan [140] that described the application of a Q-switched CO_2 laser to the measurement of vibrational relaxation in CO_2. A short sample tube was placed within the resonator of a Brewster-angle laser system much longer than the sample tube, so as to introduce only a small perturbation on the laser performance. The laser, operating on the $(00^01) \rightarrow (10^00)$ 10.6-micron transition (see Figure 3.8), induces rapid transitions between the same levels of the CO_2 in the sample tube. After the laser is pulsed, spontaneous infrared emission from the sample gas is observed with a grating monochromator equipped with an Au:Ge detector. The observed fluorescence arises from the $(00^01) \rightarrow (00^00)$ band. The level populations are found to relax exponentially to their steady-state values. The pressure dependence [141] of the volume quenching rate is found to possess a slope of 335 sec^{-1} torr^{-1}, corresponding to a relaxation time of 3.9 μsec for the asymmetric stretching vibration of CO_2 at 1 atm.

These authors also showed that for CO_2 molecules in the (00^01) state, the deactivation probability for collisions with the tube walls is 0.22 for four different wall materials investigated. This value may well be characteristic of V–T and/or V–V exchange processes involving adsorbed species (e.g., H_2O). It was also observed that the diffusion rate for vibrationally excited CO_2 in pure CO_2 gas was considerably less than that for ground-state molecules. The diffusion cross section for the excited molecules was determined to be about 90 Å^2 (as opposed to 53 Å^2 for the ground-state molecules). This important result is interpreted in terms of a large probability for single-quantum resonant vibrational-energy exchange.

Simultaneous with the publication of Hocker et al., there appeared the results of Yardley and Moore [142] on laser-excited vibrational fluorescence in CH_4. A mechanically chopped He–Ne 3.39-micron laser [143, 144] was used to excite the asymmetric stretching [$P_1 = 2948$ cm^{-1} $(36.55 \times 10^{-2}$ eV)] vibration, ν_3 (see Figure 3.17). The optical arrangement is shown in Figure 3.18. The He–Ne laser tube, 220 cm in length, is shown on the left. M_1, M_2, and M_3 are mirrors; B_1 and B_2 are baffles to eliminate stray light; L_1 and L_2 are lenses which focus the laser output into a collimated beam having a diameter of 2 mm, and thence, into a Pyrex fluorescence cell. At the focal point between L_1 and L_2 is a chopper wheel, to produce a nearly perfect square wave modulated at frequencies between 600 and 10,000 Hz. An audio oscillator and a 60-W amplifier are used to drive the synchronous chopper motor. An InSb infrared detector (response time of about 4 nsec) is used to

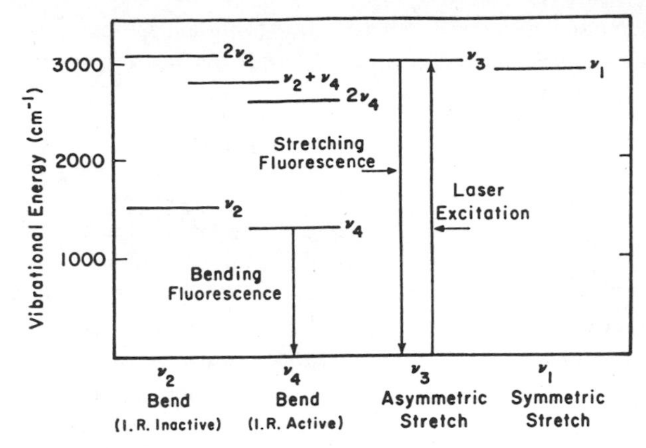

Figure 3.17 Vibrational energy-level diagram for methane (from Yardley and Moore [144]).

monitor the incident beam. A cooled Ge:Cu photoconductor views the fluorescence from the bending level, ν_4, through suitable filters (F_1 and F_2) and through a NaCl window in the cell. A phase-sensitive (lock-in) amplifier is used to measure the modulated fluorescence intensity, both in phase and 90° out of phase with the exciting beam. The ratio of these two signals is $\tan\phi$, where ϕ is the phase shift of the fluorescence [145]. The fluorescence intensity is proportional to the modulated density of CH_4 molecules with ν_4 vibrational excitation. For pure CH_4 at values of f/p between 10^5 and 10^6 Hz/atm, $\tan\phi$ increases linearly with f/p, indicating a single relaxation process corresponding to the V–T transfer for the low-frequency bending

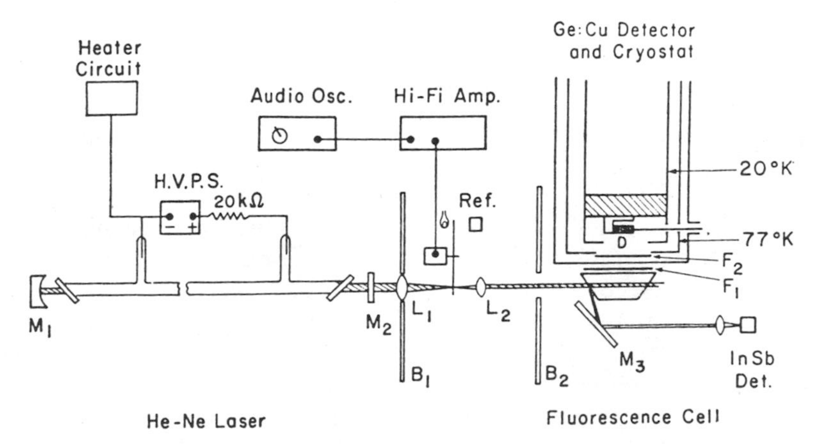

Figure 3.18 He–Ne laser system utilized by Yardley and Moore [144] to investigate vibrational energy transfer in CH_4.

mode [$\nu_4 = 1306\ \text{cm}^{-1}$ (16.19×10^{-2} eV)]. At higher f/p this dependence deviates from a straight line, indicating a second relaxation process. Yardley and Moore use the expression

$$I(f) = I_0(1 + i2\pi f\tau_1)^{-1}(1 + i2\pi f\tau_2)^{-1}, \tag{90}$$

or

$$\frac{f}{p} \cot \phi - (2\pi\ \tau_1 p)^{-1} - 2\pi\ \tau_2 p\left(\frac{f}{p}\right)^2 \tag{90'}$$

for the fluorescence intensity as a function of the chopping frequency. To obtain the relaxation time τ_2 for the faster process, they plot $(f/p) \cot \phi$ versus $(f/p)^2$. This procedure has been followed for pure methane and a number of methane–rare-gas mixtures. Thus, from measurements of the ν_4 fluorescence, Yardley and Moore were able to determine the V–T relaxation time for the lower vibrational mode ν_4 of methane and obtain a value which agreed well with the single relaxation time determined ultrasonically (see section VII.B.1). The faster relaxation process which they observe results from the short time interval required for energy to flow from the laser-excited ν_3 stretching mode to the ν_4 bending mode via such reactions as

$$CH_4(\nu_3) + CH_4 \rightarrow CH_4(2\nu_4) + CH_4 + 420\ \text{cm}^{-1} \quad (5.21 \times 10^{-2}\ \text{eV}), \tag{91a}$$

$$CH_4(\nu_3) + CH_4 \rightarrow CH_4(\nu_2 + \nu_4) + CH_4 + 196\ \text{cm}^{-1} \quad (2.43 \times 10^{-2}\ \text{eV}), \tag{91b}$$

and

$$CH_4(\nu_3) + CH_4 \rightarrow CH_4(2\nu_2) + CH_4 - 52\ \text{cm}^{-1} \quad (0.64 \times 10^{-2}\ \text{eV}), \tag{91c}$$

$$CH_4(\nu_3) + CH_4 \rightarrow CH_4(\nu_2) + CH_4(\nu_2) - 47\ \text{cm}^{-1} \quad (0.58 \times 10^{-2}\ \text{eV}), \tag{91d}$$

with rapid equilibration between the two bending modes according to

$$CH_4(\nu_2) + CH_4 \rightarrow CH_4(\nu_4) + CH_4 + 227\ \text{cm}^{-1} \quad (2.81 \times 10^{-2}\ \text{eV}). \tag{92}$$

In their later paper, Yardley and Moore [144] report also the observation of fluorescence from the ν_3 asymmetric stretching mode. The apparatus was essentially that described except for the use of a Ge:Au detector with a 1.9-μsec time constant. In this case, the phase shifts give the decay rate of the ν_3 level, excited by the laser radiation. There are many V–V processes contributing to the overall observed rate, characterized by a relaxation time of 7 nsec. This very rapid equilibration of the vibrational modes with each other occurs because of the small amounts of energy exchanged with translation as compared to that involved in the V–T process (see Section IV).

Results of Moore and his associates on V–V processes in other molecular gases and gas mixtures are given in Section VII.B.

Work that has been reported to date on the application of gas lasers to studies of molecular energy transfer is already quite impressive. Their success rests primarily on the high power density that can be attained over a very narrow range of wavelength, as compared to a continuum source of radiation. However, it is necessary that the laser wavelength exactly match the absorption line of the molecule under investigation. At present, fluorescence measurements are limited to those molecules whose levels correspond to known laser lines that can be generated with adequate power, or to mixtures with gases whose fluorescing levels may be populated by V–V transfer. Of course, stimulated Raman emission studies, such as that described above for H_2, do not require matching of frequencies. Further discussion of laser applications to molecular energy transfer phenomena may be found in the papers of Yardley and Moore [141, 142].

E. Gaseous Transport Phenomena

Thermal conductivity and thermal transpiration measurements provide two additional methods by which information may be obtained on inelastic molecular collisions. Mason and Monchick [146] have derived an expression for the thermal conductivity λ of a polyatomic gas that includes a term having an explicit dependence upon the relaxation rates of the internal degrees of freedom. Their expression can be written [147]

$$\lambda = \frac{\eta}{M}\left[\frac{5}{2} C_{\text{trans}} + \frac{\rho D_{\text{int}}}{\eta} C_{\text{int}} - \frac{2}{\pi}\left(\frac{5}{2} - \frac{\rho D_{\text{int}}}{\eta}\right)^2 \sum \frac{C_k}{Z_k}\right], \tag{93}$$

where η is the viscosity, M is the molecular weight, ρ is the density, D_{int} is the average coefficient for the diffusion of internal energy, C_{trans} and C_{int} are the translational and the internal contributions to the molar heat capacity, respectively, and Z_k is the collision number for relaxation of the kth internal mode, with associated specific heat C_k. The first two terms in equation (93) are just the modified Eucken approximation, while the third term is a higher approximation of importance whenever Z_k is small, most generally the case for rotational relaxation. Brokaw and other investigators [147, 148] have made thermal conductivity measurements in a number of polar gases and their mixtures over a range of temperatures, and have employed equation (93) to obtain rotational collision numbers. They have also found experimental evidence for the importance of rotation–rotation transfer [146] in energy transport in highly polar gases. Their results, as well as those of several other investigators, have been summarized by Zeleznik [54].

Mason and co-workers [149] have noted the importance of R–T transfer in the thermal transpiration effect, which describes the pressure gradient developing along a capillary at low pressures by virtue of a temperature difference. The theory has been utilized by several investigators (see Section VII.A.2).

F. Molecular Beams

Recent developments [150–156] in molecular-beam methods provide techniques by which inelastic collisions involving vibrational and rotational energy transfer may be studied directly at a specified value of incident particle velocity and scattering angle. These techniques deserve special attention because of their great potential and their fundamental nature.

Utilization of both ion and neutral beams for such studies has been reported. Toennies [150] has performed measurements on the inelastic collision cross section for transitions between specified rotational states using a molecular beam apparatus. TlF molecules in the state (J, M) were separated out of a beam traversing an electrostatic four-pole field by virtue of the second-order Stark effect, and were directed into a noble-gas-filled scattering chamber. Molecules which were scattered by less than $\frac{1}{2}°$ were then collected in a second four-pole field, and were analyzed for their final rotational state. The beam originated in an effusive oven source and was chopped to obtain a velocity resolution $\Delta v/v$ of about 7%. The velocity change due to the inelastic encounters was about 0.3%. Transition probabilities were calculated using time-dependent perturbation theory and the straight-line trajectory approximation. The interaction potential was taken to be purely attractive:

$$V(r, \theta) = \frac{-c}{r^6}(1 + q\cos^2\theta) - \frac{q'}{r^7}\cos^3\theta, \tag{94}$$

where θ is the angle between the molecular axis and the vector $\mathbf{r}$ connecting the centers of mass. The observed transition was $(2, 0) \rightarrow (3, 0)$, which must be attributed to the $\cos^3\theta$ term. Inelastic cross sections for TlF impacting on He, Ne, Ar, and Kr were observed to be 4, 5, 6, and 5 Å^2, respectively, with an uncertainty of about ± 2 Å^2. These values were shown to be in rough agreement with calculations based on an independently measured value of the quadrupole moment of TlF.

Blythe, Grosser, and Bernstein [151] have used crossed molecular beams to observe the $J = 2 \rightarrow 0$ rotational deexcitation process in D_2. A velocity-selected atomic beam of potassium was made to impinge on a modulated D_2 beam from an effusive ($T = 181$°K) source. The scattered K atoms were detected by surface ionization on a hot Pt–W ribbon, from which the ions were drawn into an electron multiplier equipped with lock-in amplification.

Comparison of data on the fast elastic and inelastic peaks led to an estimate of the differential inelastic cross section for the $J = 2 \rightarrow 0$ transition of

$$\left(\frac{d\sigma}{d\Omega}\right)_{2\rightarrow 0} \approx 0.05 \text{ Å}^2/\text{sr},$$

at an angle of 108° in the center-of-mass system.

More recent investigations of rotational excitation in CO_2 by a crossed potassium beam have been reported by Beck and Förster [154]. In this experiment, the K atoms were velocity analyzed before and after collision. The dependence of the scattering on center-of-mass angle from 6° to 28°, and on translational–rotational energy transfer from 10% to 50% of the initial relative collision energy, was determined. Resulting ΔJ values for the CO_2 excitation were found to range from ~6 to ~22 for the average initial value of $\bar{J} = 20$.

Recently reported measurements [152, 155, 156] on the energy loss of molecular ions scattered from gaseous targets provide information on vibrational transition probabilities. Cheng et al. [155] have examined energy loss from beams of NO^+ and O_2^+ scattered from helium, with relative energies in the range of roughly 4–25 eV. Ion beams are formed in a microwave discharge, which results in some metastable electronically excited ions (estimated at ~3%), along with the parent ground-state beam. Some vibrational excitation of the beam ions is also present. The observed energy losses showed that large changes in the vibrational quantum number occur, with $\Delta v = 10$ or more. Individual quantum changes were not resolved. Results for 180° scattering of NO^+ and O_2^+ ions by helium were compared with a modified form of equation (12):

$$\frac{\Delta E}{E_R} = \frac{4m_A m_B m_C(m_A + m_B + m_C)}{(m_A + m_B)^2(m_B + m_C)^2}\left(\frac{\pi\omega l}{v_0}\right)^2 \operatorname{csch}^2\left(\frac{\pi\omega l}{v_0}\right), \tag{95}$$

which was shown to converge properly to the impulse approximation at the high-velocity limit, and to correspond closely to Kelley and Wolfsberg's exact numerical results [19] when $m_A \ll m_B, m_C$. Good agreement was found for l values of 0.1746 and 0.1715 Å, for NO^+ and O_2^+, respectively, which are derived from equation (27) using Lennard–Jones parameters for the neutral collision partners NO–He and O_2–He. At the high energies involved in these collisions, the amount of energy transferred into vibration is less sensitive to l than to atomic masses.

Schöttler and Toennies [152] have employed a velocity-selected Li^+ beam to study inelastic collisions giving rise to vibrational excitation in H_2. Although individual vibrational levels were not completely resolved in these experiments, it was demonstrated that as the incident kinetic energy of the ions was increased from 10 to 50 eV, the probability of multiquantum transitions also increases.

Cosby and Moran [156] have described an apparatus they utilized to measure relative differential inelastic cross sections for 10–20-eV ions incident on atomic or molecular gas targets. Figure 3.19, from their paper, shows the ion source and mass spectrometer for the formation of the primary beam, which in this instance was composed of either O^+ or O_2^+. The charged reaction products are analyzed in regard to mass, energy, and angular distribution utilizing the rotatable scattering angle selector, 127° electrostatic analyzer, and quadrupole mass spectrometer. Because of the presence of O_2^+ $(a\,{}^4\Pi_u)$ in the beam at the electron impact energies employed in their ion source, these investigators also observe inelastic processes involving this metastable species. The difference between the energies of the elastically and inelastically scattered ions at a particular laboratory scattering angle θ is given by

$$E_e - E_i = \frac{m\,\Delta E}{M+m} \pm \frac{2ME_0\cos\theta}{(M+m)^2} \times \left[(m^2 - M^2\sin^2\theta)^{1/2} - \left(1 - \frac{(M+m)\,\Delta E}{mE_0}\right)^{1/2} \times \left(m^2 - \frac{M^2\sin^2\theta}{1-(M+m)\,\Delta E/mE_0}\right)^{1/2}\right], \tag{96}$$

where E_0 is the energy of the incident ion of mass M, m is the mass of the target, and ΔE is the energy absorbed due to vibrational and rotational excitation. Cosby and Moran have measured the intensity of scattered ions as a function of $E_e - E_i$ for specific values of θ and E_0, using an argon target for O_2^+ and an O_2 target for O^+. The resultant curves possess multiple peaks corresponding to different numbers of vibrational quanta excited. The maximum in the scattered ion intensity moves to progressively larger values of $E_e - E_i$ as the scattering angle is increased.

VII. SELECTED EXPERIMENTAL RESULTS

A. Diatomic Gases and Mixtures with Other Diatomic or Monatomic Species

1. Vibration–Translation Transfer

a Correlation Scheme of Millikan and White

It has been pointed out in Sections II, IV, and V that V–T transfer theory may not necessarily be applicable to any given observed vibrational relaxation

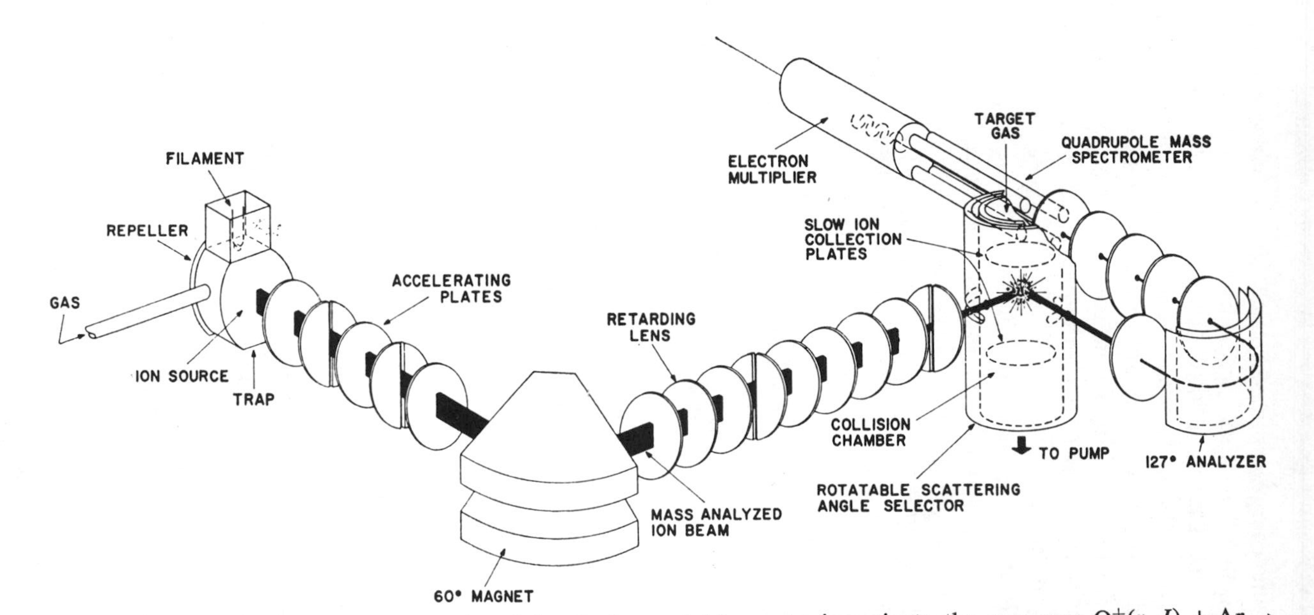

Figure 3.19 Ion-beam apparatus employed by Cosby and Moran to investigate the processes $O_2^+(v, J) + Ar \rightarrow O_2^+(v'J') + Ar$ and $O^+ + O_2(O, J) \rightarrow O^+ + O_2(v'J')$. Energy-selected product ions were measured by either a channel electron multiplier or a Cu–Be multiplier coupled to a high-speed electrometer.

time, since vibrational energy loss may proceed through other channels. Thus, it is incorrect to assume that some approximate V–T theory, such as that of Landau and Teller, has general validity. Nevertheless, for a vast number of experimental measurements, such theories have been shown to give good, and sometimes excellent, agreement. Even in several cases for which the theoretical estimates are considerably off in magnitude (such as for H_2 [157]), the exponential $T^{-1/3}$ dependence of τ_v is, with some notable exceptions, generally valid for diatomic molecules.

Millikan and White [158], in their 1963 paper on the systematics of vibrational relaxation, presented a semiempirical correlation for a considerable amount of data pertaining to diatomic species, but excluded cases such as $N_2 + CO$, in which V–V exchange is of importance. Figure 3.20, taken from their paper, shows the general validity of the exponential $T^{-1/3}$ behavior of the vibrational relaxation time for the data selected [159]. It is also seen that in most cases these data may be fitted by lines intersecting at a common point, viz., $\tau_v = 10^{-8}$ sec (at 1 atm) and $T^{-1/3} = 0.03$ (that is, $T = 37{,}000°$K). Of course, in actual cases, extrapolations to temperatures of this magnitude may be physically meaningless because of dissociation and other processes. For the cases shown of mixtures with noble gases, the quantity plotted corresponds to $p\tau_v$ for collisions between dissimilar species [τ_{AB} of equation (39)]; for example, the O_2–Ar line corresponds to the relaxation of O_2 molecules undergoing collisions with argon only, at a pressure of p atmospheres. Millikan and White constructed this figure by least-squares fitting of N_2 and O_2. The intersection of these two lines determined the common origin for all other sets of data. (The use of the ordinate $p\tau_v$ implies that the relaxation process is due to binary collisions, so that this product is a constant. When τ_v is written alone, it generally refers to the relaxation time at 1 atm.) The two cases, O_2–He and CO–He, involving light collision partners, are notable exceptions. These data clearly do not extrapolate to the common point. The dark band across the bottom of the figure indicates the variation of the collision rate with temperature; τ_c, the time between collisions, is simply $\tau_c = \mathscr{R}^{-1} = \tau_v/Z$, where Z is the collision number for vibrational deexcitation. The band is bounded by the CO–He case and the O_2–O_2 case.

A refinement of the correlation, which replaces the common point by a range in the abscissa, was attempted by Millikan and White in an effort to accommodate the cases with light collision partners. They succeeded in correlating the data according to the relation

$$p\tau_v = \exp\,[A(T^{-1/3} - 0.015\mu^{1/4}) - 18.42], \tag{97}$$

where μ is the reduced mass for the various collision partners and A is treated as an adjustable parameter. These results, while empirical, have some general usefulness in predicting trends in the temperature dependence of

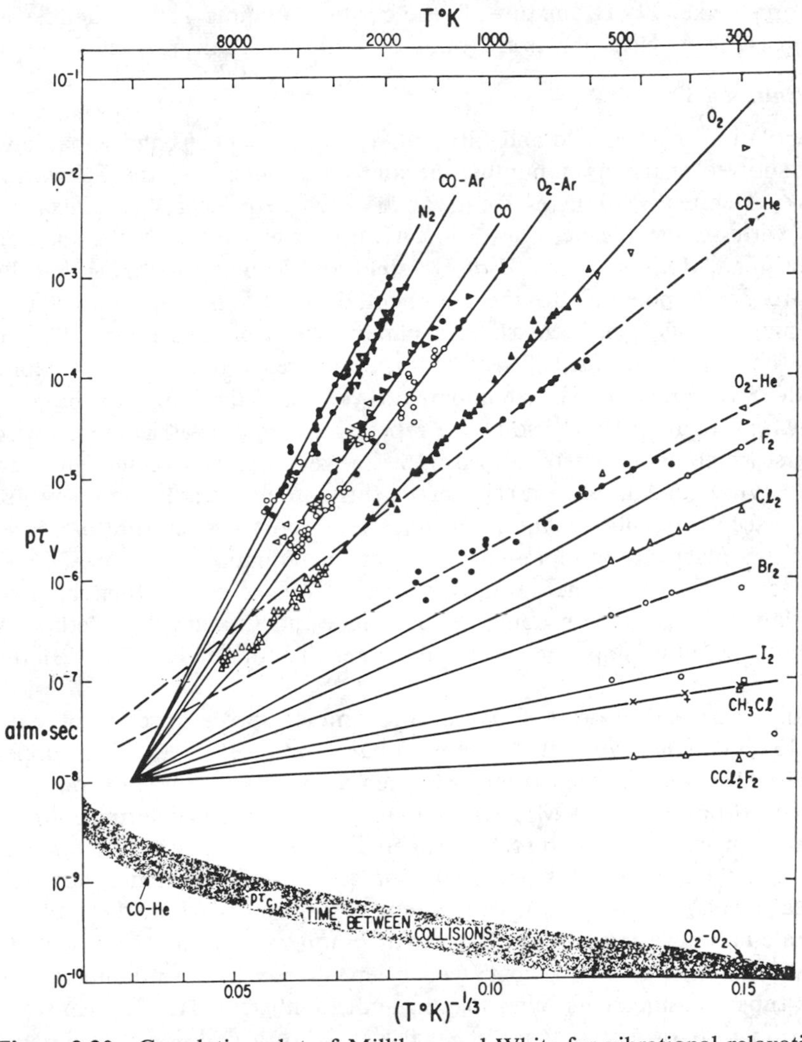

Figure 3.20 Correlation plot of Millikan and White for vibrational relaxation in a number of diatomic species and their mixtures [158]. Solid symbols refer to measurements at the General Electric Laboratory (see ref. 158 and the text for other data sources).

vibrational relaxation for those systems in which V–T transfer is the dominant deexcitation mechanism. White [160] has found that experimental observations on shocked N_2–H_2 mixtures fit the correlation quite well, while relaxation times for N_2–He mixtures are found to be about half the predicted values.

b. Results on O_2 *and* N_2

Calculations by Herzfeld and Litovitz [8] and by Dickens and Ripamonti [161] showed that, for a number of diatomic molecules, the Schwartz–Slawsky–Herzfeld (SSH) treatment can be used to predict V–T relaxation times with reasonable accuracy, depending somewhat upon the method chosen (e.g., Methods A and B or Herzfeld and Litovitz) for matching the Lennard–Jones potential to the exponential repulsions. This uncertainty from matching is also discussed in a paper by Calvert and Amme [23] in which their own semiclassical treatment is compared with experiment, along with the SSH equation. Their comparison with O_2 data [115, 130, 162–167] is shown in Figure 3.21. The solid line represents the semiclassical result with a Morse interaction potential, with the range parameter α set equal to 4.7 Å^{-1}. Very good agreement is observed throughout the entire temperature range, except possibly for the more uncertain room-temperature measurements. The dashed line represents SSH theory, equation (28), with Z_0 set equal to 3, and the range parameter selected by method B (matching of potentials at two common points) of Herzfeld and Litovitz [8]. Method A (matching potential slopes at $r = r_c$) gives a result about midway between the two curves.

Figure 3.22, taken from Calvert and Amme, shows a series of data [128, 131, 162, 169, 170] on the vibrational relaxation of N_2. The upper dashed curve was obtained from SSH theory by matching the exponential repulsion (Method A) to a Morse potential with parameters determined from transport properties. The central, broken line is for the same theory but utilizing the Lennard–Jones potential. The solid curve represents the semiclassical result, equation (28), with $\alpha = 4.7$ Å^{-1}, and is within 11% of the SSH calculation, using Method B, over the temperature range shown. Note that Method A gives a somewhat steeper dependence of Z_{10} on temperature. Shock-tube measurements by Appleton and Steinberg [131, 171] lead to a relaxation time of 5×10^{-7} sec at 9000°K; use of a collision rate derived from viscosity measurements gives an experimental value of 155 for Z_{10}. The plotted semiclassical calculation gives 176. Experimental results at temperatures below 1000°K are considerably less certain, but agreement appears to be somewhat worse.

There is fairly good agreement with theory for the dependence of Z_{10} on T in the case of O_2 deactivated by argon [167] and by helium [113, 122, 163, 164, 166] collisions.

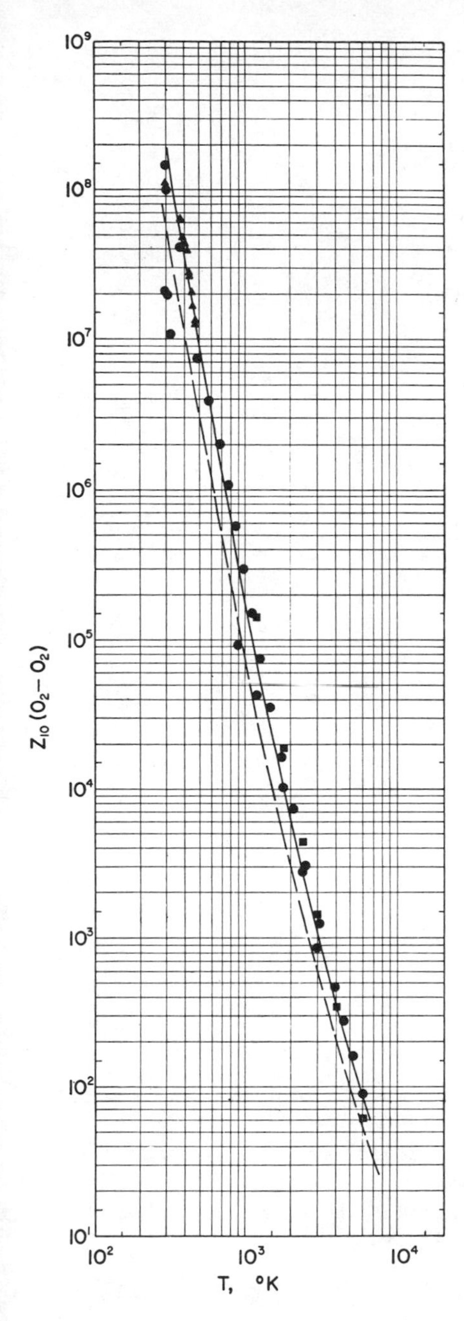

Figure 3.21 Comparison of the experimental collision number Z_{10} for O_2–O_2 with SSH theory and the semiclassical theory of Calvert and Amme using a Morse intermolecular potential (see text for data references).

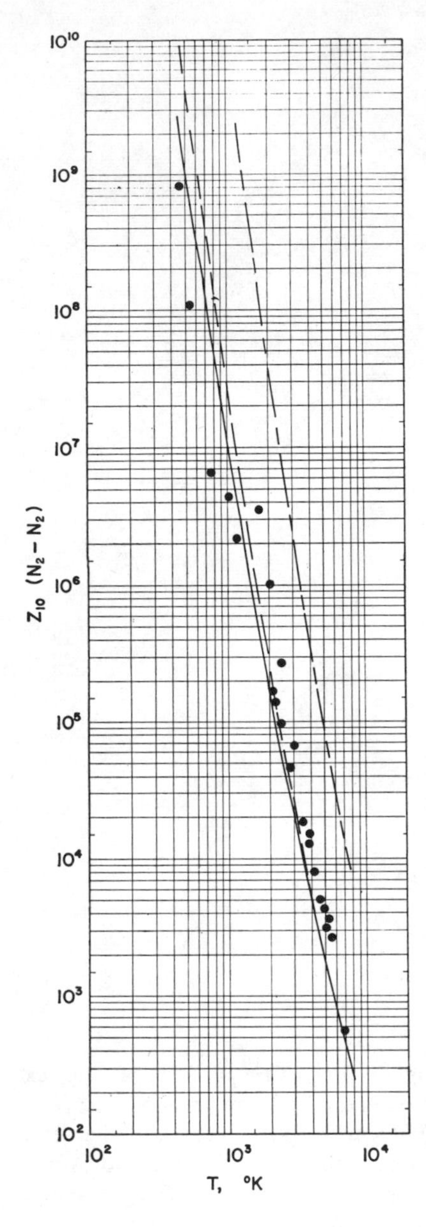

Figure 3.22 Comparison of the experimental collision number Z_{10} for N_2–N_2 with SSH theory and with the semiclassical theory of Calvert and Amme using a Morse intermolecular potential (see text for data references).

c. Results on NO *and* CO

Of the diatomic molecules that have received considerable study, nitric oxide exhibits the most interesting vibrational relaxation behavior. According to Nikitin [172], the relaxation of NO by self-collisions is nonadiabatic because of a resonance between the $^1\Sigma_g^+$ electronic state of the vibrationless $(NO)_2$ collision complex and the $^3\Sigma_g^-$ state with one NO molecule vibrationally excited. This theory predicts that at room temperature, the probability for collisional excitation or deexcitation will be many times higher than predicted on the basis of adiabatic theories, and that for temperatures less than 300°K, the probability will decrease rapidly with decreasing temperature. Shock-tube measurements by Wray [173] have confirmed that at high temperatures the relaxation of pure NO is much faster than predicted by SSH theory. Subsequent shock measurements by Breshears and Bird [174], using the laser-beam deflection method over the range $500 \lesssim T \lesssim 1900$°K, show a smoothly increasing $P_{10}(T)$ throughout that interval. Bauer and co-workers [175–177] and others [178–180] find that $P_{10} \approx 3 \times 10^{-4}$ at room temperature. Measurements by Billingsley and Callear [181, 182], using flash photolysis [180], indicate that the vibrational relaxation rate *increases* with decreasing temperature from 430°K down to 100°K; this behavior was interpreted in terms of an orientation dependence of the resonance described in Nikitin's theory. In addition, these investigators concluded that a very significant contribution to the relaxation arises from ternary collisions, with a corresponding rate coefficient, $k_2 = 3.3(\pm 0.3) \times 10^{-33}$ cm^6/molecule^{-2}/sec, which is approximately independent of temperature in the 100–300°K range. The bimolecular contribution was evaluated to be $k_1 \simeq 1.9 \times 10^{-14}$ cm^3/molecule/sec at 298°K, corresponding to a probability P_{10} of $\sim 0.7 \times 10^{-4}$ per binary collision. This value is about one-fourth of that observed by Kneser, Bauer, and Kosche [177] in their studies for which the pressures exceeded several atmospheres. Hancock and Green [183] have very recently studied the relaxation of vibrationally excited NO by self-collisions in an HF chemical laser system. The pressures involved were about 32 torr of NO and 0–3 torr of HF. The probability P_{10} for binary collisions was found to be in excellent agreement with Kneser, Bauer, and Kosche, despite the 60-fold difference in pressure. Thus, there is evidence that the thermolecular process at 300°K should not be important, in contrast to the flash photolysis findings. In any case, it appears that a minimum exists in the temperature dependence of the binary deactivation probability, at a temperature between 300°K and 400°K.

Results from shock measurements by Kamimoto and Matsui [184] over the range 900–2700°K, using infrared-emission monitoring techniques, are in excellent agreement with those of Wray (using UV absorption) and in good agreement with those of Breshears and Bird.

Electronic relaxation in NO due to the split ground state ($X\,^2\Pi_{3/2,\,1/2}$) is

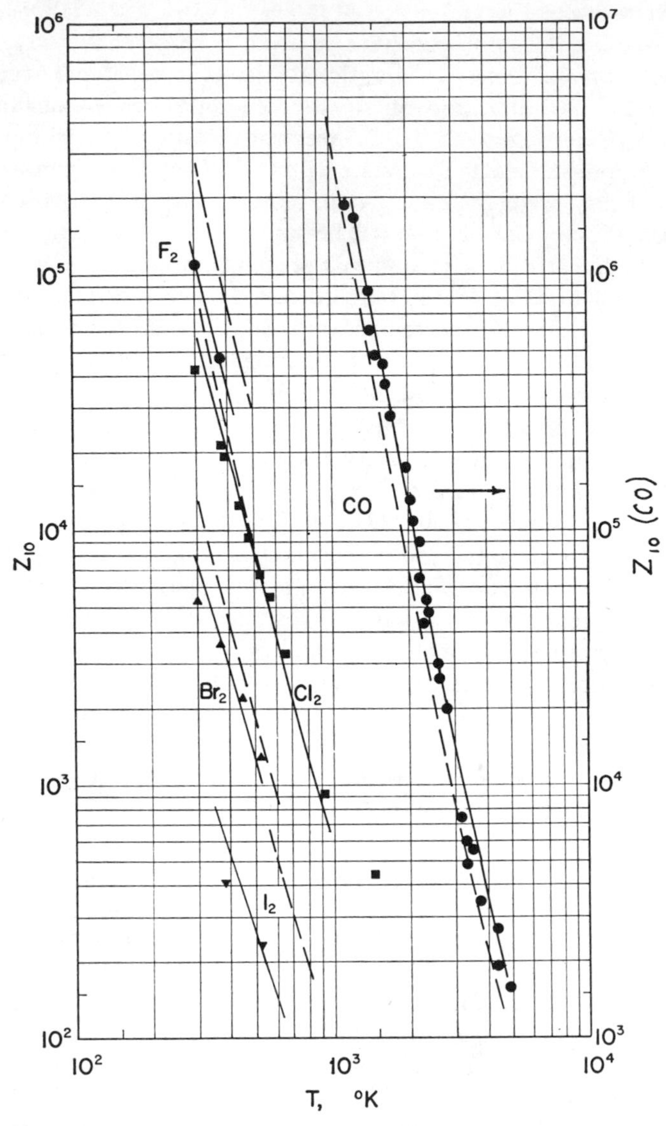

Figure 3.23 Collision numbers for CO and the halogen compared with experimental data of other investigators (see text). Solid lines: semiclassical theory with Morse potential parameter and chosen for best fit; broken lines: SSH theory, method B, using Lennard–Jones parameters and $Z_0 = 3$.

quite rapid at room temperature; equilibration appears to be independent of the vibrational and rotational degrees of freedom [177].

On the right side of Figure 3.23 is a comparison of theory and experiment [185, 186] for CO. A Morse parameter of 4.5 $Å^{-1}$ provides a good overall fit; SSH theory, with Method B and the Lennard–Jones r_0 of 3.706 Å, yields the dashed curve. A summary of some older data for CO is given by Woodmansee and Decius [187], who used a spectrophone method to measure the relaxation time of pure CO at room temperature.

A publication by McLaren and Appleton [188] describes a novel technique for the study of collisional energy transfer and chemical reaction rates in rapidly cooled gases. The apparatus employs a conventional shock tube, with helium as the driver gas, fitted near the downstream end with an expansion chamber on each side. The chambers are joined to the walls of the tube through twin opposing apertures fitted with rupture diaphragms. Measurements on the vibrational temperature of carbon monoxide were performed by monitoring the infrared emission intensity of the CO fundamental band centered at 4.65 microns. The shock-heated gas is stationary during the cooling process so that the temporal behavior can be continuously monitored. The deexcitation rate of CO in an argon heat bath was found to exhibit the same characteristic vibrational relaxation time in the expansion process as the excitation observed in a compression (shock) wave, to within experimental uncertainty, in contrast to experiments performed with the supersonic nozzle [189].

Relatively few measurements have been reported of vibrational energy transfer at low temperatures. Aside from the NO studies of Billingsley and Callear, already discussed, Miller and Millikan [190] have reported relaxation times for mixtures of CO with helium and hydrogen down to 100°K. A very clear deviation from the Landau–Teller linear extrapolation [191] was observed below 300°K for both mixtures. Shin [192] has examined the CO + He case in more detail, using the WKB approach and a Morse interaction potential. At low temperatures, he concludes, there are three important considerations:

1. The approximation $E \gg h\nu$ is no longer valid; the thermal averaging must be performed numerically.
2. The initial and final energies are significantly different after an inelastic collision; Shin uses a barrier penetration model for the incident wave and "symmetrizes" the energy.
3. The steric factor, which cannot be evaluated directly in the theory, approaches unity as T becomes small.

Considerable future work will undoubtedly be devoted to the understanding of observations for temperatures at which attractive forces play a significant role.

d. Results on the Halogens

In Figure 3.23 are shown experimental data for the halogens [6, 193, 194] compared both with the semiclassical Morse potential calculation, α being chosen for best fit, and with SSH theory (Method B), using Lennard–Jones parameters. A Morse parameter of 4.1 Å^{-1} gives reasonably good fit for Cl_2, Br_2, and I_2, while 5.0 Å^{-1} is required to fit the F_2 measurements. There is a general trend in the data, at the lower temperatures, towards lower Z_{10} than predicted by these curves; the reasons are probably similar to those for the CO + He case above. Additional measurements at still lower temperatures would be of interest. More recent studies of Cl_2 by Breshears and Bird (195) indicate that the high-temperature behavior is closer to the theoretical curve in Figure 3.23 than the earlier results indicate.

e. Results on H_2 *and* D_2

Shock-tube investigations of vibrational relaxation times in hydrogen, deuterium, and their mixtures with argon and krypton have been reported by Kiefer and Lutz [129] and by Moreno [196]. By fitting their measurements to a Landau–Teller temperature dependence, Kiefer and Lutz give, for D_2–D_2 collisions (1100–3000°K),

$$\tau_{\mathrm{DD}} = 2.7(\pm 0.3) \times 10^{-10} \exp \frac{110.5 \pm 1.5}{T^{1/3}}; \tag{98}$$

for D_2–Ar collisions (1600–3000°K),

$$\tau_{\mathrm{DA}} = 1.0(\pm 0.7) \times 10^{-9} \exp \frac{118 \pm 10}{T^{1/3}}; \tag{99}$$

for H_2–H_2 collisions (1100–2700°K),

$$\tau_{\mathrm{HH}} = 3.9(\pm 0.8) \times 10^{-10} \exp \frac{100.0 \pm 2.6}{T^{1/3}}; \tag{100}$$

and for H_2–Ar collisions (1500–2700°K),

$$\tau_{\mathrm{HA}} \approx 4.1 \tau_{\mathrm{HH}}. \tag{101}$$

Within experimental uncertainties, Moreno's measured relaxation times are in agreement with those of Kiefer and Lutz. The results of the latter two authors are shown as curves A and C, for H_2 and D_2 respectively, in Figure 3.24. Curves B and D are those calculated by Calvert, using an extension to the approach of Salkoff and Bauer [197]. The single point at room temperature as from Ducuing [137, 138].

Parker [198] has compared his classical theory with the experimental results on vibrational relaxation of hydrogen and deuterium. He finds satisfactory agreement for collisions with argon and krypton, but definite

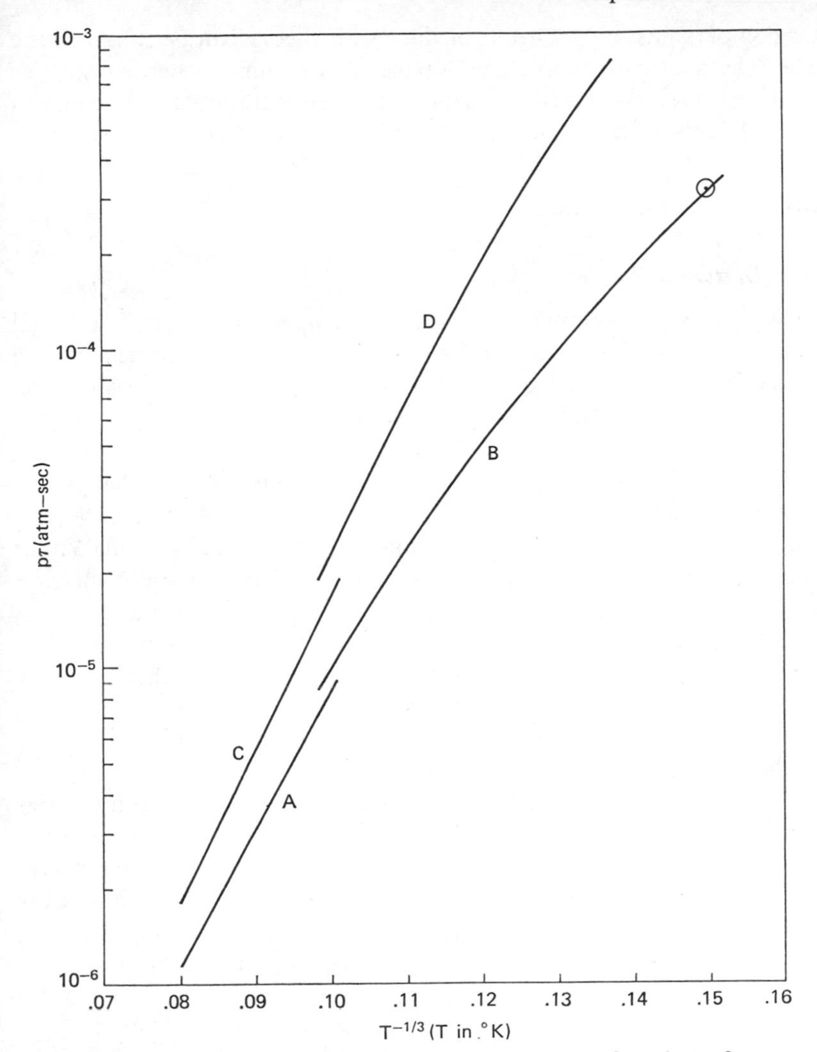

Figure 3.24 Vibrational relaxation of H_2 and D_2 as a function of temperature. Curves A and C are the data of Kiefer and Lutz on H_2 and D_2, respectively; curves B and D are those calculated by Calvert [139]. The single point at room temperature is from Ducuing [138] for H_2.

discrepancies for H_2 and D_2 self-collisions. Kiefer [199] has pointed out that the lack of symmetrization (15) (Section II.C) in Parker's formulation results is a disparity between theory and experiment for the ratio τ_{DD}/τ_{HH}. He shows further that with symmetrization to account for the change in relative motion resulting from the transfer of a vibrational quantum, Parker's theory can be brought into better agreement with the observed ratio.

Other comparisons of V–T transfer data with theory can be found in the literature. The reader is referred to the paper of Benson and Berend [48] for a comparison with the classical theory of those authors, in which both Morse and Lennard–Jones potentials have been employed.

2. Translation–Rotation Transfer

a. Heavy Diatomic Species

For the majority of diatomic gases at room temperature, many rotational states are populated (see Section III.A). If experimental measurements are interpreted in terms of a single relaxation time, τ_r, then the corresponding rotational collision number, $Z_r = \mathscr{R}\tau_r$, cannot in general be related to a single transition probability. In some instances, experimental work has demonstrated that the relaxation process is more complex, and that multi-state models must be formulated to reproduce the data. Raff and Winter [200] have treated the ultrasonic dispersion and absorption data of Winter and Hill [126] in this manner in order to reproduce their sound-dispersion curves and to explain the observed temperature dependence of the apparent inflection frequency. Parahydrogen at room temperature exhibits multiple sound dispersion, as shown by Valley and Amme [75]. Nevertheless, many measurements have been interpreted in terms of a single relaxation process, and it is interesting to compare the results with a classical theory such as Parker's [44] or that of Berend and Benson [49].

Sound-absorption and shock-tube measurements in heavy diatomic gases show that relaxation times are extremely short, with corresponding collision numbers ranging from slightly greater than unity to about 7 at room temperature. Bauer and Kosche [201] have shown that Z_r increases monotonically as a function of $\overline{\Delta E_r}/kT$, in which $\overline{\Delta E_r} = \sum n_J \Delta E_J / \sum n_J$ is a mean energy transfer for the smallest possible change of rotational quantum number:

$$\Delta E_J = \frac{\hbar^2}{2I}(2J + 2) \qquad \text{for} \quad \Delta J = 1,$$

$$= \frac{\hbar^2}{2I}(4J + 6) \qquad \text{for} \quad \Delta J = 2,$$

and

$$n_J = (2J + 1)g \exp - \frac{\hbar^2 J(J + 1)}{2IkT}. \tag{102}$$

Their result is shown in Figure 3.25. The experimental value for CO is that

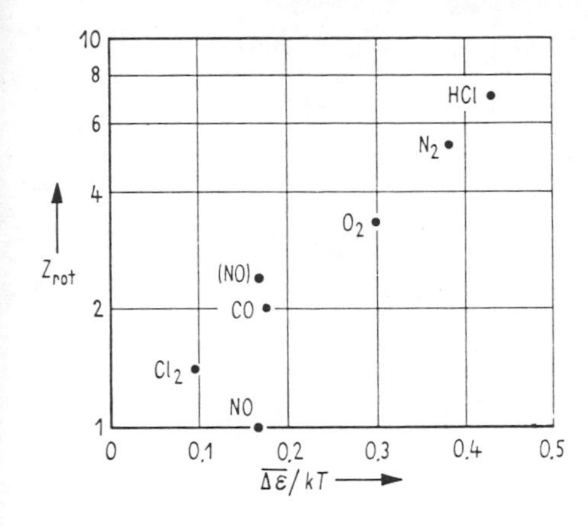

Figure 3.25 Rotational collision number for several diatomic molecules at room temperature as a function of $\overline{\Delta E_r}$ (see text), according to Bauer and Kosche.

from Bauer and Kosche; that for Cl_2 is from Sittig [194]; O_2 and N_2 results are from Sessler [202], and HCl from Breazeale and Kneser [203]. The older value for NO of unity, reported by Bauer and Sahm [176] (from a measurement in which relaxation due to electronic specific heat arising from the $^2\Pi_{1/2}$–$^2\Pi_{3/2}$ splitting was also observed, with $Z_e = 70$ collisions) was revised by Bauer and Kosche to the value shown in parentheses. The collision numbers by Sessler for N_2 and O_2 using absorption of sound are in good agreement with those reported shortly afterwards by Greenspan [105], who obtained 5.26 ± 0.05 for N_2 and 4.09 ± 0.08 for O_2, at 300°K, using absorption and dispersion of sound.

Results of Winter and Hill [126] and of Carnevale, Carey, and Larson [125] show that the equilibration rate of the rotational degree of freedom decreases, as the temperature is increased from 300°K to 1273°K, for H_2 and D_2 as well as for N_2 and O_2. Figure 3.26, from Carnevale et al., compares the temperature dependence of Z_r for N_2, from a muffle-tube experiment, with the theory of Parker [44]. Each curve is normalized to a value at 0°C. Qualitative agreement is observed.

Malinauskas and co-workers [204, 205] and also Healy and Storvich [206] have derived rotational collision numbers from thermal transpiration measurements, utilizing the theory developed by Mason and co-workers [149]. Values have been obtained for the gases N_2, CO, O_2, and CO_2, up to temperatures of 500°K. Collision numbers were also obtained by Healy and Storvick for H_2 at 444° and for CH_4 and CF_4 at 366°K. The latter three cases present difficulties for this technique, and it could only be ascertained that $Z_r(H_2)$ is greater than 100. For N_2, O_2, and CO_2, the values of Z_r obtained were in good agreement with those derived from acoustical methods, except that the increase in Z_r with increasing temperature appeared more pronounced for the thermal transpiration method.

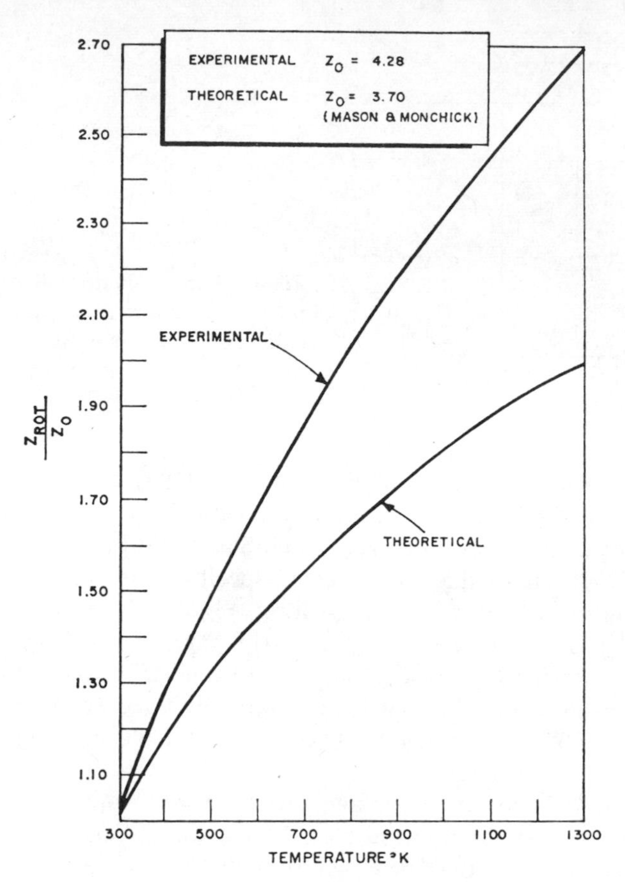

Figure 3.26 Comparison of the experimental dependence of rotational collision number Z_r on temperature with that predicted from the theory of Parker (from ref. 125).

b. Collisions Involving H_2, HD, *and* D_2

The characteristic rotational temperatures θ_{rot} for H_2, HD, and D_2 are 84.8, 63.8, and 42.6°K, respectively. The symmetry of H_2 and D_2 require that for optical transitions, $\Delta J = \pm 2$, while for HD, $\Delta J = \pm 1$. At room temperature, normal hydrogen n-H_2 is composed of 25% para-H_2 (J even) and 75% ortho-H_2 (J odd), while at lower temperatures, equilibrium hydrogen contains an increasing proportion of para-H_2. These rotational species do not change form in gaseous collisions, so that it is possible to select nearly pure p-H_2 (e.g., boil-off from liquid H_2) and perform measurements on it and its mixtures with n-H_2 over a range of temperatures, before surface catalysis on the instrument walls converts the mixture to equilibrium hydrogen (to normal-H_2 at room temperatures).

For p-H_2 at low temperatures, only two rotational states, $J = 0$ and 2, need be considered, and relaxation measurements correspond unambiguously to a single rotational transition. Thus, a number of investigations have been performed on p-H_2 and its mixtures with the noble gases at temperatures down to 77°K. Most notable are the extensive investigations of Beenakker's group at Leiden [70–73]; Sluijter, Knaap, and Beenakker [70] investigated rotational relaxation in the three isotopes over a range of temperatures using sound absorption, while Jonkman et al. [71–73] investigated mixtures of these gases with helium, neon, and argon, comparing their results with the theoretical works of Takayanagi [57], Davison [59], and Roberts [74]. (Davison had made a comparison of his calculation with the earlier observations by Rhodes [69].)

The 1969 measurements by Valley and Amme [75] demonstrated that the noble gases affect the two relaxation processes, $2 \to 0$ and $4 \to 2$ in room-temperature p-H_2, by different amounts. Double-relaxation processes with characteristic relaxation times τ_1 and τ_2 were found to persist with He, Ne, and Ar added, but in such a way that the ratio τ_1/τ_2 increased as the mole fraction of noble gas was increased. Neglecting R–R processes and the influence of $J = 0 \to 4$ transitions (see Section II), these authors applied a three-state model to p-H_2 at room temperature. The two relaxation times evolving from the rate equations were then shown to correspond closely to $\tau_{20} = [P_{20}\mathscr{R}(1 + K_2)]^{-1}$ and $\tau_{42} = [P_{42}\mathscr{R}(1 + K_4)]^{-1}$, where $K_2 = (g_2/g_0 \exp(-\varepsilon_{20}/kT)$ and $K_4 = (g_4/g_2) \exp(-\varepsilon_{42}/kT)$, with g denoting the degeneracies of the respective states and ε denoting the energies associated with the $2 \to 0$ and $4 \to 2$ transitions. P_{20} and P_{42} are the average probabilities per collision that the respective rotational transition will occur. In analyzing their dispersion curves, it was assumed that the relaxing specific heat associated with the longer relaxation time includes those contributions to the total rotational specific heat [207] due to the $J = 4$ level, and that the faster relaxation process accounts for the remainder. The collision numbers $Z = \mathscr{R}\tau$ (where $\mathscr{R}$ is the relevant collision rate computed from a hard-sphere collision model), relaxation times, and deexcitation probabilities are listed in Table 3.1. These results combined with those of other investigators [70–73, 208, 209] are plotted in Figure 3.27 as a function of temperature. In Figure 3.27a, theoretical curves derived from the work of Takayanagi [57] and the subsequent work of Davison [59] and Roberts [74] are shown for comparison. Measurements on pure p-H_2 are in the best accord, yielding the value $\beta = 0.113$ in Roberts's angle-dependent Morse potential, $\beta = 0.108$ in Davison's, or $\beta = 0.105$ in Takayanagi's (equation (54)). For the p-H_2–He interaction, the value $\beta = 0.30$ has been used in plotting the predicted T

Table 3.1
Rotational Relaxation in p-H_2 Self-Collisions and p-H_2–Noble-Gas Collisions at 300°K [75]

	H_2–H_2	H_2–He	H_2–Ne	H_2–Ar
τ_{20}	1.3×10^{-8}	0.76×10^{-8}	0.62×10^{-8}	0.92×10^{-8}
τ_{42}	3.9×10^{-8}	7.7×10^{-8}	5.6×10^{-8}	6.1×10^{-8}
Z_{20}	215	96	74	131
Z_{42}	650	970	670	870
P_{20}	2.4×10^{-3}	5.4×10^{-3}	7.0×10^{-3}	4.0×10^{-3}
P_{42}	1.5×10^{-3}	1.0×10^{-3}	1.4×10^{-3}	1.1×10^{-3}

dependence from Roberts's theory. Johnson and Secrest [210] have shown that the distorted-wave approximation, used by Roberts and others for H_2–He collisions, overestimates the cross section for the $0 \rightarrow 2$ transition by about 25%. Also, Krauss and Mies [50] indicate that Roberts' potential gives cross sections too large by the same amount, so that the results calculated by this author could be too high by $\sim$50%.

The observed temperature dependence departs from the theoretical predictions increasingly as the mass of the incident particle increases, a result which probably reflects increasing error with increasing m in the distorted-wave approximation. Helium is found to be more efficient at room temperature in deactivating the $J = 2$ level than is H_2, while neon is more efficient than either helium or argon. This behavior was also observed at 170°K by Jonkman et al. [71, 73].

Berend and Benson [49] have compared experimental rotational relaxation rates for p-H_2–p-H_2, p-H_2–He, and o-D_2–He with their classical theory utilizing the two-dimensional model of Figure 3.5, with atom-centered Morse potentials. Their conclusion that H_2–H_2 collisions are more effective than are H_2–He collisions differs from the predictions of Roberts and is not confirmed experimentally; however, there is fair agreement with regards to both magnitude and temperature dependence, particularly for the H_2–H_2 case.

Akins, Fink, and Moore [211] have made a study of the pressure dependence of the fluorescence from HD($B\,^1\Sigma_u^+$) molecules excited by the 1048- and 1066-Å argon resonance lines. The (v', J') levels of the $B\,^1\Sigma_u^+$ electronic state excited were (3, 2), (5, 2), and (5, 5). At low pressures, the fluorescence originates from transitions between these levels and the $X\,^1\Sigma_g^+$ ground state. By increasing the HD pressure, or by adding helium, neon, or D_2, these investigators were able to determine rotational energy transfer rate constants at room temperature. ^{3}He was found to be more effective than ^{4}He; processes were observed for which $\Delta E > 1.5kT$, and in fact, collisions of HD* $(v', J' = 2)$ with rare gases were found to exhibit $\Delta J = \pm 2$ changes with

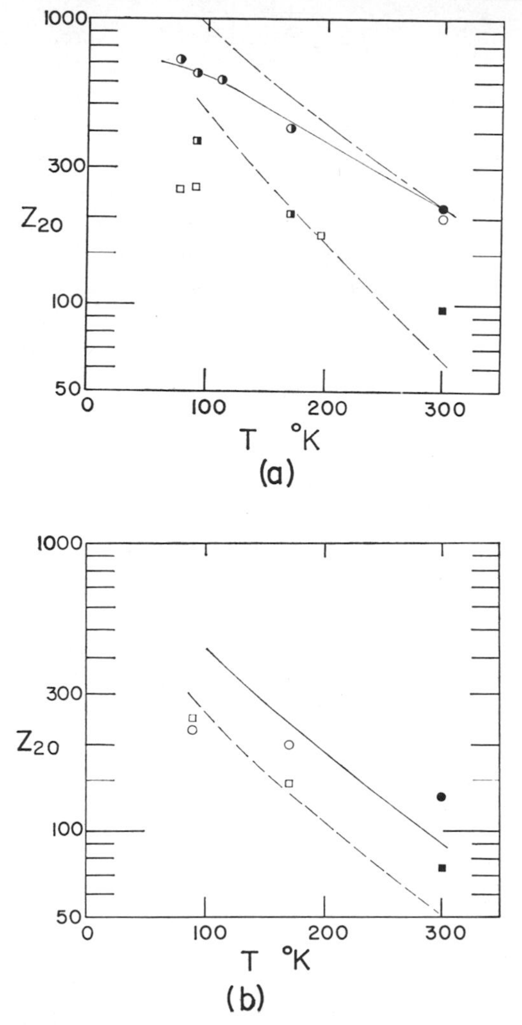

Figure 3.27 Comparison of experimental and theoretical temperature dependence of rotational collision numbers $Z_r = \mathscr{R}\tau$. (a) The cases of p-H_2–p-H_2 and p-H_2–He: Half-filled circles are the data of Jonkman et al., half-filled squares are their results for p-H_2–He. Open squares are from Bose et al. Room-temperature results are by Geide (open circle) and by Valley and Amme (solid symbols). Upper broken curve is based on Takayanagi ($\beta = 0.105$), solid line based on theory of Roberts ($\beta = 0.113$) or Davison ($\beta = 0.108$). Lower dashed line is for p-H_2–He theory of Roberts ($\beta = 0.30$). (b) Results for p-H_2–Ne (open squares, Jonkman et al.; solid square, Valley and Amme); and for p-H_2–Ar (open circles, Jonkman et al.; solid circle, Valley and Amme). Solid line is for p-H_2–Ar based on Jonkman et al., theoretical curve ($\beta = 0.13$); dashed line is for p-H_2–Ne ($\beta = 0.20$).

higher probability than for $\Delta J = \pm 1$, while for self-collisions, $\Delta J = \pm 1$ was more probable.

c. *Molecules Exhibiting Λ Doubling*

Molecules having a $^1\Pi$ electronic state possess rotational levels with doublet character (Λ doubling). Ottinger, Velasco, and Zare [212] have observed the fluorescence of the Li_2 $B\,^1\Pi_u$–$X\,^1\Sigma_g^+$ band system, induced by the irradiation of a mixture of lithium vapor and argon with an argon ion laser. A pattern of collision-induced satellite lines was found, resulting from single inelastic collisions of Ar atoms with the $^1\Pi_u$ molecules. Typically, a fluorescence line would be accompanied by six prominent satellites, resulting from inelastic collisions in which the rotational quantum number J' of the $B\,^1\Pi_u$ state is changed by ± 1 or ± 2. The relative intensities of the satellite lines were found to depend upon the component of the Λ doublet from which the collision-induced transition originates. If the upper and lower states of the doublet are denoted by c and d, respectively, then for $d \rightarrow c$ jumps an increase in J is found to be preferred over a decrease in J, while for $c \rightarrow d$ jumps a decrease in J is favored over an increase in J. On the other hand, satellite line intensities for those transitions for which $c \rightarrow c$ or $d \rightarrow d$ showed that $\pm\Delta J$ changes occur with nearly equal probabilities. These trends, labeled "propensity rules," have been interpreted by Ottinger et al. in terms of a simple classical model based upon the markedly different charge-density distributions of the upper and lower components. For a $^1\Pi$ electronic state, the charge distribution has a $\cos^2 \phi$ distribution for the upper Λ component (c states), where ϕ is the azimuthal angle about the internuclear axis measured from J. For the lower Λ component (d states) the charge distribution is proportional to $\sin^2 \phi$. Thus, the preferred collision trajectories for argon atoms striking electronically excited Li_2 molecules and causing $c \leftrightarrow d$ transitions are those having velocity components parallel to the plane separating the two lobes of the charge distributions, one lobe lying above and one below this plane. For $c \rightarrow d$ transitions, this component lies in the plane of rotation of the molecule, and the vector addition of this velocity component of the argon atom to that of the approaching Li atom results in a lowering of the rotational energy. For $d \rightarrow c$ transitions, the incident argon atom has a component of velocity perpendicular to the plane of rotation that, when added vectorially to a lithium atom velocity, leads to an increase in rotational energy, all in accordance with the propensity rule. These approximate arguments have yet to be developed in any quantitative way.

d. *Polar Diatomic Molecules*

Zeleznik [54] has summarized much of the information available on rotational relaxation in highly polar gases. For the diatomic species HCl,

DCl, HF, and DF, rotational collision numbers appear to decrease with increasing temperature over the intervals for which data are available (see Table 3.2). His comparison of acoustic-absorption and thermal-conductivity data for both HCl and H_2O indicates consistent results for these two experimental methods and their respective interpretations. However, the apparent decrease of Z_r with increasing temperature is in contradiction to Zeleznik's

Table 3.2
Experimental Rotational Collision Numbers for Polar Diatomic Gases [54].

MOLECULE	TEMP. (°K)	Z_r	REF.	METHOD
HCl	273.2	7.0	[203]	Acoustic
	300.1	6.2	[147]	Thermal Cond.
	328.5	4.6	[147]	Thermal Cond.
	374.8	3.6	[147]	Thermal Cond.
	423.1	3.2	[147]	Thermal Cond.
	471.4	3.0	[147]	Thermal Cond.
DCl	300.1	2.6	[147]	Thermal Cond.
	328.5	2.3	[147]	Thermal Cond.
	374.8	2.0	[147]	Thermal Cond.
	423.1	1.9	[147]	Thermal Cond.
	471.4	1.9	[147]	Thermal Cond.
HF	373.8	9.5	[213]	Thermal Cond.
	422.3	9.1	[213]	Thermal Cond.
DF	373.8	4.1	[213]	Thermal Cond.
	422.3	3.8	[213]	Thermal Cond.

theoretical prediction (equation (51)), as well as to that of Parker [equation (45)]. Further investigation of rotational relaxation in polar gases is desirable.

3. Vibration–Vibration Transfer

Measurements have been made by Callear [84] on the systems NO^* ($A\,^2\Sigma$) + N_2, NO^* + CO, and NO^* + N_2, in which vibration–vibration transfer occurs. The probabilities, determined for these gases at room temperature, are shown in Figure 3.4. There is at least qualitative agreement with Rapp's prediction [83] (equation (64)), based on a repulsive interaction potential.

Bauer and Roesler [111, 214] were able to determine that the probability for V–V deactivation of O_2^* by N_2 [$\gamma = 789$ cm^{-1} (9.78×10^{-2} eV)] at room temperature is $P_{10}^{01} = 5.6 \times 10^{-10}$, which is also in rough agreement with Rapp's prediction. This probability of V–V transfer is about $\frac{1}{12}$ that of V–T transfer for the same collision species (see Section VII.A.1). The reverse process, which is exothermic, occurs with a probability P_{01}^{10}, related to P_{10}^{01} through detailed balancing. P_{01}^{10} was found by Bauer and Roesler to have a value of 2×10^{-8}, which is larger than either V–T process giving rise to deexcitation of N_2^* in the mixture of N_2 and O_2.

Bauer and Roesler [111] have also studied ultrasonic absorption in mixtures of O_2 and CO at room temperature. The deactivation of vibrationally excited O_2 by V–V transfer to CO [$\gamma = 487$ cm^{-1} (6.04×10^{-2} eV)] was found to occur with a probability of 3.8×10^{-7}, while the inverse process is about an order of magnitude faster. V–T deactivation of O_2^* by CO was shown by these authors to be roughly $\frac{1}{3}$ as probable as V–V deactivation, namely, $P_{10} = 1.1 \times 10^{-7}$.

White [215] has reported shock-tube data for $N_2 + O_2$ mixtures over the range 1000–3000°K. He has concluded that the V–V exchange probability leading to N_2 deactivation increases from 1×10^{-6} to 2.3×10^{-5} over this range. As seen from Figure 3.4, the near-resonant theory of Rapp gives a probability which is higher by a factor of $\sim$12 at 3000°K.

Sato, Tsuchiya, and Kuratani [216] have solved the relaxation equation for the vibrational energies of two diatomic gases A and B diluted in an inert monatomic gas M, and have applied the solution to shock-wave relaxation profiles in order to obtain V–V transfer rates. Their solution shows that the relaxation of each of the component molecules proceeds as if it possessed two relaxation times. At the onset of the relaxation process, both components begin to relax with their respective V–T rates, whereupon the relaxation rate of that component having the smaller V–T relaxation time begins to decrease, while the relaxation rate of the other component increases. Finally, both components relax with the same rate toward their equilibrium states. By observed infrared emission from the CO fundamental behind shock waves in mixtures of CO–N_2, CO–O_2, CO–D_2, and CO–H_2, they were able to determine P_{10}^{01} as a function of temperature. Argon was used as inert buffer gas.

In addition, they analyzed the measurements of Taylor, Camac, and Feinberg [217] for CO–NO mixtures. The results are presented in Figure 3.28, taken from their paper. Also shown are the data of Hooker and Millikan [187] for V–T relaxation in CO, along with room-temperature results of Bauer and Roesler [111] and of Basco et al. [180]. The dashed lines were calculated by SSH theory [30] [see equation (59)]. The V–T probability for CO–CO collisions is shown by a solid line for comparison. The resulting V–V probabilities are found to be in general agreement with the SSH theory, except for the CO–H_2 case. For this pair, rotational energy also contributes to the transfer process, as discussed in Part 4 below.

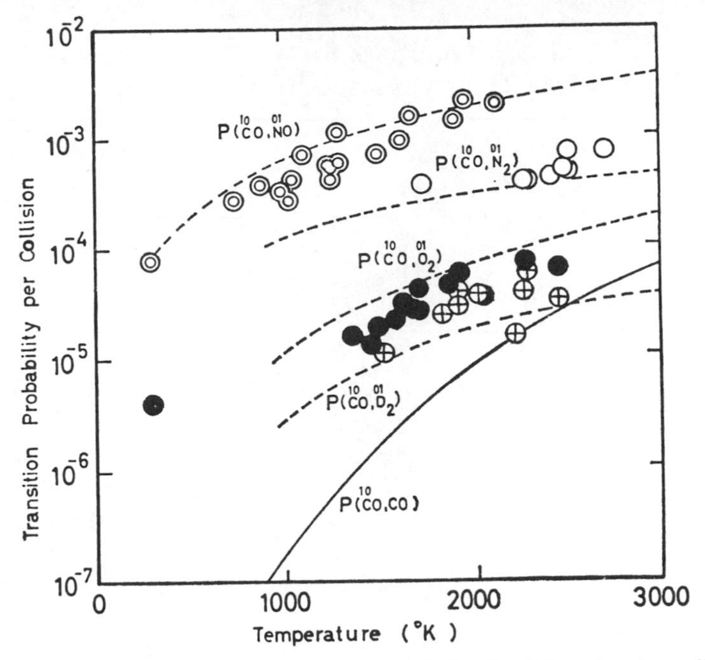

Figure 3.28 Vibrational deactivation probabilities for CO molecules colliding with various diatomic species. Data represented by double circles are those of Taylor et al. [217], and of Bauer and Roesler [111], representing V–V transfer with NO. The solid circle at $T = 300°K$ is the datum of Basco et al. [180]. The solid curve represents the result of Hooker and Millikan [186], and the dashed curves were calculated by Sato et al. [216], using SSH theory (equation (59)). All other experimental data are by Sato et al., from which this figure is taken.

V–V exchange between two hydrogen halides has been observed by Chen, Stephenson, and Moore [218] utilizing laser-excited vibrational fluorescence. The vibration-vibration transfer rate for HCl–HI collisions was determined by measurement of the exponential decay of HCl $\Delta v = 1$ fluorescence following excitation by an HCl laser pulse. The relaxation time for the process

$$\mathrm{HCl}(v = 1) + \mathrm{HI}(v = 0) \rightarrow \mathrm{HCl}(v = 0) + \mathrm{HI}(v = 1) + \Delta E, \quad (103)$$

in which $\Delta E = 653\ \mathrm{cm}^{-1}$ $(8.10 \times 10^{-2}\ \mathrm{eV})$, was found to be $(2.5 \pm 0.2) \times 10^{-7}$ sec at 1 atm.

Breshears and Bird [219] have shown that in mixtures of N_2 with HI and DI, the vibrational relaxation of N_2 is dominated by V–V transfer to the halide gas:

$$N_2^\dagger + \mathrm{HI} \rightarrow N_2 + \mathrm{HI}^\dagger + \Delta E = 101\ \mathrm{cm}^{-1} \quad (1.25 \times 10^{-2}\ \mathrm{eV}), \quad (104)$$

$$N_2^\dagger + \mathrm{HD} \rightarrow N_2 + \mathrm{HD}^\dagger + \Delta E = 731\ \mathrm{cm}^{-1} \quad (9.06 \times 10^{-2}\ \mathrm{eV}). \quad (105)$$

Table 3.3
Comparison of Collision Numbers for V–V Transfer with Values Calculated Using SSH Theory, Method B [219].

T(°K)	$Z\ (N_2^{10}, HI^{01})$		$Z\ (N_2^{10}, HD^{01})$	
	EXPT.	CALC.	EXPT.	CALC.
1200	4.77×10^3	118	4.18×10^4	1.26×10^7
2000	5.69×10^3	21	2.30×10^4	1.67×10^5
2700	6.12×10^3	10	—	—

Theoretical values for V–V transfer calculated by these authors using SSH theory (method B) were higher than experiment for the HI case and lower for the DI case, with a greater temperature dependence than that observed. Presumably, long-range forces play a more important role than the theory permits. Table 3.3 gives a comparison of collision numbers at three different temperatures, calculated from the relation $Z_{10}^{01} = \tau_{AB}^{v}\mathscr{R}_{AB}$, where the pressure is at 1 atm of B molecules, and where $\mathscr{R}_{AB}$ is the collision rate per A molecule with B molecules.

V–V transfer between HCl and DCl has been investigated by Chen and Moore [220] employing laser-excited vibrational fluorescence. An HCl laser was used to excite fluorescence in the HCl component of the mixture, and the decay rate provides information on the fast process

$$\text{HCl}(v = 1) + \text{DCl}(v = 0) \rightarrow \text{HCl}(v = 0) + \text{DCl}(v = 1) + \Delta E = 795\ \text{cm}^{-1} \quad (9.86 \times 10^{-2}\ \text{eV}), \tag{106}$$

Table 3.4
Probability per Collision of V–V Transfer for $HCl(v = 1) + M(v = 0) \rightarrow HCl(v = 0) + M(v = 1) + \Delta E$ [220] and [221]

M	$P \times 10^4$	ΔE cm^{-1} (eV)
D_2	70[a]	108 (1.34×10^{-2})
HBr	60	327 (4.05×10^{-2})
N_2	1.1	555 (6.88×10^{-2})
HI	8.0	656 (8.13×10^{-2})
CO	3.4	743 (9.21×10^{-2})
DCl	5.0	795 (9.86×10^{-2})

[a] For V–V transfer from D_2 to HCl in the exothermic direction.

as does the rate of increase in DCl fluorescence. The subsequent decay of the DCl fluorescence provides information also on the slower V $\rightarrow$ T, R processes. The V–V relaxation time was found to lie between 0.4 and 0.6 μsec at 1 atm.

Chen and Moore [221] have also examined mixtures of HCl with diatomic molecules HBr, HI, N_2, and CO to further investigate V–V processes. Their results are presented in Table 3.4. It is seen that the probability decreases rather slowly with increasing ΔE. It is particularly interesting to note that energy transfer from HCl to CO is 3 times as probable as from HCl to N_2, despite the fact that 188 cm^{-1} (2.33×10^{-2} eV) more energy must be transferred in the former case.

4. Vibration–Rotation Transfer

Two companion papers by Millikan and by Millikan and Switkes [97] describe the latest experimental results on the vibrational relaxation of CO by hydrogen. It is found that for temperatures above 600°K there is no observable difference between the deactivation efficiencies of n-H_2 and p-H_2. Below 600°K, p-H_2 is clearly more efficient, and at 300°K the ratio is approximately 2. For temperatures between 600°K and 2700°K, the relaxation time for CO–H_2 collisions is given by

$$\ln \tau_{\text{CO-H}_2} = 89.7768 T^{-1/3} - 21.8845 \tag{107}$$

for a pressure of 1 atm and with τ in seconds. Between 300°K and 100°K the results do not obey a Landau–Teller relationship. For CO–n-H_2 collisions, the result is

$$\tau = -6.81679 \times 10^{-5} + \frac{0.0584037}{T} - \frac{2.1292}{T^2}. \tag{108}$$

For CO–p-H_2,

$$\tau = 6.59328 \times 10^{-5} - \frac{0.00884759}{T} + \frac{2.3250}{T^2}. \tag{109}$$

These relaxation times, for both forms of H_2 at low temperatures, are much shorter than the Landau–Teller extrapolation from higher temperatures would predict.

Breshears and Bird [195] have reported relaxation data for Cl_2 mixed with HCl and DCl over the range 400–1100°K. $Cl_2^\ddagger$–HCl collisions were found to be ~100 times more effective than Cl_2–Cl_2 collisions at 400°K and ~2–4 times more effective than Cl_2–DCl. Since the reduced masses for Cl_2 collisions with HCl and DCl are nearly equal, and since the vibrational frequencies for these species are considerably greater than that of Cl_2, these investigators concluded that V–R exchange was highly probable in $Cl_2^\ddagger$–HCl collisions. Furthermore, since Cl_2–CO mixtures gave vibrational relaxation times indicative of longer than for Cl_2–DCl collisions, it was concluded that V–R exchange in the latter case is also highly probable.

Earlier evidence for the importance of V–R transfer in pure HCl and DCl had been reported by these and other investigators [222–225]. A theoretical treatment by Shin [226] of the energy-transfer problem in these gases, as well as in HBr and HI, shows clearly that long-range attractive forces play a significant role in the V–R exchange process.

Chen and Moore's investigation [220] of room-temperature relaxation in HCl and in DCl shows that the high-temperature shock-tube results [222, 225] do not extrapolate smoothly according to a $T^{-1/3}$ law. Figure 3.29, taken from their paper, shows that while log P versus $T^{-1/3}$ appears to be fairly linear above 1000°K, the room-temperature deactivation probability is ~15 times larger than the value expected by linear extrapolation of the shock-tube data. The upper dashed curve represents a probable interpolation between the shock-tube data and the fluorescence data; the lower dashed line is an interpolation based on SSH theory for V–T transfer. The datum represented by the square is for vibrational deactivation of DCl by HCl.

An additional result of Chen and Moore's study is that a considerable portion, perhaps nearly all, of the vibration–rotation transfer in HCl is intramolecular: The vibrator leaves the collision with the major portion of

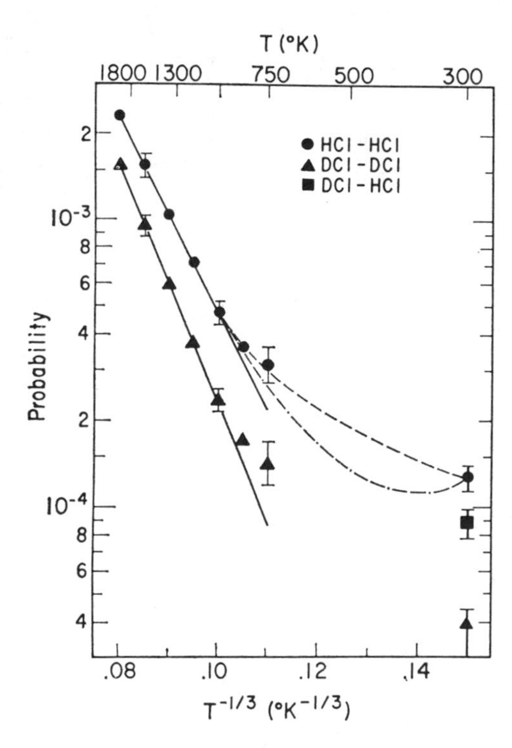

Figure 3.29 Probability of vibrational deactivation for HCl and DCl collisions versus $T^{-1/3}$. Higher-temperature data are shock-tube results of Breshears and Bird [222]. Room-temperature measurements are those by Chen and Moore, from whose paper this figure is taken [220].

the rotational energy in most of the deactivating collisions. It was concluded that the hydrogen bonding interaction between the two HCl molecules is significant; vibrational relaxation by self-collisions is very much more efficient than HCl–H_2 or HCl–rare-gas collisions. Moreover, it was found that the rotational motion of H_2 is not important in this case, since no difference between the efficiencies of p-H_2 and o-H_2 could be detected.

Vibrational relaxation studies of HF have been reported by Bott and Cohen [227], using shock waves in the temperature interval 1350–4000°K. For self-collisions, they conclude that $P\tau = 0.0102 \exp (34.39/T)^{1/3}$ μsec atm. Compared with the value $P\tau = 0.015$ μsec atm determined near room temperature by Airey and Fried [228], it is evident that the relaxation time must pass through a maximum and then decrease. A complete theoretical treatment has not yet been attempted, but it is clear that attractive forces will play a significant role, particularly at the lower temperatures.

A study of collisional relaxation of the electronically excited OH ($A\ ^2\Sigma^+$) radical has been reported by Welge, Filseth, and Davenport [229]. The excited species were formed by monochromatic photodissociation of H_2O and were created in vibrationally and rotationally excited levels v', K'. By monitoring the OH $A \rightarrow X$ emission spectrum from the UV-irradiated H_2O and mixtures of H_2O with N_2 and argon, these investigators were able to study the population of individual levels in $v' = 0$ and 1 and to observe shifts in the population due to collisions with foreign gases. The process

$$\mathrm{OH}(A\ ^2\Sigma^+, v' = 0, K' = 20) + \mathrm{Ar, N_2} \rightarrow \mathrm{OH}(A\ ^2\Sigma^+, v' = 1, K' = 15) + \mathrm{Ar, N_2} + \Delta E, \tag{110}$$

where $\Delta E = 27\ \mathrm{cm}^{-1}$ $(0.33 \times 10^{-2}\ \mathrm{eV})$, was observed to possess a rate constant of 10^{-11} cm^3/molecule/sec.

B. Polyatomic Gases and Mixtures

1. Vibrational Energy Transfer

In polyatomic molecules, vibrational relaxation usually occurs via slow T–V energy exchange into a particular mode, usually the lowest energy mode ν_{min}. V–V exchange among the remaining modes subsequently occurs with such rapidity that only one relaxation time is observed [equation (57)]. Ultrasonics measurements do not normally provide information on the V–V processes, nor do most polyatomics exhibit multiple dispersion. Of those which do, namely, SO_2, CF_3CN, C_3O_2, C_2N_2, CH_2Br_2, CH_2Cl_2, and C_2H_6, the ratio of the next lowest vibrational frequency to ν_{min} ranges from 2.2, for SO_2, to 8.7, for C_3O_2. Ethane, which has been studied by a number of investigators [230], exhibits double relaxation. Whether the two processes are both V–T or whether the slower is V–V, has not been clarified. Sulfur dioxide is the only triatomic species that has definitely been shown to exhibit two vibrational relaxation times, although there is some evidence [231] that D_2O may be doubly dispersive. CH_2Cl_2 and CH_2Br_2 are the only halogenated methanes exhibiting multiple relaxation, and both appear to exhibit double relaxation. The theoretical procedure of Tanczos [78], which provides for the replacement of the atomic masses in equation (30) by the effective masses of the normal modes, has rarely been followed in detail for molecules more complex than triatomics. His methods for calculating theoretical acoustic absorption or dispersion curves, given a set of V–T and V–V rate constants appropriate to the gas or gas mixture, have been applied to cases exhibiting multiple relaxation [232].

Earlier experimental results on a large number of polyatomic gases have been collected by Herzfeld and Litovitz [8], by Cottrell and McCoubrey [9], and more recently by Gordon, Klemperer, and Steinfeld [40]. The selected data presented in this section are confined largely to those published in the past few years. Of these, many result from the application of lasers to the study of molecular energy transfer.

a. CO_2 *and Its Mixtures*

Many of the data on carbon dioxide have been summarized by Taylor and Bitterman [88]. The most recent work of Simpson, Bridgman, and Chandler [233], using shock-wave interferometry, extends from 330 to 1600°K. The resulting relaxation times, probably among the most accurate reported for pure CO_2, are shown as a solid line in Figure 3.30. Measurements throughout this and higher ranges give evidence that neither the symmetric stretching mode nor the asymmetric stretching mode has a significantly longer relaxation time than the bending mode. The room-temperature relaxation time, from these studies, is seen to be about 7 μsec at 1 atm. The temperature dependence is less than that predicted by several theoretical treatments: Equation (28), with potential-matching methods A and B, is depicted in the figure as HLA and HLB; a modified equation by Herzfeld is labeled H; a calculation by Marriott [235] for the bending mode, M; and one by Witteman [236] is labeled W. Simpson and Chandler [237] have performed a shock-tube study of CO_2 mixtures with He, Ne, Ar, N_2, H_2, and D_2 in the range 360–1500°K. Except for Ne, there is excellent agreement with older acoustical measurements, and also with the more recent results for $CO_2 + N_2$, O_2 at room temperature by Merrill and Amme [238]. For the cases of $CO_2 + H_2$, D_2 the relaxation times are found to increase with increasing temperatures, in contrast to the other cases. This behavior results from V–R processes [equations (72)–(74)]; a subsequent investigation by Simpson and Simmie [239] over the range 344–742°K, employing mixtures with both n-H_2 and

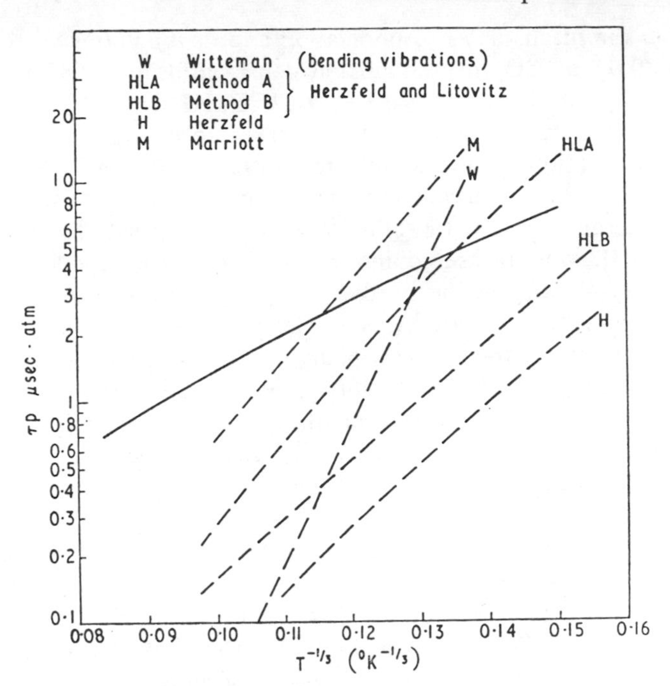

Figure 3.30 Temperature dependence of vibrational relaxation in CO_2 from Simpson et al. [233] compared with several theoretical treatments (see text).

p-H_2, agrees well with ultrasonic data at room temperature reported earlier [93]. It is concluded that both V–T and V–R processes are important in mixtures of CO_2 and H_2, over the entire temperature range.

Mixtures of CO_2 and water vapor have long been of interest because of the marked enhancement of the ν_2 relaxation by trace amounts of H_2O. Acoustic-absorption measurements by Lewis and Lee [240] and by Shields and Burks [119] on CO_2–D_2O mixtures showed that the explanation lies in a V–R process rather than a chemical affinity: D_2O is $\frac{1}{3}$ to $\frac{1}{5}$ as efficient, over the range 300–500°K, as H_2O. Sharma [241] has performed a calculation based on his earlier ideas involving long-range forces and rapid near-resonant V–R transfer. Qualitative agreement is obtained with Lewis and Lee's results; at room temperature, this corresponds to a probability per collision of about 0.007. Because of the Fermi resonance between the (100) and (020) levels of CO_2, the rapid deactivation of the bending mode by water vapor greatly enhances the vibrational relaxation of the lower laser level, as observed under laser conditions by Bulthuis and Ponsen [242].

There is additional evidence [91] that methane deactivates CO_2(010), also by a V–R process.

Interest in the mechanisms of population and depopulation of the upper laser level (001) of CO_2 has stimulated considerable experimental effort. Deactivation by collisions with N_2, He, O_2, H_2O, HDO, D_2O, NO, Cl_2, H_2, and by self-collisions has been the subject of several papers [243–249], as well as deactivation by several polyatomics [250, 251]. In Section IV, the theoretical treatment by Sharma and Brau of the nearly resonant transfer of vibrational energy from $CO_2(00^01)$ to $N_2(v = 1)$ was discussed. Taylor and Bitterman [243], who measured infrared emission from mode ν_3 behind shocks, Moore, Wood, Hu, and Yardley [244], using laser-excited vibrational fluorescence, and Rosser, Wood, and Gerry [245], also with the latter technique, have studied this process at length. The V–V exchange probability per collision decreases from a value of $\sim 2 \times 10^{-3}$ at room temperature to $\sim 5 \times 10^{-4}$, in accord with the Sharma–Brau theory. Thereafter, the probability increases rapidly to $\sim 10^{-2}$ at 2800°K. Clearly, the long-range forces account for the low-temperature behavior, while short-range forces are influential at high temperatures.

Studies of the influence of polyatomic species in deactivating $CO_2(00^01)$ by near-resonant V–V exchange have shown that both intramolecular and intermolecular processes may occur. For mixtures with deuterated hydrocarbons, Stephenson, Wood, and Moore [251] found efficiencies to be 2 to 90 times greater than for the normal hydrogen isotopes.

b. N_2O *and Its Mixtures*

Earlier investigations on N_2O have been summarized by Herzfeld and Litovitz [8] and by Calvert [252]. Simpson, Bridgman, and Chandler [253] have more recently examined vibrational relaxation in pure N_2O over the range 320–820°K using shock-tube interferometry. As in the case of CO_2, single relaxation is observed. Experimental relaxation times are depicted in Figure 3.31, and are compared with SSH theory.

Studies by Calvert [252] and by Nethery [104] of N_2O–n-H_2 and N_2O–p-H_2 at room temperature indicated the existence of a V–R process, but the distinction between p-H_2 and n-H_2 was not discernible, in spite of the close energy matching with o-H_2:

$$N_2O(01^10) + H_2(J = 1) \rightarrow N_2O(00^00) + H_2(J = 3) + 2\ \text{cm}^{-1}\ (0.02_5 \times 10^{-2}\ \text{eV}). \quad (111)$$

Calvert concluded that the ν_2 deactivation probability for n-H_2 was approximately that expected from V–T theory. From Wight's measurements [254], Calvert concluded that at room temperature $Z_{10} = 44$ for N_2O–H_2O collisions and 77 for N_2O–D_2O collisions. For self-collisions, $Z_{10} = 4500$ ($P\tau_v =$ 0.9 μsec atm). The effects of water vapor on N_2O are thus comparable to the effects seen in CO_2.

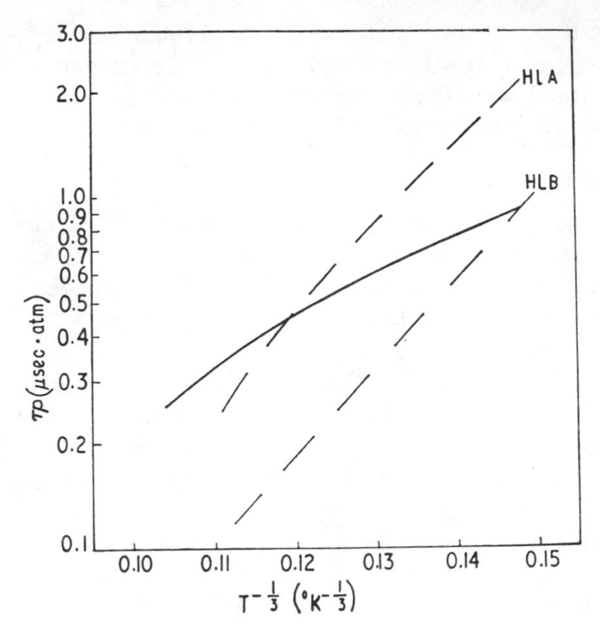

Figure 3.31 Vibrational relaxation in N_2O according to Simpson et al. [252]. Dashed lines represent SSH theory, according to methods A and B [8].

Bates, Flynn, and Ronn [255], using an N_2O–N_2–He laser to study (00^01) fluorescence quenching in N_2O, concluded that the lifetime is 1.96 μsec atm, the deactivation probably occurring through V–V transfer to nearby bending modes. Yardley [256] has observed $N_2O(00^01)$ fluorescence quenching by collisions with N_2O, the noble gases, H_2, D_2, and N_2. For self-collisions he obtained 1.74 ± 0.03 μsec atm. This value compares well with spectrophone results [135, 256]. For collisions with the noble gases, Yardley concludes that deactivation is probably through the (11^10) level, as is expected to be the case for CO_2 [249]. For collisions of $N_2O(00^00)$ with $N_2(v = 1)$, he concludes that the probability of V–V transfer to the (00^01) level is 8.7×10^{-4} per N_2–N_2O collision. Again, long-range forces appear to be quite important.

c. *Sulfur Dioxide*

Recent ultrasonic measurements on SO_2 by Bass, Winter, and Evans [258] confirms the earlier observations of multiple relaxation [259, 260] and extends the temperature range to 1090°K. As T increases, the relaxation time for the bending mode increases with increasing temperature, while that for the stretching modes decreases in the normal manner. Shields and Anderson [261], using an improved absorption apparatus, have studied pure SO_2 and its mixtures with argon. Under careful inspection, the results are shown to favor a process in which the stretching modes exchange energy with the bending mode, rather than with translation as an independent V–T process. Collisions with argon were shown to produce intramolecular V–V transfer, but with about $\frac{1}{20}$ the probability arising in self-collisions.

d. Hydrogen Sulfide

Studies by Winter and Bass [262] over the range 300–683°K show that the deactivation probability per collision has a value of about 0.003 at 300°K and at 600°K, with a minimum value of 0.002 occurring in the vicinity of 500°K. This behavior is attributed to the importance of V–R exchange, particularly at the lower temperatures.

Mixtures of H_2S and of D_2S in CO_2 have been studied by Shields and associates [119, 263], along with pure D_2S. The relaxation of D_2S was found to occur at a slower rate than in H_2S, which substantiates the V–R mechanism. As the temperature was raised, the relaxation time increased, as for H_2S. For the mixtures with CO_2, V–V transfer between the bending modes of CO_2 and D_2S was observed; as well as between the strongly coupled (03^10) and (11^10) modes of CO_2 1933 and 2077 cm^{-1} (23.97×10^{-2} and 25.75×10^{-2} eV), respectively), and the (100) and (001) levels of D_2S 1892 and 1900 cm^{-1} (23.46×10^{-2} and 23.56×10^{-2} eV).

e. Water Vapor

Many of the available data on vibrational relaxation in water vapor have been summarized by Bass, Olson, and Amme [264]. These authors, who have measured ultrasonic absorption and dispersion at temperatures ranging from 373°K to 946°K, interpret their observations in terms of a V-R transfer mechanism. Values of $p\tau_{VR}$ were found to range from 1×10^{-8} at 373°K to 1.7×10^{-9} atm sec at the higher temperature limit. These data, when compared with older absorption measurements [265], suggest that a minimum exists in the rate constant in the neighborhood of 370°K. At the higher temperatures, the vibrational and rotational degrees of freedom appear to relax together. Shin's calculation [266] indicates that V-R exchange should dominate at temperatures below about 450°K. Rotational collision numbers were found to range from about ten to twenty, increasing with temperature.

f. Ammonia

Observations in 1963 by Cottrell and Matheson [89], although unsuccessful in measuring vibrational relaxation in NH_3, showed that ND_3 is much less efficient in disposing of vibrational energy through self-collisions. These results, along with experiments on PH_3 and PD_3, were among the first to suggest that V–R exchange can be very important. Subsequent observations on NH_3 by Jones, Lambert, Saksena, and Stretton [267] showed that at

298°K $Z_v \approx Z_r = 4.9$ collisions, or a factor of 10 more efficient than ND_3 self-collisions.

g. Methane

The shock-tube data of Richards and Sigafoos [268] for pure methane, which extend over the temperature range 740–1600°K, follow the relationship

$$P\tau_v = \exp\left[\frac{40}{T^{1/3}} - 5.4\right]. \tag{112}$$

Extrapolation to room temperature yields 1.9 μsec atm, a value in good agreement with ultrasonic data by Cottrell and Matheson [269] (2.0 ± 0.2 μsec atm), and in fair agreement with Yardley and Moore [144] (1.90 ± 0.1 μsec atm), using vibrational fluorescence. As described in Section VI.D, these latter authors also observe a rapid V–V process, with a relaxation time of ~6 μsec atm. The 1962 measurements by Cottrell and Matheson were particularly significant because of the observation that $\tau(CD_4)/\tau(CH_4) \approx 2$, despite the lower value of ν_{min} for CD_4. These authors concluded that a V–R mechanism was operative in CH_4, as well as in C_2H_4 and in SiH_4 (silane), which was also investigated, as was SiD_4.

Bauer and Schotter [270] have investigated mixtures of CH_4 and CO_2 using ultrasonic absorption. V–V transfer was observed for collisions in which ν_4 of CH_4 was initially excited [1306 cm^{-1} (16.19 × 10^{-2} eV)], the energy being transferred to CO_2 levels (10^00), (02^00), and (02^20), with a probability of about 4 × 10^{-4} per collision.

Methane is known to be a very efficient partner for the deactivation of oxygen [271, 272]; the bending mode ν_2 of CH_4 (see Figure 3.17), with frequency 1526 cm^{-1} (18.92 × 10^{-2} eV), closely matches the vibrational frequency of O_2 [1556 cm^{-1} (19.29 × 10^{-2} eV)]. Since water vapor [$\nu_2 =$ 1595 cm^{-1} (19.77 × 10^{-2} eV)] can deactivate O_2 vibrations very efficiently, it is thus not surprising that vibrational relaxation in CH_4 is also greatly affected by H_2O. Monkewicz [273] observed that at 310°K a 2% H_2O mixture shifted the relaxation time by a factor of about 2.

Methane–oxygen mixtures have also been studied by laser-excited fluorescence, as have NO–CH_4 mixtures. Yardley and Moore [274] find for the V–V exchange probability from O_2 to CH_4 a value of 2.0 × 10^{-3} per collision at 297°K.

Millikan [275], employing vibrational fluorescence methods, found a V–V exchange probability of 3.0 × 10^{-5} for $CO^\dagger + CH_4 \rightarrow CO + CH_4^\dagger$ at 303°K, increasing to 4.2 × 10^{-5} at 363°K.

Yardley, Fertig, and Moore [276] have examined a series of mixtures involving methane, including noble gases. The latter observations show that the behavior is intermediate between V–T and V–R. For Ar, Kr, and Xe mixtures the behavior is closer to V–R.

h. Other Polyatomic Molecules

Vibrational relaxation measurements have been reported for a large number of other polyatomic species in recent years. Lambert [277] has reviewed the results on 75 of these gases, which include many hydrocarbons and some of their halogenated derivatives, plus a number of other species which contain no hydrogen, such as SF_6, BF_3, $GeCl_4$, $SnCl_4$, and others. Systematic variations of the collision number Z_{10} with $\nu_{\min}$ are noted and correlations are observed with the number of H atoms in each molecule. The reader is referred to that paper for an extensive list of references.

2. Rotational Energy Transfer

Among the first polyatomic molecules in which rotational relaxation was investigated is methane. Kelly [278], employing an ultrasonic interferometer, concluded that for this gas as 314°K, $Z_r \equiv \mathscr{R}\tau_r \approx 15$ collisions. More recent ultrasonic data obtained by Hill and Winter [279] provide apparent rotational relaxation times for CH_4 and C_2H_4 as a function of temperature. Their results are given in Table 3.5. The increase in Z_r with increasing temperature

Table 3.5
Effective Rotational Collision Numbers from Measurements by Hill and Winter [279].

GAS	T (°K)	Z_r
CH_4	298	12
	573	22
	773	40
	1073	30
C_2H_4	298	9
	573	14
	773	17

resembles the results from acoustic measurements on diatomic species [125, 126], and differs from the behavior of Z_r for polar diatomic molecules as derived from thermal-conductivity data (Table 3.2). Zeleznik and Svehla [280], who attempted to extract rotational collision numbers from thermal conductivities of several polyatomic molecules including SO_2, H_2O, H_2S, NH_3, and HCN, concluded that the results were highly suspect. Values for

SO_2, as an example, ranged from <0 at 450°K to a maximum value of 2.3 at 950°K. The acoustic-absorption measurements by Bass et al. [258] on SO_2 gave rotational collision numbers ranging from 3.72 at 290°K to 5.89 at 1090°K. The ratio $Z(1090°K)/Z(290°K)$ is thus 1.6, and is smaller than predicted by equation (51), which gives a ratio of 4.18.

Winter and Bass [262], in their acoustical studies on H_2S, have concluded that Z_r varies from 8.0 at 298°K to 17 at 683°K.

Malinauskas et al. [205] derived a rotational collision number for CO_2, from thermal transpiration measurements, of 1.9 at a nominal temperature of 504°K.

Oka [281] has described a series of experiments in which the technique of high-power microwave double resonance has been applied to the study of collision-induced transitions between rotational levels in a variety of molecular species, including ethylene oxide, HCN, NH_3, and mixtures of NH_3 with rare gases. This method, described earlier by Cox, Flynn, and Wilson [282], permits the investigation of the selection rules operative for collisions involving infrared-active molecules. Unland and Flygare [283] have studied rotational relaxation in OCS and its mixtures with helium, argon, and O_2 with this technique. Helium is found to have 66% of the efficiency of argon in relaxing OCS, while O_2 has 76%. These findings are in agreement with the results by Toennies [284] on TlF colliding with these same species.

REFERENCES

1. S. H. Bauer and E. Ossa, *J. Chem. Phys.*, **45,** 434 (1966).
2. A. Burcat and A. Lifshitz, *J. Chem. Phys.*, **47,** 3079 (1967).
3. A. Bar-Nun and A. Lifshitz, *J. Chem. Phys.*, **47,** 2878 (1967).
4. The radiative relaxation time τ_r of a vibrational mode ν is inversely proportional to the band strength S_n^ν. (See, for example, R. M. Goody, *Atmospheric Radiation: I. Theoretical Basis*, Oxford Univ. Press, Oxford, 1964.) Thus one finds, for example, that the radiative lifetime of the bending mode, ν_2, of H_2O is about 0.07 sec.
5. G. W. Pierce, *Proc. Am. Acad.*, **60,** 271 (1925).
6. E. F. Smiley and E. H. Winkler, *J. Chem. Phys.*, **22,** 2018 (1954).
7. J. P. Appleton, *Bull. Am. Phys. Soc.*, **13,** 784 (1968).
8. K. Herzfeld and T. Litovitz, *Absorption and Dispersion of Ultrasonic Waves*, Academic, New York, 1959.
9. T. L. Cottrell and J. C. McCoubrey, *Molecular Energy Transfer in Gases*, Butterworths, London, 1961.
10. K. Takayanagi, *Prog. Theor. Phys. Suppl.* (Japan), **25,** 1 (1963).
11. K. Takayanagi, *Advances in Atomic and Molecular Physics*, Vol. 1, Academic, New York, 1965, p. 149.

12. K. Herzfeld, in *Dispersion and Absorption of Sound by Molecular Processes*, edited by D. Sette, Academic, New York, 1963, p. 272 ff.
13. D. Rapp and T. Kassal, *Chem. Rev.*, **69,** 61 (1969).
14. C. Zener, *Phys. Rev.*, **37,** 556 (1931).
15. C. Zener, *Proc. Cambridge Phil. Soc.*, **29,** 136 (1933).
16. L. Landau and E. Teller, *Phys. Z. Sowjetunion*, **10,** 34 (1936).
17. See ref. 8, pp. 262–266.
18. See refs. 10 and 13 for a summary of recent contributions to the classical theory.
19. J. D. Kelley and M. Wolfsberg, *J. Chem. Phys.*, **44,** 324 (1966).
20. D. Rapp, *J. Chem. Phys.*, **32,** 735 (1960); **40,** 3812 (1964).
21. J. D. Kelley and M. Wolfsberg, *J. Phys. Chem.*, **71,** 2373 (1967).
22. J. D. Kelley and M. Wolfsberg, *J. Chem. Phys.*, **50,** 1894 (1969).
23. J. B. Calvert and R. C. Amme, *J. Chem. Phys.*, **45,** 4710 (1966).
24. H. K. Shin, *J. Chem. Phys.*, **49,** 3964 (1968).
25. H. K. Shin, *J. Chem. Phys.*, **47,** 3302 (1967).
26. H. K. Shin, *J. Chem. Phys.*, **42,** 59 (1965); **46,** 744 (1967); **48,** 3644 (1968). See also H. K. Shin, *J. Am. Chem. Soc.*, **90,** 3025 (1968).
27. C. F. Hansen and W. E. Pearson, *J. Chem. Phys.*, **53,** 3557 (1970).
28. J. M. Jackson and N. F. Mott, *Proc. Roy. Soc.* (London), **A137,** 703 (1932).
29. R. N. Schwartz, Z. I. Slawsky, and K. F. Herzfeld, *J. Chem. Phys.*, **20,** 1591 (1952). See also ref. 8.
30. R. N. Schwartz and K. F. Herzfeld, *J. Chem. Phys.*, **22,** 767 (1954).
31. K. Takayanagi and S. Kaneko, *Sci. Rept. Saitama Univ.* Series **A1,** 111 (1954).
32. D. Secrest and B. R. Johnson, *J. Chem. Phys.*, **45,** 4556 (1966).
33. E. Thiele and J. Weare, *J. Chem. Phys.*, **48,** 2324 (1968). See also **48,** 513 (1968) for a discussion by these authors of the reaction matrix method.
34. K. Takayanagi, *Sci. Rept. Saitama Univ.*, **IV,** 51 (1962).
35. S. L. Thompson, *J. Chem. Phys.*, **49,** 3400 (1968).
36. A. F. Devonshire, *Proc. Roy. Soc.* (London), **A158,** 269 (1937).
37. R. T. Allen and P. Feuer, *J. Chem. Phys.*, **40,** 2810 (1964).
38. H. J. Bauer, "Theory of Relaxation Phenomena in Gases," in *Physical Acoustics*, Vol. II, edited by W. P. Mason, Academic, New York, 1965.
39. H. O. Kneser, "Relaxation Processes in Gases," in *Physical Acoustics*, Vol. II, edited by W. P. Mason, Academic, New York, 1965.
40. R. G. Gordon, W. Klemperer, and J. I. Steinfeld, *Ann. Rev. Phys. Chem.*, **19,** 215 (1968).
41. B. Stevens, *Collisional Activation in Gases*, Vol. 3 of the *International Encyclopedia of Physical Chemistry and Chemical Physics*, edited by A. F. Trotman-Dickenson, Pergamon, New York, 1967.
42. H. P. Broida and T. Carrington, *J. Chem. Phys.*, **38,** 136 (1963).
43. T. Carrington, *J. Chem. Phys.*, **35,** 807 (1961).
44. J. G. Parker, *Phys. Fluids*, **2,** 449 (1959).
45. G. N. Watson, *Theory of Bessel Functions*, Cambridge Univ. Press, New York, 1952, 2nd ed., p. 181, Eq. 4.
46. L. M. Raff, *J. Chem. Phys.*, **46,** 520 (1967); **47,** 1884 (1967).

47. C. A. Brau and R. M. Jonkman, *J. Chem. Phys.*, **52,** 477 (1970).
48. S. W. Benson and G. C. Berend, *J. Chem. Phys.*, **44,** 470 (1966); 4247 (1966).
49. G. C. Berend and S. W. Benson, *J. Chem. Phys.*, **47,** 4199 (1967); **50,** 5047 (1969).
50. M. Krauss and F. H. Mies, *J. Chem. Phys.*, **42,** 2703 (1965).
51. K. Takayanagi, *Progr. Theor. Phys.*, **11,** 557 (1954).
52. K. P. Lawley and J. Ross, *J. Chem. Phys.*, **43,** 2930 (1965); **43,** 2943 (1965).
53. L. M. Raff, *J. Chem. Phys.*, **47,** 4789 (1967).
54. F. J. Zeleznik, *J. Chem. Phys.*, **47,** 3410 (1967).
55. R. Brout, *J. Chem. Phys.*, **22,** 934 (1954); **22,** 1189 (1954).
56. K. Takayanagi, *Prog. Theor. Phys.*, **8,** 497 (1952). See also ref. 51 and others cited in refs. 8 and in 11.
57. K. Takayanagi, *Sci. Rept. Saitama Univ.* Ser. **AIII,** 87 (1959).
58. A. M. Arthurs and A. Dalgarno, *Proc. Roy. Soc.* (London), **A256,** 540 (1960).
59. W. D. Davison, *Discussions Faraday Soc.*, **33,** 71 (1962).
60. C. F. Curtiss and R. B. Bernstein, *J. Chem. Phys.*, **50,** 1168 (1969).
61. W. A. Lester, Jr. and R. B. Bernstein, *J. Chem. Phys.*, **48,** 4896 (1968).
62. A. C. Allison and A. Dalgarno, *Proc. Phys. Soc.* (London), **90,** 609 (1967).
63. R. D. Levine, *Quantum Mechanics of Molecular Rate Processes*, Oxford Univ. Press, Oxford, 1969.
64. D. A. Micha, *Phys. Rev.*, **162,** 88 (1967).
65. R. D. Levine, B. R. Johnson, J. J. Muckerman, and R. B. Bernstein, *Chem. Phys. Lett.*, **1,** 517 (1968); *J. Chem. Phys.*, **49,** 56 (1968).
66. R. D. Levine, M. Shapiro, J. T. Muckerman, and B. R. Johnson, *Chem. Phys. Lett.*, **2,** 545 (1968).
67. J. T. Muckerman, *J. Chem. Phys.*, **50,** 627 (1969).
68. B. R. Johnson, M. Shapiro, and R. D. Levine, *Chem. Phys. Lett.*, **3,** 131 (1969).
69. J. E. Rhodes, *Phys. Rev.*, **70,** 932 (1946).
70. C. J. Sluijter, H. F. P. Knaap, and J. J. M. Beenakker, *Physica*, **30,** 745 (1964); **31,** 915 (1965).
71. R. M. Jonkman, G. J. Prangsma, I. Ertas, H. F. P. Knaap, and J. J. M. Beenakker, *Physica*, **38,** 441 (1968).
72. R. M. Jonkman, G. J. Prangsma, R. A. J. Keijser, R. A. Aziz, and J. J. M. Beenakker, *Physica*, **38,** 451 (1968).
73. R. M. Jonkman, G. J. Prangsma, R. A. J. Keijser, H. P. F. Knaap, and J. J. M. Beenakker, *Physica*, **38,** 456 (1968).
74. C. S. Roberts, *Phys. Rev.*, **131,** 203, 209 (1963).
75. L. M. Valley and R. C. Amme, *J. Chem. Phys.*, **50,** 3190 (1969).
76. J. D. Lambert, D. G. Parks-Smith, and J. L. Stretton, *Proc. Roy. Soc.* (London), **A282,** 380 (1964).
77. J. D. Lambert, *Quart. Rev.*, **21,** 67 (1967).
78. F. I. Tanczos, *J. Chem. Phys.*, **25,** 439 (1956).
79. D. Rapp and P. Englander-Golden, *J. Chem. Phys.*, **40,** 573 (1964).
80. B. Mahan, *J. Chem. Phys.*, **46,** 98 (1967).
81. G. C. Berend and S. W. Benson, *J. Chem. Phys.*, **51,** 1480 (1969).
82. R. Dubrow and D. J. Wilson, *J. Chem. Phys.*, **50,** 1553 (1969).

83. D. Rapp, *J. Chem. Phys.*, **43,** 316 (1965).
84. A. B. Callear, *Appl. Opt. Suppl.*, **2,** 145 (1965).
85. E. R. Fisher and R. H. Kummler, *J. Chem. Phys.*, **49,** 1075 (1968).
86. R. D. Sharma and C. A. Brau, *Phys. Rev. Lett.*, **19,** 1273 (1967).
87. R. D. Sharma and C. A. Brau, *J. Chem. Phys.*, **50,** 924 (1969).
88. R. L. Taylor and S. Bitterman, *Rev. Mod. Phys.*, **41,** 26 (1969).
89. T. L. Cottrell and A. J. Matheson, *Trans. Faraday Soc.*, **58,** 2336 (1962); **59,** 824 (1963).
90. T. L. Cottrell, R. C. Dobbie, J. McLain, and A. W. Read, *Trans. Faraday Soc.*, **60,** 241 (1964).
91. C. B. Moore, *J. Chem. Phys.*, **43,** 2979 (1965).
92. R. D. Sharma, *J. Chem. Phys.*, **50,** 919 (1969).
93. S. W. Behnen, H. L. Rothwell, and R. C. Amme, *Chem. Phys. Lett.*, **8,** 318 (1971).
94. T. L. Cottrell and M. A. Day, in *Molecular Relaxation Processes*, Chemical Society Special Publication No. 20, Academic, New York, 1966, p. 253.
95. R. D. Sharma and C. W. Kern, *J. Chem. Phys.*, **55,** 1171 (1971).
96. R. C. Millikan and L. A. Osburg, *J. Chem. Phys.*, **41,** 2196 (1964).
97. R. C. Millikan, *J. Chem. Phys.* (to be published); R. C. Millikan and E. Switkes, *J. Chem. Phys.* (to be published).
98. J. D. Kelley, *J. Chem. Phys.*, **53,** 3864 (1970).
99. A. B. Bhatia, *Ultrasonic Absorption*, Oxford Univ. Press, London, 1967.
100. *Dispersion and Absorption of Sound by Molecular Processes*, Proceedings of the International School of Physics, edited by D. Sette, Academic, New York, 1963.
101. J. D. Lambert, "Relaxation in Gases," in *Atomic and Molecular Processes*, edited by D. R. Bates, Academic, New York, 1962.
102. H. O. Kneser, *Ann. Phys.*, **11,** 761 (1931).
103. R. A. Walker, T. D. Rossing, and S. Legvold, N.A.C.A. Tech. Note 3210, Washington, D.C., 1954 (unpublished). See also the description in ref. 2, p. 36.
104. S. J. Nethery, unpublished Ph.D. thesis, University of Denver, 1968.
105. M. Greenspan, *J. Acoust. Soc. Am.*, **31,** 155 (1959).
106. R. C. Amme and B. E. Warren, *J. Acoust. Soc. Am.*, **41,** 419 (1968).
107. J. V. Martinez, J. G. Strauch, Jr., and J. C. Decius, *J. Chem. Phys.*, **40,** 186 (1964).
108. J. G. Strauch, Jr. and J. C. Decius, *J. Chem. Phys.*, **44,** 3319 (1966).
109. J. K. Hancock and J. C. Decius, *J. Chem. Phys.*, **51,** 5374 (1969).
110. E. Meyer and G. M. Sessler, *Z. Phys.*, **149,** 15 (1957).
111. See, for example, H.-J. Bauer and H. Roesler, in *Molecular Relaxation Processes*, Academic, New York, 1966, and H. Roesler, *Acustica*, **17,** 73 (1966).
112. H. Sell, *Z. Tech. Physik*, **18,** 3 (1937).
113. F. A. Angona, *J. Acoust. Soc. Am.*, **25,** 1111 (1953).
114. J. W. L. Lewis and F. D. Shields, *J. Acoust. Soc. Am.*, **36,** 1593 (1964).
115. F. D. Shields and K. P. Lee, *J. Chem. Phys.*, **40,** 737 (1964).
116. F. D. Shields, *J. Acoust. Soc. Am.*, **31,** 248 (1959).

117. F. D. Shields, *J. Acoust. Soc. Am.*, **29,** 450 (1957).
118. J. W. L. Lewis and F. D. Shields, *J. Acoust. Soc. Am.*, **41,** 100 (1967).
119. F. D. Shields and J. A. Burks, *J. Acoust. Soc. Am.*, **43,** 510 (1968).
120. F. D. Shields, *J. Acoust. Soc. Am.*, **45,** 481 (1969).
121. F. D. Shields and R. T. Lagemann, *J. Acoust. Soc. Am.*, **29,** 470 (1957).
122. J. G. Parker, *J. Chem. Phys.*, **34,** 1763 (1961).
123. H. Knötzel and L. Knötzel, *Ann. Phys.*, **2,** 393 (1948).
124. M. C. Henderson and G. J. Donnelly, *J. Acoust. Soc. Am.*, **34,** 779 (1962). See also Henderson and Peselnck, **29,** 769 (1957) for a description of the apparatus.
125. E. H. Carnevale, C. Carey, and G. Larson, *J. Chem. Phys.*, **47,** 2829 (1967).
126. T. G. Winter and G. L. Hill, *J. Acoust. Soc. Am.*, **42,** 848 (1967).
127. A. G. Gaydon and I. R. Hurle, *The Shock Tube in High Temperature Chemical Physics*, Chapman and Hall, London, and Reinhold, New York, 1963, p. 162.
128. R. C. Millikan and D. R. White, *J. Chem. Phys.*, **39,** 98 (1963).
129. J. H. Kiefer and R. W. Lutz, *J. Chem. Phys.*, **44,** 658, 668 (1966).
130. V. Blackman, *J. Fluid Mech.*, **1,** 61 (1956).
131. J. P. Appleton, *J. Chem. Phys.*, **47,** 3231 (1967).
132. P. W. Huber and A. Kantrowitz, *J. Chem. Phys.*, **15,** 275 (1947).
133. I. R. Hurle, A. L. Russo, and J. G. Hall, *J. Chem. Phys.*, **40,** 2076 (1964).
134. C. W. von Rosenberg, R. L. Taylor, and J. D. Teare, *J. Chem. Phys.*, **48,** 5731 (1968).
135. T. L. Cottrell, I. M. Macfarlane, and A. W. Read, *Trans. Faraday Soc.*, **63,** 2093 (1967).
136. R. C. Millikan, *J. Chem. Phys.*, **38,** 2855 (1963).
137. F. DeMartini and J. Ducuing, *Phys. Rev. Lett.*, **17,** 117 (1966).
138. J. Ducuing, reported at Cambridge Conference on Molecular Energy Transfer, Cambridge, England, July 1971 (unpublished).*
139. J. B. Calvert, *J. Chem. Phys.*, **56,** 5071 (1972).
140. L. O. Hocker, M. A. Kovacs, C. K. Rhodes, G. W. Flynn, and A. Javan, *Phys. Rev. Lett.*, **17,** 233 (1966). See also G. W. Flynn, M. A. Kovacs, C. K. Rhodes, and A. Javan, *Appl. Phys. Lett.*, **8,** 63 (1966).
141. M. A. Kovacs, D. R. Rao, and A. Javan, *J. Chem. Phys.*, **48,** 3339 (1968).
142. J. T. Yardley and C. B. Moore, *J. Chem. Phys.*, **45,** 1066 (1966).
143. C. B. Moore, "Laser Excited Vibrational Fluorescence and its Application to Vibration-Vibration Energy Transfer in Gases," in *Fluorescence*, edited by G. Guilbault, Marcel Dekker, New York, 1967, p. 133.
144. J. T. Yardley and C. B. Moore, *J. Chem. Phys.*, **49,** 1111 (1968).
145. In the more detailed paper by Yardley and Moore (ref. 144), equations are presented for obtaining the vibrational energy transfer rate constants from the measured phase shifts, assuming first-order processes. This procedure is valid when the nonequilibrium concentration of excited states is small.
146. E. A. Mason and L. Monchick, *J. Chem. Phys.*, **36,** 1622 (1962).
147. C. E. Baker and R. S. Brokaw, *J. Chem. Phys.*, **40,** 1523 (1964); **43,** 3519 (1965).

* See also J. Ducuing, C. Joffrin, and J. P. Coffinet, Opt. Commun. (Netherlands) *2*, 245 (1970).

148. See, for example, C. O. O'Neil and R. S. Brokaw, *Phys. Fluids*, **5,** 567 (1962); **6,** 1675 (1963).
149. E. A. Mason, *J. Chem. Phys.*, **39,** 522 (1963). See also refs. 204 and 206 for additional background publications.
150. J. P. Toennies, *Z. Physik*, **182,** 257 (1965); **193,** 76 (1966). See also H. Pauly and J. P. Toennies, *Advan. Atomic Mol. Phys.*, **1,** 195 (1965).
151. A. R. Blythe, A. E. Grosser, and R. B. Bernstein, *J. Chem. Phys.*, **41,** 1917 (1964).
152. J. Schöttler and J. P. Toennies, *Z. Physik*, **214,** 472 (1968); *Ber. Bunsen—Ges. Physik. Chem.*, **72,** 979 (1968). See also W. D. Held, J. Schöttler, and J. P. Toennies, *Chem. Phys. Lett.*, **6,** 304 (1970).
153. P. F. Dittner and S. Datz, *J. Chem. Phys.*, **49,** 1969 (1968).
154. D. Beck and H. Förster, *Z. Physik*, **240,** 136 (1970).
155. M. H. Cheng, M. H. Chiang, E. A. Gislason, B. H. Mahan, C. W. Tsao, and A. S. Werner, *J. Chem. Phys.*, **52,** 6150 (1970). See also W. R. Gentry, E. A. Gislason, B. H. Mahan, and C. W. Tsao, *ibid.*, **49,** 3058 (1968).
156. P. C. Cosby and T. F. Moran, *J. Chem. Phys.*, **52,** 6157 (1970).
157. J. H. Kiefer and R. W. Lutz, *J. Chem. Phys.*, **44,** 668 (1966).
158. R. C. Millikan and D. R. White, *J. Chem. Phys.*, **39,** 3209 (1963).
159. References for the data presented in Fig. 3.20 are given in the paper by Millikan and White (158).
160. D. R. White, *J. Chem. Phys.*, **46,** 2016 (1967); **48,** 525 (1968).
161. P. G. Dickens and A. Ripamonti, *Trans. Faraday Soc.*, **57,** 735 (1961).
162. M. C. Henderson, *J. Acoust. Soc. Am.*, **34,** 349 (1962). See also M. C. Henderson, A. V. Clark, and P. R. Lintz, *ibid.*, **37,** 457 (1965).
163. D. R. White and R. C. Millikan, *J. Chem. Phys.*, **39,** 1803 (1963).
164. N. A. Generalov, *Dokl. Akad. Nauk. SSSR*, **148,** 552 (1963) [English transl: *Soviet Phys.—Doklady*, **8,** 60 (1963)].
165. J. G. Parker and R. H. Swope, *J. Acoust. Soc. Am.*, **37,** 718 (1965).
166. R. Holmes, F. A. Smith, and W. Tempest, *Proc. Phys. Soc.* (London), **81,** 311 (1963).
167. M. Camac, *J. Chem. Phys.*, **34,** 448 (1961).
168. H. Roesler, *Acustica*, **17,** 73 (1966).
169. S. J. Lukasik and J. E. Young, *J. Chem. Phys.*, **27,** 1149 (1957).
170. A. G. Gaydon and I. R. Hurle, Symp. Combust. 8th, Pasadena, Calif., 1960, p. 309 (1962).*
171. J. P. Appleton and M. Steinberg, *J. Chem. Phys.*, **46,** 1521 (1967).
172. E. E. Nikitin, *Opt. Spectry.*, **9,** 8 (1960).
173. K. L. Wray, *J. Chem. Phys.*, **36,** 2597 (1962).
174. W. D. Breshears and P. F. Bird, *Nature*, **224,** 268 (1969).
175. H.-J. Bauer, H. O. Kneser, and E. Sittig, *J. Chem. Phys.*, **30,** 1191 (1959).
176. H.-J. Bauer and K. F. Saum, *J. Chem. Phys.*, **42,** 3400 (1965).
177. H. O. Kneser, H.-J. Bauer, and H. Kosche, *J. Acoust. Soc. Am.*, **41,** 1029 (1967).

* Proc. Eighth Symp. on Combustion, 1961 Academic Press, Inc., N. Y. (1962).

178. For a review of earlier studies of NO, see A. B. Callear, *Discussions Faraday Soc.*, **33,** 28 (1962), and K. Takayanagi, ref. 11.
179. C. J. Hochanadel, J. A. Ghormley, and P. J. Ogren, *J. Chem. Phys.*, **50,** 3075 (1969).
180. N. Basco, A. B. Callear, and R. G. W. Norrish, *Proc. Roy. Soc.* (London), **A260,** 459 (1961); **A269,** 180 (1962).
181. J. Billingsley and A. B. Callear, *Nature*, **221,** 1136 (1969).
182. J. Billingsley and A. B. Callear, *Trans. Faraday Soc.*, **67,** 257 (1971).
183. J. K. Hancock and W. H. Green, U.S. Naval Research Laboratories (private communication).
184. G. Kamimoto and H. Matsui, *J. Chem. Phys.*, **53,** 3987 (1970).
185. D. L. Matthews, *J. Chem. Phys.*, **34,** 639 (1961).
186. W. J. Hooker and R. C. Millikan, *J. Chem. Phys.*, **38,** 214 (1963).
187. W. E. Woodmansee and J. C. Decius, *J. Chem. Phys.*, **36,** 1831 (1962).
188. T. I. McLaren and J. P. Appleton, *J. Chem. Phys.*, **53,** 2850 (1970).
189. A. L. Russo, *J. Chem. Phys.*, **47,** 5201 (1967).
190. D. J. Miller and R. C. Millikan, *J. Chem. Phys.*, **53,** 3384 (1970).
191. R. C. Millikan, *J. Chem. Phys.*, **40,** 2594 (1964).
192. H. K. Shin, *J. Chem. Phys.*, **55,** 5233 (1971).
193. F. D. Shields, *J. Acoust. Soc. Am.*, **32,** 180 (1960); **34,** 271 (1961).
194. E. Sittig, *Acustica*, **10,** 81 (1960).
195. W. D. Breshears and P. F. Bird, *J. Chem. Phys.*, **51,** 3660 (1969).
196. J. B. Moreno, *Phys. Fluids*, **9,** 431 (1966).
197. M. Salkoff and E. Bauer, *J. Chem. Phys.*, **29,** 26 (1958).
198. J. G. Parker, *J. Chem. Phys.*, **45,** 3641 (1966).
199. J. H. Kiefer, *J. Chem. Phys.*, **48,** 2332 (1968).
200. L. M. Raff and T. G. Winter, *J. Chem. Phys.*, **42,** 848 (1967).
201. H.-J. Bauer and H. Kosche, *Acustica*, **17,** 96 (1966).
202. G. Sessler, *Acustica*, **8,** 395 (1958).
203. M. A. Breazeale and H. O. Kneser, *J. Acoust. Soc. Am.*, **32,** 885 (1960).
204. A. P. Malinauskas, *J. Chem. Phys.*, **44,** 1196 (1966).
205. A. P. Malinauskas, J. W. Gooch, Jr., B. K. Annis, and R. E. Fuson, *J. Chem. Phys.*, **53,** 1317 (1970).
206. R. N. Healy and T. S. Storvick, *J. Chem. Phys.*, **50,** 1419 (1969).
207. J. E. Meyer and M. G. Meyer, *Statistical Mechanics*, Wiley, New York, 1940.
208. K. Geide, *Acustica*, **13,** 31 (1963).
209. T. K. Bose, H. Zink, and A. van Itterbeek, *Phys. Lett.*, **19,** 642 (1966).
210. B. R. Johnson and D. Secrest, *J. Chem. Phys.*, **48,** 4682 (1968).
211. D. L. Akins, E. H. Fink, and C. B. Moore, *J. Chem. Phys.*, **52,** 1604 (1970).
212. C. Ottinger, R. Velasco, and R. N. Zare, *J. Chem. Phys.*, **52,** 1636 (1970).
213. C. E. Baker, *J. Chem. Phys.*, **46,** 2846 (1967).
214. H-J. Bauer and H. Roesler, *Z. Naturforsch.*, **19a,** 656 (1964).
215. D. R. White, *J. Chem. Phys.*, **49,** 5472 (1968).
216. Y. Sato, S. Tsuchiya, and K. Kuratani, *J. Chem. Phys.*, **50,** 1911 (1969).
217. R. L. Taylor, M. Camac, and R. M. Feinberg, Symp. Combus. 11th, 1966, 49 (1967).

218. H. Chen, J. C. Stephenson, and C. B. Moore, *Chem. Phys. Lett.*, **2,** 593 (1968).
219. W. D. Breshears and P. F. Bird, *J. Chem. Phys.*, **54,** 2968 (1971).
220. H.-L. Chen and C. B. Moore, *J. Chem. Phys.*, **54,** 4072 (1971).
221. H.-L. Chen and C. B. Moore, *J. Chem. Phys.*, **54,** 4080 (1971).
222. W. D. Breshears and P. F. Bird, *J. Chem. Phys.*, **50,** 333 (1969).
223. J. H. Kiefer, W. D. Breshears, and P. F. Bird, *J. Chem. Phys.*, **50,** 3641 (1969).
224. C. C. Chow and E. F. Greene, *J. Chem. Phys.*, **43,** 324 (1965).
225. C. T. Bowman and D. J. Seery, *J. Chem. Phys.*, **50,** 1904 (1969).
226. H. K. Shin, *J. Phys. Chem.*, **75,** 1079 (1971); *Chem. Phys. Lett.*, **6,** 494 (1970).
227. J. F. Bott and N. Cohen, *J. Chem. Phys.*, **55,** 3698 (1971).
228. J. R. Airey and S. F. Fried, *Chem. Phys. Lett.*, **8,** 23 (1971).
229. K. H. Welge, S. V. Filseth, and J. Davenport, *J. Chem. Phys.*, **53,** 502 (1970).
230. R. C. Amme and B. E. Warren, *J. Acoust. Soc. Am.*, **44,** 419 (1968), and references cited therein.
231. J. R. Olson, R. C. Amme, and J. Fanchi (to be published).
232. B. Anderson, F. D. Shields, and H. Bass, *J. Chem. Phys.*, **56,** 1147 (1972).
233. C. J. S. M. Simpson, K. B. Bridgman, and T. R. D. Chandler, *J. Chem. Phys.* **49,** 513 (1968).
234. K. Herzfeld, *Discussions Faraday Soc.*, **33,** 22 (1962).
235. R. Marriott, *Proc. Phys. Soc.* (London), **84,** 877 (1964).
236. W. J. Witteman, *J. Chem. Phys.*, **35,** 1 (1961).
237. C. J. S. M. Simpson and T. R. D. Chandler, *Proc. Roy. Soc.* (London), **A317,** 265 (1970).
238. K. M. Merrill and R. C. Amme, *J. Chem. Phys.*, **51,** 844 (1969).
239. C. J. S. M. Simpson and J. M. Simmie, *Proc. Roy. Soc.* (London), **A325,** 197 (1971).
240. J. W. L. Lewis and K. P. Lee, *J. Acoust. Soc. Am.*, **38,** 813 (1965).
241. R. D. Sharma, *J. Chem. Phys.*, **54,** 810 (1971).
242. K. Bulthuis and G. J. Ponsen, *Phys. Lett.*, **36A,** 123 (1971).
243. R. L. Taylor and S. Bitterman, *J. Chem. Phys.*, **50,** 1720 (1969).
244. C. B. Moore, R. E. Wood, B. Hu, and J. T. Yardley, *J. Chem. Phys.*, **46,** 4222 (1967).
245. W. A. Rosser, Jr., A. D. Wood, and E. T. Gerry, *J. Chem. Phys.*, **50,** 4996 (1969).
246. W. A. Rosser, Jr. and E. T. Gerry, *J. Chem. Phys.*, **51,** 2286 (1969); **54,** 4131 (1971).
247. W. A. Rosser, Jr., R. D. Sharma, and E. T. Gerry, *J. Chem. Phys.*, **54,** 1196 (1971).
248. D. F. Heller and C. B. Moore, *J. Chem. Phys.*, **52,** 1005 (1970).
249. J. T. Yardley and C. B. Moore, *J. Chem. Phys.*, **46,** 4491 (1967).
250. J. C. Stephenson and C. B. Moore, *J. Chem. Phys.*, **52,** 2333 (1970).
251. J. C. Stephenson, R. E. Wood, and C. B. Moore, *J. Chem. Phys.*, **56,** 4813 (1972).
252. J. B. Calvert, unpublished Ph.D. thesis, University of Colorado, Boulder, Colorado, 1963.

253. C. J. S. M. Simpson, K. B. Bridgman, and T. R. D. Chandler, *J. Chem. Phys.* **49**, 509 (1968).
254. H. M. Wight, *J. Acoust. Soc. Am.*, **28**, 459 (1956).
255. R. D. Bates, G. W. Flynn, and A. M. Ronn, *J. Chem. Phys.*, **49**, 1432 (1968).
256. J. T. Yardley, *J. Chem. Phys.*, **49**, 2816 (1968).
257. P. V. Slobodskaya and Tkachenko, *Opt. Spektrosk.*, **23**, 480 (1967) [English transl: *Opt. Spectrosc.*, **23**, 256 (1967)].
258. H. E. Bass, T. G. Winter, and L. B. Evans, *J. Chem. Phys.*, **54**, 644 (1971).
259. J. D. Lambert and R. Salter, *Proc. Roy. Soc.* (London), **A243**, 78 (1957).
260. F. D. Shields, *J. Chem. Phys.*, **46**, 1063 (1967).
261. F. D. Shields and B. Anderson, *J. Chem. Phys.*, **55**, 2636 (1971).
262. T. G. Winter and H. E. Bass, *J. Acoust. Soc. Am.*, **49**, 110 (1971).
263. F. D. Shields and G. P. Carney, *J. Acoust. Soc. Am.*, **47**, 1269 (1970).
264. H.-J. Bauer, *J. Acoust. Soc. Am.*, **44**, 285 (1968).
265. See H. Roesler and K. F. Sahm, *J. Acoust. Soc. Am.*, **37**, 386 (1965).
266. H. K. Shin, J. Phys. Chem. **77**, 346 (1973).
267. D. G. Jones, J. D. Lambert, M. P. Saksena, and J. L. Stretton, *Trans. Faraday Soc.*, **65**, 965 (1969).
268. L. W. Richards and D. H. Sigafoos, *J. Chem. Phys.*, **43**, 492 (1965).
269. T. L. Cottrell and A. J. Matheson, *Trans. Faraday Soc.*, **58**, 2336 (1962).
270. H.-J. Bauer and R. Schotter, *J. Chem. Phys.*, **51**, 3261 (1969).
271. U. E. Schnaus, *J. Acoust. Soc. Am.*, **37**, 1 (1965).
272. J. G. Parker and R. H. Swope, *J. Chem. Phys.*, **43**, 4427 (1965).
273. A. A. Monkewicz, *J. Acoust. Soc. Am.*, **42**, 258 (1967).
274. J. T. Yardley and C. B. Moore, *J. Chem. Phys.*, **48**, 14 (1968).
275. R. C. Millikan, *J. Chem. Phys.*, **43**, 1439 (1965).
276. J. T. Yardley, M. N. Fertig, and C. B. Moore, *J. Chem. Phys.*, **52**, 1450 (1970).
277. J. D. Lambert, *J. Chem. Soc. Faraday Trans.*, II, **68**, 364 (1972).
278. B. T. Kelly, *J. Acoust. Soc. Am.*, **29**, 1005 (1957).
279. G. L. Hill and T. G. Winter, *J. Chem. Phys.*, **49**, 440 (1968).
280. F. J. Zeleznik and R. A. Svehla, *J. Chem. Phys.*, **53**, 632 (1970).
281. T. Oka, *J. Chem. Phys.*, **45**, 752, 754 (1966); **47**, 13, 4852 (1967); **48**, 4919 (1968); **49**, 3135 (1968).
282. A. P. Cox, G. W. Flynn, and E. B. Wilson, *J. Chem. Phys.*, **42**, 3094 (1965).
283. M. L. Unland and W. H. Flygare, *J. Chem. Phys.*, **45**, 2421 (1966).
284. J. P. Toennies, *Discussions Faraday Soc.*, **33**, 96 (1962).

CHAPTER FOUR

SENSITIZED FLUORESCENCE AND QUENCHING

L. Krause

Department of Physics, University of Windsor, Windsor, Ontario, Canada

Contents

I. INTRODUCTION

The study of interactions between excited atoms and ground-state atoms or molecules and also electrons is attracting the interest of a growing group of physicists and chemists. Excitation transfer and quenching of resonance radiation were studied during the first 30 years of this century and, after a period of dormancy, interest in these collisional phenomena was revived in the 1950s by the work of R. Seiwert in Berlin and of the author's group in Windsor. Experimental and theoretical investigations of low-energy collisions involving excited atoms are now being pursued in several research institutes.

Until recently very little was known about the interactions of heavy particles. The subject is very difficult to treat exactly, because it is essentially a many-body problem. Nevertheless, theoreticians working in this field have recently had some gratifying success, often with the help of semiclassical models. Studies of such interactions provide an important check on our understanding of atomic structure and of quantum-mechanical calculations. The results are also found useful by scientists and engineers working in a variety of fields, both pure and applied. It has been suggested, for instance [1, 2], that collisional mixing between fine-structure states in O, C^+, and Si^+ may constitute a possible cooling mechanism in interstellar space. Collisional effects are even more important in processes that take place in the earth's atmosphere, especially collisions involving metastable ^1D oxygen atoms and $^1\Delta_g$ oxygen molecules. Other interactions dominate the atmospheres of other planets, which are now being probed by satellites. The considerable effort which is being put into laser research and development draws heavily on the information furnished by basic research on low-energy atomic and molecular collisions, information which is also crucial to the understanding of chemical reactions in the gas phase. Many phenomena are studied under conditions such that the distribution functions of local thermodynamic equilibrium are not applicable, so that atomic cross sections, rate constants, and lifetimes are needed for the detailed computation of occupation numbers.

For fairly obvious reasons, the resonance states of mercury and of the alkali atoms are frequently chosen for studying the processes of excitation transfer. These states are spectroscopically accessible, emit spectral lines with relatively large oscillator strengths, and are well enough separated from other states so that experiments need not be complicated by cascade effects. The fine-structure splittings in the alkali atoms are sufficiently large that (with the exception of lithium) the resonance doublets can be easily separated and yet the energy defects between them can be spanned collisionally by drawing energy from the kinetic-energy continuum. The processes that have been studied most extensively and that will be discussed below, are the collisional transfer of excitation between fine-structure states of the same atom or of different atoms (this process manifests itself by the emission of sensitized fluorescence), and the quenching of resonance radiation. Excitation transfer may be caused in a collision between an excited atom and a ground-state atom or molecule but quenching, which causes the excited atom to decay to its ground state without the accompanying emission of radiation, takes place only as a result of collisions with molecules whose various vibrational, rotational, and translational degrees of freedom may absorb some of the released atomic excitation energy.

The early work dealing with sensitized fluorescence and quenching has been discussed by Mitchell and Zemansky [3] and by Pringsheim [4]. Reviews of more recent experiments on collisions between excited and unexcited atoms have been provided by Ermisch [5], by Seiwert [6, 7], Krause [8, 9], and by Kraulinya [10,11], whose two monographs (in Russian) give particular emphasis to the experimental work carried out in the Soviet Union. Gilmore, Bauer, and McGowan [12] cover collisions with both atoms and molecules in their review of atomic and molecular excitation mechanisms, while comprehensive accounts of experiments dealing with interactions between excited atoms and molecules may be found in Callear's [13, 14] two articles. Although collisions of excited atoms at thermal energies have been investigated largely in transparent fluorescence vessels, considerable work has also been done in flames. A detailed account of the flame-spectrometric methods together with the appropriate results may be found in a recent book of Alkemade and Zeegers [15]. A survey of literature dealing with inelastic processes involving the alkali resonance doublets has recently been carried out by Lijnse [16].

This chapter is intended to provide an overview of the results obtained from recent experimental studies of sensitized fluorescence and quenching of resonance radiation in alkali and mercury vapors, up to 1974. The many inconsistencies and discrepancies that still exist between experiment and theory as well as among the experimental results themselves, and that are a source of concern to those working in the field, will no doubt continue to be resolved as new ideas are put forward and as both theoretical and experimental techniques advance in precision and sophistication.

II. RESONANCE FLUORESCENCE AND SENSITIZED FLUORESCENCE

When a pure sample of metal vapor at low density is irradiated with the appropriate resonance radiation, the atoms become excited to their resonance states and, in decaying, emit resonance fluorescence which, except for a small

Doppler shift, is of the same frequency as the exciting radiation. The polarization and angular distribution of the emitted radiation depend on various factors, such as polarization of the exciting radiation, the atomic hyperfine structure, the pressure, the strength of ambient magnetic field, and atomic collision processes. The various properties of atomic fluorescence arising from its polarization are a powerful tool in the radio-frequency spectroscopy of excited and ground-state atoms, and they form the basis of effects such as optical pumping, optical double resonance, and level crossing [17], which are beyond the scope of this discussion. A full treatment of polarization of resonance fluorescence may be found in Feofilov's monograph [18] and in Happer's [19] review article.

If resonance fluorescence is excited in a vapor at a somewhat higher density (sodium or potassium above 5×10^{-7} torr), the fluorescent light may not reach the observer directly but may become reabsorbed and re-emitted by the ground-state atoms in the fluorescence vessel. This reabsorption gives rise to a phenomenon known as imprisonment or trapping of resonance radiation, which manifests itself by an increase in the observed mean lifetime of the resonance state. In all experiments dealing with resonance fluorescence, the problem of radiation trapping must be resolved, otherwise the experimental results will be subject to serious error. One way of avoiding the issue is to work at very low vapor densities, at which radiation is not imprisoned, provided that the resulting very low fluorescent intensities can be tolerated. On the other hand, if the geometry of the experiment is well known, the experimental measurements can be corrected for radiation trapping with the help of Holstein's [20, 21] and Walsh's [22] theory. The problem has been investigated experimentally for sodium by Kibble, Copley, and Krause [23, 24] and for potassium by Copley and Krause [25, 26], who carried out a series of lifetime determinations in relation to the appropriate alkali vapor density. The results for potassium are shown in Figure 4.1, from which the extent of radiation trapping that occurs at already low densities may be seen quite clearly. The experiments were carried out by the method of delayed coincidences [27], which was also used for similar studies in mercury in which not only radiation but also polarization was found to be imprisoned [28].

At still higher atomic densities (above 10^{-5} torr), inelastic collisions between the excited and ground-state atoms begin to influence the properties of the fluorescent light. During such a collision, excitation may be transferred, producing an excited atom of the same or different species, so that the resulting fluorescence may include frequency components additional to the component present in the exciting light. These additional components are commonly known as sensitized fluorescence, and their presence may be due to collisions in which the ground-state atoms received part of the transferred energy, became excited, and decayed optically, or they may be due to

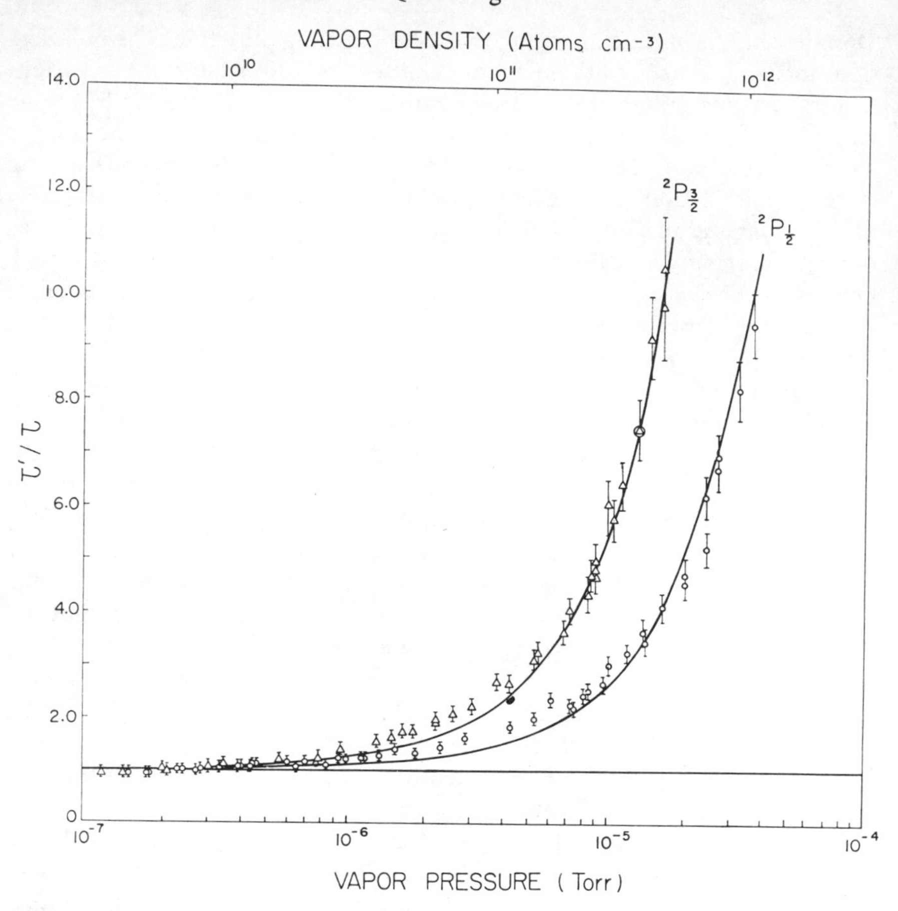

Figure 4.1. The variation of the effective lifetimes τ' with potassium vapor pressure. $\tau = 27.7$ nsec (for the "natural" lifetime) and the encircled point on the $^2P_{3/2}$ curve were used to fit the theoretical curves, representing Holstein's [20, 21] relation $1/g = \tau'/\tau$ to the experimental results.

collisions with atoms or molecules which, serving merely as carriers of kinetic energy, remained in their electronic ground states and caused the primarily excited atom to be transferred to another excited state. Both cases are illustrated in Figure 4.2, which shows the energy levels involved in the sensitized fluorescence of potassium and rubidium, induced in collisions between excited potassium and ground-state rubidium atoms [29, 30]. When only the 4 $^2P_{1/2}$ or the 4 $^2P_{3/2}$ state in potassium is excited by irradiation with monochromatic light, the excited potassium atom, when colliding with a ground-state rubidium atom, may either be transferred to the other

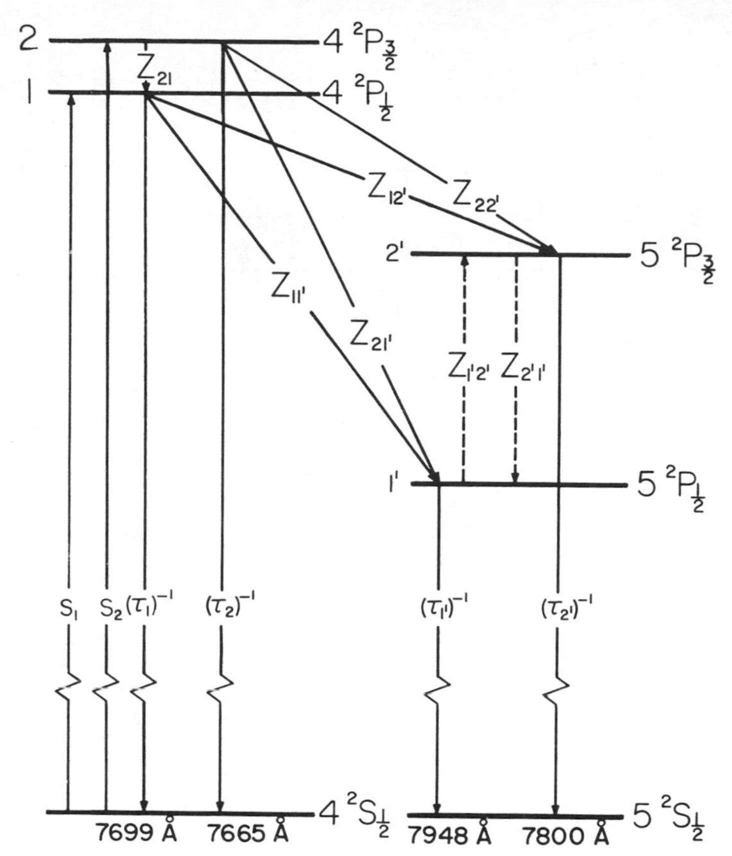

Figure 4.2. Energy levels involved in the sensitized fluorescence of potassium and rubidium, induced in collisions between excited potassium and ground-state rubidium atoms. The *s* coefficients refer to optical excitation, τ^{-1} to radiative decay, and Z to radiationless transfer of energy by inelastic collisions.

fine-structure state with the excess energy supplied from or released to the kinetic energy of relative motion, or the excitation may be transferred to the rubidium atom, leaving it in the $5\ ^2P_{1/2}$ or $5\ ^2P_{3/2}$ state. The three processes take place simultaneously, though with different probabilities, and they give rise to sensitized fluorescence that consists of three spectral components.

Sensitized fluorescence, especially when emitted in the decay of resonance states, has properties similar to those of resonance fluorescence, being equally subject to imprisonment and being sensitive to the presence of polarization and of ambient magnetic fields. It must thus be studied under conditions as well controlled as those necessary for studies of resonance fluorescence.

III. EXCITATION TRANSFER IN COLLISIONS BETWEEN EXCITED AND UNEXCITED ATOMS

A. Apparatus and Experimental Methods

The basic experimental method of investigating sensitized fluorescence in atomic vapors has not altered much since the early experiments of Wood [31] and Lochte-Holtgreven [32], even though considerable improvement in the various techniques has been made possible because of technological advances. A fluorescence vessel containing the metallic vapor (pure or in mixture with another vapor or gas) is mounted in an oven whose temperature can be measured and controlled accurately, and the gaseous mixture is irradiated with the appropriate resonance radiation emitted by a spectral lamp. The resulting fluorescence is monitored at right angles to the direction of irradiation by a suitable light detector. A representative arrangement of such apparatus is shown in Figure 4.3 and is discussed below in some detail.

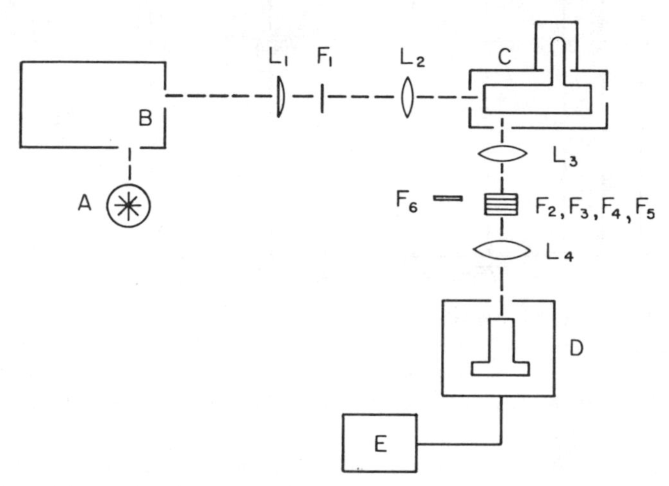

Figure 4.3. Schematic diagram of apparatus used in sensitized fluorescence studies. A, spectral lamp; B, monochromator; C, fluorescence cell in oven; D, photomultiplier in cryostat; E, preamplifier and scaler; L_{1-4}, condensing lenses; F_{1-5}, interference filters; F_6, neutral density filter.

1. Spectral Lamps

It is of crucial importance to the success of the experiments that the resonance lines used to excite the fluorescence be of high and constant intensity and of small breadth and that they not be self-reversed. When the gaseous mixture in the fluorescence vessel is kept at pressures of at most a few torr, as is frequently done in fluorescence experiments, the resonance absorption lines are extremely narrow. For maximal efficiency of excitation, the profiles of the lines in the exciting light should match the absorption profiles as closely as possible, otherwise much of the light emitted by the spectral lamp will be wasted. Even more serious consequences result from self-reversal of the exciting lines because changes in the pressure of the fluorescing mixture will cause the absorption profile to be broadened or shifted along the steep trough in the exciting line, with a consequent rapid variation in the fluorescent intensity.

Many of the usual discharge arcs containing electrodes, whether with or without admixed gases to carry the discharges, tend not to satisfy all these criteria. The early lamps of this type have been described by Mitchell and Zemansky [3]. More recently, spectral lamps emitting intense lines of various elements have been produced commercially, notably by the Osram and Philips Companies. Although very convenient and easy to use, they still suffer from some of the above disadvantages, mainly broadening and self-reversal [33]. Some workers have used such lamps with the outer jackets removed and with additional cooling arrangements, and have experienced satisfactory performance [34]. There seems little doubt, however, that electrodeless discharges and hollow cathode discharges are preferable as light sources for fluorescence experiments. An electrodeless discharge in a bulb containing a small amount of the appropriate metal with 1–2 torr of a noble gas (to carry the discharge) can be excited directly with a radio frequency oscillator, producing intense and unreversed resonance lines whose profiles can be controlled by varying the temperature of the small reservoir at the bottom of the bulb and thus the vapor pressure of the metal [35–37]. Such a discharge can also be excited by microwaves, either directed onto the discharge bulb from a horn or by mounting the bulb in a tunable microwave cavity [38, 39]. Of late, hollow cathode discharges have enjoyed considerable popularity as sources of resonance radiation. Their special advantage is due to the extremely small breadth of the emitted lines and to their adaptability to refractory materials, which makes them particularly useful in experiments involving the hyperfine structure of many elements. However, the integrated intensities of the emitted resonance lines tend to be not very high [40], thus decreasing their usefulness for many fluorescence experiments. A comprehensive review of recent designs of spectral lamps has been given by Budick, Novick, and Lurio [41].

The advent of tunable dye lasers has opened a new era in light sources for the excitation of atomic fluorescence. An increasing number of descriptions of these may be found in the literature [42], and populations of alkali atoms excited to their ^2P resonance states, approaching the theoretical limit equilibrium of 50%, have been reported [43].

2. Fluorescence Cells

The fluorescent cells in use today tend to be variations on Wood's original design. A typical cell is shown in Figure 4.4. The two windows at right angles to one another make possible the excitation and observation of fluorescence such that the paths of the exciting and fluorescent light in the vapor are extremely short, thus reducing the reabsorption and trapping of resonance radiation. The cell terminates in a Wood's horn to decrease the

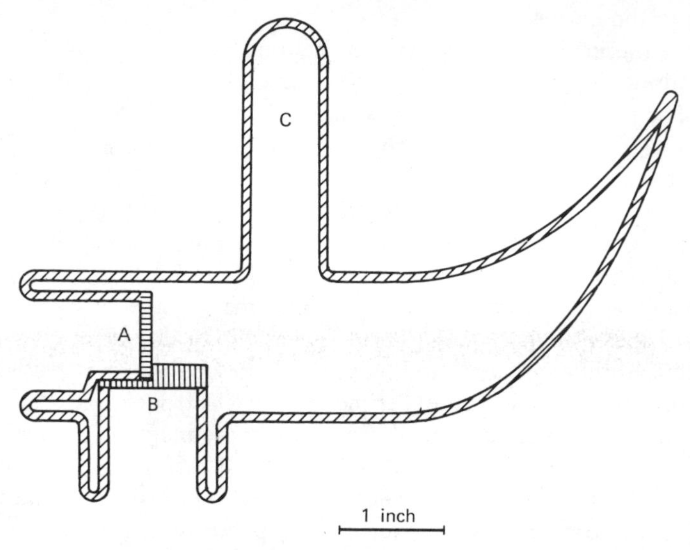

Figure 4.4. Horizontal section of the fluorescence cell. A, front (entrance) window; B, side (exit) window; C, side-arm.

stray scattered light. The side-arm that contains the liquid metal, which has a temperature lower than that of the main cell, is relatively short and thick so that the density of the metal atoms in the cell may be represented by the usual temperature–vapor-pressure relation [44, 45, 46]. Although completely sealed fluorescence cells have been used in many experiments, there is a significant risk that the metal sample in a sealed tube may become contaminated by gaseous impurities adsorbed on the walls of the cell. Experience has shown that a fluorescence cell connected to a pumping system by a narrow-bore (2-mm) tube does not suffer from such disadvantages.

3. Detection Systems

The various wavelength components present both in the exciting and in the fluorescent light must be resolved to ensure that only the desired component is present in the exciting radiation and that the intensities of each fluorescent component may be measured separately. This wavelength separation may be accomplished by monochromators and spectrometers (or spectrographs) or by interference filters. When only a few components are present in the fluorescence, interference filters are preferable since their high transmission ensures a relatively large light flux reaching the detector. There are filters available which can separate the Na–D lines, giving a transmission of about 70% at the desired wavelength and of less than 1% at the other wavelength. Two or more filters in tandem can produce a monochromatic light beam with a spectral purity of 1 p.p.m.

Photographic methods of detection and recording are still being used occasionally, but they have largely been superseded by photoelectric techniques. The most common detector is the photomultiplier tube, which must be properly selected as to the spectral response of the photocathode and to speed. Because fluorescent light is often very faint, considerations of signal-to-noise ratio are paramount and, when dealing with signals in the red region of the spectrum, it is necessary to refrigerate the photomultiplier (the whole photomultiplier rather than just the photocathode). This decreases the dark noise in the tube sometimes by six or seven orders of magnitude and, in favorable cases, it is possible to reach dark currents of 10^{-14} A, or less than one noise count per second.

The output signal of the photomultiplier can be measured with a good electrometer amplifier, which should have low noise, fast response, and a zero-suppression feature. If there is a large background component such as black-body radiation from an oven, it is necessary to resort to lock-in techniques. If the actual fluorescent light is extremely faint, photon-counting methods are preferable, provided that the photomultiplier tube has suitable dark-noise characteristics and that a discriminator is included in the counting circuit. The photon-counting method lends itself well to automation of the whole experiment [47]. The reader is referred to recent assessments of the photon-counting techniques [48] and of the relative merits of the photon-counting method as compared with the lock-in method [49]. Several of the advanced optical and electronic techniques have been applied recently by Pace and Atkinson [50] to fluorescence experiments in alkali vapors. The authors used a pulsed N_2-pumped dye laser and gated the output pulses from the photomultipliers, obtaining a high signal-to-noise ratio and a higher sensitivity than was available from previous methods.

B. Excitation Transfer between Fine-Structure States in Alkali Atoms

Most, though not all, studies of collisional excitation transfer in alkali atoms have dealt with the $^2P_{1/2}$ and $^2P_{3/2}$ resonance substates. The experimental procedure usually involves the excitation of one ^2P fine-structure state by the appropriate component of the resonance doublet and the measurement of the relative intensities of both components present in the fluorescent light. The appearance of the second component, generally known as sensitized fluorescence, is due to inelastic collisions between the excited and unexcited atoms present in the mixture.

The excitation transfer may be considered as proceeding according to the following equation:

$$\mathrm{M}(n\,^2P_{1/2}) + \mathrm{X} + \Delta E \leftrightarrow \mathrm{M}(n\,^2P_{3/2}) + \mathrm{X}, \tag{1}$$

where M represents the excited alkali atom and X is an alkali or other atom in its ground state. ΔE is the energy defect between the resonance substates, which is drawn from or converted into kinetic energy of relative motion of the collision partners and which equals 17 cm^{-1} for sodium, 57 cm^{-1} for potassium, 238 cm^{-1} for rubidium and 554 cm^{-1} for cesium. It is usually assumed that the fluorescing vapor or vapor–gas mixture exists in a dynamic equilibrium that involves only optical excitation, spontaneous decay, and the inelastic collisional processes. The steady-state conditions may be represented by two pairs of rate equations, each pair corresponding to optical excitation with one component of the resonance doublet. When the $^2P_{1/2}$ state is being optically excited, the following equations apply:

$$\frac{dN_1}{dt} = -\left(\frac{1}{\tau_1}\right)N_1 - Z_1N_1 + Z_2N_2 + s_1 = 0, \tag{2}$$

$$\frac{dN_2}{dt} = -\left(\frac{1}{\tau_2}\right)N_2 + Z_1N_1 - Z_2N_2 = 0, \tag{3}$$

where N_1 is the density of the atoms excited optically to the $^2P_{1/2}$ state, N_2 is the density of the collisionally excited $^2P_{3/2}$ atoms, Z_1 and Z_2 are the collision numbers (collision frequencies per excited atom) corresponding to the processes $^2P_{1/2} \rightarrow {}^2P_{3/2}$ and $^2P_{1/2} \leftarrow {}^2P_{3/2}$, respectively, s_1 is the density of atoms excited per second to the $^2P_{1/2}$ state by the incident radiation, and τ_1 and τ_2 are the lifetimes of the two resonance states. Equations (2) and (3) may be solved, together with a similar set corresponding to excitation with the other resonance component, to give the collision numbers Z_1 and Z_2 in terms of η_1 and η_2, the experimentally determined sensitized-to-resonance fluorescent intensity ratios:

$$Z_1 = \left(\frac{1}{\tau_1}\right)\frac{1+\eta_1}{(1/\eta_2) - \eta_1} \tag{4}$$

$$Z_2 = \left(\frac{1}{\tau_2}\right)\frac{1+\eta_2}{(1/\eta_1) - \eta_2} \tag{5}$$

The total collision cross sections for excitation transfer, $Q_1(^2P_{1/2} \rightarrow {}^2P_{3/2})$ and $Q_2(^2P_{1/2} \leftarrow {}^2P_{3/2})$ are defined analogously to the gas-kinetic cross section:

$$Z_{1,2} = NQ_{1,2}v_r, \tag{6}$$

where N is the density of the ground state atoms and v_r is the average relative velocity of the colliding partners. The principle of detailed balancing predicts that the two cross sections Q_1 and Q_2 should be in the ratio

$$\frac{Q_1}{Q_2} = \frac{g_2}{g_1} \exp\left(-\frac{\Delta E}{kT}\right) \tag{7}$$

(g_2 and g_1 are the statistical weights of the $^2P_{3/2}$ and $^2P_{1/2}$ states, respectively); it thus provides a useful check for the consistency of experimental results.

It is assumed in the above argument that neither imprisonment nor quenching of resonance radiation takes place in the system. Quenching of resonance radiation, which will be considered in a subsequent section, takes place when excited atoms are collisionally deexcited to the ground state, their decay not being accompanied by the emission of radiation. Contrary to some early opinions, collisions between excited alkali and ground-state noble-gas atoms do not result in quenching of alkali resonance radiation to any measurable extent. Nevertheless, even small amounts of molecular impurities present in the system may cause significant "quenching," especially if they are chemically reactive. Such impurities should be eliminated by baking out the fluorescence cell with the alkali metal in vacuo and by thoroughly purifying the noble gases; the noble gases should be of spectroscopic grade and should be subjected to prolonged gettering with hot alkali vapor or to cataphoresis.

1. Collisions between Alkali Atoms of the Same Species

Extensive experimental studies of mixing between ^2P resonance substates in alkali atoms, induced in collisions between excited and unexcited atoms of the same species, have been carried out by the groups of Seiwert in Berlin and of Krause in Windsor, and by Thangaraj in Toronto. All the experiments of the Windsor group were carried out at very low alkali vapor pressures, at which trapping of resonance radiation could be avoided. Seiwert worked at considerably higher vapor densities and employed a modification of Holstein's theory [20] to correct his measurements for radiation trapping, while Thangaraj, who worked at still higher densities, used Milne's theory [51] of radiation diffusion for this purpose. The various experimental collision cross sections are summarized in Table 4.1 and are compared with theoretical values calculated by Nikitin and co-workers [52, 53]. It may be seen that the calculated cross sections are much smaller than the experimental values and also that there are significant discrepancies between the experimental results

Table 4.1
Cross Sections for Excitation Transfer in Collisions between Alkali Atoms of the Same Species

COLLISIONAL PROCESS	COLLISION CROSS SECTIONS ($Å^2$) EXPERIMENTAL (TEMPERATURE IN °K)	THEORETICAL
Na–Na($3\,^2P_{1/2} \rightarrow 3\,^2P_{3/2}$)	532(424°)[a]; 170(560°)[b]	170[k]
Na–Na($3\,^2P_{1/2} \leftarrow 3\,^2P_{3/2}$)	283(424°)[a]; 100(560°)[b]	101[c]; 100[k]
K–K($4\,^2P_{1/2} \rightarrow 4\,^2P_{3/2}$)	370(358°)[d]; 120(493°)[e]; 330(550°)[f]	
K–K($4\,^2P_{1/2} \leftarrow 4\,^2P_{3/2}$)	250(358°)[d]; 60(493°)[e]; 165(550°)[f]	20[c]
Rb–Rb($5\,^2P_{1/2} \rightarrow 5\,^2P_{3/2}$)	53(360°)[g]; 74(543°)[f]	
Rb–Rb($5\,^2P_{1/2} \leftarrow 5\,^2P_{3/2}$)	68(360°)[g]; 46(543°)[f]	4.8[c]
Rb–Rb($6\,^2P_{1/2} \rightarrow 6\,^2P_{3/2}$)	212(433°)[m]	
Rb–Rb($6\,^2P_{1/2} \leftarrow 6\,^2P_{3/2}$)	140(433°)[m]	
Cs–Cs($6\,^2P_{1/2} \rightarrow 6\,^2P_{3/2}$)	6.4(323°)[h]; 6(473°)[i]	
Cs–Cs($6\,^2P_{1/2} \leftarrow 6\,^2P_{3/2}$)	31(323°)[h]; 13(473°)[i]	4.2[c]
Cs–Cs($8\,^2P_{1/2} \rightarrow 8\,^2P_{3/2}$)	190(420°)[n]	
Cs–Cs($8\,^2P_{1/2} \rightarrow 7\,^2D_{3/2}$)	21(420°)[n]	
Cs–Cs($8\,^2P_{1/2} \rightarrow 7\,^2D_{5/2}$)	20(420°)[n]	56[o]
Cs–Cs($8\,^2S_{1/2} \leftarrow 8\,^2P_{1/2}$)	$\leq$50(525°)[j]	
Cs–Cs($8\,^2P_{1/2} \rightarrow 4F$)	$\leq$50(525°)[j]	
Cs–Cs($8\,^2S_{1/2} \rightarrow 4F$)	850(525°)[j]	
Cs–Cs($8\,^2S_{1/2} \leftarrow 8\,^2P_{3/2}$)	120(525°)[j]	
Cs–Cs($4F \leftarrow 8\,^2P_{3/2}$)	$<$12(525°)[j]	

[a] Pitre and Krause [54];
[b] Seiwert [55];
[c] Dashevskaya et al. [52];
[d] Chapman and Krause [56];
[e] Hoffmann and Seiwert [57];
[f] Thangaraj [58];
[g] Rae and Krause [47];
[h] Czajkowski and Krause [59];
[i] Bunke and Seiwert [60];
[j] Berlande [61], private communication;
[k] Vdovin et al. [62];
[m] Pace and Atkinson [63];
[n] Pimbert [64];
[o] Zembekov and Nikitin at 525°K [65].

of the Windsor group and of Seiwert, whose cross sections Q_1 and Q_2 are not in very good agreement with the prediction of the principle of detailed balancing. Thangaraj's [58] cross sections are in surprisingly good agreement with those of the Windsor group, bearing in mind that he worked at high alkali vapor pressures in the region of "surface fluorescence," even though his two cross sections for rubidium are in a relation opposite to that predicted by the principle of detailed balancing.

Quite recently the group of Berlande at Saclay has been investigating collisional excitation transfer between higher fine-structure states in cesium. Taking advantage of the nearly exact coincidence between the energies of the $6\,^2S_{1/2}$–$8\,^2P_{1/2}$ transition in cesium and the $2\,^3S_1$–$3\,^3P_1$ transition in helium, they used a helium spectral lamp to excite the Cs $8\,^2P_{1/2}$ state and, by measuring the relative intensities of fluorescent components emitted from closely lying levels, determined the cross sections for the processes $8\,^2P_{1/2} \rightarrow 8\,^2P_{3/2}$, $8\,^2P_{1/2} \rightarrow 7\,^2D_{3/2,5/2}$ [64], and $8\,^2S_{1/2} \rightarrow 4\,^2F_{5/2,7/2}$ [66], and estimated values for the transitions $8\,^2P_{1/2} \rightarrow 8\,^2S_{1/2}$, $8\,^2P_{1/2} \rightarrow 4\,^2F$, $8\,^2P_{3/2} \rightarrow 8\,^2S_{1/2}$ and $8\,^2P_{3/2} \rightarrow 4\,^2F$ [61]. The collisional mixing between the $6\,^2P_{1/2}$ and $6\,^2P_{3/2}$ states in rubidium has been studied by Pace and Atkinson [63] who used a N_2 laser-pumped dye laser to excite each state in turn. The results pertinent to these higher excited states are also quoted in Table 4.1.

The cross sections for mixing between the resonance fine-structure states in potassium, rubidium, and cesium were found to vary inversely as ΔE, the

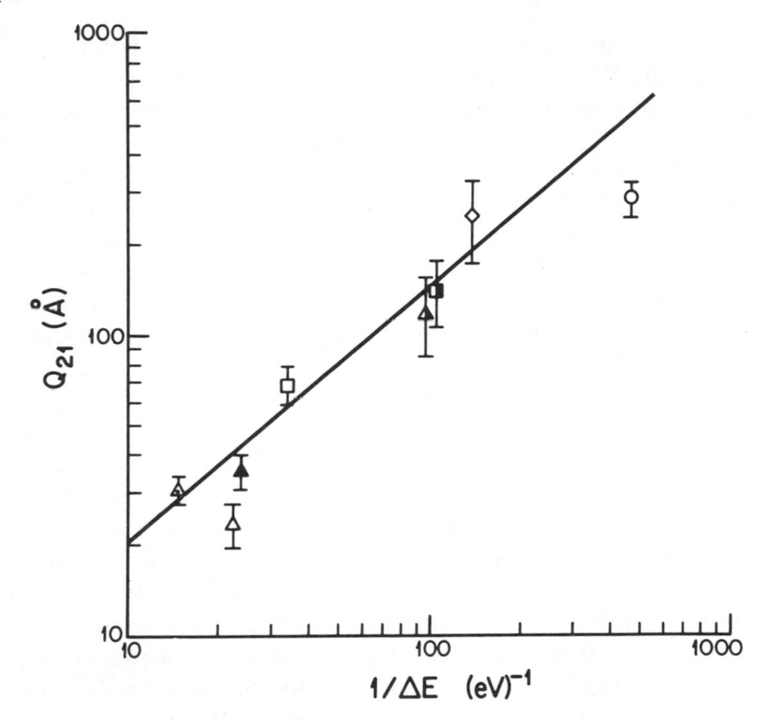

Figure 4.5. The relationship between the cross section Q_{21} ($^2P_{1/2} \leftarrow {}^2P_{3/2}$) and the fine-structure splitting ΔE for collisions between alkali atoms of like species. △, Cs($6\,^2P_{1/2} \leftarrow 6\,^2P_{3/2}$) [59]; △, Cs ($8\,^2P_{1/2} \leftarrow 7\,D_{5/2}$) [64] (not included in least-squares analysis); ▲, Cs ($8\,^2P_{1/2} \leftarrow 7\,^2D_{3/2}$) [64]; □, Rb ($5\,^2P_{1/2} \leftarrow 5\,^2P_{3/2}$) [47]; ▲, Cs ($8\,^2P_{1/2} \leftarrow 8\,^2P_{3/2}$) [64]; ◨, Rb ($6\,^2P_{1/2} \leftarrow 6\,^2P_{3/2}$) [63]; ◇, K ($4\,^2P_{1/2} \leftarrow 4\,^2P_{3/2}$) [56]; ○, Na ($3\,^2P_{1/2} \leftarrow 3\,^2P_{3/2}$) [54] (not included in least-squares analysis).

resonance defect, and thus to obey Franck's rule [67]. In the case of sodium, ΔE is so small compared with kT that it no longer limits the collision cross section. The cross sections for $8\,^2P_{1/2} \rightarrow 8\,^2P_{3/2}$ mixing in cesium and $6\,^2P_{1/2} - 6\,^2P_{3/2}$ mixing in rubidium do, however, appear to obey the same empirical relation as may be seen in Fig. 4.5.

2. Collisions between Dissimilar Alkali Atoms

When an excited alkali atom collides with a ground-state alkali atom of a different species, electronic excitation energy may be transferred either between fine-structure states of a single atom or from an excited state of one atom to that of another, as represented by the following typical equations:

$$\mathrm{K}(4\,^2P_{3/2}) + \mathrm{Rb}(5\,^2S_{1/2}) \leftrightarrow \mathrm{K}(4\,^2P_{1/2}) + \mathrm{Rb}(5\,^2S_{1/2}) + \Delta E, \qquad (8)$$

$$\mathrm{K}(4\,^2P_{3/2}) + \mathrm{Rb}(5\,^2S_{1/2}) \leftrightarrow \mathrm{K}(4\,^2S_{1/2}) + \mathrm{Rb}(5\,^2P_{3/2}) + \Delta E', \qquad (9)$$

where ΔE is the fine-structure splitting in potassium and $\Delta E'$ is the energy interval between the $\mathrm{K}(4\,^2P_{3/2})$ and $\mathrm{Rb}(5\,^2P_{3/2})$ states (see Figure 4.2). Here the interaction is somewhat different than in the case of collisions between identical partners [52, 53] and the cross sections are smaller. Equation (8) represents mixing between the ^2P resonance states of an atom, induced in a collision with an atom of a different species, while equation (9) describes excitation transfer from one species to the other. The former process has been studied experimentally for Na–K mixtures by Seiwert [55] and for K–Rb mixtures by Hrycyshyn and Krause [30]. The latter processes have been investigated in K–Rb mixtures by Thangaraj [58], Hrycyshyn and Krause [29], Ornstein and Zare [68], and Stacey and Zare [69]. The theoretical aspects of excitation transfer between two dissimilar alkali atoms were considered by Nikitin and co-workers [52, 53].

The collision cross sections reported by the various authors are summarized in Table 4.2. It may be seen that there is considerable divergence between the theoretical and experimental results and also that there is no complete agreement between the various experimental cross sections reported for the K–Rb system. It is not surprising that the experimental values agree only within order of magnitude with the calculated cross sections, bearing in mind the difficulties associated with such calculations; even this degree of agreement should be regarded as gratifying. The discrepancies between the experimental results are rather more unexpected. Thangaraj's cross section was obtained at very high vapor density, under conditions of surface fluorescence, and was corrected for trapping of resonance radiation using Milne's theory [51]. Of the four K–Rb cross sections determined both by Hrycyshyn and Krause [29] and by Stacey and Zare [69], two are in good agreement and the other two are not. There are no obvious reasons to account for the

Table 4.2
Cross Sections for Excitation Transfer in Collisions between Dissimilar Alkali Atoms

Collision partners and process	$\Delta E'$ (cm^{-1})	Collision cross sections ($Å^2$): Experimental (temperature in °K)	Theoretical
Na–K(Na 3 $^2P_{1/2} \rightarrow$ Na 3 $^2P_{3/2}$)	17	65(550°)[a]	
Na–K(Na 3 $^2P_{1/2} \leftarrow$ Na 3 $^2P_{3/2}$)	17	45(550°)[a]	
K–Rb(K 4 $^2P_{1/2} \rightarrow$ K 4 $^2P_{3/2}$)	57	260(355°)[b]	<67[c]
K–Rb(K 4 $^2P_{1/2} \leftarrow$ K 4 $^2P_{3/2}$)	57	175(355°)[b]	
K–Rb(K 4 $^2P \rightarrow$ Rb 5 ^2P)		~3.2(550°)[d]	
K–Rb(K 4 $^2P_{1/2} \rightarrow$ Rb 5 $^2P_{3/2}$)	168	40(368°)[e]; 5.3(370°)[f]	10[c]
K–Rb(K 4 $^2P_{3/2} \rightarrow$ Rb 5 $^2P_{3/2}$)	225	27(368°)[e]; 5.5(370°)[f]	
K–Rb(K 4 $^2P_{1/2} \rightarrow$ Rb 5 $^2P_{1/2}$)	409	2.7(368°)[e]; 2.3(370°)[f]	4[c]
K–Rb(K 4 $^2P_{3/2} \rightarrow$ Rb 5 $^2P_{1/2}$)	466	1.9(368°)[e]; 2.5(370°)[f]	
Rb–Cs(Rb 5 $^2P_{1/2} \rightarrow$ Cs 6 $^2P_{3/2}$)	847	1.5(348°)[g]	
Rb–Cs(Rb 5 $^2P_{1/2} \rightarrow$ Cs 6 $^2P_{1/2}$)	1401	0.5(348°)[g]	
Rb–Cs(Rb 5 $^2P_{3/2} \rightarrow$ Cs 6 $^2P_{3/2}$)	1084	0.9(348°)[g]	
Rb–Cs(Rb 5 $^2P_{3/2} \rightarrow$ Cs 6 $^2P_{1/2}$)	1638	0.3(348°)[g]	

[a] Seiwert [55].
[b] Hrycyshyn and Krause [30].
[c] Dashevskaya *et al.* [52, 53].
[d] Thangaraj [58].
[e] Hrycyshyn and Krause [29].
[f] Stacey and Zare [69].
[g] Czajkowski *et al.* [70].

discrepancies, but it should be noted that the four cross sections quoted by Hrycyshyn and Krause decrease with increasing energy differences $\Delta E'$ as do also the cross sections for Rb–Cs collisions [70], while no such dependence on $\Delta E'$ is borne out in the results of Stacey and Zare. Czajkowski, Skardis, and Krause [71] who recently studied collisional excitation transfer from Hg(6 $^3P_1$) atoms to many states in sodium, found that the cross sections for Rb–Cs excitation transfer [70] as well as those for K–Rb transfer [29] fit very well into the general scheme of cross sections for inelastic collisions between dissimilar partners.

3. Collisions between Alkali and Noble-Gas Atoms

Studies of mixing between fine-structure states in alkali atoms, induced in collisions with ground-state noble-gas atoms, have been carried out by several authors. The process may be represented by the equation

$$M(n\,^2P_{3/2}) + X(^1S_0) \leftrightarrow M(n\,^2P_{1/2}) + X(^1S_0) + \Delta E. \quad (10)$$

The noble gas acts as a carrier of kinetic energy, which is interconverted with the electronic excitation energy of the alkali atoms.

The first semiquantitative studies of sensitized fluorescence in sodium, induced in collisions with various gas atoms and molecules, were carried out by Lochte-Holtgreven [32]. On the basis of his experimental data for the Na–He system, Mitchell and Zemansky [3] calculated the mixing cross sections for the $3\,^2P_{1/2} \leftrightarrow 3\,^2P_{3/2}$ collisional transitions, but they did not take into consideration the reabsorption and trapping of the sodium resonance radiation and obtained results which were in disagreement with the principle of detailed balancing. Seiwert [55] was able to improve considerably on these results and his values, derived from Lochte-Holtgreven's experiments upon proper consideration of reabsorption, are in reasonable agreement with the more recent measurements, results of which are listed in Table 4.3 together with the available theoretical values. There is fairly good agreement among the experimentally determined cross sections quoted by various authors, and the theoretical values, where available, tend to agree with the experimental results at least within an order of magnitude.

On the whole, the experimental cross sections are internally consistent and the Q_1/Q_2 ratios agree with the values predicted from the principle of detailed balancing. A notable exception to this is provided by some of the cross sections for rubidium and cesium. Gallagher [34], who carried out an extensive study of the mixing cross sections in rubidium and cesium and their variation with temperature, suggested that the discrepancies probably arose from a pressure broadening of the alkali resonance lines, which caused an error in the Q_2 values. This contention has since been confirmed [87], and the erroneous Q_2 values have been omitted from the table. Gallagher's results, shown in Figure 4.6, indicate a rapid variation of the cross sections with temperature, according to a power law $Q \propto T^n$, where $2 \leq n \leq 4.5$.

In the cases of potassium, rubidium, and cesium the cross sections given in Table 4.3 correlate very well with the elastic electron scattering cross sections for the noble gases. Figure 4.7 shows this correlation for potassium [56] and, as in rubidium and cesium, the electron scattering cross sections are exactly one order of magnitude smaller than the mixing cross sections. Even though it has not been corroborated theoretically, this result suggests that the alkali–noble-gas atomic interaction takes place between the noble-gas atom and the orbital electron in the alkali atom, which acts as a quasifree particle. A similar correlation has been observed with the $6\,^2P$ mixing cross sections in rubidium [88], and with the $8\,^2P$ mixing cross sections in cesium [64], but not with sodium, where the mixing collisions are entirely non-adiabatic.

Although most of the experimental and theoretical work on collisional mixing has been done with the resonance fine-structure substates, several

Table 4.3

Cross Sections Q for Mixing between Alkali Resonance Fine-Structure States, Induced in Collisions with Noble-Gas Atoms (See Also Fig. 4.6)

COLLISION PARTNERS	$Q_1(^2P_{1/2} \rightarrow {^2P_{3/2}})$ (Å²) EXPERIMENTAL	Q_1 (Å²) THEORETICAL	$Q_2(^2P_{1/2} \leftarrow {^2P_{3/2}})$ (Å²) EXPERIMENTAL	Q_2 (Å²) THEORETICAL	RATIO Q_1/Q_2 EXPERIMENTAL	RATIO Q_1/Q_2 $(g_2/g_1)\exp(-\Delta E/kT)$
Na–He	86[a]	3.1[b]; 78[c,p,r]; 90[q]	45[a]; 41[d]	35[d]; 22[e]; 19[f]	1.9[a]	1.9[a]
Na–Ne	67[a]	64[b]; 68[r]	35[a]; 36[d]	46[d]	1.9[a]	1.9[a]
Na–Ar	110[a]; 111[g]	45–120[h]; 120[i]; 62[r]	56[a]; 66[g]; 65[d]	70[d]	2.0[a]	1.9[a]
Na–Kr	85[a]	57[r]	43[a]; 68[d]	74[d]	2.0[a]	1.9[a]
Na–Xe	90[a]	55[r]	46[a]; 62[d]	83[d]	2.0[a]	1.9[a]
K–He	60[j]	30[i]	41[j]; 53[k]	36[k]	1.5[j]	1.6[j]
K–Ne	14[j]	7[i]	9[j]; 14[k]	41[k]	1.5[j]	1.6[j]
K–Ar	37[j]	6[i]	22[j]; 34[k]	91[k]	1.6[j]	1.6[j]
K–Kr	61[j]	7[i]	41[j]		1.5[j]	1.6[j]
K–Xe	104[j]	12[i]	72[j]		1.4[j]	1.6[j]
Rb–He	(7.6×10^{-2})[l]; 0.1[m]	0.1[i]; (0.93×10^{-2})[n]	0.1[l]; 0.12[m]		0.74[l]	0.73[l]
Rb–Ne	(1.7×10^{-3})[l]	(8×10^{-4})[i]	(2.3×10^{-3})[l]		0.74[l]	0.73[l]
Rb–Ar	(10^{-3})[l]	(2.7×10^{-4})[n]	(1.6×10^{-3})[l]		0.63[l]	0.73[l]
Rb–Kr	(6.4×10^{-4})[l]					0.73[l]
Rb–Xe	(7.9×10^{-4})[l]	(1.1×10^{-3})[n]				0.73[l]
Cs–He	(5.7×10^{-5})[o]	(7×10^{-6})[i]; (4.6×10^{-5})[n]	(3.9×10^{-4})[o]		0.15[o]	0.15[o]
Cs–Ne	(1.9×10^{-5})[o]					0.15[o]
Cs–Ar	(1.6×10^{-5})[o]					0.15[o]
Cs–Kr	(8.3×10^{-5})[o]					0.15[o]
Cs–Xe	(7.2×10^{-5})[o]					0.15[o]

[a] Pitre and Krause [72] at 397°K.
[b] Moskovitz and Thorson [73].
[c] Reid and Dalgarno [74].
[d] Jordan and Franken [75] at 400°K.
[e] Callaway and Bauer [76].
[f] Kumar and Callaway [77].
[g] Seiwert. [55]
[h] Nikitin [78].
[i] Dashevskaya and Nikitin [79].
[j] Chapman and Krause [56] at 368°K.
[k] Jordan and Franken [75] at 340°K.
[l] Pitre, Rae, and Krause [80] at 340°K.
[m] Beahn, Condell, and Mandelberg [81] at 373°K.
[n] Dashevskaya, Nikitin, and Resnikov [82].
[o] Czajkowski, McGillis, and Krause [83] at 311°K.
[p] Reid [84];
[q] Lewis and McNamara [85];
[r] Masnou-Seeuws [86].

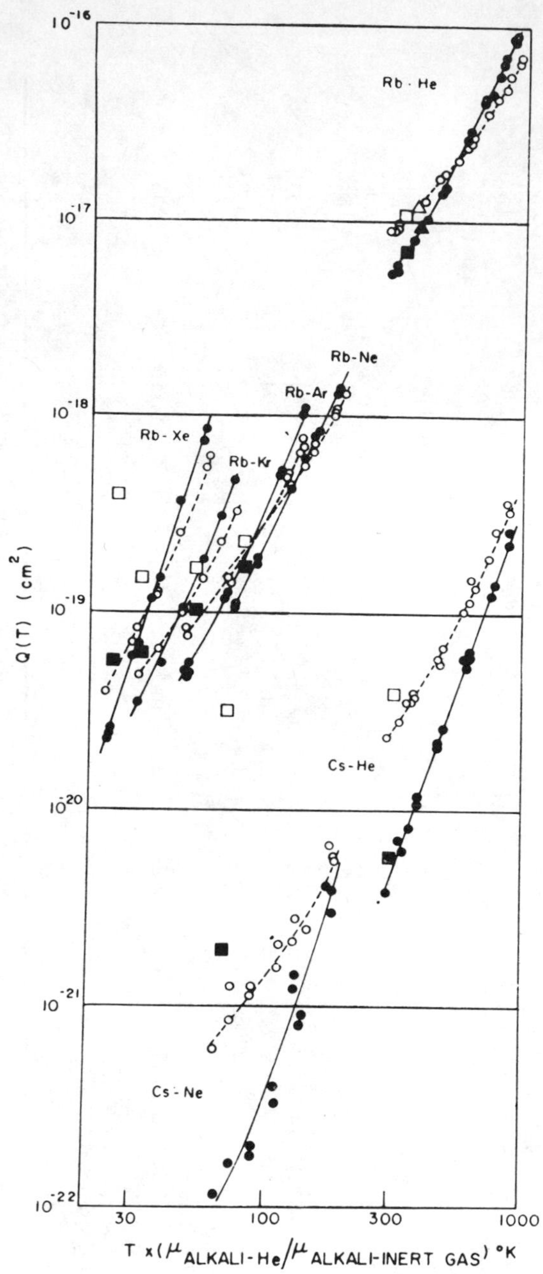

Figure 4.6. Plots of measured ^2P transfer cross sections $Q(T)$ against reduced temperature that describes the distribution of collision velocities. The cross sections for $^2P_{1/2} \leftarrow {}^2P_{3/2}$ are represented by broken lines and open circle points, those for $^2P_{1/2} \rightarrow {}^2P_{3/2}$ by solid lines and circles [34]. Similarly, square points represent the results quoted by Krause [8] and triangles, by Beahn, Condell, and Mandelberg [81].

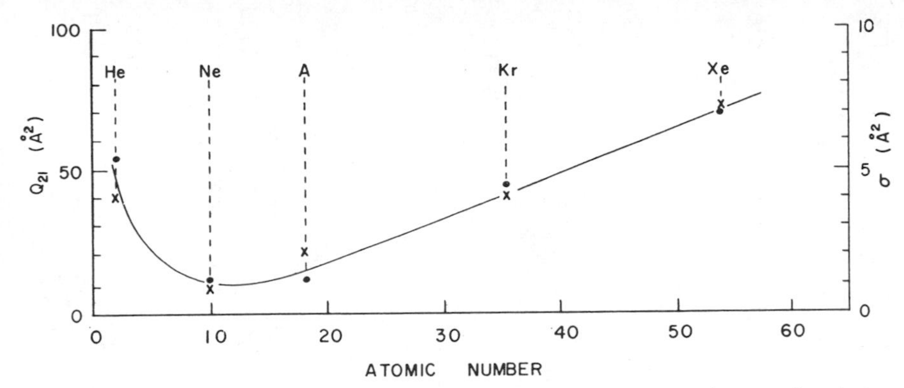

Figure 4.7 A comparison of the inelastic cross sections $Q_{21}(^2P_{1/2} \leftarrow {}^2P_{3/2})$ for potassium–noble-gas collisions, with the elastic electron–noble-gas scattering cross sections (x, Q_{21}; ●, σ).

studies of mixing in more highly excited alkali atoms have also been reported recently. Siara, Hrycyshyn, and Krause [88] determined the cross sections for $6\,^2P_{1/2} - 6\,^2P_{3/2}$ mixing in rubidium, Pimbert [64] investigated $8\,^2P_{1/2} \rightarrow 8\,^2P_{3/2}$ and $8\,^2P_{1/2} \rightarrow 7\,^2D_{3/2}$ processes in cesium and Cuvellier et al. [64a] studied Cs $7\,^2P$ mixing. Table 4.4 shows the cross sections for $8\,^2P_{1/2} \rightarrow 8\,^2P_{3/2}$ and $7\,^2P_{1/2} - 7\,^2P_{3/2}$ excitation transfer in cesium and for $6\,^2P_{1/2} - 6\,^2P_{3/2}$ mixing in rubidium, compared with the mixing cross sections appropriate to the potassium and rubidium resonance doublets. It appears that the cross sections depend largely on the value of the energy defect and decrease with increasing ΔE irrespectively whether the first, second or third ^2P doublet is involved in the process of excitation transfer. The $8\,^2P_{1/2} \rightarrow 8\,^2P_{3/2}$ cross sections do not show the rapid variation with temperature found by Gallagher for the Rb and Cs resonance states [34]; there seems to be at most a linear temperature dependence [64]. On the other hand, the results of a recent investigation by Siara, Kwong and Krause [89] shown in Fig. 4.8, indicate that the cross sections for $7\,^2P_{1/2} - 7\,^2P_{3/2}$ mixing in cesium ($\Delta E = 181$ cm^{-1}), exhibit a temperature dependence of the form $Q_{21} \propto T^{\alpha}$, where $0.1 \leq \alpha \leq 2.2$.

C. Transfer of Excitation from Mercury to Other Metallic Atoms

Mercury may be easily excited to the $6\,^3P_1$ state by 2537-Å radiation or by electron impact, and the excitation energy may be removed partly or totally in collisions with unexcited atoms or molecules. It is generally recognized that collisions with noble-gas atoms do not result in excitation transfer to the $6\,^3P_0$ metastable state or in quenching to the $6\,^1S_0$ ground state [90–92] and that such processes can be accomplished only in collisions with molecules

Table 4.4
Cross Sections for 8 $^2P_{1/2} \rightarrow$ 8 $^2P_{3/2}$ Excitation Transfer in Cesium, Induced in Collisions with Noble-Gas Atoms [64].

MIXING PROCESS	ΔE (cm^{-1})	T (°K)	CROSS SECTION	He(0.2)	Ne(0.4)	Ar(1.6)	Kr(2.5)	Xe(4.0)
				VALUES FOR VARIOUS COLLISION PARTNERS (Å²)				
$K(4\,^2P_{1/2}-4\,^2P_{3/2})$[a]	57	368	$Q_1(^2P_{1/2}\rightarrow{}^2P_{3/2})$	59.5 ± 1.5	14.3 ± 1.3	36.7 ± 1.4	61.4 ± 2.6	104 ± 2
			$Q_2(^2P_{1/2}\leftarrow{}^2P_{3/2})$	$40.8\pm2{\cdot}1$	9.4 ± 0.7	22.4 ± 1.6	40.7 ± 1.8	72.3 ± 2.2
$Rb(6\,^2P_{1/2}-6\,^2P_{3/2})$[b]	77	420	$Q_1(^2P_{1/2}\rightarrow{}^2P_{3/2})$	$29.3\pm9\%$	$10.3\pm12\%$	$24.0\pm10\%$	$23.2\pm10\%$	$43.9\pm9\%$
			$Q_2(^2P_{1/2}\leftarrow{}^2P_{3/2})$	$19.0\pm9\%$	$6.4\pm12\%$	$14.9\pm10\%$	$14.6\pm10\%$	$27.7\pm9\%$
$Cs(8\,^2P_{1/2}-8\,^2P_{3/2})$[c]	83	420	$Q_1(^2P_{1/2}\rightarrow{}^2P_{3/2})$	$34\pm30\%$	$4.4\pm30\%$	$5.5\pm30\%$	$4.5\pm30\%$	$15\pm30\%$
$Cs(7\,^2P_{1/2}-7\,^2P_{3/2})$[e]	181	448	$Q_1(^2P_{1/2}\rightarrow{}^2P_{3/2})$	12 ± 2	0.18 ± 0.03	0.12 ± 0.02	0.091 ± 0.020	0.65 ± 0.10
			$Q_2(^2P_{1/2}\leftarrow{}^2P_{3/2})$	11 ± 2	0.16 ± 0.03	0.10 ± 0.02	0.077 ± 0.017	0.57 ± 0.10
$Rb(5\,^2P_{1/2}-5\,^2P_{3/2})$[d]	238	340	$Q_1(^2P_{1/2}\rightarrow{}^2P_{3/2})$	$7.6\times10^{-2}\pm10\%$	$1.7\times10^{-3}\pm10\%$	$1.0\times10^{-3}\pm10\%$	$6.4\times10^{-4}\pm10\%$	$7.9\times10^{-4}\pm10\%$
			$Q_2(^2P_{1/2}\leftarrow{}^2P_{3/2})$	$10.3\times10^{-2}\pm10\%$	$2.3\times10^{-3}\pm10\%$	$1.6\times10^{-3}\pm10\%$	$1.5\times10^{-3}\pm10\%$	$2.1\times10^{-3}\pm10\%$

[a] Chapman and Krause [56].
[b] Siara, Hrycyshyn and Krause [88].
[c] Pimbert [64].
[d] Pitre, Rae, and Krause [80].
[e] Cuvellier et al. [64a].

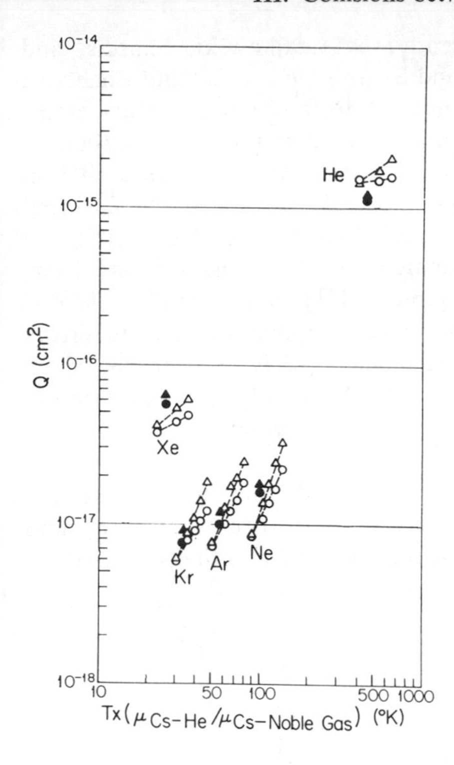

Figure 4.8 Plots of the $7\,^2P$ mixing cross sections Q_{12} and Q_{21} against normalized temperature $T' = T(\mu_{Cs-He}/\mu)$. △ points correspond to $Q_{12}(^2P_{1/2} \rightarrow {}^2P_{3/2})$ and ○ points to $Q_{21}(^2P_{1/2} \leftarrow {}^2P_{3/2})$. The solid data points represent measurements performed by Cuvellier et al. [64a].

as will be described in a subsequent section, even though Waddell and Hurst [93] suggested that Hg($6\,^3P_1$) atoms can be transferred to the $6\,^3P_0$ state in binary collisions with ground-state mercury atoms (with a cross section in the range 3–10 × 10^{-2} Å^2). On the other hand, the occurrence of collisional excitation transfer from the Hg($6\,^3P_1$) state to closely lying levels of other atoms has been known for many years and the effect has been studied extensively with various atoms as collision partners. The first experiments dealing with mercury–thallium collisions were performed by Cario and Franck [94] and mercury–sodium collisions were investigated by Beutler and Josephy [95]. Most of the early results have been discussed by Pringsheim [4], and some more recent experiments will be mentioned in the following paragraphs.

1. Mercury–Sodium Collisions

The excitation transfer has been investigated by studying the sensitized fluorescence of sodium, arising from the decay of several S, P, and D states that have energies close to the excitation energy of the $6\,^3P_1$ (and $6\,^3P_0$) state in mercury. Experiments of this type have been reported by the groups of

Frish and Kraulinya and, most recently, by Czajkowski, Skardis, and Krause [71] and Czajkowski, Krause and Skardis [96]. Frish and Bochkova [97, 98] studied excitation transfer from the 6 $^3P_1$ and 6 $^3P_0$ mercury atoms excited by collisions with electrons in a discharge, to various states in sodium. Kraulinya [99] optically excited the Hg(6 $^3P_1$) state and followed the excitation transfer to sodium by monitoring the intensities of the collisionally sensitized sodium lines. Her results which are quoted within $\pm 30\% - 50\%$ are summarized in Table 4.5 and are compared with the cross sections determined by Czajkowski, Skardis and Krause [71]. The considerable discrepancies between the two sets of results are apparently due to errors arising from the trapping of mercury resonance radiation [100, 28] which must have particularly affected Kraulinya's results, and from the uncertainty in the determination of the mercury and sodium vapor densities in the binary mixture.

The cross sections determined by Czajkowski, Skardis, and Krause [71] are plotted in Fig. 4.9 against the energy defect ΔE between the Hg 6 $^3P_1$ state and the appropriate states in sodium. The plot also includes the cross sections

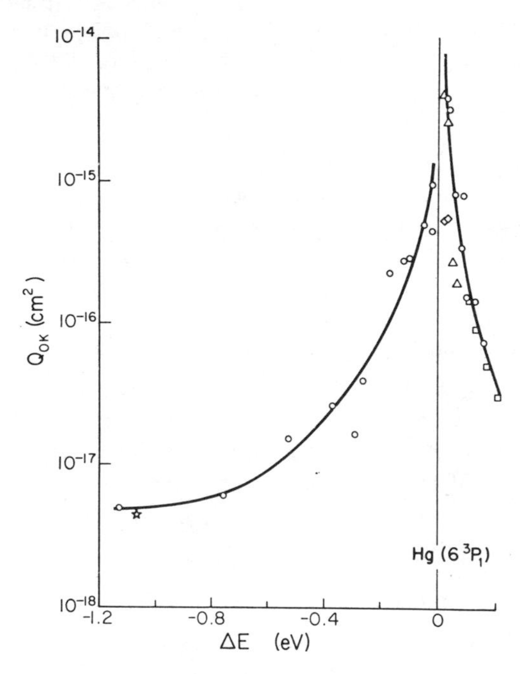

Figure 4.9 A plot against ΔE of the cross sections for excitation transfer to S, P and D states in sodium. ○, Hg → Na transfer [71]; □, Rb → Cs transfer [70]; △, K → Rb transfer [29]; ◇, K → Rb transfer [69]; *Hg → Cd transfer [101].

Table 4.5
Cross Sections Q for Excitation Transfer from Hg 6 $^3P_1$ to Various Levels in Sodium; ΔE Denotes the Energy Difference between the Appropriate Sodium Level and Hg 6 $^3P_1$

Hg → Na TRANSFER PROCESS	OBSERVED SENSITIZED FLUORESCENCE OF Na DESIGNATION	λ (Å)	ΔE (eV)	Q (Å²) (a)	(b)
$6\,^3P_1 \to 10S$	$10S \to 3P$	4340/43	0.07	0.2	3.42
$6\,^3P_1 \to 9S$	$9S \to 3P$	4420/23	0.02	2.7	38.50
$6\,^3P_1 \to 8S$	$8S \to 3P$	4542/45	−0.05	0.9	4.92
$6\,^3P_1 \to 7S$	$7S \to 3P$	4748/52	−0.17	0.7	2.30
$6\,^3P_1 \to 6S$	$6S \to 3P$	5149/54	−0.37	0.5	0.26
$6\,^3P_1 \to 5S$	$5S \to 3P$	6154/64	−0.76	0.3	0.06
$6\,^3P_1 \to 9D$	$9D \to 3P$	4321/24	0.09	0.5	8.06
$6\,^3P_1 \to 8D$	$8D \to 3P$	4390/93	0.04	3.0	31.36
$6\,^3P_1 \to 7D$	$7D \to 3P$	4494/98	−0.03	2.4	4.50
$6\,^3P_1 \to 6D$	$6D \to 3P$	4665/69	−0.13	2.0	2.87
$6\,^3P_1 \to 5D$	$5D \to 3P$	4979/83	−0.29	1.8	0.17
$6\,^3P_1 \to 4D$	$4D \to 3P$	5683/88	−0.60	1.7	~0.02
$6\,^3P_1 \to 12P$	$12P \to 3S$	2464	0.16	0.1	0.74
$6\,^3P_1 \to 11P$	$11P \to 3S$	2475	0.13	0.1	1.47
$6\,^3P_1 \to 10P$	$10P \to 3S$	2490	0.09	0.1	1.56
$6\,^3P_1 \to 9P$	$9P \to 3S$	2512	0.06	0.2	8.06
$6\,^3P_1 \to 8P$	$8P \to 3S$	2543	−0.01	0.2	9.35
$6\,^3P_1 \to 7P$	$7P \to 3S$	2593	−0.10	0.2	2.90
$6\,^3P_1 \to 6P$	$6P \to 3S$	2680	−0.25	0.1	0.40
$6\,^3P_1 \to 5P$	$5P \to 3S$	2853	−0.53	0.1	0.16
$6\,^3P_1 \to 4P$	$4P \to 3S$	3302	−1.10	0.1	≤0.05

(a) Kraulinya [99].
(b) Czajkowski, Skardis, and Krause. [71].

for excitation transfer in K–Rb, Rb–Cs, and Hg–Cd mixtures, in all of which the relative speeds of the colliding partners are sufficiently close to one another to justify a direct comparison [71]. It appears that the various cross sections for transfer of excitation between dissimilar metallic atoms may be represented in a single such plot from which it may be inferred that the interaction mechanisms are similar to one another. The cross sections for excitation transfer between atoms of the same species, which are shown in Fig. 4.5 do not lie on this resonance curve, nor do the cross sections for collisions with noble gas atoms. There is at present no comprehensive theory which would account for this behaviour of the cross sections.

2. Mercury–Thallium Collisions

Thallium has five excited states ($7\,^2P_{1/2}$, $7\,^2P_{3/2}$, $6\,^2D_{3/2}$, $6\,^2D_{5/2}$, and $8\,^2S_{1/2}$) whose energies are close to those of the $6\,^3P_1$ and $6\,^3P_0$ states of mercury, and it might be expected that collisional transfer of energy from mercury to thallium should proceed with ease. The Hg–Tl system has been studied by several authors, but absolute values of collision cross sections have only recently been determined by Hudson and Curnutte [102] and by Kraulinya [103, 104]. The cross sections are summarized in Table 4.6. Methods of sensitized fluorescence were employed in both investigations, but Kraulinya did not appear to take into account cascade transitions among the thallium levels, while Hudson and Curnutte allowed for these on the basis of theoretical transition probabilities. It seems, however, that both sets of results are subject to experimental error caused by the trapping of mercury resonance radiation, which becomes significant even at mercury densities lower than those prevailing in the experiments [28].

It is obvious from Table 4.6 that the problem of excitation transfer from mercury to thallium is in a very unsatisfactory state. There is an apparent lack of consistency in the results of Kraulinya et al. (104), whose cross sections for excitation transfer to the $8\,^2S_{1/2}$, $6\,^2D_{3/2}$, and $7\,^2S_{1/2}$ levels in thallium seem to depend on the wavelength of the observed fluorescent component. The results of the two groups (Hudson and Curnutte and Kraulinya et al.) do not agree well with each other, and there is no consistent dependence of the measured cross sections on temperature. Finally, one would expect that the cross sections should decrease in some manner with increasing energy gap $\Delta E'$, but the results seem to indicate the opposite. It is manifest that considerable additional experimental work is needed to overcome these difficulties.

3. Collisions between Mercury and Other Metallic Atoms

Several investigations have been reported of excitation transfer from excited mercury atoms to ground-state atoms of silver, bismuth, cadmium, chromium, copper, indium, lead, and zinc. Most of these experiments which had been completed some time ago were surveyed by Seiwert [6, 7]. Perhaps of particular interest is an investigation by Gough [105], who studied excitation transfer from mercury to cadmium and concluded that not only excitation energy but also coherence was transferred in the collisions. A similar conclusion was reached by Kraulinya, Sametis, and Bryukhovetskii [106] as the result of their study of the Hg–Tl system. Cross sections for Hg–Cd

Table 4.6

Cross Sections Q for Excitation Transfer from Hg $6\,^3P_1$ to Various Levels in Thallium; $\Delta E'$ Denotes the Energy Difference between the Appropriate Thallium Level and Hg $6\,^3P_1$; Temperatures Are Given in °K

Hg → Tl TRANSFER PROCESS	OBSERVED SENSITIZED FLUORESCENCE OF Tl DESIGNATION	λ (Å)	$\Delta E'$ (eV)	Q (Å²) (KRAULINYA *et al.* [104]) 710°	770°	930°	Q (Å²) (HUDSON AND CURNUTTE [102]) 1070°	1120°	1170°
$6\,^3P \rightarrow 9\,^2S_{1/2}$	$9\,^2S_{1/2} \rightarrow 6\,^2P_{3/2}$	2826	0.45	—	—	1.3	—	—	—
$6\,^3P \rightarrow 7\,^2D_{5/2}$	$7\,^2D_{5/2} \rightarrow 6\,^2P_{3/2}$	2918	0.33	—	—	0.23	—	—	—
$6\,^3P \rightarrow 7\,^2D_{3/2}$	$7\,^2D_{3/2} \rightarrow 6\,^2P_{3/2}$	2921	0.32	—	—	1.9	—	—	—
$6\,^3P \rightarrow 8\,^2S_{1/2}$	$8\,^2S_{1/2} \rightarrow 6\,^2P_{3/2}$	3230	−0.08	3.4	2.9	11	2.2 ± 1.3	1.6 ± 0.6	1.3 ± 0.3
$6\,^3P \rightarrow 8\,^2S_{1/2}$	$8\,^2S_{1/2} \rightarrow 6\,^2P_{1/2}$	2580	−0.08	—	—	0.5			
$6\,^3P \rightarrow 6\,^2D_{5/2}$	$6\,^2D_{5/2} \rightarrow 6\,^2P_{3/2}$	3519	−0.40	8.5	6.5	17			
$6\,^3P \rightarrow 6\,^2D_{3/2}$	$6\,^2D_{3/2} \rightarrow 6\,^2P_{3/2}$	3529	−0.41	5.9	13	38	9.4 ± 3.1	11.3 ± 3.5	8.5 ± 2.5
$6\,^3P \rightarrow 6\,^2D_{3/2}$	$6\,^2D_{3/2} \rightarrow 6\,^2P_{1/2}$	2768	−0.41	—	0.21	0.06			
$6\,^3P \rightarrow 7\,^2S_{1/2}$	$7\,^2S_{1/2} \rightarrow 6\,^2P_{3/2}$	5350	−1.60	26	28	83	—	—	—
$6\,^3P \rightarrow 7\,^2S_{1/2}$	$7\,^2S_{1/2} \rightarrow 6\,^2P_{1/2}$	3776	−1.60	18	4.7	1.6	—	—	—
$6\,^3P \rightarrow 7\,^2P$	—	—	−0.6	—	—	—	17.6 ± 6.0	20.8 ± 6.6	23.3 ± 7.2
$6\,^3P \rightarrow$ all	Quenching	—	—	—	—	—	108 ± 41	99 ± 25	80 ± 19

collisions were also measured by Morozov and Sosinskii [107], who reported a value of 0.9 ± 0.3 Å^2 for the Hg $6\,^3P_1 \rightarrow$ Cd $5\,^3P_1$ transition, and values of 4 ± 1 Å^2 for each of the transitions from Hg $7\,^3S_1$ to Cd $5\,^3D_1$, $^3D_2$, and $^3D_3$. These are unexpectedly large cross sections, considering the huge energy gaps between the respective levels (8756 cm^{-1} between Hg $6\,^3P_1$ and Cd $5\,^3P_1$, and 2700 cm^{-1} between Hg $7\,^3S_1$ and Cd $5\,^3D$). It also seems that the measurements may have been perturbed by radiation trapping. Kraulinya and Arman [108], who obtained a cross section of about 0.2 Å^2 for the transfer Hg $6\,^3P_1 \rightarrow$ Cd $5\,^3P_1$, suggested that the large cross sections may be explained by the formation of molecular complexes Hg $6\,^3P_0$ + Hg $6\,^1S_0 \rightarrow Hg_2(^3O_u^-)$ whose vibrational levels are in near resonance with the Cd 5^3P_1 level (a similar argument was proposed in the case of the Zn $4\,^3P_1$ level). The Hg–Zn system has also been studied by Kraulinya and Arman (108) and by Sosinskii and Morozov (109), both reporting a cross section of approximately 2×10^{-2} Å^2 for the process Hg $6\,^3P_1 \rightarrow$ Zn $4\,^3P_1$. Kraulinya and Yanson [110] investigated excitation transfer from the mercury $6\,^3P_1$ state to several states in indium, obtaining again larger than expected collision cross sections which they interpreted on the basis of an intermediate Hg_2 molecular state.

On the other hand, the results of more recent studies by Chéron [111] and by Czajkowski and Krause [101] who investigated Hg($6\,^3P_1$) $\rightarrow$ Cd($5\,^3P_1$) excitation transfer and obtained a cross section of 4.6×10^{-2} Å^2, suggest that the transfer proceeds by ordinary 2-body collisions. Czajkowski's cross section, which is considerably smaller than Morozov's and Kraulinya's values, lies well on the general resonance curve shown in Fig. 4.9, together with other similar cross sections for excitation transfer between dissimilar partners.

D. Quenching in Collisions between Atoms

Some mention should be made of the suggestion that atomic fluorescence might be subject to quenching by collisions with ground-state atoms, in the course of which the excited atoms decay to the ground state without emitting radiation and the excitation energy is entirely converted into kinetic energy of relative motion. This quenching process should be distinguished from others, in which the primarily excited atom decays not to the ground state but to some intermediate (sometimes metastable) state, or in which the excitation energy is partly or wholly used to excite the collision partner.

The existence of quenching was inferred by Demtröder [112] on the basis of his measurements of sodium resonance state lifetimes in relation to argon pressure. His conclusions appeared to be corroborated by the theoretical study of Nikitin and Bykhovskii [113], but a subsequent extensive investigation by Copley, Kibble, and Krause [114], who measured the lifetime of the sodium resonance state in relation to the pressures of all the noble gases, showed conclusively that no quenching took place. The results are illustrated in Figure 4.10. Corroboration of these conclusions was provided by the shock-tube experiments of Tsuchiya and Kuratani [115] and by the theoretical study of Stamper [116]. Further experiments failed to discover any quenching by noble gases of potassium [25, 26] or mercury [90, 92] resonance radiation. It has also been found recently that helium, neon, and argon do not quench the 6 $^2P_{3/2}$ metastable state in thallium, or that the quenching cross section does not exceed 2×10^{-3} Å^2 [117], although Pickett and Anderson [118] measured a cross section of 5.4 ± 0.5 Å^2 for the self-quenching of the metastable state, which they interpreted on the basis of the process $6\,^2P_{3/2} + 6\,^2P_{1/2} \rightarrow 2(6\,^2P_{1/2}) + \Delta E$. Yet another investigation of quenching by noble gases of sodium, potassium, rubidium, and cesium resonance radiation, based on the photodissociation technique, showed quenching to be insignificant [119, 119a].

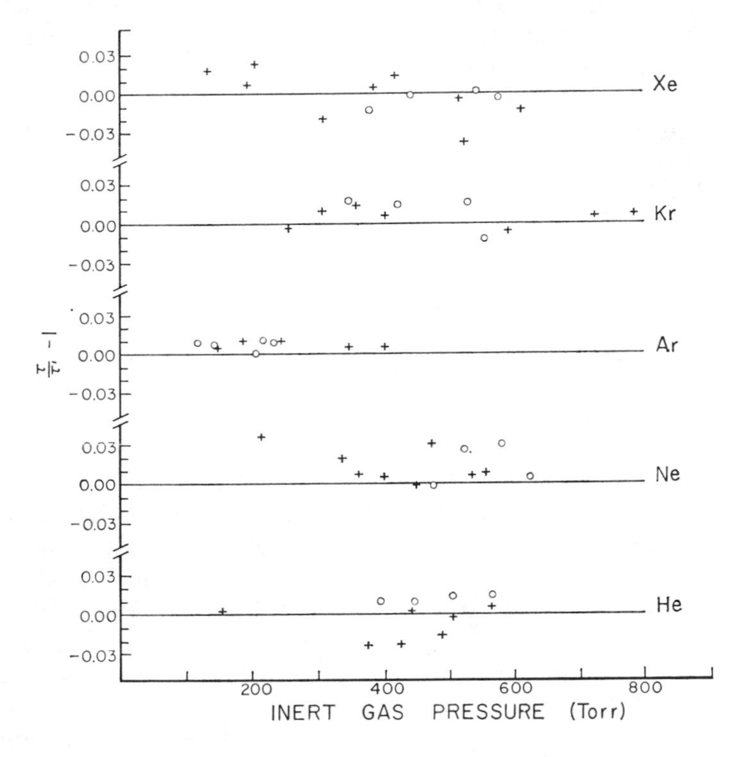

Figure 4.10 The variations of $(\tau/\tau' - 1)$ with pressures of He, Ne, Ar, Kr, and Xe [see equation (14)]. x, τ' at 130° C; 0, τ' at 210° C.

IV. COLLISIONS BETWEEN EXCITED ATOMS AND UNEXCITED MOLECULES

When an excited atom collides inelastically with a molecule in its ground state and the collision takes place at low (thermal) energy, the resulting interaction may give rise either to quenching or to a transfer of excitation energy. Quenching takes place if all the atomic excitation energy is removed and distributed among the available translational, vibrational, and rotational degrees of freedom, thus causing the atom to decay to its ground state without the emission of radiation. The effect manifests itself by the decrease in the intensity of the atomic resonance fluorescence as the pressure of the quenching gas is increased, or by the equivalent decrease in the observed mean lifetime of the atomic resonance state. The alternative process of excitation transfer involves the interconversion of electronic, translational, vibrational, and rotational energy, as a result of which the atom may find itself in another higher or lower excited state. Such excitation transfer causes the appearance of sensitized fluorescence emitted in the spontaneous decay of the state to which the atom was transferred during the collision.

Quenching and excitation transfer may be represented respectively by the following two equations, which apply to the fine-structure components of the sodium 3 ^2P resonance state:

$$\mathrm{Na}(3\,^2P_{1/2}) + \mathrm{M} \rightarrow \mathrm{Na}(3\,^2S_{1/2}) + \mathrm{M}^*, \tag{11}$$

$$\mathrm{Na}(3\,^2P_{1/2}) + \mathrm{M} \rightleftharpoons \mathrm{Na}(3\,^2P_{3/2}) + \mathrm{M}^*, \tag{12}$$

where M* represents a molecule that, as the result of the collision, has become modified in its content of translational, vibrational, and rotational energy. A third process that sometimes results from an atom–molecule collision and that may be considered as intermediate between the two others is the transfer of the atom to a metastable state from which spontaneous transition to the ground state may be forbidden; the atom then decays by further collisions with molecules or with the walls of the vessel.

The detailed mechanism of these interactions is not yet fully understood, and divergences of opinion still exist as to the relative contribution and importance of the translational, vibrational, and rotational energies in the collisional processes. It has recently been suggested, for example, that the cross sections for quenching of alkali resonance radiation by nitrogen should be insensitive to the relative kinetic energy of the colliding partners, since most of the released atomic excitation energy is channeled into the vibrational states of the nitrogen molecule [120, 121]. On the other hand it is stated elsewhere with equal confidence that the conversion of electronic to vibrational energy is inefficient and that the translational energy component and the chemical reactivity of the quenching gas are much more important considerations [12–14, 122].

A. Quenching of Atomic Fluorescence

The quenching of atomic resonance radiation by various molecular gases has been the subject of investigations for over 50 years. In the early experimental work, which has been summarized by Pringsheim [4], the metallic vapor was excited in a glass cell by the appropriate resonance radiation and was subjected to varying pressures of the quenching gases. The relative decrease in the fluorescent intensity was monitored in relation to the gas pressure and the experimental data were substituted in the Stern–Volmer relation [123], which yielded the appropriate cross sections for the quenching collisions. Many of these experiments were subject to errors arising from imprisonment of the resonance radiation or from the presence of impurities in the quenching gases. It was not realized that alkali resonance radiation becomes significantly trapped already at vapor pressures of 10^{-5} torr, at which the fluorescent intensities are relatively low and difficult to measure without the use of photomultiplier tubes. Thus many experiments were carried out at pressures higher by some orders of magnitude, at which the resonance radiation was imprisoned and the effective lifetime was much longer than the "natural" lifetime of the resonance state. According to the Stern–Volmer equation,

$$\frac{I_0}{I} = 1 + \tau N v_r Q, \tag{13}$$

where I_0 and I are the intensities of resonance radiation in pure metal vapor and in the presence of quenching gas of density N, respectively, Q is the quenching cross section, v_r is the average relative speed of the colliding partners, and τ is the lifetime of the atomic resonance state. Thus, if τ is of uncertain value, so is the cross section Q obtained from the experiment. Some quite recent determinations of quenching cross sections have been carried out at vapor pressures below 10^{-5} torr to avoid radiation trapping [124, 125]. Similar experiments have been performed with flames at much higher temperatures, at which the relative kinetic energies of the collision partners and the populations of the molecular vibrational and rotational states are quite different. A full account of quenching in flames and of the appropriate experimental techniques may be found in recent reviews by Alkemade and Zeegers [13] and Lijnse [16]. Finally, quenching cross sections are also being obtained from measurements of effective lifetimes of atomic resonance states in mixtures of metallic vapors and quenching gases [24].

A typical arrangement of the apparatus used for fluorescent intensity measurements in mixtures of alkali vapors and quenching gases is shown in

Figure 4.11 [126]. The experimental techniques are similar to those described in Section III. The alkali resonance doublet emitted by a radiofrequency lamp is separated with a spectral purity of one part in 10^6 by a grating monochromator in series with an interference filter or filters. The monochromatic beam is then split into two parts: One is condensed into a fluorescence cell

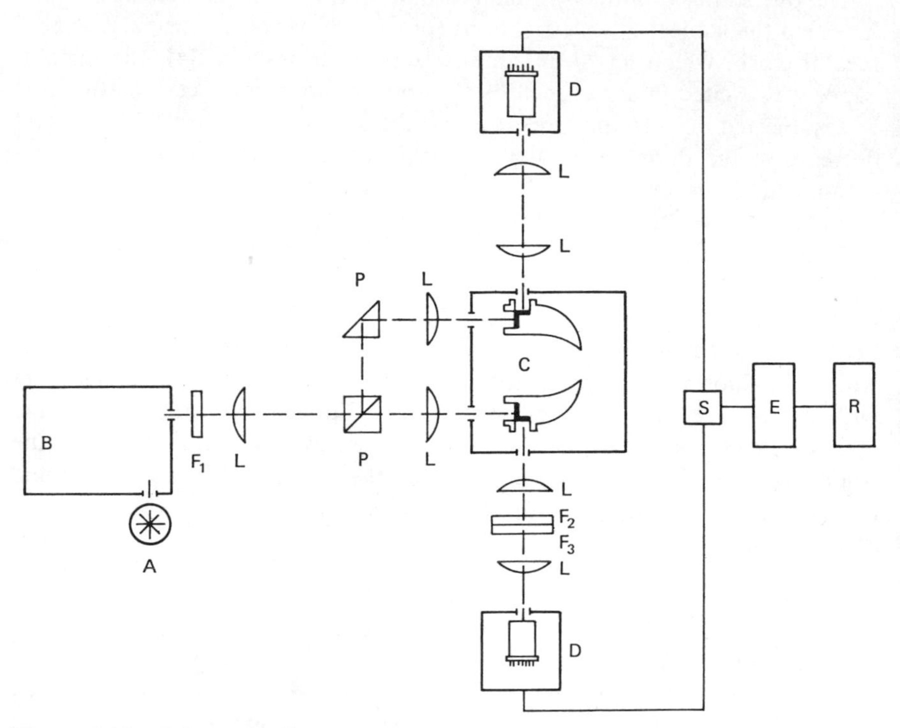

Figure 4.11 Schematic diagram of apparatus used for studies of quenching of resonance radiation. A, spectral lamp; B, monochromator; C, fluorescence cells in oven; D, photomultiplier tubes in cryostats; E, electrometer; R, recorder; S, coaxial switch; F_{1-3}, interference filters; L, lenses; P, prisms.

containing the pure alkali vapor and the other into a similar cell containing the vapor–gas mixture. The fluorescent light emerging from the latter cell is again resolved into its two spectral components and is focused onto the cathode of a photomultiplier. A second photomultiplier is used to detect the total (unresolved) fluorescence from the cell containing the alkali vapor alone. The output signals of both photomultipliers are directed, through a coaxial switch, to an electrometer amplifier and are registered with a strip-chart recorder. Both fluorescence cells are mounted in a single oven at constant temperature, and their side-arms containing the liquid metal are enclosed in separate small ovens kept at constant and equal temperatures. Each cell is connected by a narrow-bore tube (1–2 mm I.D.) to a vacuum and gas-filling system. In this way, it is possible to obtain the intensity of the fluorescence in the gas-filled cell that would have been produced in the absence of the gas. This method, employing two fluorescence cells, is preferable to the more usual arrangement with a single cell, because experimental errors due to fluctuations in temperature and lamp intensity are eliminated. Typical experimental results that show the quenching of cesium resonance radiation by nitrogen are presented in Figure 4.12. Effects due to pressure broadening are negligible at nitrogen pressures below 1 torr. Although the observed decrease in the fluorescent intensities is due in principle to collisional transitions both to the ground state and to the other resonance fine-structure state, the effect of ^2P mixing cannot be seen in the plots because the Cs–N_2 mixing cross sections are much smaller than the quenching cross sections.

These methods were applied extensively in investigations of quenching of alkali resonance radiation by nonpolar diatomic and polyatomic molecules

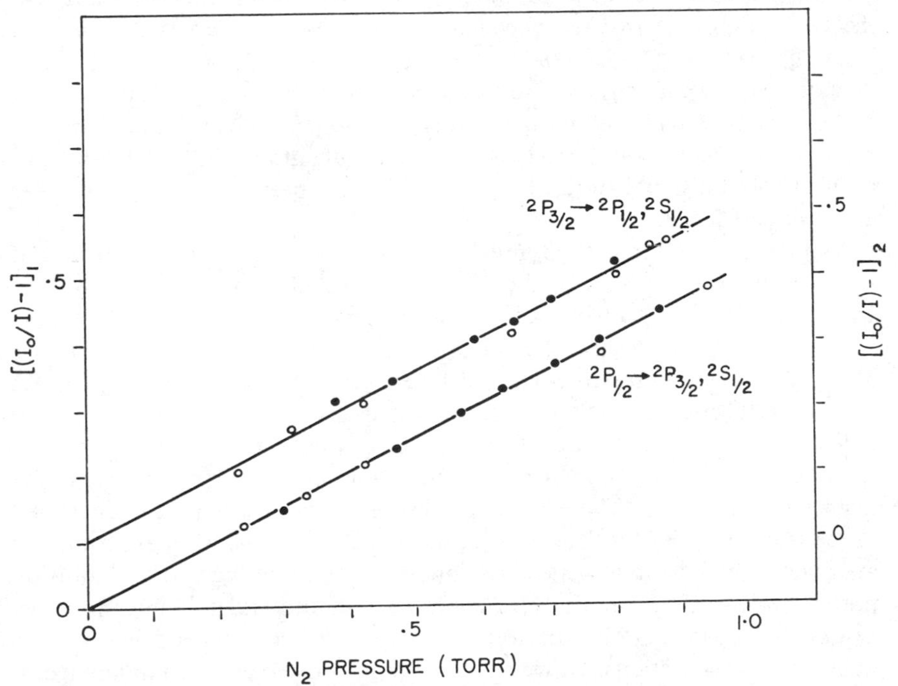

Figure 4.12 Stern–Volmer plots of $[(I_0/I) - 1]_1$ for the Cs 8944-Å resonance component and $[(I_0/I) - 1]_2$ for the 8521-Å component against N_2 pressure, showing the quenching of cesium resonance radiation. ○, results obtained with rising gas pressure; ●, results obtained with failing pressure.

[124–129]. Similarly, determination of effective quenching cross sections have been reported for the Rb $6\,^2P_{1/2,3/2}$ states [130], the Cs $8\,^2P_{1/2,3/2}$ and $7\,^2D_{3/2,5/2}$ states [131] and, also of the cross sections for quenching by N_2 of fifteen S, P, and D states in sodium [96]. The latter cross sections appear to exhibit resonance properties with respect to upward vibrational transitions in N_2. Measurements of relative fluorescent intensities were also used to study the quenching of the $7\,^3S_1$ and $6\,^3D_1$ states [132] and of the $6\,^3P_1$ state [133] in mercury, of the $6\,^2D_{3/2}$ state in thallium [134], the $4\,^2P$ resonance states in copper [135] and of the $5\,^3P_1$ state in cadmium [101]. The deexcitation of the Hg $6\,^3P_1$ state is complicated by the fact that atoms may be collisionally transferred either to the $6\,^1S_0$ ground state or to the $6\,^3P_0$ metastable state with varying degrees of efficiency. In addition, $6\,^3P_1$ and $6\,^3P_0$ atoms combine with the $6\,^1S_0$ ground-state atoms to form excited Hg_2 molecules that decay emitting band fluorescence [136, 137]. To determine separately the relative densities of the $6\,^3P_1$ and $6\,^3P_0$ atoms, which are needed to obtain the separate cross sections for the $6\,^3P_1 \rightarrow 6\,^1S_0$, $6\,^3P_1 \rightarrow 6\,^3P_0$, and $6\,^3P_0 \rightarrow 6\,^1S_0$ processes, it is necessary to measure not only the fluorescent intensities but also the absorption at a frequency which corresponds to an upward transition from the $6\,^3P_0$ state [138] ($n\,^3S_1$–$6\,^3P_0$ or $n\,^3D_1$–$6\,^3P_0$), since spontaneous decay of the $6\,^3P_0$ state is forbidden. Alternatively, the density of the Hg $6\,^3P_0$ atoms may be determined by measuring the flux of electrons ejected from a metallic electrode upon impact by the metastable atoms [139]. The difficult problem of collisional processes in mercury has been reviewed by Callear and Lambert [14].

A somewhat similar method, in which the excited metal atoms are produced by photodissociation, was used to study the quenching of sodium [140, 141] and potassium [142] resonance radiation and of the thallium metastable state [117, 118]. In the photodissociation experiments, the metallic halides are irradiated with ultraviolet light of sufficiently high frequency to dissociate the halide, producing excited metal atoms as well as ground-state halogen atoms. Some of the radiative energy absorbed by the halide is used to dissociate the molecule, and the excess is converted into translational energy of the dissociation products. The excited atoms emit fluorescence that is quenched in the usual manner by the addition of quenching gases. This experimental approach, which permits some variation in the relative velocity of the collision partners, depending on the frequency of the ultraviolet light, might be expected to provide information about the velocity dependence of the quenching cross sections. However, the attempts to obtain such data have so far been only partly successful [140–142].

Another experimental method available for the measurement of quenching cross sections hinges on the determination of the effective lifetime of the appropriate atomic resonance state in relation to the pressure of the quenching gas. If τ' is the effective lifetime at a density N of the quenching gas and τ is the "natural" mean lifetime, then

$$\frac{1}{\tau'} = \frac{1}{\tau} + NQv_r, \tag{14}$$

where Q is the quenching cross section and v_r is the average relative velocity of the colliding atom–molecule pair. Careful determinations of τ' in relation to N provide the best source of quenching cross sections, as long as the pressure of the atomic vapor is kept low enough to avoid trapping of resonance radiation. Because the experiments involve measurements of time rather than of intensity, fluctuations in the brightness of the exciting light source (or in the output of the electron gun if electrons are used to excite the atoms) do not give rise to experimental error, nor does the possible broadening of the atomic resonance lines. For similar reasons the measurements are not affected by the presence of small concentrations of reactive impurities that might combine with ground-state atoms and decrease their density. The advantages of this experimental approach have been recognized in recent years, and several quenching cross sections have been obtained in this way. The method of delayed coincidences [27] has been used to study the quenching of sodium [23, 24, 143], potassium [25, 26, 144], rubidium [144], and mercury [92] resonance radiation. The technique has been employed as well to determine the cross section for $6\,^3P_0 \rightarrow 6\,^3P_1$ transfer in Hg, induced in collisions with N_2 molecules. The results of this determination showed that collisions with N_2 lead only to $^3P_0$–$^3P_1$ mixing but do not quench $^3P_0$ or $^3P_1$ atoms to the $^1S_0$ ground state [145]. Quenching in sodium has also been extensively investigated using the technique of phase-shift fluorometry [112, 146, 147] to measure the effective lifetimes of the Na–D states. Quenching in thallium was studied in both delayed coincidence [143] and Hanle [134] experiments.

Some cross sections for the quenching of atomic resonance radiation are quoted in Table 4.7. There are obvious disparities between the results obtained at low and at high temperatures. In the cases of sodium, potassium, and rubidium the high-temperature cross sections measured in flames are consistently smaller than the low-temperature cross sections measured in fluorescence vessels. Figure 4.13 shows a plot against temperature of the cross sections for quenching of sodium resonance radiation by nitrogen. The Na–N_2 system has been studied perhaps more extensively than others, and quite recently Lijnse (122a) measured the quenching cross section over a range of flame temperatures. It may be seen in Figure 4.13 that, as temperatures increase, there is an initial steep decrease in the cross sections, followed by a levelling-off at higher temperatures. This behaviour of the cross sections for sodium as well as other alkali resonance states, has been successfully interpreted by Lijnse [122, 155] on the basis of a model for the interaction

Table 4.7
Cross Sections for Quenching of Atomic Fluorescence

QUENCHING PROCESS	MOLECULE	TEMPERATURE (°K)	CROSS SECTION (Å²)	EXPERIMENTAL METHOD	REF.
$Na(3\,^2P_{1/2,3/2}) \rightarrow Na(3\,^2S_{1/2})$	N_2	388	40.3	Delayed coinc.	a
	N_2	403	45.6	Fluor. vessel	b
	N_2	473	43	Phase-shift fluorom.	c
	N_2	523	36.6	Phase-shift fluorom.	d
	N_2	538; 638	30.7; 28.7	Phase-shift fluorom.	e
	N_2	1400–1800	21.5	Flame fluorescence	f
	N_2	1900	21	Flame fluorescence	g
	H_2	388	16.2	Delayed coinc.	a
	H_2	403	23.2	Fluor. vessel	b
	H_2	573	12.2	Phase-shift fluorom.	e
	H_2	1400–1800	9.0	Flame fluorescence	f
	H_2	1900	8	Flame fluorescence	g
	H_2	—	9.5	Photodissociation	h
	HD	388	11.6	Delayed coinc.	a
	HD	573	11.5	Phase-shift fluorom.	e
	D_2	388	9.8	Delayed coinc.	a
	D_2	573	10.2	Phase-shift fluorom.	e
	CO	403	88	Fluor. vessel	b
	CO	573	41	Phase-shift fluorom.	e
	CO	1400–1800	37.4	Flame fluorescence	f
	CO	1900	41	Flame fluorescence	g
	CO_2	1400–1800	53.5	Flame fluorescence	f
	CO_2	1900	50	Flame fluorescence	g
	CO_2	—	48–103	Photodissociation	h
	H_2O	1400–1800	1.6	Flame fluorescence	f
	H_2O	2100	2.2	Flame fluorescence	g
	CH_4	403	0.34	Fluor. vessel	b
	CH_4	573	0.7	Phase-shift fluorom.	e
$Na(9S \rightarrow 3S)$	N_2	503	54.8	Fluor. vessel	v
$Na(8S \rightarrow 3S)$	N_2	503	94.7	Fluor. vessel	v
$Na(7S \rightarrow 3S)$	N_2	503	15.7	Fluor. vessel	v
$Na(6S \rightarrow 3S)$	N_2	503	41.0	Fluor. vessel	v
$Na(5S \rightarrow 3S)$	N_2	503	27.4	Fluor. vessel	v
$Na(8D \rightarrow 3S)$	N_2	503	25.2	Fluor. vessel	v

Table 4.7 (Continued)

QUENCHING PROCESS	MOLECULE	TEMPERATURE (°K)	CROSS SECTION (Å²)	EXPERIMENTAL METHOD	REF.
Na(7$D \rightarrow$ 3S)	N_2	503	45.1	Fluor. vessel	v
Na(6$D \rightarrow$ 3S)	N_2	503	10.2	Fluor. vessel	v
Na(5$D \rightarrow$ 3S)	N_2	503	7.3	Fluor. vessel	v
Na(4$D \rightarrow$ 3S)	N_2	503	37.7	Fluor. vessel	v
Na(7$P \rightarrow$ 3S)	N_2	503	~380	Fluor. vessel	v
Na(6$P \rightarrow$ 3S)	N_2	503	3.1	Fluor. vessel	v
Na(5$P \rightarrow$ 3S)	N_2	503	7.6	Fluor. vessel	v
Na(4$P \rightarrow$ 3S)	N_2	503	23.4	Fluor. vessel	v
Na(3$P \rightarrow$ 3S)	N_2	503	40.8	Fluor. vessel	v
K(4 $^2P_{1/2,3/2}$)$\rightarrow$	N_2	353	34.0	Delayed coinc.	i
K(4 $^2S_{1/2}$)	N_2	1400	17.6	Flame fluorescence	j
	N_2	1900	19	Flame fluorescence	g
	N_2	—	20.5	Photodissociation	k
	H_2	353	9.4	Delayed coinc.	i
	H_2	1400–1800	3.2	Flame fluorescence	j
	H_2	1900	3.4	Flame fluorescence	g
	HD	353	12	Delayed coinc.	i
	D_2	353	8	Delayed coinc.	i
	CO	1400	37.9	Flame fluorescence	j
	CO	1900	44	Flame fluorescence	g
	CO_2	1400	67.2	Flame fluorescence	j
	CO_2	1900	66	Flame fluorescence	g
	CO_2	—	60	Photodissociation	k
	H_2O	1400–1800	2.8	Flame fluorescence	j
	H_2O	2100	2.6	Flame fluorescence	g
Rb(5 $^2P_{1/2,3/2}$)$\rightarrow$	N_2	1400	19	Flame fluorescence	j
Rb(5 $^2S_{1/2}$)	N_2	1900	25	Flame fluorescence	l
	H_2	1400	1.9	Flame fluorescence	j
	H_2	1900	3.6	Flame fluorescence	l
	CO	1400	37.0	Flame fluorescence	j
	CO_2	1400	75.4	Flame fluorescence	j
	H_2O	1400	4.0	Flame fluorescence	j
	H_2O	2100	4.0	Flame fluorescence	l
Rb(5 $^2P_{1/2}$)$\rightarrow$	N_2	340	58	Fluor. vessel	m
Rb(5 $^2S_{1/2}$)	N_2	?	37	Delayed coinc.	n

Table 4.7 (Continued)

QUENCHING PROCESS	MOLECULE	TEMPERATURE (°K)	CROSS SECTION (Å²)	EXPERIMENTAL METHOD	REF.
	H_2	340	6	Fluor. vessel	*m*
	HD	340	6	Fluor. vessel	*m*
	D_2	340	3	Fluor. vessel	*m*
	CH_4	340	<1	Fluor. vessel	*m*
	CD_4	340	2	Fluor. vessel	*m*
$Rb(5\,^2P_{3/2}) \rightarrow$	N_2	340	43	Fluor. vessel	*m*
$Rb(5\,^2S_{1/2})$	N_2	?	36	Delayed coinc.	*n*
	H_2	340	3	Fluor. vessel	*m*
	HD_2	340	5	Fluor. vessel	*m*
	D_2	340	5	Fluor. vessel	*m*
	CH_4	340	3	Fluor. vessel	*m*
	CD_4	340	3	Fluor. vessel	*m*
$Rb(6\,^2P_{1/2} \rightarrow$	H_2	430	36	Fluor. vessel	*w*
$^2S_{1/2})$	HD	430	47	Fluor. vessel	*w*
	D_2	430	28	Fluor. vessel	*w*
	N_2	430	128	Fluor. vessel	*w*
	CH_4	430	129	Fluor. vessel	*w*
	CD_4	430	82	Fluor. vessel	*w*
$Rb(6\,^2P_{3/2} \rightarrow$	H_2	430	31	Fluor. vessel	*w*
$^2S_{1/2})$	HD	430	38	Fluor. vessel	*w*
	D_2	430	21	Fluor. vessel	*w*
	N_2	430	126	Fluor. vessel	*w*
	CH_4	430	114	Fluor. vessel	*w*
	CD_4	430	76	Fluor. vessel	*w*
$Cs(6\,^2P_{1/2,3/2}) \rightarrow$	N_2	1400	78.5	Flame fluorescence	*j*
$Cs(6\,^2S_{1/2})$	H_2	1400	5.3	Flame fluorescence	*j*
	H_2O	1400	17.3	Flame fluorescence	*j*
$Cs(6\,^2P_{1/2}) \rightarrow$	N_2	315	77	Fluor. vessel	*o*
$Cs(6\,^2S_{1/2})$	H_2	315	7	Fluor. vessel	*o*
	HD	315	4	Fluor. vessel	*o*
	D_2	315	8	Fluor. vessel	*o*
$Cs(6\,^2P_{3/2}) \rightarrow$	N_2	315	69	Fluor. vessel	*o*
$Cs(6\,^2S_{1/2})$	H_2	315	5	Fluor. vessel	*o*
	HD	315	3	Fluor. vessel	*o*
	D_2	315	7	Fluor. vessel	*o*

Table 4.7 (Continued)

QUENCHING PROCESS	MOLECULE	TEMPERATURE (°K)	CROSS SECTION (Å^2)	EXPERIMENTAL METHOD	REF.
$Tl(7\,^2S_{1/2}) \rightarrow Tl(6\,^2P_{1/2,3/2})$	N_2	1400	20.1	Flame fluorescence	p
	H_2	1400	≤ 0.1	Flame fluorescence	p
	CO	1400	42.7	Flame fluorescence	p
	CO_2	1400	102	Flame fluorescence	p
	H_2O	1400	5.5	Flame fluorescence	p
$Cu(4\,^2P_{1/2}) \rightarrow Cu(4\,^2S_{1/2})$	N_2	?	14	Fluor. vessel	q
	M_2	?	22	Fluor. vessel	q
	CO_2	?	50	Fluor. vessel	q
$Cu(4\,^2P_{3/2}) \rightarrow Cu(4\,^2S_{1/2})$	N_2	?	19	Fluor. vessel	q
	H_2	?	23	Fluor. vessel	q
	CO_2	?	36	Fluor. vessel	q
$Hg(6\,^3P_1) \rightarrow Hg(6\,^3P_0, 6\,^1S_0)$	N_2	248	<1	Hanle effect	r
	N_2	296	0.73	Delayed coinc.	s
	N_2	1400	3.0	Flame fluorescence	t
	H_2	248	25.2	Hanle effect	s
	H_2	273	16.6	Fluor. vessel	u
	H_2	296	24.6	Delayed coinc.	r
	H_2	1400	15.7	Flame fluorescence	t
	D_2	273	15.7	Fluor. vessel	u
	D_2	296	22.7	Delayed coinc.	r
	CO	248	20.4	Hanle effect	s
	CO	296	21.7	Delayed coinc.	r
	CO	1400	16.3	Flame fluorescence	t
	CO_2	248	15.7	Hanle effect	s
	CO_2	296	10.2	Delayed coinc.	r
	CO_2	1400	20.4	Flame fluorescence	t

[a] Kibble, Copley, and Krause [24].
[b] Norrish and Smith [148].
[c] Demtröder [112].
[d] Hulpke, Paul, and Paul [146].
[e] Bästlein, Baumgartner, and Brosa [147].
[f] Jenkins [149].
[g] Hooymayers and Lijnse [150].
[h] Hanson [140, 141].
[i] Copley and Krause [25, 26].
[j] Jenkins [151].
[k] Gatzke [142].
[l] Hooymayers and Nienhuis [152].
[m] Hrycyshyn and Krause [129].
[n] Bellisio, Davidovits, and Kindlmann [144].
[o] McGillis and Krause [125].
[p] Jenkins [153].
[q] Bleekrode and van Benthem [135].
[r] Barrat *et al.* [90].
[s] Deech, Pitre, and Krause [92].
[t] Jenkins [154].
[u] Yang [133].
[v] Czajkowski [96].
[w] Siara [130].

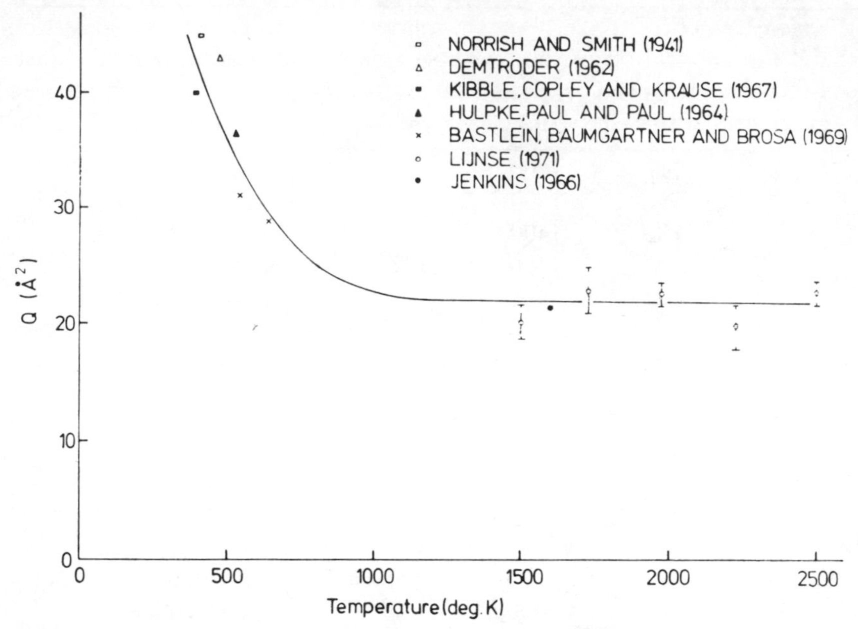

Figure 4.13 The variation with temperature of the cross sections for the quenching of sodium resonance radiation by nitrogen. □, Norrish and Smith [148]; △, Demtröder [112]; ■, Kibble, Copley and Krause [24]; ▲, Hulpke, Paul and Paul [146]; X, Bästlein, Baumgartner and Brosa [147]; ○, Lijnse [122]; ●, Jenkins [146].

involving an intermediate ionic state, in which attractive van der Waals forces play an essential role.

Mention should also be made of the "inverse quenching" process in which vibrationally excited molecules collide with ground-state atoms, effecting a conversion of vibrational energy to atomic excitation energy. The excited atoms then decay to the ground state with the emission of resonance fluorescence. Such transfer of energy from N_2 to Na has been observed in a flowing afterglow of a microwave discharge by Starr [156] and from N_2 to K by Starr and Shaw [157]. Mentall, Krause, and Fite [158] studied the N_2–Na system using crossed beams, and Tsuchiya and Suzuki [159] investigated energy transfer from CO to Na in a shock tube. The results of all these experiments point to the conclusion that collisions between vibrationally excited molecules and ground-state alkali atoms result in the efficient production of excited alkali atoms, thus verifying the contention that the sodium emission in the aurora is caused by a transfer of excitation from N_2 molecules (excited to the $v = 8$ state by energetic electrons) to sodium atoms [160]. The crossed-beams experiment [158] yielded a cross section of 20 Å^2 for this process, which is of the correct magnitude to explain the auroral phenomena and which also correctly correlates with the Na–N_2 quenching cross section through the application of the principle of detailed balancing. In contrast, the cross sections for the purely kinetic excitation by collisions with nitrogen molecules in their ground vibrational states have been quoted as 12 Å^2 for sodium and 19 Å^2 for potassium at beam energies of 5 eV [161, 162] and estimated as about 1 Å^2 at threshold. The results of the "inverse quenching" experiments thus indicate that, in quenching collisions, it is likely that the internal molecular energy plays a more important role than does kinetic energy.

B. Excitation Transfer between Fine-Structure States

Until recently, except for Lochte–Holtgreven's [32] early experiments, all studies of interactions between excited metal atoms and ground-state molecules dealt with the quenching process (in the case of mercury also with the transfer to the 6 $^3P_0$ metastable state). Yet, when investigating the quenching of alkali resonance radiation, it should be borne in mind that significant transfer of excitation does take place between the ^2P fine-structure states, often with a cross section larger than the quenching cross section.

During the past few years, several experimental studies have been reported, dealing with excitation transfer between the resonance fine-structure states of alkali atoms, induced in collisions with molecules. The ^2P mixing process in sodium has been investigated by Stupavsky and Krause [127, 128], in potassium by McGillis and Krause [124] and by Lijnse and Hornman [163], in rubidium by Bellisio, Davidovits, and Kindlmann [144], Hrycyshyn and Krause [129] and by Lijnse, Zeegers, and Alkemade [164], and in cesium by McGillis and Krause [125] and by Lijnse, Zeegers, and Alkemade [165]. Siara and Krause [130] also investigated fine-structure mixing in the 6 ^2P states of rubidium and Rocchiccioli [131] studied transfers from 8 $^2P_{1/2}$ to 8 $^2P_{3/2}$ and 7 $^2D_{3/2,5/2}$ in cesium. Most of the results were obtained using methods of sensitized fluorescence. In order to extract the mixing cross sections from the fluorescent intensity measurements, it is necessary to solve rate equations that include quenching terms as well as mixing terms, because in the vapor-gas mixture quenching and mixing processes occur simultaneously, though with different probabilities [126]. The various measured cross sections are summarized in Table 4.8. It should be noted that the two cross sections Q_2 for Rb–N_2 collisions, reported by Bellisio, Davidovits, and Kindlmann [144] and by Hrycyshyn and Krause [129], disagree by a factor of about 3; there is no obvious explanation for this large discrepancy.

Studies of excitation transfer induced in collisions with molecules were not limited to fine-structure states of alkali atoms. Excitation transfer from the 6 $^2D_{3/2}$ to the 6 $^2D_{5/2}$ state in thallium was investigated in a Hanle experiment

Table 4.8

Cross Sections for Mixing between Alkali and between Mercury Fine-Structure States: The Values in Brackets Are Theoretical, the Mercury Cross Sections Are Relative; the Ratios Q_1/Q_2 Are Experimental

DESCRIPTION OF PROCESS	CROSS SECTIONS (Å²) AND THEIR RATIOS FOR VARIOUS MOLECULES									
	N_2	H_2	HD	D_2	CH_4	CD_4	C_2H_2	C_2H_4	C_2H_6	CO
$Na(3\ ^2P_{1/2} \rightarrow 3\ ^2P_{3/2})$	144[a]	80[a]	84[a]	98[a]	148[b]	151[b]	182[b]	178[b]	182[b]	
	76[a]	42[a]	44[a]	52[a]	77[b]	81[b]	96[b]	94[b]	95[b]	
$Na(3\ ^2P_{1/2} \leftarrow 3\ ^2P_{3/2})$	(66.6)[c]	(33.1)[c]	(35.6)[c]	(37.5)[c]	(76.4)[c]	(78.4)[c]	(92.5)[c]	(95.9)[c]	(98.1)[c]	
$Q_1/Q_2(T = 398°K)$	1.89	1.90	1.91	1.88	1.92	1.86	1.90	1.89	1.91	
$(g_2/g_1)\exp(-\Delta E/kT)$	1.88	1.88	1.88	1.88	1.88	1.88	1.88	1.88	1.88	
$K(4\ ^2P_{1/2} \rightarrow 4\ ^2P_{3/2})$	100[d]	76[d]	74[d]	72[d]						
$K(4\ ^2P_{1/2} \leftarrow 4\ ^2P_{3/2})$	66[d]	53[d]	49[d]	50[d]						
$Q_1/Q_2(T = 363°K)$	1.52	1.43	1.51	1.44						
$(g_2/g_1)\exp(-\Delta E/kT)$	1.59	1.59	1.59	1.59						

$Rb(5\,^2P_{1/2} \rightarrow 5\,^2P_{3/2})$	16^e	11^e	18^e	22^e	30^e	28^e		23^e	57^e	
	23^e; 7^f	15^e	25^e	30^e	42^e	38^e		32^e	77^e	
$Rb(5\,^2P_{1/2} \leftarrow 5\,^2P_{3/2})$	(8)g									
$Q_1/Q_2(T = 340°K)$	0.73	0.71	0.74	0.75	0.73	0.73		0.74	0.74	
$(g_2/g_1)\exp(-\Delta E/kT)$	0.72	0.72	0.72	0.72	0.72	0.72		0.72	0.72	
$Rb(6\,^2P_{1/2} \rightarrow 6\,^2P_{3/2})$	107^j	41^j	42^j	42^j	38^j	52^j				
$Rb(6\,^2P_{1/2} \leftarrow 6\,^2P_{3/2})$	70^j	26^j	27^j	27^j	24^j	34^j				
$Q_1/Q_2(T = 430°K)$	1.53	1.58	1.55	1.55	1.58	1.52				
$(g_2/g_1)\exp(-\Delta E/kT)$	1.55	1.55	1.55	1.55	1.55	1.55				
$Cs(6\,^2P_{1/2} \rightarrow 6\,^2P_{3/2})$	4.7^h	6.7^h	4.8^h	4.2^h						
	25^h	44^h	32^h	28^h						
$Cs(6\,^2P_{1/2} \leftarrow 6\,^2P_{3/2})$	$(8.4)^g$									
$Q_1/Q_2(T = 315°K)$	0.19	0.15	0.15	0.15						
$(g_2/g_1)\exp(=\Delta E/kT)$	0.16	0.16	0.16	0.16						
$Hg(6\,^3P_2 \rightarrow 6\,^3P_1)$	0.71^i	0.42^i		0.10^i	0.62^i	0.21^i				1.00^i

[a] Stupavsky and Krause [127].
[b] Stupavsky and Krause [128].
[c] Values calculated from formula of Callaway and Bauer [76].
[d] McGillis and Krause [124].
[e] Hrycyshyn and Krause [129].
[f] Bellisio, Davidovits, and Kindlmann [144].
[g] Andreev and Voronin [166].
[h] McGillis and Krause [125].
[i] Doemeny, van Itallie, and Martin [167].
[j] Siara and Krause [130].

by Rityn, Chaika, and Cherenkovskii [134], who obtained cross sections of 12 Å^2 for collisions with nitrogen and 16 Å^2 with hydrogen. Doemeny, van Itallie, and Martin [167] observed fluorescence from crossed molecular beams and measured cross sections for the energy transfer Hg($6\,^3P_2 \rightarrow 6\,^3P_1$), induced in collisions with various molecules; their results are also quoted in Table 4.8.

A comparison of the results shown in Tables 4.3 and 4.8 leads to the immediate conclusion that in the cases of adiabatic collisions involving rubidium and cesium, the ^2P mixing cross sections for collisions with molecules are very much larger than for collisions with noble-gas atoms. A theoretical model for the collisional process, which involves the formation of an intermediate ionic complex [166], successfully explained these large differences in the cases of the Cs–N_2 and Rb–N_2 interactions and permitted the calculation of cross sections that agreed with experimental results. The applicability of the ionic complex model to interactions involving N_2 molecules has recently been confirmed by Goddard [168] who investigated the velocity dependence of the cross sections for ^2P mixing in potassium in a crossed molecular beam. The cross sections for the nonadiabatic mixing process in sodium are, on the other hand, in good agreement with values calculated on the basis of Callaway and Bauer's [76] theory for collisions between alkali and noble-gas atoms, in which the alkali fine-structure splitting is assumed to be negligible and the collision partner is considered as a polarizable sphere.

The general mechanism of excitation transfer in collisions with molecules, and particularly the role of the molecular vibrational and rotational levels in the interaction, are not yet fully understood. It is difficult to explain the large sizes of the cross sections on the basis of a resonant energy transfer to the vibrational states, which would require significantly more energy than is released in the atomic transition [169]. On the other hand, correlations between the sizes of the cross sections and possible resonances with specific molecular rotational transitions are rather uncertain [125]. Considerable additional light has been shed upon this problem by a recent series of investigations in which ^2P mixing in cesium [170, 171] and rubidium [87], induced in collisions with various deuterated methane molecules, were studied in relation to temperature. The temperature-dependence of the cross sections, shown in Fig. 4.14, differs markedly from that observed with noble gas atoms as collision partners (see Fig. 4.6) and an isotope effect is apparent, which has also been observed in the cross sections for collisions between cesium and isotopic hydrogens and ethanes [172]. The isotope effect as well as the large sizes of the cross sections have been shown to arise as the result of coupling between the atomic fine-structure transitions and the molecular rotation through the interaction of the third spatial derivative of the atomic

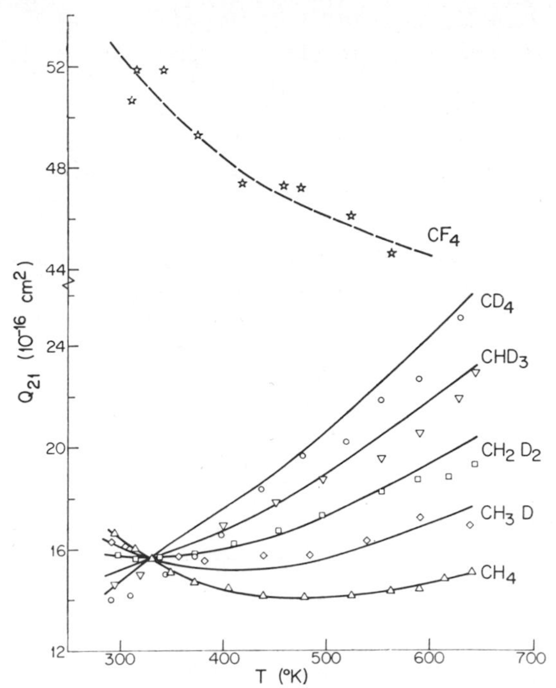

Figure 4.14 The variation with temperature of the cross sections for $6^2P_{1/2} \leftarrow 6^2P_{3/2}$ mixing in cesium. △, Cs-CH_4; ◇, Cs-CH_3D; □, Cs-CH_2D_2; ▽, Cs-CHD_3; 0, Cs-CD_4; ★, Cs-CF_4. The points are experimental and the solid curves represent theoretical calculations [170]. The theory does not apply to the case of CF_4 where only resonant electronic-to-vibrational energy transfer is likely. No specific functional relation is implied by the dashed curve.

electrostatic potential with the rotating methane octupole moment [170]. It is expected that this theory which was the first to provide evidence for strong rotational enhancement of collisionally induced nonadiabatic electronic transitions, will constitute a basis for the interpretation of other similar inelastic phenomena.

REFERENCES

1. A. Dalgarno, *Rev. Mod. Phys.*, **39,** 850 (1967).
2. A. Dalgarno, *Rev. Mod. Phys.*, **39,** 858 (1967).
3. A. C. G. Mitchell and M. W. Zemansky, *Resonance Radiation and Excited Atoms*, Cambridge Univ. Press, London, 1934.
4. P. Pringsheim, *Fluorescence and Phosphorescence*, Interscience, New York, 1949.

5. W. Ermisch, *Beitr. Plasmaphysik.*, **6**, 73 (1966).
6. R. Seiwert, *Habilitationsschrift* (Techn. Universität, Berlin, 1965) (unpublished).
7. R. Seiwert, *Ergeb. Exakt. Naturw.*, **47**, 143 (1968).
8. L. Krause, *Appl. Opt.*, **5**, 1375 (1966).
9. L. Krause, *Physics of Electronic and Atomic Collisions*, edited by T. R. Govers and F. J. DeHeer (North-Holland Publishing Company, Amsterdam), p. 65 (1972).
10. E. K. Kraulinya, *Sensitized Fluorescence in Mixtures of Metal Vapors*, Vol. I, Latvian State University Press, Riga, 1968 (in Russian).
11. E. K. Kraulinya, *Sensitized Fluorescence in Mixtures of Metal Vapors*, Vol. II Latvian State University Press, Riga, 1969 (in Russian).
12. F. R. Gilmore, E. Bauer, and J. Wm. McGowan, *J. Quant. Spectr. Radiative Transfer*, **9**, 157 (1969).
13. A. B. Callear, *Appl. Opt. Suppl. Chemical Lasers*, 145 (1965).
14. A. B. Callear and J. D. Lambert, *Comprehensive Chemical Kinetics*, Vol. 3, Elsevier, Amsterdam, 1969.
15. C. Th. J. Alkemade and P. J. Th. Zeegers, *Spectrochemical Methods of Analysis: Quantitative Analysis of Atoms and Molecules*, Wiley, New York, 1971.
16. P. L. Lijnse, Report i398, Fysisch Laboratorium, Rijksuniversiteit, Utrecht, Netherlands (1972). Unpublished.
17. G. zu. Putlitz, *Atomic Physics*, Plenum, New York, 1969.
18. P. P. Feofilov, *The Physical Basis of Polarized Emission*, Consultants Bureau, New York, 1961.
19. W. Happer, *Rev. Mod. Phys.* **44**, 169 (1972).
20. T. Holstein, *Phys. Rev.*, **72**, 1212 (1947).
21. T. Holstein, *Phys. Rev.*, **83**, 1159 (1951).
22. P. J. Walsh, *Phys. Rev.*, **116**, 511 (1959).
23. B. P. Kibble, G. Copley, and L. Krause, *Phys. Rev.*, **153–1**, 9 (1967).
24. B. P. Kibble, G. Copley, and L. Krause, *Phys. Rev.*, **159–1**, 11 (1967).
25. G. Copley and L. Krause, *Can. J. Phys.*, **47**, 533 (1969).
26. G. Copley and L. Krause, unpublished, 1969.
27. E. E. Habib, B. P. Kibble, and G. Copley, *Appl. Opt.*, **7**, 673 (1968).
28. J. S. Deech and W. E. Baylis, *Can. J. Phys.*, **49**, 90 (1971).
29. E. S. Hrycyshyn and L. Krause, *Can. J. Phys.*, **47**, 215 (1969).
30. E. S. Hrycyshyn and L. Krause, *Can. J. Phys.*, **47**, 223 (1969).
31. R. W. Wood, *Phil. Mag.*, **27**, 1018 (1914).
32. W. Lochte-Holtgreven, *Z. Physik.*, **47**, 362 (1928).
33. J. L. Snider, *J. Opt. Soc. Am.*, **57**, 1394 (1967).
34. A. Gallagher, *Phys. Rev.*, **172–1**, 88 (1968).
35. R. J. Atkinson, G. D. Chapman, and L. Krause, *J. Opt. Soc. Am.*, **55**, 1269 (1965).
36. D. H. Burling, M. Czajkowski, and L. Krause, *J. Opt. Soc. Am.*, **57**, 1162 (1967).
37. W. Berdowski, T. Shiner, and L. Krause, *Appl. Opt.*, **6**, 1683 (1967).

38. F. C. Fehsenfeld, K. M. Evenson, and H. P. Broida, *Rev. Sci. Instr.*, **36,** 294 (1965).
39. R. M. Dagnall and T. S. West, *Appl. Opt.*, **7,** 1287 (1968).
40. G. zu. Putlitz, *Ann. Physik.*, **11,** 284 (1963).
41. B. Budick, R. Novick, and A. Lurio, *Appl. Opt.*, **4,** 229 (1965).
42. A. J. Gibson, *J. Phys. E.*, **2,** 802 (1969).
43. H. Walther and J. L. Hall, *Appl. Phys. Lett.*, **17,** 239 (1970).
44. M. Rozwadowski and E. Lipworth, *J. Chem. Phys.*, **43,** 2347 (1965).
45. A. Gallagher and E. L. Lewis, *J. Opt. Soc. Am.* **63,** 864 (1973).
46. A. N. Nesmeyanov, *Vapor Pressures of the Chemical Elements*, (Elsevier, New York) 1963.
47. A. G. A. Rae and L. Krause, *Can. J. Phys.*, **43,** 1574 (1965).
48. G. A. Morton, *Appl. Opt.*, **7,** 1 (1968).
49. J. K. Nakamura and S. E. Schwartz, *Appl. Opt.*, **7,** 1073 (1968).
50. P. W. Pace and J. B. Atkinson, *J. Phys. E.* **7,** 556 (1974).
51. E. Milne, *J. London Math. Soc.*, **1,** 1 (1926).
52. E. I. Dashevskaya, E. E. Nikitin, A. I. Voronin, and A. A. Zembekov, *Can. J. Phys.*, **47,** 981 (1969).
53. E. I. Dashevskaya, A. I. Voronin, and E. E. Nikitin, *Can. J. Phys.*, **47,** 1237 (1969).
54. J. Pitre and L. Krause, *Can. J. Phys.*, **46,** 125 (1968).
55. R. Seiwert, *Ann. Physik*, **18,** 54 (1956).
56. G. D. Chapman and L. Krause, *Can. J. Phys.*, **44,** 753 (1966).
57. K. Hoffmann and R. Seiwert, *Ann. Physik*, **7,** 71 (1961).
58. M. A. Thangaraj, Ph.D. Thesis, University of Toronto, 1948.
59. M. Czajkowski and L. Krause, *Can. J. Phys.*, **43,** 1259 (1965).
60. H. Bunke and R. Seiwert, *Optik u. Spektroskopie aller Wellenlängen*, Akademie Verlag, Berlin, 1962.
61. J. Berlande, private communication, 1970.
62. Yu. A. Vdovin, V. M. Galitskii, and N. A. Dobrodeev, *Soviet Phys. J.E.T.P.*, **29,** 722 (1969).
63. P. W. Pace and J. B. Atkinson, *Can. J. Phys.* **52,** 1635 (1974).
64. M. Pimbert, *J. Physique*, **33,** 331 (1972).
64a. J. Cuvellier, P. R. Fournier, F. Gounand, and J. Berlande, *Compt. Rend.*, **B276,** 855 (1973).
65. A. A. Zembekov and E. E. Nikitin, *Chem. Phys. Lett.* **9,** 213 (1971).
66. M. Pimbert, J. Rocchiccioli, and J. Cuvellier, *Compt. Rend.*, **B270,** 684 (1970).
67. J. Franck, *Naturwiss.*, **14,** 211 (1929).
68. M. H. Ornstein and R. N. Zare, *Phys. Rev.*, **181–1,** 214 (1969).
69. V. Stacey and R. N. Zare, *Phys. Rev.* A, **1,** 1125 (1970).
70. M. Czajkowski, D. A. McGillis, and L. Krause, *Can. J. Phys.*, **44,** 741 (1966).
71. M. Czajkowski, G. Skardis, and L. Krause, *Can. J. Phys.*, **51,** 334 (1973).
72. J. Pitre and L. Krause, *Can. J. Phys.*, **45,** 2671 (1967).
73. J. W. Moskovitz and R. T. Thorson, *J. Chem. Phys.*, **38,** 1848 (1963).
74. R. H. G. Reid and A. Dalgarno, *Chem. Phys. Lett.*, **6,** 85 (1970).
75. J. A. Jordan and P. A. Franken, *Phys. Rev.*, **142,** 20 (1966).

76. J. Callaway and E. Bauer, *Phys. Rev.*, **140,** A1072 (1965).
77. L. Kumar and J. Callaway, *Phys. Lett.*, **28A,** 385 (1968).
78. E. E. Nikitin, *Opt. Spectry.*, **19,** 91 (1965).
79. E. I. Dashevskaya and E. E. Nikitin, *Opt. Spectry.*, **22,** 866 (1967).
80. B. Pitre, A. G. A. Rae, and L. Krause, *Can. J. Phys.*, **44,** 731 (1966).
81. T. J. Beahn, W. J. Condell, and H. I. Mandelberg, *Phys. Rev.*, **141–1,** 83 (1966).
82. E. I. Dashevskaya, E. E. Nikitin, and A. I. Resnikov, *J. Chem. Phys.*, **53,** 1175 (1970).
83. M. Czajkowski, D. A. McGillis, and L. Krause, *Can. J. Phys.*, **44,** 91 (1966).
84. R. H. G. Reid, *J. Phys. B.* **6,** 2018 (1973).
85. E. L. Lewis and L. F. McNamara, *Phys. Rev. A* **5,** 2643 (1972).
86. F. Masnou-Seeuws, *J. Phys. B.* **3,** 1437 (1970).
87. R. Phaneuf, Ph.D. Thesis, University of Windsor, 1973.
88. I. Siara, E. S. Hrycyshyn, and L. Krause, *Can. J. Phys.* **50,** 1826 (1972).
89. I. Siara, H. S. Kwong, and L. Krause, *Can. J. Phys.* **52,** 945 (1974).
90. J. P. Barrat, D. Casalta, J. L. Cojan, and J. Hamel, *J. Phys. Rad.*, **27,** 608 (1966).
91. A. I. Voronin and V. A. Kvlividze, *Theor. Chim. Acta*, **8,** 334 (1967).
92. J. S. Deech, J. Pitre, and L. Krause, *Can. J. Phys.* **49,** 1976 (1971).
93. B. V. Waddell and G. S. Hurst, *J. Chem. Phys.*, **53,** 3892 (1970).
94. G. Cario and J. Franck, *Z. Physik.*, **17,** 202 (1923).
95. B. Beutler and H. Josephy, *Z. Physik.*, **53,** 747 (1929).
96. M. Czajkowski, L. Krause, and G. Skardis, *Can. J. Phys.* **51,** 1582 (1973).
97. S. E. Frish and O. P. Bochkova, *Soviet Phys. J.E.T.P.*, **16,** 237 (1963).
98. S. E. Frish and O. P. Bochkova, *Bull. Acad. Sci. U.S.S.R.*, **27,** 1038 (1963).
99. E. K. Kraulinya, *Opt. Spectry.*, **17,** 250 (1964).
100. S. E. Frish and E. K. Kraulinya, *Dokl. Akad. Nauk S.S.S.R.*, **101,** 837 (1955).
101. M. Czajkowski and L. Krause, *Can. J. Phys.* to be published.
102. B. C. Hudson and B. Curnutte, *Phys. Rev.*, **152–1,** 56 (1966).
103. E. K. Kraulinya and A. E. Lezdin, *Opt. Spectry.*, **20,** 304 (1966).
104. E. K. Kraulinya, A. E. Lezdin, and O. S. Sametis, in *Proceedings of the Fifth International Conference on Electronic and Atomic Collisions*, Nauka, Leningrad, 1967.
105. W. Gough, *Proc. Phys. Soc.* (London), **90,** 287 (1967).
106. E. K. Kraulinya, O. S. Sametis, and A. P. Bryukhovetskii, *Opt. Spectry.*, **29,** 227 (1970).
107. E. N. Morozov and M. L. Sosinskii, *Opt. Spectry.*, **25,** 282 (1968).
108. E. K. Kraulinya and M. G. Arman, *Opt. Spectry.*, **26,** 285 (1969).
109. M. L. Sosinskii and E. N. Morozov, *Opt. Spectry.*, **23,** 475 (1967).
110. E. K. Kraulinya and M. L. Yanson, *Opt. Spectry.*, **29,** 239 (1970).
111. B. Chéron, *Opt. Communications* **3,** 437 (1971).
112. W. Demtröder, *Z. Physik.*, **166,** 42 (1962).
113. E. E. Nikitin and V. K. Bykhovskii, *Opt. Spectry.*, **17,** 444 (1965).
114. G. Copley, B. P. Kibble, and L. Krause, *Phys. Rev.*, **163–1,** 34 (1967).
115. S. Tsuchiya and K. Kuratani, *Combust. Flame.*, **8,** 299 (1964).
116. J. H. Stamper, *J. Chem. Phys.*, **43,** 759 (1965).

117. J. A. Bellisio and P. Davidovits, *J. Chem. Phys.*, **53,** 3474 (1970).
118. R. C. Pickett and R. Anderson, *J. Quant. Spectr. Radiative Transfer*, **9,** 697 (1969).
119. M. G. Edwards, *J. Phys. B*, **2,** 719 (1969).
119a. J. N. Dodd, E. Enemark, and A. Gallagher, *J. Chem. Phys.* **50,** 4838 (1969).
120. E. R. Fisher and G. K. Smith, R.I.E.S. Report 70-10 (Wayne State University, Detroit, Mich.) (1970).
121. E. R. Fisher and G. K. Smith, *Chem. Phys. Lett.*, **13,** 448 (1972).
122. P. L. Lijnse, *J. Quant. Spectr. Radiative Transfer*, to be published.
122a. P. L. Lijnse and R. J. Elsenaar, *J. Quant. Spectr. Radiative Transfer*, **12,** 1115 (1972).
123. O. Stern and M. Volmer, *Physik. Z.*, **20,** 183 (1919).
124. D. A. McGillis and L. Krause, *Can. J. Phys.*, **46,** 25 (1968).
125. D. A. McGillis and L. Krause, *Can. J. Phys.*, **46,** 1051 (1968).
126. D. A. McGillis and L. Krause, *Phys. Rev.*, **153–1,** 44 (1967).
127. M. Stupavsky and L. Krause, *Can. J. Phys.*, **46,** 2127 (1968).
128. M. Stupavsky and L. Krause, *Can. J. Phys.*, **47,** 1249 (1969).
129. E. S. Hrycyshyn and L. Krause, *Can. J. Phys.*, **48,** 2761 (1970).
130. I. Siara and L. Krause, *Can. J. Phys.* **51,** 257 (1973).
131. J. Rocchiccioli, *Compt. Rend.*, **B274,** 787 (1972).
132. J. L. Cojan and M. Huet, *Compt. Rend.*, **263,** 1223 (1966).
133. K. Yang, *J. Am. Chem. Soc.*, **87,** 5294 (1965).
134. E. Rityn, M. Chaika, and V. Cherenkovskii, *Opt. Spectry.*, **28,** 344 (1970).
135. R. Bleekrode and W. van Benthem, *J. Chem. Phys.*, **51,** 2757 (1969).
136. S. Mrozowski, *Z. Physik.*, **104,** 228 (1937).
137. R. A. Phaneuf, J. Skonieczny, and L. Krause, *Phys. Rev.* **A8,** 2980 (1973).
138. H. Horiguchi and S. Tsuchiya, 1970, private communication.
139. A. C. Vikis and H. C. Moser, *J. Chem. Phys.*, **53,** 2333 (1970).
140. H. G. Hanson, *J. Chem. Phys.*, **23,** 1391 (1955).
141. H. G. Hanson, *J. Chem. Phys.*, **27,** 491 (1957).
142. J. Gatzke, *Z. Physik. Chem.*, **223,** 321 (1963).
143. L. E. Brus, *J. Chem. Phys.*, **52,** 1716 (1970).
144. J. A. Bellisio, P. Davidovits, and P. J. Kindlmann, *J. Chem. Phys.*, **48,** 2376 (1968).
145. J. Pitre, K. Hammond, and L. Krause, *Phys. Rev.* **A6,** 2101, (1972).
146. E. E. Hulpke, E. Paul, and W. Paul, *Z. Physik*, **177,** 257 (1964).
147. C. Bästlein, G. Baumgartner, and B. Brosa, *Z. Physik*, **218,** 319 (1969).
148. G. W. Norrish and W. M. Smith, *Proc. Roy. Soc.* (London), **A176,** 295 (1941).
149. D. R. Jenkins, *Proc. Roy. Soc.* (London), **A293,** 493 (1966).
150. H. P. Hooymayers and P. L. Lijnse, *J. Quant. Spectr. Radiative Transfer*, **9,** 995 (1969).
151. D. R. Jenkins, *Proc. Roy. Soc.* (London), **A303,** 453 (1968).
152. H. P. Hooymayers and G. Nienhuis, *J. Quant. Spectr. Radiative Transfer*, **8,** 955 (1968).
153. D. R. Jenkins, *Proc. Roy. Soc.* (London), **A303,** 467 (1968).
154. D. R. Jenkins, 1971, private communication.

155. P. L. Lijnse, *Chem. Phys. Lett.* **18,** 73 (1973).
156. W. L. Starr, *J. Chem. Phys.*, **43,** 73 (1965).
157. W. L. Starr and T. M. Shaw, *J. Chem. Phys.*, **44,** 4181 (1966).
158. J. E. Mentall, H. F. Krause, and W. L. Fite, *Discussions Faraday Soc.* **44,** 157 (1967).
159. S. Tsuchiya and I. Suzuki, *J. Chem. Phys.*, **51,** 5725 (1969).
160. D. M. Hunten, *J. Atmos. Terrest. Phys.*, **27,** 538 (1965).
161. V. Kempter, W. Mecklenbrauck, M. Menzinger, G. Schuller, D. R. Herschbach, and C. Schlier, *Chem. Phys. Lett.*, **6,** 97 (1970).
162. W. Mecklenbrauck, Diplomarbeit, Freiburg University, 1971 (unpublished).
163. P. L. Lijnse and J. C. Hornman, *J. Quant. Spectr. Radiative Transfer*, **14,** 1079 (1974).
164. P. L. Lijnse, P. J. Th. Zeegers, and C. Th. J. Alkemade, *J. Quant. Spectr. Radiative Transfer*, **13,** 1033 (1973).
165. P. L. Lijnse, P. J. Th. Zeegers, and C. Th. J. Alkemade, *J. Quant. Spectr. Radiative Transfer*, **13,** 1301 (1973).
166. E. A. Andreev and A. I. Voronin, *Chem. Phys. Lett.*, **3,** 488 (1969).
167. L. J. Doemeny, F. J. van Itallie, and R. M. Martin, *Chem. Phys. Lett.*, **4,** 302 (1969).
168. T. P. Goddard, Ph.D. Thesis, University of California at Santa Cruz (1973). Unpublished.
169. D. A. McGillis and L. Krause, *Can. J. Phys.*, **47,** 473 (1969).
170. W. E. Baylis, E. Walentynowicz, R. A. Phaneuf, and L. Krause, *Phys. Rev. Lett.* **31,** 741 (1973).
171. E. Walentynowicz, R. A. Phaneuf, W. E. Baylis, and L. Krause, *Can. J. Phys.* **52,** 584 (1974).
172. E. Walentynowicz, R. A. Phaneuf, and L. Krause, *Can. J. Phys.*, **52,** 589 (1974).

CHAPTER FIVE

THEORY OF NONADIABATIC COLLISION PROCESSES INCLUDING EXCITED ALKALI ATOMS

E. E. Nikitin

Institute of Chemical Physics, Academy of Sciences, Moscow, U.S.S.R.

Contents

I. INTRODUCTION

A large number of elementary molecular collision processes proceeding via (or in) excited electronic states are known at present. A prominent feature of all these is that as a rule they can not be interpreted (even at a very low kinetic energy of nuclei) in terms of the motion of a representative point over a multidimensional potential-energy surface. The breakdown of the Born–Oppenheimer approximation, which manifests itself in the so-called nonadiabatic coupling of electronic and nuclear motion, induces transitions between electronic states that remain still well defined at infinitely large intermolecular distances.

The problem of cross-section calculation for various inelastic collisions is mathematically equivalent to the solution of a set (in principle, infinite) of coupled wave equations for nuclear motion [1]. Machine calculations have been done recently to obtain information about nonadiabatic coupling in some representative processes. Although very successful, these calculations do not make it easy to interpret particular transitions in terms of a particular interaction. It is here that the relatively simple models of nonadiabatic coupling still play an important part in the detailed interpretation of a mechanism, thus contributing to our understanding of the dynamic interaction between electrons and nuclei in a collision complex.

The simplicity of the models that makes them analytically tractable is due mainly to the following:

1. Semiclassical treatment of the relative motion of collision partners.
2. Small number of strongly coupled electronic states.
3. Rather small extent ("localization") of regions where nonadiabatic coupling is large.

It will be seen in what way these conditions facilitate the consideration of nonadiabatic coupling. Following these three points we are now able to do the following:

1. Replace the set of second-order coupled equations in the nuclear configuration space by a set of first-order equations appropriate to the time-dependent Schrödinger equation.
2. Separate out a small number of equations in an attempt to find a simple solution.
3. Consider only the small part of the trajectory that is important for nonadiabatic coupling.

These simplifications can be made by introducing several approximations, some of which are very difficult to assess. The most important questions that arise are the following:

1. Is the conception of a single trajectory determining the relative motion of partners over several electronic surfaces still useful?
2. Will the solution of a limited number of equations give a result which is not sensitive to the selected electronic basic set?
3. How does the "overlap" of coupling regions influence the net result; or how and when can known solutions for localized regions be used to construct a solution for "overlapping" regions?

At this stage it is not always easy to estimate possible errors introduced by the nature of these approximations. More work is needed to answer definitely these questions.

In this chapter some simple collision processes proceeding via electronic excited states will be discussed. It is not intended to present another review on energy transfer, but rather to demonstrate the present status of the theory dealing with nonadiabatic processes in slow collisions of heavy particles. Also, we shall not discuss questions that arise in the calculation of differential inelastic scattering since these have been reviewed recently [2–5].

For obvious reasons the theory of processes occurring in atom–atom collisions is developed much better than that for atom–molecule or molecule–molecule collisions. Thus the emphasis will be on the former. To present a more unified approach and to escape dealing with many tricky questions of interatomic interactions and construction of potential surfaces, we confine ourselves to the simplest collision processes exhibiting the main features that could be found in other processes. These "simplest" processes are inelastic collisions of alkali atoms in their first excited states with unexcited atoms under gas-kinetic conditions. Historically, such processes were the first studied both experimentally and theoretically [1], and many recent papers deal with the subject. The systems chosen offer a convenient classification of possible inelastic events according to the amount of electronic energy converted to translational energy of nuclei. In general, the smaller the energy change the larger would be the interatomic distance where the mixing of electronic states occurs. As the interaction is better known at large interatomic separations, these processes receive more attention, not necessarily proportional to their role in the kinetics of energy transfer. Nevertheless, it is believed that some general approach that proved successful for relatively simple processes might eventually be used for more complicated inelastic collisions.

II. SEMICLASSICAL TREATMENT OF NONADIABATIC COUPLING IN THE SCATTERING PROBLEMS

It is supposed that a collision can be characterized by the Hamiltonian $\mathscr{H}$ written as a sum of the Hamiltonian of noninteracting partners $\mathscr{H}_0(\mathbf{r})$ and the interaction $\mathscr{V}(\mathbf{r}, \mathbf{R})$ that vanishes at large interatomic distance $\mathbf{R}$ ($\mathbf{r}$ represents a set of electron coordinates). Given the classical trajectory $\mathbf{R} = \mathbf{R}(t)$, the problem of inelastic scattering is equivalent to the solution of the time-dependent Schrödinger equation:

$$i\hbar \frac{\partial \psi(\mathbf{r}, t)}{\partial t} = \mathscr{H}(\mathbf{r}, t)\psi(\mathbf{r}, t) = \{\mathscr{H}_0(\mathbf{r}) + \mathscr{V}(\mathbf{r}, \mathbf{R}[t])\}\psi(\mathbf{r}, t). \tag{1}$$

It is usual for the scattering theory to expand ψ in terms of eigenfunctions of $\mathscr{H}_0$, and then to solve the time-dependent equations for the corresponding probability amplitudes. An alternative approach would be to use an adiabatic basic set ϕ_k obtained by diagonalization of $\mathscr{H}$ at each configuration of the nuclei:

$$\mathscr{H}(\mathbf{r}, \mathbf{R})\phi_k(\mathbf{r}, \mathbf{R}) = U_k(\mathbf{R})\phi_k(\mathbf{r}, \mathbf{R}), \tag{2}$$

where $U_k(\mathbf{R})$ is just an adiabatic potential surface for the kth electronic state of interacting partners. If ψ is now expanded in the form

$$\psi(\mathbf{r}, t) = \sum_k c_k(t)\phi_k(\mathbf{r}, \mathbf{R}) = \sum_k a_k(t)\phi_k(\mathbf{r}, \mathbf{R}) \exp\left(-\frac{i}{\hbar}\int^t U_k(\mathbf{R})\, dt\right),$$

$$\mathbf{R} = \mathbf{R}(t) \tag{3}$$

then the amplitudes $a_k(t)$ are to be found from the following set of equations:

$$i\hbar\dot{a}_k = \sum_m \langle\phi_k| - i\hbar\frac{\partial}{\partial t}|\phi_m\rangle \exp\left(-\frac{i}{\hbar}\int^t (U_m - U_k)\, dt\right)a_m, \tag{4}$$

which describe the nonadiabatic coupling between adiabatic states ϕ_k. The two- (or N-) state approximation essentially consists of truncating this set of equations and reducing it to two (or N) coupled equations. It is assumed further that the two-state approximation is a good one in the sense that it is not sensitive to the basis set of electronic functions used. If so, two adiabatic electronic functions $\phi_i(\mathbf{r}, \mathbf{R})$ ($i = 1, 2$) that depend on electronic and nuclear coordinates can be reasonably approximated by a linear combination of two diabatic electronic functions $\phi_i^0(\mathbf{r})$ which themselves do not depend on nuclear coordinates. It is the $\mathbf{R}$ dependence of the expansion coefficients that is now responsible for the nonadiabatic coupling. Generally, if ψ is represented in a form

$$\psi(r, t) = \sum_k b_k(t)\phi_k^0(\mathbf{r}) \exp\left(-\frac{i}{\hbar}\int^t H_{kk}\, dt\right), \tag{5}$$

then the $b_k(t)$ are to be found from

$$i\hbar \dot{b}_k = \sum_m H_{km}(t) \exp\left(-\frac{i}{\hbar}\int^t (H_{mm} - H_{kk})\, dt\right) b_m. \tag{6}$$

Here $H_{km}(\mathbf{R})$ are matrix elements of $\mathscr{H}$ calculated in the diabatic basis set.

A further, very important, assumption that is usually valid for low-energy collisions is that nonadiabatic coupling occurs only in effectively localized regions (so-called nonadiabatic regions), so that an explicit approximation to $H(\mathbf{R})$ is needed only within these regions. This implies that away from the nonadiabatic region, coefficients $a_k(t)$ tend to constant values—which is, of course, only an approximation. In particular, at $R \to \infty$ the adiabatic basis gives stationary atomic orbitals, while the collision problem corresponds to atomic orbitals centered at moving nuclei. This shortcoming of the adiabatic electronic set is partly compensated for by using traveling atomic orbitals [6], but this method is not general enough to be recommended for interatomic distances that correspond to the strong interaction $\mathscr{V}(\mathbf{R})$. However, for low-energy collisions this effect is negligible, and the typical situation corresponds to distinctly separated nonadiabatic regions, located at positions $\mathbf{R}_p^{(k)}$ on the trajectory $\mathbf{R} = \mathbf{R}(t)$ or at moments $t_p^{(k)}$ on the time scale.

Passing over the coupling region at the moment $t = t_p$ will result in a transformation between coefficients a_k described by the matrix of nonadiabatic transitions (N matrix). The scattering matrix (S matrix) connecting coefficients c_n before and after collision can be constructed by successive multiplication of N matrices and diagonal matrices describing adiabatic evolution of $c_n(t)$ (A matrices). Thus for N regions of nonadiabaticity,

$$\mathscr{S} = \mathscr{A}(\infty, t_p^{(N)})\mathscr{N}_N \cdots \mathscr{N}_2\mathscr{A}(t_p^{(2)}, t_p^{(1)})\mathscr{N}_1\mathscr{A}(t_p^{(1)}, -\infty), \tag{7}$$

where

$$A_{km}(t_p, t_p') = \delta_{km} \exp\left(-\frac{i}{\hbar}\int_{t_p'}^{t_p} U_m[\mathbf{R}(t)]\, dt\right) \tag{8}$$

and $\mathscr{N}_k$ are to be found from the set of equations (4) solved for the kth coupling region. This last step is precisely the basic problem of the theory of nonadiabatic transitions.

It must also be mentioned that the $\mathbf{R}(t)$ appearing in diagonal elements of A need not all be the same, but each can be related to the corresponding potential $U_n(\mathbf{R})$ governing the nuclear motion on an adiabatic region of the potential surface. This constitutes an essential improvement over the one-trajectory approximation, and it is widely used in the theory of collision spectroscopy.

For a diatomic system the operator of nonadiabatic coupling $\mathscr{C} = -i\hbar\, \partial/\partial t$ can be written as

$$\mathscr{C} = -i\hbar \frac{\partial}{\partial t} = -i\hbar \dot{R} \frac{\partial}{\partial R} - i\hbar \dot{\phi} \frac{\partial}{\partial \phi}, \tag{9}$$

where the two terms on the right-hand side are contributions from radial and rotational relative motion of the colliding system.

This partitioning of $\mathscr{C}$ is useful because adiabatic functions are classified according to a symmetry character with respect to the fixed molecular axis. Thus, $\partial/\partial R$ couples electronic states of similar angular symmetry, and $\partial/\partial \phi$ couples states differing in Ω (projection of electronic angular momentum on the molecular axis) by ± 1. More detailed discussion of selection rules is given in ref. 7, though the general approach is similar to that for predissociation [8, 9].

As discussed earlier, matrix elements $C_{km}(t)$ of $\mathscr{C}$ are supposed to be essentially different from zero near $t = t_p$ and to vanish as $|t - t_p| \to \infty$. Let τ_{km} be the characteristic time of changing C_{km} near t_p, and $\hbar\omega_{km}$ be the difference in adiabatic terms at t_p. Then the N matrix will be quite different for the two limiting cases $\tau_{km}\omega_{km} \gg 1$ and $\tau_{km}\omega_{km} \ll 1$. If either of these conditions is fulfilled, then general methods can be suggested for the calculation of the matrix of nonadiabatic transitions.

A. Almost Adiabatic Perturbations; $\tau\omega \gg 1$

For two terms of the same symmetry, this condition is always satisfied at low velocity as, $\tau \sim 1/v$ and the noncrossing rule implies that $\omega \neq 0$. Terms of different symmetry may cross, but the most efficient nonadiabatic coupling need not be localized at the crossing point, though this is usually the case. If v is a small parameter, a perturbation treatment of equation (2) can be attempted. It was shown by Landau [10] that for noncrossing terms the matrix elements N_{km} are exponentially small, that is, $N_{km} \sim \exp(-\omega_{km}\tau_{km})$. As for the preexponential factor B_{km}, more detailed treatment shows two possibilities: Either B_{km} depends on velocity v and is small for low v, or it is independent of v. Only the first case, which is realized for terms of different symmetries, can be treated by the usual perturbation method. Thus, to first order we have

$$N_{km} = \int_{-\infty}^{\infty} \frac{C_{km}(t)}{i\hbar} \exp\left[-\frac{i}{\hbar}\int^{t} (U_m - U_k)\, dt\right] dt$$
$$= B_{km} \exp(-\omega_{km}\tau_{km}). \tag{10}$$

Higher orders will yield higher powers of B_{km} and the same exponential.

The second case, which arises for terms of the same symmetry, cannot be exactly solved by the usual perturbation technique. All orders give comparable contribution to the preexponential, so that the evaluation of N_{km} requires an infinite summation of leading contributions from all orders, even for $N_{km} \ll 1$ [11]. If so, how can the large value of $\omega\tau$ be used? This is the case for which the idea of integrating equation (4) in the complex t plane, put forward by Landau [10], proves to be very useful. The essence of the approach is to consider all molecular terms $U_n(R)$ of the same symmetry as different branches of one function $U(R)$ that is analytical on the Riemann plane of R. Continuous transition from U_n to U_m can be accomplished by passing through branch cuts neither of which crosses the real axis R: for real R all $U_n(R)$ are analytical functions of R. A transformation from R to t representations gives the integration contour l in the t plane that begins and ends at real times t'' and t'. The small value of the transition amplitude N_{km}, which is interpreted as a result of the destructive interference of two wave packets moving during the time interval $t' - t''$, now emerges as an increment of the trivial phase factor

$$\exp\left(-\frac{i}{\hbar}\int_{l} U\, dt\right)$$

when moving from t'' to t' over l. It can be shown that the main contribution to B_{km} in integration of equation (4) over l comes from a small region Δl near the singularity of $U(t)$. This means that the actual integration path is Δl, and not l, so that C_{km} can be approximated on Δl by a simple function of t that is determined by the type of singularity. The exact solution of the emerging approximate equations near the singularity of $U(t)$ and the continuation of this solution for the rest of l, making use of the adiabatic approximation, is equivalent to partial (still infinite) summation of the perturbation series. In a sense the theory of nearly adiabatic perturbations for two electronic states is close to the quasiclassical approximation for one wave equation: The isolated turning point for the latter plays the same part as the isolated branching point of $U(R)$ for the former. This model case seems to be the only one for which N_{km} can be found without explicit solution of equations, just using different circumventions [12–14] of singularity points. Attempts to generalize the circumvention method for more complicated situations as first suggested by Stueckelberg [15] fail and do not give correct results [16].

For an isolated branch point at $t = t_c$ the off-diagonal matrix elements of N_{km} are [9]

$$N_{km} = \exp\left[-\frac{i}{\hbar}\int_{t_p}^{t_c}(U_m - U_k)\,dt\right],$$
$$N_{mk} = -\exp\left[-\frac{i}{\hbar}\int_{t_p}^{t_c^*}(U_k - U_m)\,dt\right]. \tag{11}$$

Although this semiclassical result seems to be very simple, still there is no generalization for the quantal treatment except for the linear model (vide infra).

If the coupling region is passed twice, the S matrix will be

$$\mathscr{S} = \mathscr{A}(R > R_p)\mathscr{N}^{\dagger}\mathscr{A}(R < R_p)\mathscr{N}\mathscr{A}(R > R_p). \tag{12}$$

Substituting equation (11) in equation (12) we obtain for the two-state system

$$S_{12} = \exp(i\delta_1 + i\delta_2)2i\sin s'\exp(-s''). \tag{13}$$

Here $s' + is''$ is the complex phase difference for two adiabatic terms $U_1(R)$ and $U_2(R)$ calculated from the complex branch point R_c to corresponding turning points, and δ_1 and δ_2 are adiabatic phases for potentials U_1 and U_2. Equation (13) is valid when two coupling regions do not overlap, that is, when $s' \gg s'' \gg 1$. If, however, $s' \sim 1$, two (or more) branching points should be taken into account simultaneously while integrating over l.

Conditions of nearly adiabatic coupling are frequently met in atomic collisions. This will be exemplified in Section V for the collisionally induced intramultiplet mixing in alkalies.

Consider now the coupling between terms of different symmetry. If two terms U_n and U_m cross, they can be considered near the crossing point as being degenerate for a time interval τ_{nm}. It follows then from equations (9) and (10) that N_{nm} (and B_{nm}) will be proportional to the angle of rotation of the collision axis during the time interval τ_{nm}. If τ_{nm} is short enough, the linear approximation for the terms near the crossing point will suffice, and the range of the crossing region will be determined by the difference in slopes $F_n - F_m$ of terms $U_n - U_m$. For this case N_{nm} is [9, 17]

$$N_{mn} = \frac{1}{\hbar}(J_x)_{mn}\dot{\phi}_c\left(\frac{2\pi\hbar}{|F_n - F_m|\,v_R}\right)^{1/2}\exp\frac{i\pi}{4}, \tag{14}$$

where $\dot{\phi}_c$ is the angular velocity of a colliding pair at a distance $R = R_c$ corresponding to term crossing, v_R is the radial velocity at this distance, and J_x is a component of the electronic angular momentum operator perpendicular to the collision plane. This type of coupling will be discussed in Section VII.

When terms of different symmetry do not cross, the transition amplitude is very low; It is proportional to the exponential with a small additional pre-exponential factor. Nevertheless, under certain conditions this coupling may prevail compared to others (Section V).

These seem to be the most important cases of weak nonadiabatic coupling between terms of different symmetry. Strong coupling is expected either at $R \to \infty$ or at $R \to 0$, where molecular terms merge into a degenerate atomic term of a separated or united atom. The first case is discussed in Section VI in connection with collisionally induced depolarization. The second case is a consequence of high-energy collisions and thus will not be discussed here.

B. Sudden Perturbations; $\omega\tau \ll 1$

When the nonadiabatic perturbation is strong but sudden, the lowest-order approximation consists of reexpansion of one adiabatic set into another. Let ϕ_m be adiabatic functions of the Hamiltonian $\mathscr{H}(R)$ at $R > R_p$ and ϕ_μ be adiabatic functions of $\mathscr{H}(R)$ at $R < R_p$. Denote by $l_m(\mu)$ the components of ϕ_μ in the ϕ_m representation. Then

$$N_{\mu m} = l_m(\mu). \tag{15}$$

For a double crossing of the coupling region equation (12) gives

$$S_{mn} = \exp(i\Delta_n + i\Delta_m) \sum_\mu l_m(\mu) \exp(i\Delta\mu) l_n(\mu), \tag{16}$$

where Δ_n are phase shifts for adiabatic evolution at $R > R_p$ and $\Delta\mu$ are phase shifts for adiabatic evolution at $R < R_p$. In terms of the interaction $\mathscr{V}$ and zero-order functions ϕ_m^0, equation (16) can be approximated as

$$S_{nm} = \langle\phi_n^0| \exp\left(-\frac{i}{\hbar}\int_{-\infty}^{\infty} \mathscr{V}\, dt\right) |\phi_m^0\rangle \tag{17}$$

provided the contribution from the commutator $[\mathscr{V}(\mathbf{R}), \mathscr{V}(\mathbf{R}')]$ is neglected. The exact result will be obtained by replacing the integral in equation (17) by the series of integrals from commutators [18, 19]. In attempting to generalize this approach for transitions between not too closely spaced levels, it was suggested [20] to use equation (17) replacing $\mathscr{V}$ by its counterpart in the interaction representation:

$$S_{nm} = \langle\phi_n^0| \exp\left(-\frac{i}{\hbar}\int_{-\infty}^{\infty} \exp\left(\frac{i}{\hbar}\mathscr{H}_0 t\right)\mathscr{V} \exp\left(-\frac{i}{\hbar}\mathscr{H}_0 t\right) dt\right) |\phi_m^0\rangle. \tag{18}$$

This gives the S matrix that is correct also for nonsudden but weak perturbations.

For the subset of a degenerate state, equations (17) and (18) give identical results.

III TWO-STATE MODELS OF STRONG NONADIABATIC COUPLING

Even now that computers are available for solving collision problems, two-state models of the strong nonadiabatic coupling still play an important part. This is substantiated by the development of collision spectroscopy, which successfully interprets the complicated differential scattering in terms of these models. In such cases the main object of the theory of nonadiabatic processes consists in the explicit formulation of the 2×2 scattering matrix as a function of the parameters of the accepted model.

For the two-state Hamiltonian the adiabatic functions can be easily transformed into diabatic functions. In the latter representation the Hamiltonian is not diagonal, and the model is completely specified by the matrix H_{ik} as a function of $\mathbf{R}$. If, besides, the semiclassical approximation is adopted there will be only two essential functions: $\Delta H(t) \equiv H_{11}(t) - H_{22}(t)$ and $H_{12}(t)$.

Semiclassical problems that permit an analytical solution must be described by a set of two first-order time-dependent equations that reduce to the well-known equation of the second order, namely, to a hypergeometric equation. Thus, the maximum number of free parameters is equal to the number of parameters of a hypergeometric function (three, less the argument). Because the adiabatic splitting is $\Delta U = [(\Delta H)^2 + 4\,|H_{12}|^2]^{1/2}$, it is convenient to consider the coordinate plane $(\Delta H, 2H_{12})$ on which different models are characterized by curves scanned by the vertex of the vector $(\Delta H, 2H_{12})$ during passage over the coupling region. The length of the vector is just $\Delta U(R)$. It is expected that $\Delta H(t)$ and $H_{12}(t)$ are related in some simple way, so that with a change of time the tip of the vector will draw a curve that is not too complicated. This is summarized in Figure 5.1, which shows characteristic curves for model problems permitting a semiclassical analytical solution; these models have been considered by Zener [21] (straight line 1), Rosen and Zener [22] (loop 3), Nikitin [23, 24], Ellison and Borovitz [25] (straight line 4), and Kunicke and Demkov [26] (parabola 5 and straight line 6). Curve 2 of Figure 5.1 corresponds to the so-called linear model for which semiclassical and quantal solutions over a wide range of characteristic parameters are known [27]. The explicit time dependence of ΔH and H_{12} is not depicted in Figure 5.1. This leaves room for construction of new models, replacing t by a new variable.

If the characteristic time of changing H_{ik} is fixed, the evolution of the system proceeds almost adiabatically beyond the coupling region (the region outside the circle in Figure 5.1). In this region the N matrix can be calculated using a theory of almost adiabatic perturbations. If, on the other hand, H_{12}

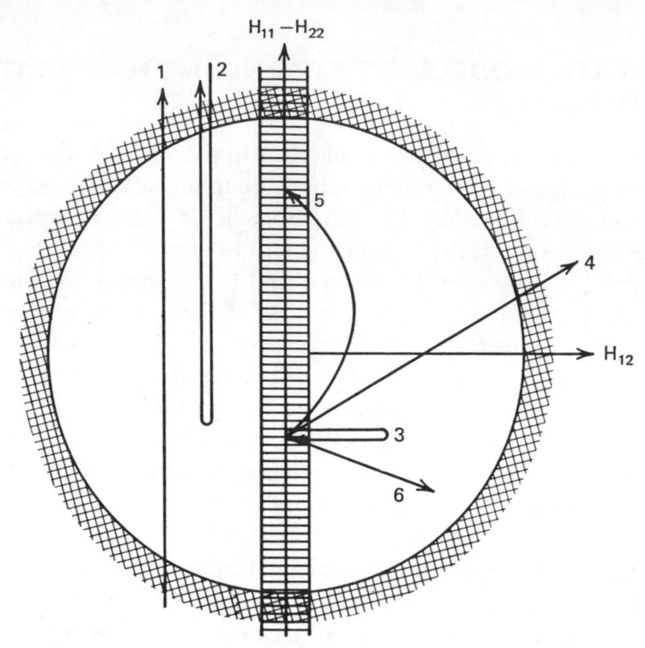

Figure 5.1 Graphic representation of two-state models of the strong nonadiabatic coupling.

is sufficiently low (the vertically striped band in Figure 5.1) the usual perturbation treatment will provide the result. The region inside the circle and outside the strip corresponds to strong coupling, and this is where the models present new results.

For a two-state problem the general N matrix is

$$N = \begin{pmatrix} \cos\chi \exp(i\phi) & \sin\chi \exp(i\psi) \\ -\sin\chi \exp(-i\psi) & \cos\chi \exp(-i\phi) \end{pmatrix}, \tag{19}$$

with three parameters χ, ϕ, and ψ. The value of R_p can be defined in different ways, but the S matrix will be independent of it. For instance, R_p may be found either from the constraint $\psi = 0$, or it may be identified as the real part of a complex internuclear distance R_c at which the two adiabatic terms cross. It is convenient, however, to have R_p defined in such a manner that phases ϕ and ψ are not large for any possible variations of the model parameters.

Now, the two-state S matrix is built from the N matrix using equation (12). Simple calculations give

$$S = \begin{pmatrix} \exp(2i\delta_1 + i\Gamma)(1 - \Pi^2)^{1/2} & \exp(i\delta_1 + i\delta_2)i\Pi \\ \exp(i\delta_1 + i\delta_2)i\Pi & \exp(2i\delta_2 - i\Gamma)(1 - \Pi^2)^{1/2} \end{pmatrix}, \tag{20}$$

where the three real parameters of the S matrix are $2\delta_1 + \Gamma$, $2\delta_2 - \Gamma$, and Π. Here δ_1 and δ_2 are adiabatic phase shifts for potentials U_1 and U_2; Γ and Π depend on the parameters of the N matrix:

$$\begin{aligned} \Pi &= \sin 2\chi \sin \Phi, \\ \Gamma &= \arg \{\exp (2i\phi)[1 - 2i \sin^2 \chi \sin \Phi \exp (-i\phi)]\}, \\ \Phi &= \int_{R_{t1}}^{R_r} \frac{p_1}{\hbar}\, dR - \int_{R_{t2}}^{R_p} \frac{p_2}{\hbar}\, dR + \psi + \phi. \end{aligned} \tag{21}$$

If the center of the coupling region R_p is defined by the condition $\psi = 0$, then Φ is interpreted as the phase difference s' at $R < R_p$ (adiabatic contribution) plus the correction term ϕ (nonadiabatic contribution).

In the past, most attention was paid to the mean transition probability $\bar{P} = \bar{\Pi}^2 = (\sin 2\chi)^2/2$ averaged over oscillations of the factor $\sin^2 \Phi$. The probability $\bar{P}$ is simply expressed in terms of the transition probability $p = (\sin \chi)^2$ for one passage of the coupling region:

$$\bar{P} = 2p(1 - p). \tag{22}$$

It has been realized recently that the parameter Φ is of importance in calculating differential scattering. This, in turn, stimulated calculation of ϕ for the earliest model of nonadiabatic coupling, the Landau–Zener–Stueckelberg model [28–30]. The model is formulated for a Hamiltonian with ΔH linear in R:

$$\begin{aligned} \Delta H(R) &= -(F_1 - F_2)(R - R_0), \\ H_{12} &= a. \end{aligned} \tag{23}$$

Parameters of the N matrix are

$$\begin{aligned} &\psi = 0 \quad \text{if} \quad R_p = R_0, \\ &\phi = \frac{\pi}{4} + \delta \ln \delta - \delta - \arg \Gamma(1 + i\delta), \\ &(\sin \chi)^2 = p = \exp(-2\pi\delta), \\ &\delta = \frac{a^2}{\hbar v_p \,|F_1 - F_2|}, \end{aligned} \tag{24}$$

where v_p is the velocity at R_p. From equations (20), (21), and (24), the Landau–Zener formula follows:

$$P = \Pi^2 = 4 \sin^2 [s' + \phi] \exp(-2\pi\delta)[1 - \exp(-2\pi\delta)], \tag{25}$$

which is valid at $s' \gg 1$, that is, on the condition that the crossing point R_0 is far from at least one turning point. Relaxation of the condition $s' \gg 1$ and the region of applicability of the model are discussed elsewhere [27, 31].

It follows from equation (24) that at the maximum of $\bar{P}$ ($2\pi\delta = \ln 2$) the phase ϕ is close to $\pi/4$, and at $\pi\delta \gg 1$ the phase ϕ vanishes as would be expected according to the theory of almost adiabatic perturbations.

Another model approximates the Hamiltonian near the coupling region by exponential functions of R. If R_p is defined as Re (R_c), then $\Delta H(R)$ and $H_{12}(R)$ are given in this model by

$$\begin{aligned} \Delta H(R) &= \Delta\varepsilon\{1 - \cos\theta \exp[-\alpha(R - R_p)]\}, \\ H_{12}(R) &= \frac{\Delta\varepsilon}{2} \sin\theta \exp[-\alpha(R - R_p)]. \end{aligned} \tag{26}$$

Parameters ψ and ϕ for the corresponding N matrix have not yet been found. As for $\sin\chi$, it is [32]

$$(\sin\chi)^2 = p = \exp\left[-\frac{\pi\xi}{2}(1 - \cos\theta)\right]\frac{\sinh[\frac{1}{2}\pi\xi(1 + \cos\theta)]}{\sinh[\pi\xi]}, \tag{27}$$

with

$$\xi = \frac{\Delta\varepsilon}{\hbar\alpha v_p},$$

if the trajectory at the coupling region is approximated by $R = R_p + v_p t$. If another approximation is accepted for the trajectory, expression for p will be different. This and related problems are discussed in a review article [32].

A great deal of research has been done recently on approximate solutions of strongly coupled semiclassical equations. Some solutions have been extensively used for cross-section calculations, often without any estimation of possible errors. Final expressions for the nonadiabatic transition probability P between two electronic states will be written down for the most frequently used approximations.

It is suggested that the two electronic states ϕ_1^0 and ϕ_2^0 of noninteracting partners provide a good basis set to be used for constructing orthonormal adiabatic electronic functions ϕ_1 and ϕ_2. At $R \to \infty$ the functions ϕ_1 and ϕ_2 adiabatically correlate with ϕ_1^0 and ϕ_2^0, so that the nonadiabatic transition probability calculated for a particular trajectory $\mathbf{R}(t)$ refers to the collision-induced transition between the two states of the partners.

Let us put

$$\begin{aligned} \Delta H &= \hbar\omega + \hbar\delta\omega(t), \\ H_{12} &= \hbar f(t). \end{aligned} \tag{28}$$

In 1932 Rosen and Zener [22] found the analytical solution for a certain interaction $f(t)$ and then postulated a general formula for any interaction with $\delta\omega = 0$:

$$P = \sin^2\left(\int_{-\infty}^{\infty} f\,dt\right)\left|\frac{\int_{-\infty}^{\infty} f \exp(i\omega t)\,dt}{\int_{-\infty}^{\infty} f\,dt}\right|^2. \tag{29}$$

In 1957 Gurnee and Magee [33] suggested that for the case with $\delta\omega = 0$ the transition probability can be calculated according to the following expression:

$$P = \sin^2\left[\int_{-\infty}^{\infty} f(t) \exp(i\omega t)\,dt\right]. \tag{30}$$

In 1962 Vainstein, Presnyakov, and Sobelman [34] considered a more general case with $\delta\omega \neq 0$ and obtained the expression

$$P = \sin^2\left[\int_{-\infty}^{\infty} f(t) \exp\left(i\int_0^t [(\omega + \delta\omega)^2 + 4\,|f|^2]^{1/2}\,dt\right)dt\right]. \tag{31}$$

To complete the list of approximate expressions for P we give here the result based on equation (18):

$$P = \sin^2\left[\left(\left|\int_{-\infty}^{\infty} \tfrac{1}{2}\delta\omega\,dt\right|^2 + \left|\int_{-\infty}^{\infty} f \exp(i\omega t)\,dt\right|^2\right)^{1/2}\right] \times \frac{\left|\int_{-\infty}^{\infty} f \exp(i\omega t)\,dt\right|^2}{\left|\int_{-\infty}^{\infty} \tfrac{1}{2}\delta\omega\,dt\right|^2 + \left|\int_{-\infty}^{\infty} f \exp(i\omega t)\,dt\right|^2}. \tag{32}$$

Expressions (29)–(32) give correct results for the case of weak coupling ($\int_{-\infty}^{\infty} f\,dt \ll 1$) and for exact resonance ($\omega = 0,\ \delta\omega = 0$). Neither is correct at the adiabatic limit [$\int_{-\infty}^{\infty} f\,dt \gg 1$, $\int_{-\infty}^{\infty} f \exp(i\omega t)\,dt \ll 1$] when P must be found from equation (15). For intermediate cases these expressions have been investigated mainly with respect to certain interactions $f(t)$. Skinner [35] discussed equations (29) and (30) for an exponential interaction and compared them with computed values of P. It was found that equation (29) is more reliable than equation (30), but a collision must not be too adiabatic ($\omega\tau \leq \frac{1}{2}$). A comparison of equations (29), (30), and (31) for the dipole–dipole interaction was made by Nakamura [36], who showed that equation (31) gave the lowest transition probability under adiabatic conditions ($\omega\tau \gg 1$). Equation (30) has been discussed also by Mori and Fujita [37].

At present it is difficult to estimate which approximate solution would be better for a particular problem, so that each case requires an independent estimate of possible errors introduced by the nature of the approximation.

IV. MOLECULAR TERMS FOR ALKALI–NOBLE-GAS AND ALKALI–ALKALI PAIRS

A. Adiabatic Potentials for Alkali–Noble-Gas Pair M* + X

For low-energy collisions we are interested in the interaction for rather large interatomic distances R. This suggests that the spin-orbit coupling even for the outer electron of M is still unaffected by the interaction, which is of an electrostatic nature. Thus, the Hamiltonian $\mathscr{H}_{\mathrm{MX}}$ of a quasimolecule MX can be put in the form

$$\mathscr{H}_{\mathrm{MX}}(R) = \mathscr{H}^0_{\mathrm{MX}}(R) + \mathscr{H}^{\mathrm{SO}}_{\mathrm{M}}, \tag{33}$$

where $\mathscr{H}^0_{\mathrm{MX}}(R)$ is the Hamiltonian of MX less the magnetic interaction, and $\mathscr{H}^{\mathrm{SO}}_{\mathrm{M}}$ is the spin-orbit coupling in the free M atom. The hyperfine interaction will be ignored completely.

If eigenvalues $V_k(R)$ of $\mathscr{H}^0_{\mathrm{MX}}$ are known, then eigenvalues $U_k(R)$ of $\mathscr{H}_{\mathrm{MX}}$ are obtained by diagonalization of the matrix containing $V_k(R)$ and atomic constants of the spin-orbit interaction in M. In particular,[1] molecular terms $A\frac{1}{2}$, $A\frac{3}{2}$, and $B\frac{1}{2}$, which correlate at $R \to \infty$ with atomic states ${}^2P_j(\mathrm{M}) + {}^1S_0(\mathrm{X})$ are given by [38]

$$\begin{aligned}
U(A\tfrac{1}{2}) &= \tfrac{1}{2}[V_\sigma + V_\pi + \Delta\varepsilon - \Delta U],\\
U(A\tfrac{3}{2}) &= V_\pi + \Delta\varepsilon,\\
U(B\tfrac{1}{2}) &= \tfrac{1}{2}[V_\sigma + V_\pi + \Delta\varepsilon + \Delta U],\\
\Delta U &= [(\Delta\varepsilon)^2 + \tfrac{2}{3}\Delta\varepsilon\,\Delta V + (\Delta V)^2]^{1/2},\\
\Delta V &= V_\sigma - V_\pi,
\end{aligned} \tag{34}$$

where V_σ and V_π are adiabatic potentials corresponding to Σ and Π molecular terms, and $\Delta\varepsilon$ is the atomic spin-orbital splitting of the ${}^2P$ doublet.

[1] Here and below we follow the usual spectroscopic notations for molecular terms [8, 9]. In the case of the Hund coupling scheme **a** the notation ${}^{2s+1}\Lambda_w$ refers to a molecular term with total electron spin S, the projection $\pm\Lambda$ ($\Lambda > 0$) of electronic orbital angular momentum on the molecular axis, and parity w.

In the case of the Hund coupling scheme **c** the notation Ω_w means a term with projection of the total electronic (orbital plus spin) angular momentum Ω on the molecular axis and parity w. The superscript σ in Σ^σ or 0^σ taking + or − refers to the character of the wave function upon reflection in the collision plane. In the notation, Λ is identified as Σ, Π, $\Delta, \ldots$, w as g or u (*gerade* or *ungerade* states). In equations, $\Sigma, \Pi, \Delta, \ldots$ mean $0, 1, 2, \ldots$, **g** or **u** mean +1 or −1, and + or − for σ mean +1 or −1. Capital letters $X, A, B, \ldots$ preceding the term notation show the term ordering in the energy scale, beginning with the ground state. An adiabatic potential $U(R)$ referred to a given electronic term will bear the term notation (or an essential part of it) either as a subscript, or as an argument, say $U({}^{2s+1}\Lambda_w, R)$

There are different approximate methods for calculation of V_σ and V_π. The direct variational approach [39] shows that for the $B\Sigma$ term of the Li–He system, the van-der-Waals attraction is overpowered by the exchange repulsion of the $p\sigma$ electron of M from the X core at rather large distances ($R > 10$ a.u.). This is substantiated by recent calculations made by Krauss [40] and Baylis [41], who demonstrated besides that for the $p\pi$ electron of M, the noble-gas atom penetrates into the excited atom M* until it experiences the core–core repulsion M^+–X. Before the repulsion region for the A term there is net attraction that is due to the polarization of X in the field of the charged core M^+.

This general pattern is in line with the asymptotic estimation of interaction [42, 43] for large R, which uses the simple additive scheme for the exchange and the van-der-Waals interactions. For a M*–X system the former contribution is given by

$$
\begin{aligned}
V_\sigma^{\text{ex}} &= C_\sigma R^{4/\alpha a_0 - 2} \exp(-\alpha R), \\
V_\pi^{\text{ex}} &= C_\pi R^{4/\alpha a_0 - 4} \exp(-\alpha R),
\end{aligned}
\tag{35}
$$

where $\alpha/2$ is the orbital exponent of the excited electron in M. The constants C_σ and C_π depend on parameters of free atoms M and X. Estimates of C_σ for different M*–X pairs give positive values around 0.1 a.u. varying within one order of magnitude, in qualitative agreement with Baylis' results. As for C_π, its value is of no importance for what follows, because in the asymptotic region $V_\sigma \gg V_\pi$.

Another approach for estimating C_σ is based on the negative-energy extrapolation of the phase-shift $\delta(E)$ for the s scattering of low-energy electrons on X. Use of the standard procedure [42] will give negative values of C_σ for heavier atoms X due to the negative scattering lengths for these atoms. The same result has been obtained [45] in the direct calculation of C_σ replacing atom X by the pseudopotential with long-range polarization attraction and hard-core repulsion, which yields correctly the Ramsauer effect. It might be that the adiabatic treatment of noble-gas polarization in the field of the moving electron at not too large distances greatly underestimates the attraction, so that negative values of C_σ are an artifact of the approximation. If C_σ were negative, the first excited state of a MX pair would have a very deep potential well, much deeper than that needed to explain the satellite structure of the $M^* \to M$ optical transition. This argument and also the previous explanation of intramultiplet mixing in M*–X collisions [44] suggest that positive values of C_σ, as given by traditional quantum-chemistry calculations, are correct. This is the view adopted here. We add also that the cross sections discussed later are not very sensitive to the actual values of C_σ: the uncertain factor of 3 is in line with other approximations that have to be made.

As for the van-der-Waals (polarization) part of the potential for the M*–X system, it can be put in a simple form

$$V_\sigma^{\mathrm{pol}} = -\frac{7}{5}\frac{\alpha_p e^2\langle r^2\rangle}{R^6},$$
$$V_\pi^{\mathrm{pol}} = -\frac{4}{5}\frac{\alpha_p e^2\langle r^2\rangle}{R^6}, \tag{36}$$

which is valid when the ionization potential of X much exceeds that of M* [38, 46]. Here α_p is the polarizability of X and $\langle r^2\rangle$ is the mean square distance of the excited electron in M* from the M nucleus.

Summarizing, the qualitative picture of the lower molecular terms for a M–X pair is presented in Figure 5.2. The van-der-Waals attraction is too

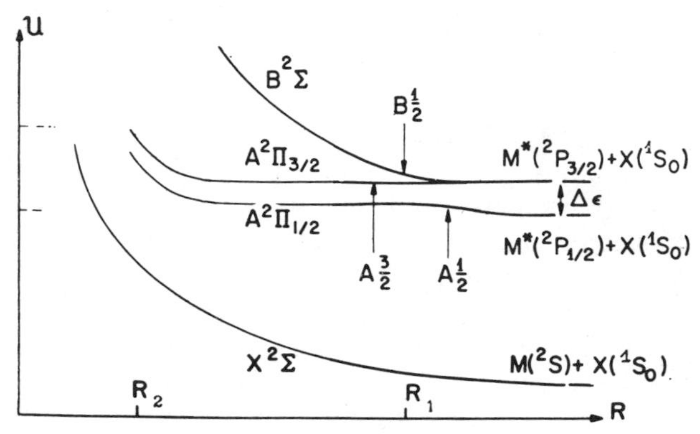

Figure 5.2 Qualitative pattern of lower molecular terms for a pair consisting of an alkali atom and a noble-gas atom.

weak to be shown in the figure. The distances R_1 and R_2 characterize coupling regions where nonadiabatic transitions are most likely to occur.

B. Adiabatic Potentials for a Pair of Similar Alkali Atoms M′* + M″

Similar to equation (33) the molecular Hamiltonian is

$$\mathscr{H}_{\mathrm{M'M''}}(R) = \mathscr{H}^0_{\mathrm{M'M''}}(R) + \mathscr{H}^{\mathrm{SO}}_{\mathrm{M'}} + \mathscr{H}^{\mathrm{SO}}_{\mathrm{M''}}, \tag{37}$$

where $\mathscr{H}^0_{\mathrm{M'M''}}(R)$ is the electrostatic Hamiltonian, and $\mathscr{H}^{\mathrm{SO}}_{\mathrm{M'}}$ and $\mathscr{H}^{\mathrm{SO}}_{\mathrm{M''}}$ are the spin-orbital interactions in atoms M′ and M″. For the M′*–M″ pair the limited molecular basis set is built from atomic functions $^2P + {}^2S$, and resulting molecular functions $^{2s+1}\Lambda_w$ ($S = 0, 1; \Lambda = \Sigma, \Pi, w = g, u$) are used to calculate the interaction $V(^{2s+1}\Lambda_w)$.

To first order the main contribution for large R comes from the resonance dipole–dipole interaction [47]:

$$V^{\mathrm{dip}}(^{2S+1}\Lambda_w, R) = (-1)^S w(-1)^\Lambda[2 - \Lambda]V_0^{\mathrm{dip}}(R),$$
$$V_0^{\mathrm{dip}} = \frac{d^2}{R^3}. \quad (38)$$

Here d is the z projection of the dipole matrix elements for the spinless S–P_z transition of an outer electron. The factor $(-1)^S w$ in equation (38) characterizes the symmetry in the arrangement of atomic dipoles in mixed S–P atomic states. Spin S and parity w of a molecular state are relevant either to the exchange of electrons (with atomic orbital fixed at nuclei) or to the exchange both of electrons and atomic orbitals. The exchange of orbitals (with atomic electrons attached to corresponding nuclei) is accomplished by the product of these two transformations. This explains the appearance of the spin quantum number in (38).

Adiabatic terms in equation (38) are degenerate in quantum numbers S and w that yield the same value of the factor $(-1)^S w$. After diagonalization of the matrix $\mathscr{H}_{\mathrm{M'M''}}$, with $\mathscr{H}^0_{\mathrm{M'M''}}$, $\mathscr{H}^{\mathrm{SO}}_{\mathrm{M'}}$, and $\mathscr{H}^{\mathrm{SO}}_{\mathrm{M''}}$ taken into account, the degeneracy is partly lifted but virtually remains for certain terms at the limit where $V_0^{\mathrm{dip}} \gg \Delta\varepsilon$. At this limit the spin-orbital coupling can be calculated to the first order, giving

$$U(^{2S+1}\Lambda_w, R) = V^{\mathrm{dip}}(^{2S+1}\Lambda_w) + V^{\mathrm{SO}}(^{2S+1}\Lambda_{w,\Omega}), \quad (39)$$

$$V^{\mathrm{SO}}(^{2S+1}\Lambda_{w,\Omega}) = \tfrac{2}{3}\Delta\varepsilon\Lambda(\Omega - 1). \quad (40)$$

As seen from equations (39) and (40), degenerate terms are those with $\Omega = 1$ and with the same factor $(-1)^S w$. The above is summarized in the correlation diagram (Figure 5.3) drawn between states of free atoms (FA region) and molecular states with resonance dipole–dipole interaction (RDI region). The four bunches of states $(^3\Sigma_u^+, {}^1\Sigma_g^+)$, $(^1\Sigma_u^+, {}^3\Sigma_g^+)$, $(^3\Pi_g, {}^1\Pi_u)$, $(^1\Pi_g, {}^3\Pi_u)$ in the former are those for which the dipole–dipole energies are $\pm 2d^2/R^3$ and $\pm d^2/R^3$.

At smaller distances the exchange interaction comes into play. In the Heitler–London approximation there will be two types of exchange integrals: without and with excitation transfer. These are

$$Y_\Lambda = \langle s^{\mathrm{M'}}(1)p_\Lambda^{\mathrm{M''}}(2)|\, \mathscr{H}^0_{\mathrm{M'M''}}\, |s^{\mathrm{M'}}(2)p_\Lambda^{\mathrm{M''}}(1)\rangle,$$
$$Y_\Lambda^* = \langle s^{\mathrm{M'}}(1)p_\Lambda^{\mathrm{M''}}(2)|\, \mathscr{H}^0_{\mathrm{M'M''}}\, |s^{\mathrm{M''}}(2)p_\Lambda^{\mathrm{M'}}(1)\rangle. \quad (41)$$

The contribution of these integrals to the exchange interaction is

$$V^{\mathrm{ex}}(^{2S+1}\Lambda_w) = (-1)^S Y_\Lambda - wY_\Lambda^*. \quad (42)$$

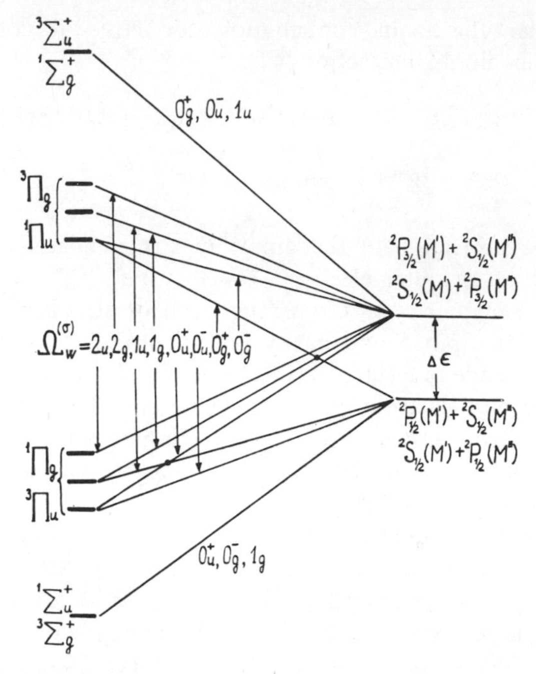

Figure 5.3 Correlation diagram of molecular terms for a pair of similar (excited and ground-state) alkali atoms. The right part of the figure corresponds to free atoms, the left part corresponds to the strong dipole–dipole interaction ($V^{\text{dip}} \gg \Delta\varepsilon$). Dots mark crossings of terms coupled by the Coriolis interactions.

The calculation of Y and Y^* and the relation between them are discussed in ref. 47.

Thus, in the region where $V^{\text{ex}} \gg \Delta\varepsilon$ or $V^{\text{dip}} \gg \Delta\varepsilon$ the adiabatic terms are

$$U(^{2S+1}\Lambda_{w,\Omega}) = V^{\text{ex}}(^{2S+1}\Lambda_w) + V^{\text{dip}}(^{2S+1}\Lambda_w) + V^{\text{SO}}(^{2S+1}\Lambda_{w,\Omega}), \qquad (43)$$

with individual contributions given by equations (42), (38), and (40). This expression is valid if terms of the same symmetry are well separated from each other. Otherwise there will be an interaction resulting in pseudocrossings for such terms in equation (43).

Figure 5.4 gives the correlation diagram of terms between the RDI region and that where the exchange interaction strongly exceeds the dipole–dipole interaction (EI region). One pseudocrossing of O_g^+ terms can be seen to arise from the crossing of $^1\Sigma_g^+$ and $^3\Pi_{1g}$ terms at $R = R_1$. At all other crossings the interaction is zero because of the corresponding selection rules.

At still smaller distances the asymptotic estimation of exchange integrals and the leading dipole–dipole term of the Coulomb interaction cannot be

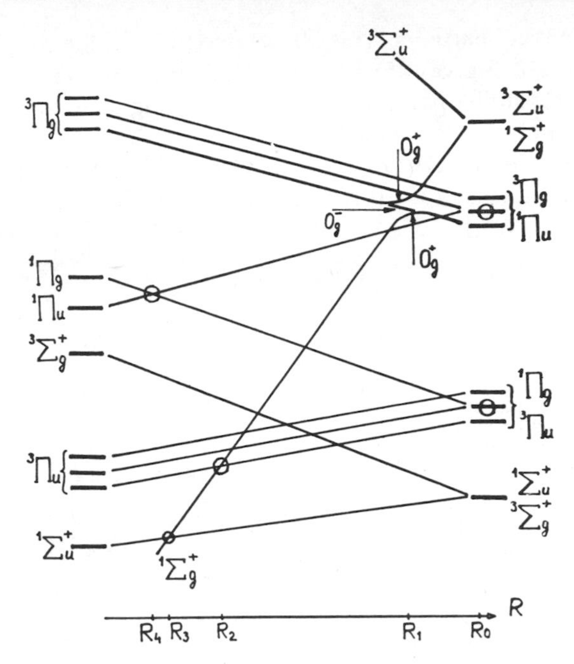

Figure 5.4 Correlation diagram of molecular terms for a pair of similar (excited and ground-state) alkali atoms. The right part of the figure corresponds to the strong dipole–dipole interaction ($V^{\mathrm{dip}} \gg \Delta\varepsilon \gg V^{\mathrm{ex}}$). The left part corresponds to the strong exchange interaction ($V^{\mathrm{ex}} \gg V^{\mathrm{dip}} \gg \Delta\varepsilon$). Encircled crossings or mergings become pseudocrossings or splitting in the case of dissimilar atoms.

relied upon. Unfortunately, there is no virtually independent information on the interaction at small distances except for the variational calculation for Li_2 [48]. Comparison of variational and asymptotic calculations show that both yield the same term ordering at $R > 7$ a.u. This value gives the approximate lower limit of applicability of the asymptotic approach, which is difficult to estimate otherwise.

C. Adiabatic Potentials for a Pair of Dissimilar Alkali Atoms $M_1^* + M_2$

For a pair of dissimilar alkali atoms M_1^* and M_2 there is no degeneracy between $M_1^*M_2$ and $M_1M_2^*$ states, so that asymptotically, as $R \to \infty$, the interaction is proportional to R^{-6}. However, because the energy difference ΔE between doublet centers of different alkali atoms is not high, the dipole–dipole interaction exceeds ΔE at distances where the exchange interaction is still negligible. In this region the interaction again varies as R^{-3}, being similar to that for identical partners. This allows one to draw the simple correlation diagram shown in Figure 5.5. The main difference between Figures 5.3 and 5.5 is that for dissimilar partners there is no crossing of adiabatic terms in the region where exchange is very weak.

The approximate pattern of adiabatic terms $U(^{2S+1}\Lambda_\Omega)$ at smaller separations can be established resorting to equation (43). If V^{ex}, $V^{\text{dip}} \gg \Delta\varepsilon_1, \Delta\varepsilon_2$,

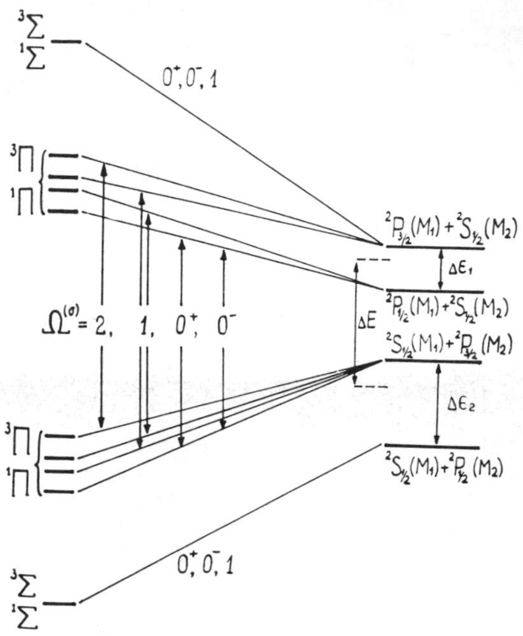

Figure 5.5 Correlation diagram of molecular terms for a pair of dissimilar (excited and ground-state) alkali atoms. The right part of figure corresponds to free atoms, the left part corresponds to the strong dipole–dipole interaction ($V^{\text{dip}} \gg \Delta\varepsilon$, $\Delta\varepsilon_1$, $\Delta\varepsilon_2$).

ΔE, the well-separated molecular terms for a pair of dissimilar atoms are close to those for a pair of similar atoms, equation (43), except with d^2 replaced by $d_1 d_2$ and Y_Λ replaced by $\frac{1}{2}(Y_\Lambda^{(12)} + Y_\Lambda^{(21)})$, where

$$Y_\Lambda^{(12)} = \langle s^{M_1}(1) p_\Lambda^{M_2}(2) | \, H^0_{M_1M_2} \, | s^{M_1}(2) p_\Lambda^{M_2}(1) \rangle,$$
$$Y_\Lambda^{(21)} = \langle s^{M_2}(1) p_\Lambda^{M_1}(2) | \, H^0_{M_1M_2} \, | s^{M_2}(2) p_\Lambda^{M_1}(1) \rangle. \tag{44}$$

However, numerous crossings and mergings that exist for terms $U(^{2S+1}\Lambda_{\Omega,w})$ become now pseudocrossing for terms $U(^{2S+1}\Lambda_\Omega)$ with splitting depending on "asymmetry" parameters ΔE, $\Delta\varepsilon_1 - \Delta\varepsilon_2$, and $(Y_\Lambda^{(12)} - Y_\Lambda^{(21)})$.

The most important pseudocrossings are at R_0, R_1 (already present for similar partners), R_2, R_3, and R_4 (Figure 5.4). These distance locate possible regions of nonadiabatic coupling for intramultiplet mixing and nonresonant exitation transfer in collisions of dissimilar alkalies.

V. INTRAMULTIPLET MIXING IN EXCITED ALKALI ATOMS UPON COLLISIONS WITH NOBLE-GAS ATOMS

The theory of the intramultiplet mixing in alkalies $M^*(^2P)$ in collisions with noble-gas atoms $X(^1S_0)$,

$$M^*(^2P_{1/2}) + X(^1S_0) \rightarrow M^*(^2P_{3/2}) + X(^1S_0), \tag{45}$$

has been discussed in many papers. Calculations of the cross sections were based either on the van-der-Waals [46, 50, 53–55] or on the exchange interactions [38, 44, 50–52, 56, 57]. Mathematically, the problem is formulated as a set of three coupled equations. Their semiclassical version has been discussed in [38]. In general the one-trajectory approximation will not be good because the kinetic energy is low compared to the difference of adiabatic potentials in question. Thus it is helpful to find the more or less localized nonadiabaticity regions and use one-trajectory approximations within each of these. Referring to Figure 5.2 we can see two regions that could be nonadiabaticity regions: one located near R_1 of a characteristic length $1/\alpha_1$ and another near R_2 of a range $1/\alpha_2$. Though these parameters are not yet well defined, we introduce for each region two dimensionless ratios $\xi_k = \Delta\varepsilon/\hbar\alpha_k v$ and $\xi' = \Delta\varepsilon R_k/\hbar v$ giving the degree of adiabaticity for relative radial and rotational motion.

Now, if $\xi_1' \gg 1$, the two coupling regions do not overlap [38] and it is possible to treat them separately in a sense discussed in Section II. Moreover, on this condition the Coriolis coupling in the region $R \approx R_1$ can be neglected [38]. This is in fact the case for all alkalies at typical experimental temperatures except Li* and Na*. Leaving the latter case for awhile and ignoring all interference effects, the probability $P(\frac{1}{2} \rightarrow \frac{3}{2})$ of intramultiplet mixing can be expressed in terms of probability p_1 of transition $A\frac{1}{2} \rightarrow B\frac{1}{2}$ at R_1 for a single passage of the coupling region and the total probability P_2 of transition $A\,^2\Pi_{1/2} \rightarrow A\,^2\Pi_{3/2}$ near R_2. The summation of fluxes gives

$$P(\tfrac{1}{2} \rightarrow \tfrac{3}{2}) = 2\bar{p}_1(1 - \bar{p}_1) + P_2(1 - \bar{p}_1)^2, \tag{46}$$

where the first term on the right-hand side of equation (46) is recognized as the total probability P_1 for double passage of the coupling region at R_1:

$$P_1 = 2p_1(1 - p_1). \tag{47}$$

For small p_1, which is the case for K*, Rb*, and Cs*, equation (46) simplifies to

$$P(\tfrac{1}{2} \rightarrow \tfrac{3}{2}) = \bar{p}_1 + P_2, \tag{48}$$

implying independent contributions from two coupling regions. The pattern of adiabatic terms shown in Figure 5.2 suggests the following mechanism of reorientation of the orbital momentum **l** and the spin **s** of the valence electron in atom M* upon collision with X. At large interatomic separations $R \approx R_1$ the atomic spin-orbit coupling (that gives rise to the total angular momentum j) is partly destroyed due to competition between electrostatic and magnetic intra-atomic interactions. This decoupling together with the dynamic electron–nucleus interaction leads to nonadiabatic transitions between molecular terms of the same symmetry (given below in mechanism 1).

At smaller distances the orbital momentum is always strongly coupled to the molecular axis **n**, and the spin **s** follows the rotation if the frequency of the fine-structure splitting between $A\,^2\Pi_{1/2}$ and $A\,^2\Pi_{3/2}$ components is appreciably higher than the angular velocity $\dot{\phi}$ of **n**.

When X impinges on the core of M* at $R \approx R_2$ the angular velocity increases and this causes transitions between the components, thus leading finally to transitions between $^2P_{1/2}$ and $^2P_{3/2}$ atomic states (mechanism 2). This mechanism is operative at smaller distances compared to those involved in mechanism 1. The transition probability for mechanism 2 is proportional to ratio $(\dot{\phi}/\omega)^2$, which does not appear in mechanism 1. Both factors tend to decrease the importance of mechanism 2 compared to mechanism 1. Nevertheless, the difference in steepness of the repulsive interaction between X and M* in $B\,^2\Sigma$ and $A\,^2\Pi$ states can lead to preference of mechanism 2 for nearly adiabatic collisions when the mixing cross section is very small.

A. Mechanism 1

According to Section II, the location of the region of nonadiabatic coupling between states $A\frac{1}{2}$ and $B\frac{1}{2}$ is determined by the condition of term crossing,

$$U(B\tfrac{1}{2}, R_c) - U(A\tfrac{1}{2}, R_c) = 0. \tag{49}$$

The real part R_1 of R_c corresponds to the center of the nonadiabatic region. From equation (34) we get

$$\Delta V(R_c) = \Delta\varepsilon \exp(\pm i\theta), \qquad \cos\theta = -\tfrac{1}{3}. \tag{50}$$

To find R_c we adopt a simple approximation to both the exchange and the van-der-Waals contributions to V_σ and V_π. First, V_π^{ex} is neglected in comparison with V_σ^{ex}, and the latter is replaced by a simple exponential near R_1:

$$\begin{aligned}\Delta V^{\text{ex}}(R) &\approx V_\sigma^{\text{ex}}(R) = C_\sigma R^{4/\alpha a_0 - 2} \times \exp(-\alpha R)\\ &\approx C_\sigma R_1^{4/\alpha a_0 - 2} \exp(-\alpha_1 R), \qquad \alpha_1 = \alpha - \frac{4/\alpha a_0 - 2}{R_1}.\end{aligned} \tag{51}$$

Second, the van-der-Waals contribution $\Delta V^{\text{pol}}(R)$ is replaced by a constant $\Delta V^{\text{pol}}(R_1)$. With a low value of the ratio $V^{\text{pol}}(R_1)/\Delta\varepsilon$, we can neglect the contribution $\Delta V^{\text{pol}}(R)$ in $\Delta V(R)$ altogether if $\Delta\varepsilon$ in equation (50) is replaced by $\Delta\varepsilon_{\text{eff}}$ (51):

$$\Delta\varepsilon_{\text{eff}}(R_1) = \Delta\varepsilon - \frac{3}{5}\frac{\alpha_p e^2\langle r^2\rangle}{R_1^6}. \tag{52}$$

The final equation for R_1 is

$$C_\sigma R_1^{4/\alpha a_0 - 2} \exp(-\alpha_1 R_1) = \Delta\varepsilon_{\text{eff}}(R_1). \tag{53}$$

When R_1 is found, we get

$$R_c = R_1 \pm \frac{i\theta}{\alpha_1}. \tag{54}$$

Now, the two-state Hamiltonian with approximations made can be simulated near R_1 by the model Hamiltonian equation (26) with parameters $\cos\theta \to \frac{1}{3}$, $\alpha \to \alpha_1$, $\Delta\varepsilon \to \Delta\varepsilon_{\text{eff}}$. The transition probability $\bar{P}_1$ can be obtained from equations (27) and (22) provided interference is neglected:

$$\bar{P}_1 = 2\exp\left(-\frac{\pi}{3}\xi_{1R}\right)\frac{\sinh(\frac{1}{3}\pi\xi_{1R})\sinh(\frac{2}{3}\pi\xi_{1R})}{\sinh^2(\pi\xi_{1R})}, \tag{55}$$

$$\xi_{1R} = \frac{\Delta\varepsilon_{\text{eff}}}{\hbar\alpha_1 v_R},$$

where v_R is the radial velocity at R_1:

$$v_R = v\left(1 - \frac{b^2}{R_1^2}\right)^{1/2}.$$

At $\xi_{1R} \gg 1$, equation (55) gives a result that follows from the theory of almost adiabatic perturbations applied to the model given by equation (26). However, in this limit the semiclassical quantity v_R can be corrected for the change in relative velocity due to the energy transfer [51]. This gives the following expression for the probability of upward transition:

$$\bar{P}_1(v, b) = 2\exp\left[-\tfrac{4}{3}\pi\xi_{1R}\left(1 + \frac{\Delta\varepsilon}{2\mu v_R^2}\right)\right]. \tag{56}$$

We see that head-on collisions are most effective in bringing about transition in region 1, and the probability depends on R_1 rather weakly, only via $\Delta\varepsilon_{\text{eff}}$ (equation 52). The mean cross section σ_1 corresponding to equation (56) is

$$\sigma_1 = \pi R_1^2 4(\tfrac{1}{3}\pi)^{1/2}\left(\frac{\gamma_1}{2}\right)^{1/3} \exp\left[-3\left(\frac{\gamma_1}{2}\right)^{2/3} - \frac{\Delta\varepsilon}{2kT}\right],$$

$$\gamma = \frac{4\pi}{3}\frac{\Delta\varepsilon_{\text{eff}}}{\alpha_1}\left(\frac{\mu}{2kT}\right)^{1/2}. \tag{57}$$

B. Mechanism 2

Since the Coriolis coupling is largest at the distance of closest approach, the nonadiabatic region is localized near the turning point of the radial motion. In this region, close to R_2, the two parallel terms $A\,^2\Pi_{1/2}$ and $A\,^2\Pi_{3/2}$ are approximated by exponential functions $\sim\exp(-\alpha_2 R)$, and the average potential governing the relative motion is taken as (44)

$$U(R) = B\exp(-\alpha_2 R) - U_0. \tag{58}$$

The shift $-U_0$ simulates the attraction just before the repulsion begins.

The first-order perturbation treatment in the nearly adiabatic limit gives (44)

$$P_2(v, b) = \frac{(\alpha_2 b)^2}{(1 - b^2/R_2^2)^{1/2}}\frac{\pi^2}{\nu^2}[\tfrac{1}{2} + F(y)]^2 \exp\left[-\tfrac{4}{3}\pi\xi_{2R}\left(1 + \frac{\Delta\varepsilon}{2\mu v_R^2} - \frac{U_0}{\mu v_R^2}\right)\right],$$

$$\xi_{2R} = \frac{\Delta\varepsilon}{\hbar\alpha_2 v_R}, \qquad y = \frac{\tfrac{4}{3}\zeta_{2R}}{\nu}, \qquad \nu = \exp\frac{\alpha_2 R_2}{2}. \tag{59}$$

The plot of $F(y)$ is given in Figure 5.6. It is seen from equation (59) that most effective are the collisions with nonzero (but still small) impact parameters, and the probability strongly depends on R_2, decreasing with increasing R_2. The mean cross section corresponding to equation (59) is

$$\sigma_2 = \pi R_2^2\frac{\pi^2}{\nu^2}\frac{(\alpha_2 R_2)^2}{2}[\tfrac{1}{2} + F(y^*)]^2 4(\tfrac{1}{3}\pi)^{1/2}\left(\frac{\gamma_2}{2}\right)^{-1/3} \exp\left[-3\left(\frac{\gamma_2}{2}\right)^{2/3} - \frac{\Delta\varepsilon}{2kT} + \frac{U_0}{kT}\right]$$

$$\gamma_2 = \tfrac{4}{3}\pi\frac{\Delta\varepsilon}{\hbar\alpha_2}\left(\frac{\mu}{2kT}\right)^{1/2}, \qquad y^* = 2\left(\frac{\gamma_2}{2}\right)^{2/3}\frac{\pi}{\nu}. \tag{60}$$

The point of interest is the comparison between corresponding factors (exponential and pre-exponential) in equations (57) and (60). While the exponential factor in equation (57) is smaller than that in equation (60) because of the greater steepness of the repulsive potential at $R \approx R_2$, the

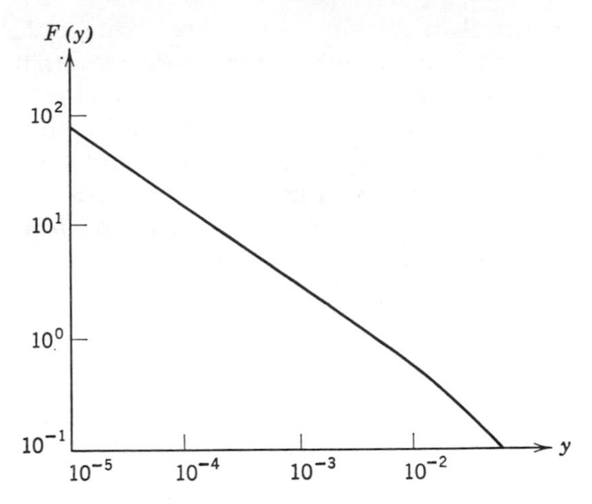

Figure 5.6 Plot of the function $F(y)$; see equation (59).

opposite is valid for pre-exponentials. Thus, various relations between σ_1 and σ_2 are possible and depend on the parameters.

C. Nonlocalized Transitions

If $\xi_1' \lessdot 1$, the Coriolis coupling between $A\,^2\Pi_{1/2}$ and $A\,^2\Pi_{3/2}$ states is effective at all $R < R_1$. This means that region 2 actually extends to region 1 although the latter is still fairly localized extending over interval $1/\alpha_1$ near R_1. Taking advantage of the "small overlap," some insight into the mechanism can be gained using equation (46). From equation (55) the maximum value of P_1 is $\frac{4}{9}$, corresponding to $p_1 = \frac{1}{3}$. Then

$$\sigma(\tfrac{1}{2} \to \tfrac{3}{2})_{\max} \approx \tfrac{4}{9}\pi R_1^2[1 + \langle P_2 \rangle], \tag{61}$$

where $\langle P_2 \rangle$ is the average transition probability per collision with target area πR_1^2. For the two limiting cases $\langle P_2 \rangle = 0$ and $\langle P_2 \rangle = \frac{1}{2}$, equation (61) for the maximum value of $\sigma(\frac{1}{2} \to \frac{3}{2})$ ranges between $\frac{4}{9}\pi R_1^2$ and $\frac{2}{3}\pi R_1^2$. With increasing velocity the cross section $\sigma(\frac{1}{2} \to \frac{3}{2})$ will pass through its maximum and eventually drop, as must be the case for any quasiresonance process. At these energies it will be necessary to consider both types of coupling (due to the radial and rotational motion) simultaneously.

This problem has recently been solved numerically by Masnou–Seeuws [57] for rectilinear trajectories, assuming ΔV in the form equation (51) with $R_1 = 13.8$ a.u., neglecting the van-der-Waals contribution and ignoring repulsive core-core interaction. If the cross section $\sigma(\frac{1}{2} \to \frac{3}{2})$ is written in the form

$$\sigma(\tfrac{1}{2} \to \tfrac{3}{2}) = \pi R_1^2 S(\gamma, \alpha_1 R_1), \tag{62}$$

the results of numerical calculations allow one to infer the function S. Though S depends on two parameters, γ and $\alpha_1 R_1$, the variation of γ with collision partners is expected to be much larger than that of $\alpha_1 R_1$. If such is the case, then S can be considered a universal function relevant to all collision processes of the type represented by equation (45). The plot of $S(\gamma)$, based on numerical results [57], is given in Figure 5.7. For $\gamma > 3$ it matches the asymptotic expression [57] because at this limit mechanism 1 is the only operative mechanism of mixing.

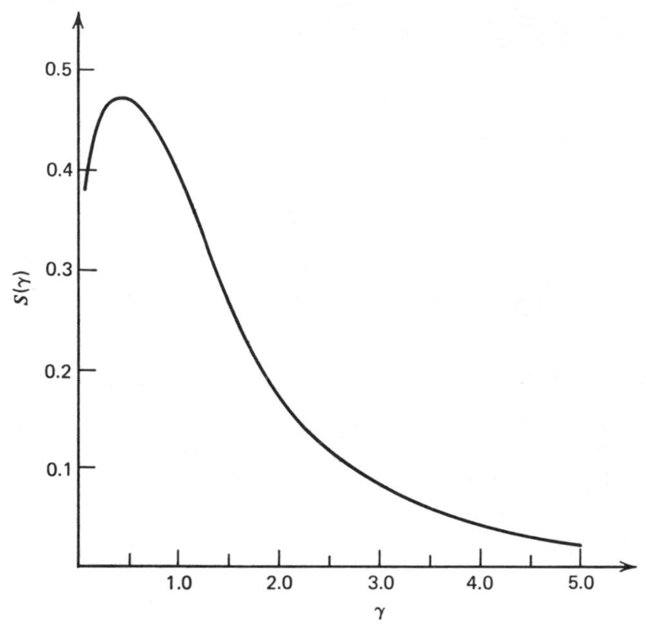

Figure 5.7 Plot of the function $S(\gamma)$, see equation (62). Recently (F. Masnou-Seeuws, These, Faculté des Sciences de Paris, Paris, 1973) a mistake was found in calculation of $S(\gamma)$ in (57). Corrected $S(\gamma)$ near its maximum should be about 15% higher.

This completes the description of different types of nonadiabatic coupling in collisions represented by equation (45). Equations (57) and (60) and the plot $S(\gamma)$ will now be used for the interpretation of experimental results. Parameters needed for calculations are either atomic constants 5($\Delta\varepsilon$, α, α_p, $\langle r^2 \rangle$) or are taken from Baylis' results [41] (R_1, R_2, α_2, U_0).

1. Na* + X (X = He, Ne, Ar, Kr, Xe)

Numerical calculations by Masnou–Seeuws [57] give a monotonic decrease of $\sigma(\frac{1}{2} \rightarrow \frac{3}{2})$ with increasing atomic number of X. This is due to the monotonic

Table 5.1
Intramultiplet-Mixing Cross Sections (in Å^2) for Na* (3 $^2P_{1/2}$) + X ($^1S_0$) Collisions at 400°K

	PARTNERS				
	Na*–He	Na*–Ne	Na*–Ar	Na*–Kr	Na*–Xe
γ	0.550	0.973	1.178	1.27	1.32
$\sigma_{\text{theor.M}}$ [a]	78	68	62	57	55
γ_{eff}	0.516	0.866	0.684	0.584	0.396
$\sigma_{\text{theor.}}$	78	71.5	75.5	77	78
σ_{expt} [b]	86	67	56	85	89

[a] Reference 57.
[b] Reference 58.

increase of γ (the first and second line in Table 5.1). If the polarization correction is taken into account replacing $\Delta\varepsilon$ by $\Delta\varepsilon_{\text{eff}}$, then γ_1 goes through its maximum, and σ reaches minimum when passing from He to Xe. This is in qualitative agreement with experimental findings [58]. In regard to absolute values, the discrepancy is not too large. Better agreement can be achieved varying R_1.

2. K* + X (X = He, Ne, Ar, Kr, Xe)

The main contribution to $\sigma(\frac{1}{2} \rightarrow \frac{3}{2})$ is from region 1. Theoretical values calculated according to equation (57) are lower than experimental ones (59), see Table 5.2. The latter are of the same order of magnitude as the depolarization cross sections in the $^2P_{1/2}$ state of K* [60]. The theory (Section VI)

Table 5.2
Intramultiplet-Mixing Cross Sections (in Å^2) for K* (4 $^2P_{1/2}$) + X ($^1S_0$) Collisions at 368°K

	PARTNERS				
	K*–He	K*–Ne	K*–Ar	K*–Kr	K*–Xe
γ_{eff}	2.12	3.94	4.03	4.08	3.50
σ_{theor}	24	6.0	6.0	6.0	9.2
σ_{expt} [a]	60	14.3	36.7	61.4	104

[a] Reference 59.

predicts that the mixing cross sections should be considerably lower, thus casting some doubt on experimental data. Further study of K*–X collisions will be helpful in accounting for the discrepancy.

3. Rb* + X (X = He, Ar, Xe)

For X = He the main contribution comes from region 1. For X = Ar and Xe, region 2 is responsible for mixing. Competition between the two mechanisms of coupling is illustrated in Figure 5.8, which together with Table 5.3 shows general agreement between theory [44] and experiment [61] both for absolute values and for the temperature dependence of cross sections. Some

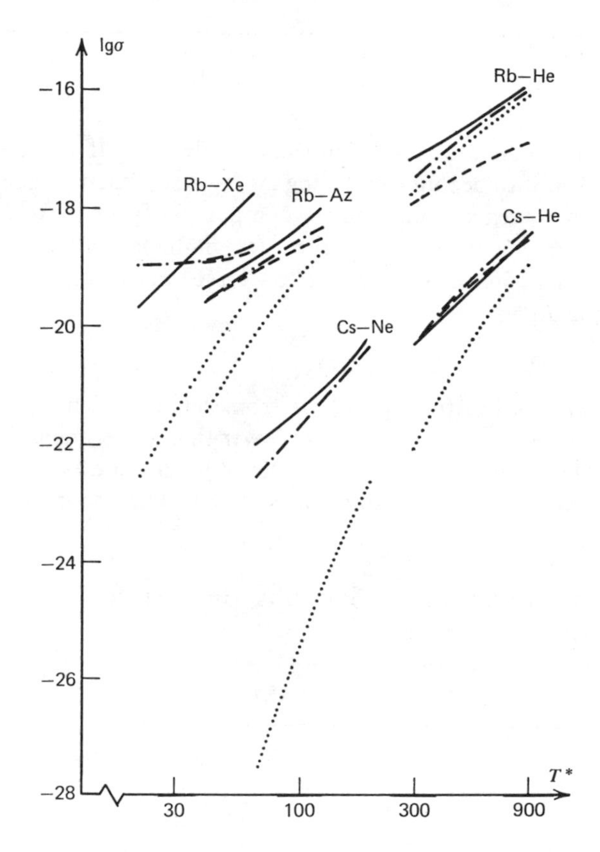

Figure 5.8 Intramultiplet-mixing cross sections $\sigma(\frac{1}{2} \rightarrow \frac{3}{2})$ (in Å^2) for different M*–X pairs as a function of the effective temperature $T^* = T\ (\mu_{\text{M-He}}/\mu_{\text{M-X}})$. Experimental data [61]: ———. Theory [50]: mechanism 1, · · · · · · ; mechanism 2, – – – –; sum of both mechanisms, – · – · – · – ·.

Table 5.3
Intramultiplet-Mixing Cross Sections (in $Å^2$) for Rb* ($5\,^2P_{1/2}$) and Cs* ($6\,^2P_{1/2}$) at 600° K[a]

	PARTNERS								
	R_1 (a.u.)	α_1 (a.u.)	$\gamma_{1,\mathrm{eff}}$	R_2 (a.u.)	α_2 (a.u.)	γ_2	$\sigma_{1,\mathrm{theor}}$	$\sigma_{2,\mathrm{theor}}$	σ_{expt}[b]
Rb*–He	13	0.67	9.10	5.0	2.3	2.68	0.16	0.047	0.28
Rb*–Ar	13	0.67	22.8	5.7	3.3	4.98	2.3×10^{-4}	1.3×10^{-3}	2.3×10^{-3}
Rb*–Xe	13	0.67	27.7	5.7	3.3	6.87	3.5×10^{-5}	1.3×10^{-3}	3.3×10^{-3}
Cs*–He	12	0.63	22.9	5.4	2.3	6.26	1.0×10^{-4}	7.6×10^{-4}	5.8×10^{-4}
Cs*–Ne	12	0.63	48.2	5.4	2.3	13.4	7.0×10^{-9}	7.6×10^{-6}	1.0×10^{-5}

[a] Reference 44.
[b] Reference 61.

discrepancy for the Rb*–Xe case is explained on the ground that equation (60) overestimates the effect of attraction by using a shifted exponential interaction.

4. Cs* + X (X = He, Ne)

For X = He and Ne, which has been studied experimentally [61], the main contribution comes from region 2. Agreement between theory and experiment is satisfactory (see Figure 5.8 and Table 5.3).

Thus, the intramultiplet mixing in heavy alkalies can be accounted for by transitions in two nonadiabaticity regions that give competing contributions to the mixing cross sections. It seems clear now that an attempt [62] to interrelate experimental data resorting only to equation (57) cannot be justified.

VI. DEPOLARIZATION OF EXCITED ALKALI ATOMS UPON COLLISIONS WITH NOBLE-GAS ATOMS

Several recent papers [45, 63–69] are devoted to the theoretical study of depolarizing collisions of alkalies $M^*(^2P_j)$ with noble gases, that is, to processes

$$M^*(^2P_{j,m}) + X(^1S_0) \rightarrow M^*(^2P_{j,m'}) + X(^1S_0). \tag{63}$$

Here m stands for the projection of j on the z axis of a laboratory frame.

There are at present several distinct approaches to the treatment of collisional reorientation of electronic angular momentum j of an atom. We shall discuss these by comparing them with known "exact" results which, though scarce, provide the possibility of verifying our understanding of the mechanism involved. These "exact" results are based on numerical solution of the collision dynamics for an assumed form of interatomic interaction and for coupling of states $^2P_{1/2}$ and $^2P_{3/2}$ of alkalies. Thus it would be helpful to discuss first different types of coupling in the light of results given in the preceding section.

A. Energetically Isolated States $^2P_{1/2}$ and $^2P_{3/2}$: $\Delta V \ll \Delta\epsilon$, $\xi_1 = \Delta\epsilon/\hbar\alpha_1 v \gg 1$

If the interaction ΔV [equation (34)] responsible for the coupling between states $^2P_{1/2}$ and $^2P_{3/2}$ remains small for a given trajectory $\mathbf{R}(t)$ compared to $\Delta\varepsilon$, then there is no appreciable adiabatic mixing of these states during a collision. The condition $\xi_1 \gg 1$ also prohibits nonadiabatic mixing of the states. However, the interaction $\Delta V(R)$ induces a splitting of the $^2P_{3/2}$ term (equal to $\frac{2}{3}\,\Delta V$), thus giving rise to "dephasing" of different components of the $^2P_{3/2}$ state, which ultimately results in depolarization. A large phase difference for terms $A\frac{3}{2}$ and $B\frac{1}{2}$ (see Figure 5.2) can build up for large impact parameters b, so that the main contribution to ΔV will be due to polarization interaction, equation (36). This suggests that the van-der-Waals interaction alone would be sufficient for interpreting depolarization of $^2P_{3/2}$ states of Rb*, Cs* and, approximately, K*, completely ignoring the $^2P_{1/2}$–$^2P_{3/2}$ mixing. As for state $^2P_{1/2}$ (and for any state with $j = \frac{1}{2}$), it is not split by interaction, so that there will be no ordinary adiabatic dephasing. Nonadiabatic dephasing, however, can occur if there is some mixture of states with $j > \frac{1}{2}$ to the state with $j = \frac{1}{2}$. This is best illustrated by depolarization of the ground state of alkalies where the collisionally induced mixing is actually very low and the depolarization cross section amounts to 10^{-22}–10^{-26} cm^2 [70]. The situation changes dramatically for the excited state $^2P_{1/2}$, for which typical values of depolarization cross sections are of the order of 100 a.u. [60]. This is due to an efficient mixing of $^2P_{1/2}$ and $^2P_{3/2}$ states, that is, to the departure from the scheme of energetically isolated states.

B. Adiabatically Isolated States $^2P_{1/2}$ and $^2P_{3/2}$: $\Delta V \sim \Delta\epsilon$, $\xi_1 = \Delta\epsilon/\hbar\alpha_1 v \gg 1$

Under this condition a collision practically does not induce intramultiplet mixing, but during a collision there is an appreciable distortion of atomic functions. The mere fact of the existence of the molecular term $A\,^2\Pi_{1/2}$ described by the Hund coupling scheme **a** means that the electronic spin follows rotation of the molecular axis. To this approximation there will be depolarization of the $^2P_{1/2}$ state although the relevant molecular term

$A\ ^2\Pi_{1/2}$ is degenerate. The depolarization probability depends on the angle of rotation of the molecular axis for the part of the trajectory where $R < R_1$ rather than on the interaction as such.

Approximation of adiabatically isolated states can be used to interpret depolarization of $^2P_{1/2}$ states of Rb*, Cs* and, approximately, K*.

C. Quasidegenerate States $^2P_{1/2}$ and $^2P_{3/2}$: $\Delta V \ll \Delta\epsilon$, $\xi_1' = \Delta\epsilon R_1/\hbar v \ll 1$

If the collision time R_1/v is short compared to the period of the Larmor precession of the electronic spin in the intra-atomic magnetic field, spin orientation can be considered to be fixed during the collision. Then the scattering matrix $S(jm, j'm')$ connecting $^2P_{jm}$ and $^2P_{j'm'}$ states will be expressed via the matrix $S(1M, 1M')$ connecting components of a spinless P state [71, 72]. This interrelation for depolarization probabilities is given in Table 5.4 (last column).

An estimation of ξ_1' gives $\xi_1' \approx \frac{1}{4}\gamma_1(\alpha_1 R_1)$ with γ_1 defined by equation (57). Taking $\alpha_1 R_1 \approx 10$ and referring to Table 5.1, we find that even for Na*–He collisions the condition $\xi_1' \ll 1$ is not fulfilled, although it is often assumed [68]. As for K*, Rb*, and Cs*, this approximation is not valid for ordinary temperatures.

Table 5.4

Relations between Population Transfer and Multipole-Relaxation Transition Probabilities

j	$P^{(j)}_{mm'} = \sum_{\chi} a_{\chi} P^{(j)}_{\chi}$	Quasidegenerate States $^2P_{1/2}$ and $^2P_{3/2}$
$\frac{1}{2}$	$P^{(1/2)}_{1/2,-1/2}{}^{a} = \frac{1}{2}P^{(1/2)}_1$	$\frac{10}{27}(P^{(1)}_2 - \frac{5}{3}P^{(1)}_1)^b$
1	$P^{(1)}_{0,1} = \frac{1}{3}P^{(1)}_2$	$P(j = \frac{1}{2} \to j = \frac{3}{2}) =$
	$P^{(1)}_{1,-1} = \frac{1}{2}P^{(1)}_1 - \frac{1}{6}P^{(1)}_2$	$\frac{2}{3}P^{(1)}_1$
$\frac{3}{2}$	$P^{(3/2)}_{3/2,1/1} = -\frac{3}{20}P^{(3/2)}_1 + \frac{1}{4}P^{(3/2)}_2 + \frac{3}{20}P^{(3/2)}_3$	$\frac{2}{9}P^{(1)}_2$
	$P^{(3/2)}_{3/2,-1/2} = \frac{3}{20}P^{(3/2)}_1 + \frac{1}{4}P^{(3/2)}_2 - \frac{3}{20}P^{(3/2)}_3$	$\frac{1}{6}P^{(1)}_1 - \frac{1}{18}P^{(1)}_2$
	$P^{(3/2)}_{1/2,-1/2}{}^{c} = \frac{1}{20}P^{(3/2)}_1 - \frac{1}{4}P^{(3/2)}_2 + \frac{9}{20}P^{(3/2)}_3$	$\frac{10}{27}(P^{(1)}_2 - \frac{3}{5}P^{(1)}_1)^b$
	$P^{(3/2)}_{3/2,-3/2}{}^{c} = \frac{9}{20}P^{(3/2)}_1 - \frac{1}{4}P^{(3/2)}_2 + \frac{1}{20}P^{(3/2)}_3$	0

[a] Zero for an energetically isolated state $j = \frac{1}{2}$.
[b] Zero in the exponential or adiabatic approximation for an energetically isolated state $j = 1$.
[c] Zero in the exponential approximation for an energetically isolated state $j = \frac{3}{2}$.

This concludes the discussion of possible coupling schemes, showing the significance of at least two physical parameters rather than one as sometimes believed [73].

We turn now to the discussion of the symmetry of a scattering matrix for depolarizing collisions. Calculation of cross sections includes, inter alia, the averaging of bilinear combinations of the scattering matrix elements over different orientations of the collision plane. This can be accomplished in a general form giving invariant expressions for cross sections valid for any orientation of the collision plane, for example, for a plane normal to the z axis of a laboratory frame (the so-called "standard collision"). Averaging also reveals that not all cross sections $\sigma_{mm'}^{jj'}$ characterizing transitions $j, m \to j', m'$ are independent, because of the additional constraints that arise due to the invariance of the description of $m \to m'$ transitions in spaced-fixed frames differing in orientation of axis. Two different approaches to the formulation of these constraints are possible. The first implies an expansion of the S matrix in irreducible tensors and a calculation of a rotationally invariant expression for the squared moduli of S (64, 65). The second involves expansion of the density matrix of a state with momentum j in irreducible tensors, and bilinear combinations of scattering matrix elements are rearranged in such a manner as to describe transitions between irreducible components of the density matrix [74, 75]. The two approaches differ in the coupling scheme of the four angular momenta j in constructing scalars and in obtaining either Zeeman (population transfer) cross sections $\sigma_{mm'}^{(j)}$, or those $\sigma_{\chi}^{(j)}$ for relaxation of 2^{χ}-pole moments of an atom in a state with angular momentum j (magnetic moments for odd χ and electric moments for even χ). Different multipole moments relax to zero independently of each other, and all components of a 2^{χ}-pole moment relax at the same rate $v\sigma_{\chi}^{(j)}$.

Due to these constraints, the number of independent parameters for transitions within the ^2P doublet is eight (65). If, however, the discussion concerns only depolarization of $^2P_{1/2}$ and $^2P_{3/2}$ states independently, and total excitation transfer between sublevels $j = \frac{1}{2}$ and $j = \frac{3}{2}$, this number reduces to five: one cross section for depolarization of the $^2P_{1/2}$ state, three cross sections for depolarization of the $^2P_{3/2}$ state, and the remaining cross section for the total intramultiplet mixing (see Section V).

The invariant expression for probabilities[1] $P_{\chi}^{(j)}$ is of the form

$$P_{\chi}^{(j)}(b) = 1 - \sum_{\substack{r,r' \\ s,s',p}} (-1)^{s-r} \begin{pmatrix} j & j & \chi \\ r & -r' & p \end{pmatrix} \begin{pmatrix} j & j & \chi \\ s & -s' & p \end{pmatrix} S_{rs}^{jj} S_{r's'}^{*jj}, \tag{64}$$

where the S matrix refers to any space-fixed frame.

[1] The $P_{\chi}^{(j)}(b)$ are related to the $\sigma_{\chi}^{(j)}$ by $\sigma_{\chi}^{(j)} = 2\pi \int_0^{\infty} b\, db\, P_{\chi}^{(j)}(b)$. $P_{\chi}^{(j)}$ may exceed unity because of the interference effects incorporated in equation (64).

Two such frames that will be used are those with the z axis directed perpendicular to the collision plane for the "standard collision" and with the z axis along the relative velocity vector before the collision. The z projections of j in these two frames are denoted by m and n; μ stands for the projection of j on the rotating molecular axis.

The relation between $P_\chi^{(j)}$ and $P_{mm'}^{(j)}$ is given by (66)

$$\sum_\chi (2\chi + 1)\begin{pmatrix} j & j & \chi \\ m & -m & 0 \end{pmatrix}\begin{pmatrix} j & j & \chi \\ m' & -m' & 0 \end{pmatrix} P_\chi^{(j)} = (-1)^{m-m'} P_{mm'}^{(j)}. \quad (65)$$

From equation (64) we obtain

$$P_0^{(j)} = 1 - \frac{1}{(2j+1)} \sum_{\tau,s} |S_{\tau s}^{jj}|^2, \quad (66)$$

whence the probability of total depletion of the state j upon one collision. Thus for the ^2P doublet we have

$$P_0^{(1/2)} = 1 - P(\tfrac{1}{2} \to \tfrac{3}{2}) \quad (67)$$

The relations between $P_\chi^{(j)}$ and $P_{mm'}^{(j)}$ are summarized in Table 5.4 for a general case and for the case of quasidegenerate states, $^2P_{1/2}$ and $^2P_{3/2}$.

We discuss now approximate methods for calculation of the S matrix with the aim of obtaining more information on individual transition probabilities.

D. Adiabatic Approximation

It is assumed that the atomic angular momentum is strongly coupled to the collision axis so that its projection on this axis is conserved during the collision [76]. Then for a rectilinear trajectory one has

$$S(jn, jn') = (-1)^{j+n}\, \delta_{n,-n'} \exp(i\Delta_n), \qquad \Delta_n = \Delta_{-n}, \quad (68)$$

which is actually the result of rotating the collision axis by an angle π with additional build-up of phases Δ_n, different for various molecular states. From equations (64) and (68) we get

$$P_\chi^{(j)} = 1 - \frac{(-1)^\chi}{2\chi + 1}. \quad (69)$$

For a spinless P state this gives $P_1^{(1)} = \frac{4}{3}$, $P_2^{(1)} = \frac{4}{5}$. The depolarization probability does not depend on b, so that calculation of the cross section requires a cutoff parameter defining the region of the adiabatic approximation.

E. Uniform Mixing Approximation

This is based on the statistical approximation to the scattering matrix [77, 78]

$$S(j, n, jn') = \frac{1}{(2j + 1)^{1/2}} \exp(i\Delta_{nn'}), \tag{70}$$

where phases $\Delta_{nn'}$ are not correlated. Equations (64) and (70) give

$$P_\chi^{(j)} = 1. \tag{71}$$

As before, the cutoff parameter $b_{\max}$ is needed to calculate cross sections.

F. Exponential Approximation

This approximation uses equation (18). The mathematical problem reduces to the matrix exponentiation. A computer program was worked out recently to calculate $P_\chi^{(j)}$ for rectilinear trajectories and the polarization interaction in the isolated-state approximation (66). A similar approach has been used taking into account the coupling between ${}^2P_{1/2}$ and ${}^2P_{3/2}$ states for a rectilinear trajectory (63) and for a trajectory corresponding to specular reflection from a hard sphere (69). The S matrix calculated according to equation (18) satisfies the symmetry relation

$$S(j, m, j, m') = (-1)^{2j-m-m'} S(j, -m', j, -m). \tag{72}$$

This relation can be traced back to the fact that the operator $\int_{-\infty}^{\infty} \mathscr{V}\, dt$ is a real matrix although $\mathscr{V}$ itself is not. From equation (72) follows a linear relation

$$\sum_\chi (2\chi + 1)(-1)^\chi P_\chi^{(j)} = 0. \tag{73}$$

This and the relation between $P_{mm'}^{(j)}$ and $P_\chi^{(j)}$ lead to the selection rule $m \nleftrightarrow -m$. Cross sections can be calculated directly by integration of $P_\chi^{(j)}(b)$ over impact parameters since at large b the exponential approximation goes over to the perturbation result.

G. Matching Approximation

In this approximation the evolution of perturbed atomic states during collision is described by two time-independent Hamiltonians, each taking into account only one part of the interaction. The simplest way is to neglect $\mathscr{V}(R)$ altogether outside the radius R_m, and consider it to be so strong at $R < R_m$ that the adiabatic approximation be valid. The basic formula is similar to equation (16). The suggested form of the S matrix, with different trajectories $\mathbf{R}_\mu(t)$ for different adiabatic potentials $U_\mu(R)$ at $R < R_m$, is (79)

$$S_{mm'}^{(j)} = \sum_\mu e^{-im(\pi-\chi_\mu)} D_{\mu m}(0, \tfrac{1}{2}\pi, \alpha) e^{i\Delta_\mu} D_{\mu m'}(0, \tfrac{1}{2}\pi, \alpha), \tag{74}$$

where the D functions are recognized as $l_m(\mu)e^{i\Delta_m}$ in equation (16), χ_μ is the scattering angle (deflection function) for the potential $U_\mu(R)$, and $\sin\alpha = b/R_m$. The symmetry relation equation (72) is now replaced by

$$S(j, m, jm'; \alpha, \chi_1, \ldots, \chi_n) = (-1)^{2j-m-m'} S(j, -m', j, -m; -\alpha, -\chi_1, \ldots, -\chi_n), \tag{75}$$

which does not impose any special selection rules prohibiting, for example, transitions $m \to -m$.

In spite of being very simple, this approximation provides the possibility of accounting in a simple way for motion of the collision partners in different adiabatic potentials. This is of special importance for depolarizing collisions where the angle of rotation of the molecular axis is an essential parameter of the S matrix (in particular, of those matrix elements that connect atomic states correlating to a degenerate molecular state).

Comparing now the different approximate schemes, it is astounding to see from Table 5.4 how various approximations yield spurious, physically unwarranted selection rules.

Only two selection rules can be explained. Absence of transition $m = \frac{1}{2} \to m = -\frac{1}{2}$ for an energetically isolated term ${}^2P_{1/2}$ is due to lack of splitting and failure to build up the ${}^2\Pi_{1/2}$ molecular term. Absence of transition $m = \frac{3}{2} \to m = -\frac{3}{2}$ for ${}^2P_{3/2}$ under quasidegenerate conditions is due to the impossibility of a $\Delta m = 3$ transition without spin flip.

However for the term $j = \frac{3}{2}$ there are extra selection rules for energetically isolated or quasidegenerate states, provided the probabilities are calculated in the exponential approximation. For the energetically isolated state the selection rule is due to the symmetry relations given by equation (72), while for the quasidegenerated state it follows from the particular value of the ratio $P_1^{(1)}/P_2^{(1)} = \frac{5}{3}$. This value is also responsible for vanishing of the cross section $\sigma_1^{(1/2)}$.

On these grounds it is suggested that the matching method is superior to others, and it will be used now for interpretation of depolarization in process (63).

1. Term ${}^2P_{1/2}$

The matching method provides the following simple expression for $\sigma_1^{(1/2)}$:

$$\sigma_1^{(1/2)} = 2\sigma_{1/2,-1/2}^{(1/2)} = 2\pi \int_0^{R_1} b\, db\, \tfrac{2}{3}\{1 + \cos[2\alpha(b) + \chi(b)]\}, \tag{76}$$

where $\sin \alpha = b/R_1$, and $\chi(b)$ is to be found from the dynamics of elastic scattering of $M^*(^2P_{1/2})$ on $X(^1S_0)$ in the potential $U(A\ ^2\Pi_{1/2})$. For a rectilinear trajectory $\chi(b) = 0$, and equation (76) gives $\sigma_1^{(1/2)} = \frac{2}{3}\pi R_1^2$. The effect of trajectory distortion is best demonstrated for a simple hard-sphere model:

$$U(A\ ^2\Pi_{1/2}, R) = \begin{cases} 0, & R > R_2 \\ \infty, & R < R_2 \end{cases} \tag{77}$$

that corresponds qualitatively to the real situation (see Figure 5.2). From equation (76) we get

$$\sigma_1^{(1/2)} = \frac{2}{3}\pi R_1^2 Q\left(\frac{R_2}{R_1}\right) \tag{78}$$

and the function $Q(\lambda)$ is plotted in Figure 5.9. It is seen that the larger the core, the smaller the cross section. At $R_2 = R_1$ the cross section vanishes. This

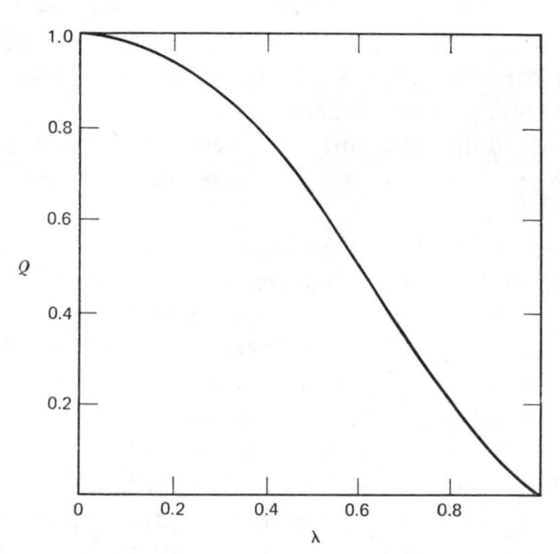

Figure 5.9 Plot of the function $Q(\lambda)$; see equation (78).

is due to the vanishing extension of the region where spin follows the molecular rotation axis. More sophisticated treatment is needed at this limit to find the nonzero but very small value of the cross section [80].

For R_2/R_1 not too close to unity, equation (78) gives a good approximation to more exact calculations (45) for the same potential. Now, application of equation (78) to experimental findings does not lead to satisfactory results. If experimental data (60) on $K^*(^2P_{1/2}) + X$ (X = He, Ne, Ar, Kr, Xe) with cross sections $\sigma_{1/2,-1/2}^{(1/2)}$ ranging from 46 to 107 Å^2 are to be reproduced by equation (78) the value of R_1, found even without core effect, must change from 14 to 20 a.u. If the core with $R_2 \approx 5$ a.u. is introduced, R_1 must increase by an additional 20–30%. For the K*–Xe pair this means that at $R \approx 25$ a.u. the exchange interaction is of the order of 60 cm^{-1}. This is not in line either with Baylis' results (41) or with other simple estimations.

For Rb*–X collisions, measured values of $\sigma^{1/2}_{1/2-1/2}$ are 9, 6, 9.7, and 10.6 Å^2 for X = He, Ne, Ar, Kr [81]. This order of magnitude can be obtained from equation (78) with $R_1 \approx 10$ a.u., $R_2 \approx 5$ a.u. If a larger value of R_1 is accepted, the hard-core radius has to be larger too, for example, $R_1 \approx$ 13–14 a.u., $R_2 \approx 10$ a.u. These figures are close to those obtained recently (69) using the exponential approximation for the manifold of states belonging to the ^2P doublet.

For Cs*–X collisions the low experimental values of $\sigma^{(1/2)}_{1/2-1/2} = 1$–$2$ Å^2, found in ref. 81, can hardly be rationalized on the basis of equation (78). It is possible, however, that for large values of $\Delta\varepsilon$ the "step" in the potential energy curve $A\frac{1}{2}$ near R_1 will distort the trajectory, thus diminishing the cross section.[1]

It follows from equation (76) that $\sigma_1^{(1/2)}$ depends only on the velocity via $\chi(b)$. In general for soft interactions $\chi(b)$ will decrease with increasing energy so that $\sigma_1^{(1/2)}$ might be expected to increase slightly with temperature, at least until the $^2P_{1/2}$ state remains adiabatically isolated. At present there seem to be no experimental data on this effect.

2. Term $^2P_{3/2}$

It is reasonable to attempt an interpretation of depolarization for the state $^2P_{3/2}$ within the energetically-isolated-state approximation. As the contribution to $\sigma_\chi^{(3/2)}$ comes from large impact parameters, one may assume a rectilinear trajectory ($\chi_\mu = 0$) and use the random-phase approximation for $R < R_m$, R_m being the matching distance. The value of R_m itself is determined by the condition that the Coriolis coupling between molecular terms $A\frac{3}{2}$ and $B\frac{1}{2}$ is equal to the adiabatic splitting $U(B\frac{1}{2}) - U(A\frac{3}{2})$. From equation (34) on the condition that $\Delta V \ll \Delta\varepsilon$, we get $U = U(B\frac{1}{2}) - U(A\frac{3}{2}) = \frac{2}{3}\,\Delta V(R)$, and using the polarization interaction equation (36) we finally obtain

$$\Delta U(R) = \frac{C'}{R^6}, \qquad C' = \tfrac{2}{5}\alpha_p e^2\langle r^2\rangle. \tag{79}$$

With this interaction and the simplifications stated above the cross sections

[1] Bulos and Happer [Phys. Rev. **A 4**, 849 (1971)] have recently shown that due to nonzero nuclear spin the experimental values [81] for Cs*–X depolarization cross sections are smaller by about a factor of 3 than the corresponding values would be in the absence of nuclear spin. In the present chapter, effects of nuclear spin have not been considered.

are

$$\sigma_1^{(3/2)} = \tfrac{13}{15}\pi\left(\frac{C'}{2\hbar v}\right)^{2/5},$$
$$\sigma_2^{(3/2)} = \pi\left(\frac{C'}{2\hbar v}\right)^{2/5}. \tag{80}$$

Numerical solution (67) of semiclassical equations gives 0.8 instead of $\frac{13}{15}$ in the first line, and 0.985 instead of 1 in the second. Again, the matching procedure is satisfactory.

Optical pumping experiments [82] with Rb give for $\sigma_2^{(3/2)}$ values ranging from 100 to 300 Å^2 for He, Ne, Ar, Kr, Xe. Calculation gives close agreement with theory except for He, for which case the theoretical value is about 57 Å^2. It was suggested that this discrepancy might be due to the unaccounted exchange interaction. Calculations along this line for $K^*(^2P_{3/2}) + X$ collisions show that the repulsive exchange interaction gives a larger contribution to depolarization cross sections for He and Ne than the van-der-Waals interaction. However, theoretical results are very rough approximations, because the adoption of one particular coupling scheme to simplify the mathematics cannot be substantiated, and the use of rectilinear trajectories may introduce some errors, as for example for the $^2P_{1/2}$ state.

The latter point deserves more comment. Referring to equation (76) we see that for head-on collisions $\alpha = 0$ and χ is equal to either π or 0 for cases without or with a hard core, respectively. The corresponding depolarizing probabilities will be either zero or $\frac{4}{3}$, which is of course a large difference. The appearance of the scattering angle and slow oscillations of P with b mean that the dephasing occurs along the entire trajectory within the region of appreciable interaction. This is quite different from intramultiplet mixing transitions, which are actually localized (under almost adiabatic conditions) near R_1 and R_2 and which depend only on the motion of atoms in the vicinity of R_1 and R_2. This makes the theory of intramultiplet mixing in a sense less complicated than that of depolarization.

VII. INTRAMULTIPLET MIXING AND EXCITATION TRANSFER IN COLLISIONS OF EXCITED AND GROUND-STATE ALKALI ATOMS

The resonant and nonresonant excitation transfer in collisions of excited and ground-state alkali atoms proceed with very large cross sections. This is primarily due to the long-range dipole–dipole interaction mixing electronic states of colliding partners at large interatomic separations. The peculiarity of the interaction is that its time dependence along the typical trajectory is very slow, thus giving very low transition probabilities for many channels. The number of adiabatic molecular states strongly coupled by atomic motion is rather small, and this makes it possible to make a more or less detailed analysis of the actual energy-transfer mechanism.

In this section we consider first the resonant and nonresonant energy transfer for the spinless S–P transitions

$$A^*(P) + B(S) \to A(S) + B^*(P). \tag{81}$$

This energy-transfer reaction simulates collisions of alkali atoms M_1 and M_2 under conditions when the spin-orbital splittings $\Delta\varepsilon_1$ and $\Delta\varepsilon_2$ in isolated atoms are very weak. Then we discuss collisions of similar alkali atoms

$$M^*(^2P_j) + M(^2S) \to M^*(^2P_{j'}) + M(^2S) \tag{82}$$

and collisions of dissimilar atoms

$$M_1^*(^2P_j) + M_2(^2S) \begin{cases} \nearrow M_1^*(^2P_{j'}) + M_2(^2S) \\ \searrow M_1(^2S) + M_2^*(^2P_j) \end{cases} \tag{83}$$

The process (82) occurs by both resonant and nonresonant channels, while all channels of equation (83) are of a nonresonant character.

A. Resonant S–P Excitation Transfer

If atoms A and B are similar, adiabatic molecular terms are classified as $^1\Lambda_w$. If only the dipole–dipole interaction is taken into account, adiabatic energies are given by equation (38) with $S = 0$. According to selection rules the only coupling between states will be due to the Coriolis interaction between Σ and Π terms of the same parity.

To obtain a rough estimate of the excitation transfer cross section, consider first the adiabatic approximation, neglecting the Coriolis coupling (the so-called "rotating-atom approximation"). The transfer is then associated with mere interference between *gerade* and *ungerade* states. As an example, the $^1\Sigma$ time-dependent wave function can be written as

$$\Psi(^1\Sigma, t) = \phi(^1\Sigma_g, R) \exp\left[-\frac{i}{\hbar}\int_{-\infty}^{t} U(^1\Sigma_g, R)\, dt\right] + \phi_2(^1\Sigma_u, R) \exp\left[-\frac{i}{\hbar}\int_{-\infty}^{t} U(^1\Sigma_u, R)\, dt\right], \tag{84}$$

giving as an initial condition at $t = -\infty$, the localization of excitation on one of the partners, say partner A. The A* → B* excitation transfer probability is

$$P_{AB}(\Sigma) = \sin^2 \Delta(\Sigma, b), \tag{85}$$

where

$$\Delta(\Sigma, b) = \frac{1}{2\hbar}\int_{-\infty}^{\infty} [V^{\text{dip}}(^1\Sigma_g) - V^{\text{dip}}(^1\Sigma_u)]\, dt = \frac{d^2}{\hbar}\int_{-\infty}^{\infty} \frac{dt}{R^3}. \tag{86}$$

Integrating (86) over a rectilinear trajectory and using the Firsov cutoff parameter b^* in calculating the excitation transfer cross section we get

$$\sigma_{AB}(\Sigma) = 2\pi\int_0^{\infty} P_{AB}(\Sigma) b\, db = \tfrac{1}{2}\pi(b_\sigma^*)^2, \tag{87}$$

where b_σ^* is defined by $\Delta(\Sigma, b_\sigma^*) = 1/\pi$ (83). A similar approach gives b_π^*. Both $b_\sigma^*(v)$ and $b_\pi^*(v)$ are given in Table 5.5. The average cross section $\bar{\sigma}_{AB}$, calculated in the adiabatic approximation, is

$$\bar{\sigma}_{AB} = \tfrac{1}{3}[\sigma_{AB}(\Sigma) + 2\sigma_{AB}(\Pi)] = \tfrac{4}{3}\pi\, \pi p_0^2 = 4.2\pi p_0^2, \qquad p_0 = \frac{d^2}{\hbar v}. \tag{88}$$

Numerical solutions of equations allowing for the Coriolis coupling, give, instead of the factor 4.2, in equation (88), considerably lower values of 2.25–2.28 [84–8].

To understand this discrepancy let us estimate the characteristic distance R_m at which the off-diagonal element of the Coriolis interaction between Σ and Π terms prevails over the difference in adiabatic energies:

$$\left|\frac{V^{\text{dip}}(\Sigma) - V^{\text{dip}}(\Pi)}{2}\right| = \frac{\hbar v}{R_m}. \tag{89}$$

Table 5.5
Characteristic Lengths Associated with Dipole–Dipole Resonant S–P Excitation Transfer ($\rho_0^2 = d^2/\hbar v$)

Firsov impact parameters for adiabatic terms	$b_\sigma^* = \sqrt{4\pi}\,\rho_0$ $b_\pi^* = \sqrt{2\pi}\,\rho_0$
Boundary distance between prevailing Coriolis and electrostatic interactions	$R_m = \sqrt{\frac{3}{2}} \cdot \rho_0$
Firsov impact parameters for nonadiabatic terms	$b_{1,2}^* = \sqrt{\pi}\,\rho_0$ $b_3^* = \sqrt{2\pi}\,\rho_0$

It can be inferred from Table 5.5 that R_m is less than $b_\sigma^*(v)$ and $b_\pi^*(v)$. This means that the adiabatic approximation is totally unacceptable for impact parameters close to the cutoff value b^*.

A better approximation may be obtained if the Σ and one component of the Π functions be at first completely mixed and new functions used to calculate the effective interaction V_k^{dip}. Thus constructed, the new functions and terms are

$$\begin{aligned} \phi_{w,1} &= \frac{1}{\sqrt{2}}[\phi(^1\Sigma_w) + \phi'(^1\Pi_w)], & V_1^{\mathrm{dip}} &= \frac{w}{2} V_0^{\mathrm{dip}}(R), \\ \phi_{w,2} &= \frac{1}{\sqrt{2}}[\phi(^1\Sigma_w) - \phi'(^1\Pi_w)], & V_2^{\mathrm{dip}} &= \frac{w}{2} V_0^{\mathrm{dip}}(R), \\ \phi_{w,3} &= \phi''(^1\Pi_w), & V_3^{\mathrm{dip}} &= -wV_0^{\mathrm{dip}}(R), \end{aligned} \tag{90}$$

where ϕ' and ϕ'' are components of the Π state, symmetric and antisymmetric with respect to reflection in the collision plane. Corresponding Firsov parameters for these terms are listed in Table 5.5. Since $R_m < b_k^*$, it is safe to use b_k^* as a maximum impact parameter for excitation transfer between nonadiabatic *gerade* and *ungerade* states. Thus, we obtain

$$\bar{\sigma}_{\mathrm{AB}} = \tfrac{1}{3}(\tfrac{1}{2}\pi(b_1^*)^2 + \tfrac{1}{2}\pi(b_2^*)^2 + \tfrac{1}{2}\pi(b_3^*)^2] = \tfrac{2}{3}\pi\ \pi p_0^2 = 2.1\pi p_0^2. \tag{91}$$

The factor 2.1 in (7.10) is rather close to the mentioned "exact" results, 2.25–2.28. This demonstrates the possibilities of a simple approach based on the estimation of nonadiabatic coupling.

It should be noted that result given by equation (91) coincides with that calculated under the assumption that the orientation of atomic orbitals remains fixed during collision ("fixed-atom approximation") [84, 87]. This is to be expected, because the strong Coriolis mixing of molecular functions in the rotating frame prevents rotation of orbitals in a space fixed frame.

B. Nonresonant S–P Excitation Transfer

In the approximation of the dipole–dipole interaction, the adiabatic terms of a system A + B* are

$$\begin{aligned} U(\Lambda, R) &= \pm\{(\Delta\varepsilon)^2 + 4(2-\Lambda)^2(V_0^{\mathrm{dip}})^2\}^{1/2}, \\ V_0^{\mathrm{dip}}(R) &= \frac{d^{\mathrm{A}}d^{\mathrm{B}}}{R^3}, \end{aligned} \tag{92}$$

where $\Delta\varepsilon$ is the energy difference between the initial and final states in equation (81).

Nonadiabatic coupling mixes all terms. However, if $\Delta\varepsilon$ is sufficiently high, the location of Σ–Σ and Π–Π nonadiabatic coupling due to radial motion will be different from that of a Σ–Π coupling due to rotational motion. This argument, and also the fact that the probability of nonresonant transition is small, permit depolarization and transition with energy transfer to be treated separately. Let us discuss now the latter. According to the theory of nearly adiabatic perturbations (Section II),

$$P_{AB}(\Lambda, b) = 2 \exp\left[-\frac{2}{\hbar v}\,\mathrm{Im}\int_{|R_c^\Lambda|}^{R_c^\Lambda} 2U(\Lambda, R)\,\frac{dR}{(1-b^2/R^2)^{1/2}}\right], \tag{93}$$

where R_c^Λ is found from $\Delta U(\Lambda, R_c) = 0$, that is,

$$R_c^\Lambda = \left(\frac{2(2-\Lambda)d^A d^B}{\Delta\varepsilon}\right)^{1/3} \exp\left(\frac{i\pi}{6}\right). \tag{94}$$

Calculations give (88)

$$P_{AB}(\Lambda, b) = 2 \exp\left[-\frac{\Delta\varepsilon\,|R_c^\Lambda|}{3\hbar v}\left(3.41 + \frac{1.36b^2}{|R_c^\Lambda|^2} + \cdots\right)\right], \tag{95}$$

$$\sigma_{AB}(\Lambda, v) = \pi\,|R_c^\Lambda|^2 \left(\frac{3\hbar v}{1.36\Delta\varepsilon\,|R_c^\Lambda|}\right) \exp\left(-\frac{3.41\Delta\varepsilon\,|R_c^\Lambda|}{3\hbar v}\right). \tag{96}$$

The numerical coefficients 3.41 and 1.36 in equation (95) are not very different from π and $\frac{1}{2}\pi$ that would have been obtained had the power interaction $(1/R)^3$ been approximated in the coupling region by

$$|R_c|^{-3} \exp\left(-3\,\frac{R - |R_c|}{|R_c|}\right).$$

This shows that the exponential model (see Section III) can be used to match nonresonant [equation (95)] and resonant [equation (85)] probabilities within the two-state close-coupling approximation.

Equation (96) differs from those found earlier by Stueckelberg [15] and Bates [89]. Stueckelberg [15] used the perturbation theory to calculate $P_{AB}(b)$ at $b > |R_c|$ and applied a Landau–Zener type of formula for $b < R_c$. It was then found that the main contribution to σ_{AB} comes from distant collisions with $b > |R_c|$.

This was essentially due to the high pre-exponential factor, entering an equation similar to equation (95). However, the high pre-exponential factor results in a first-order perturbation theory that is incorrect under strong but slowly varying perturbation. Bates [89] corrected this, applying a solution [equation (29)] conjectured by Rosen and Zener [22] to the interaction of the type $A/(a^2 + R^2)^{3/2}$ with an extra cutoff parameter a that is supposed to be less than $|R_c|$. If this parameter is set equal to zero, neither the transition probability nor the cross section will be exponentially small, when $\Delta\varepsilon R_c/\hbar v \gg 1$. This is because the conjectured solution erroneously considers the internal region $R < R_c$ as contributing to transitions between molecular states. Table 5.6 summarizes results obtained by various approaches.

Table 5.6
Comparison of Various Approximations for Cross Sections of Nonresonant Dipole–Dipole *S–P* Excitation Transfer

APPROXIMATION FOR THE CASE $\xi = \dfrac{\Delta\varepsilon\,\|R_c\|}{\hbar v} \gg 1$	INTERATOMIC DISTANCES CONTRIBUTING TO σ	IMPACT PARAMETERS CONTRIBUTING TO σ	ORDER OF MAGNITUDE OF REDUCED CROSS SECTION $\sigma/\pi\,\|R_c\|^2$
Born approximation[a]	$R \approx \|R_c\|$	b varies near R_c in the interval $\|\Delta b\| \sim \|R_c\|/\xi$	$\xi^2 \exp(-\xi)$
Conjectured Rosen–Zener solution[b]	$R \approx a$	$0 \leqslant b \leqslant \|R_c\|(\alpha/\xi)^{1/2}$ $\alpha = 2a/\|R_c\| < 1$	$\alpha^2 \exp(-\alpha\xi)$
Theory of almost adiabatic pertubation[c]	$R \approx \|R_c\|$	$0 < b < \|R_c\|/\xi^{1/2}$	$\xi^{-2} \exp(-\xi)$

[a] Reference 15.
[b] Reference 89.
[c] Reference 88.

Considering the last column of reduced cross sections, in which all non-essential numerical factors have been omitted to facilitate comparison, we see how large the difference between various approaches is, and how large the error can be when nonadiabatic coupling is not properly treated. Some examples along this line can he found in Gallagher's paper [90].

C. Collisions of Similar Alkali Atoms

For the process (82) there are both resonant and nonresonant channels (Table 5.7). From the above we have the following estimates for the resonant

Table 5.7
Intramultiplet-Mixing Cross Sections (in Å^2) for $M^*(^2P_{3/2})$ + $M(^2S_{1/2})$ Collisions

PARTNERS	$\Delta\varepsilon \times 10^4$ (a.u.)	d^2 (a.u.)	T (°K)	R_c (a.u.)	σ_{theor}	σ_{capt}	σ_{expt}
Na* + Na	0.78	6.36	423	43	101	78.4	280[a]
K* + K	2.65	8.36	360	31	20.4	84	252[b]
Rb* + Rb	10.8	9.16	330	20	4.8	118	67.2[b]
Cs* + Cs	25.2	10.66	310	15	4.2	134	30.8[b]

[a] Reference 93.
[b] Reference 59.

cross section and the maximum value of the nonresonant cross section:

$$\sigma^{\text{res}} \approx \frac{\pi d^2}{\hbar v},$$
$$\sigma^{\text{nonres}}_{\text{max}} \approx \pi\,|R_c^2| \approx \pi\left(\frac{d^2}{\Delta\varepsilon}\right)^{2/3}. \tag{97}$$

Combining these we obtain

$$\frac{\sigma^{\text{res}}}{\sigma^{\text{nonres}}_{\text{max}}} \approx \frac{\Delta\varepsilon\,|R_c|}{\hbar v}. \tag{98}$$

Thus, if the ratio $\Delta\varepsilon\,|R_c|/\hbar v = \Delta\varepsilon^{2/3}\,d^{2/3}/\hbar v$ is large, the two processes can be discussed separately. A simple estimate shows that for thermal collisions of a light alkali (e.g., sodium, with $\Delta\varepsilon = 17\ \text{cm}^{-1}$), R_c is about 50 a.u., and the ratio $\Delta\varepsilon R_c/\hbar v$ is approximately equal to 40. For heavier alkalies this ratio increases still more. For this reason the discussion of the excitation-transfer mechanism is simplified considerably.

Resonant excitation transfer $^2P_{1/2}$–$^2S_{1/2}$ and $^2P_{3/2}$–$^2S_{1/2}$ can be considered in the same way as the S–P process. The results of cross section calculations are [91]

$$\sigma^{\text{res}}(^2P_{1/2} - {}^2S_{1/2}) = 1.13\,\frac{\pi d^2}{\hbar v},$$
$$\sigma^{\text{res}}(^2P_{3/2} - {}^2S_{1/2}) = 1.66\,\frac{\pi d^2}{\hbar v}. \tag{99}$$

Nonresonant excitation transfer, that is, intramultiplet mixing, requires discussion of possible nonadiabatic regions. As seen from Figure 5.8 there are several pseudocrossings, crossings, and mergings, but selection rules greatly simplify the coupling scheme. *Gerade* and *ungerade* manifolds are not coupled, and none of the merging 0_w^σ terms is coupled. We are interested now in nonadiabatic coupling between terms correlating at $R \to \infty$ with different atomic states. Terms of the same symmetry are coupled by radial motion, and terms of different symmetry are coupled by relative rotation of the pair. For the first group there will be a pseudocrossing situation, so that transition probabilities $P(\Omega, b)$ are given by an equation similar to equation (95) with a change of numerical factors in the exponent. Now the characteristic length entering equation (95) is $|R_c|/3$ instead of $1/\alpha$ as for alkali–noble-gas collisions. Thus the exponent for dipole–dipole M*–M interactions is 15 times higher than that for M*–X exchange interactions, and collisions M*–M are strongly adiabatic for all alkali atoms with respect to transitions between terms of the same symmetry.

There are two kinds of pairs of terms with different Ω. As discussed in Section II, if the terms do not cross, the transition probability will be exponentially small (as for case discussed above) with an extra small pre-exponential factor. If the terms do cross, there will be no exponential, and thus this case will give the main contribution to the excitation-transfer cross section. As seen from Figure 5.3 there are two crossing points R_S, each corresponding to pairs 1_g, 0_g^+ and 1_u, 0_u^+. Corresponding partial cross sections σ_{if} calculated using equation (14) are

$$\sigma_{if}(\varepsilon) = \frac{16\sqrt{2}\,\pi^2\,|\langle i|\,J_x\,|f\rangle|^2}{3\mu^{1/2}\,\Delta F_{if}^S\hbar}\,\frac{(E - U_i^s)^{3/2}}{E}, \qquad E = \frac{\mu v^2}{2}, \tag{100}$$

where U_i^S is the value of a molecular term U_i at $R = R_s$,

$$\Delta F_{if}^S = \frac{d}{dR}(U_i - U_f)$$

and $\langle i|\,J_x\,|f\rangle$ is the matrix element of the projection of total angular momentum on the relative angular momentum, calculated with adiabatic molecular functions $|i\rangle$ and $|f\rangle$. It can be inferred from the correlation diagrams that $\langle 1_g|\,J_x\,|0_g^*\rangle$ should be small compared to $\hbar$, but $\langle 1_u|\,J_x\,|0_u^*\rangle$ may be as large as its asymptotic "molecular" (the Hund coupling case **a**) value, $\hbar/\sqrt{2}$. Thus the "*gerade–gerade*" possibility virtually does not contribute to the total cross section, and $\sigma(\frac{3}{2} \to \frac{1}{2})$ is given just by equation (100) with $|i\rangle = |0_u^+\rangle$, $|f\rangle = |1_u\rangle$ and $\langle 1_u|\,J_x\,|0_u^+\rangle = \hbar/\sqrt{2}$. The intramultiplet mixing cross section is then

$$\sigma(j \to j') = \frac{1}{4(2j + 1)}\,\sigma_{if}, \tag{101}$$

where the factor $4(2j + 1)$ accounts for the degeneracy of the initial state [2 is due to parity, $2(2j + 1)$ is due to the total rotational degeneration of the $M(^2S_{1/2}) + M^*(^2P_J)$ pair]. Results of more detailed calculations (47, 92) are summarized in Table 5.7. Theoretical results are in qualitative agreement with experimental data only in the dependence of $\sigma(\frac{3}{2} \rightarrow \frac{1}{2})$ on $\Delta\varepsilon$. Theory predicts $\sigma(\frac{3}{2} \rightarrow \frac{1}{2}) \sim |\Delta F^c_{if}|^{-1} \sim (\Delta\varepsilon/R_S)^{-1} \sim (\Delta\varepsilon)^{-4/3} d^{1/3}$, whereas experimental data can be satisfactorily correlated by the $\Delta\varepsilon^{-1}$ dependence [59]. As for absolute values of $\sigma(\frac{3}{2} \rightarrow \frac{1}{2})$ the disagreement is the largest for heavy alkalis. However, even for the Na*–Na collision the discrepancy is significant because $\sigma_{\text{theor}}(\frac{3}{2} \rightarrow \frac{1}{2})$ is very close to its maximum value $\pi R_S^2/8$. It is difficult to see how the theory can explain large experimental values of $\sigma(\frac{3}{2} \rightarrow \frac{1}{2})$, which are even larger than the orbiting capture cross section σ_{capt} in the field of the resonant dipole–dipole potential [47]. Clearly, more work is needed to clarify the situation.

D. Collisions of Dissimilar Alkali Atoms

For dissimilar alkali atoms possible inelastic channels correspond either to intramultiplet mixing or to nonresonant excitation transfer. In general, the predicted pattern of molecular states is very sensitive to approximations adopted in the calculations. In particular, this refers to certain pseudocrossings for which the splitting depends on the difference between exchange integrals rather than on atomic parameters such as $\Delta\varepsilon_1 - \Delta\varepsilon_2$ and ΔE. Another difficulty is that uncertainty in one coupling region can affect many or all cross sections, because transition from an initial to a final atomic state proceeds via many intermediate transitions between molecular states. This makes analysis of the processes [83] very complicated.

Since there are no crossings of molecular terms at large distances where the dipole–dipole interaction is operative (Figure 5.4), no appreciable contribution from this region—either to intramultiplet mixing or to transfer of excitation between the partners—takes place. Thus the region of strong exchange interaction becomes increasingly important. Some results obtained for the $K^*(4\,^2P)$–$Rb(5\,^2S)$ collisions [49] will be cited as an illustration. The schematic diagram of molecular terms relevant to the problem is shown in Figure 5.10. Distances R_0, R_1, R_2, R_3, and R_4 correspond to those in Figure 5.4. Transition probabilities P_1–P_4 are calculated according to the Landau–Zener formula (25), P_5 is given by equation (14), and P_0 describes the transition between terms approximated by the exponential model (26).

Calculated cross sections for different channels are given in Table 5.8.

Intramultiplet mixing in K* was found to be due mainly to nonadiabatic coupling near $R_0 \approx 15$–22 a.u. between terms $d1$ and $c1$, which arise from merging terms $^3\Pi_{1g}$ and $^1\Pi_{1u}$ of similar partners under the influence of a perturbation proportional to $\Delta\varepsilon_k - \Delta\varepsilon_{Rb}$. The upper bound of the estimated

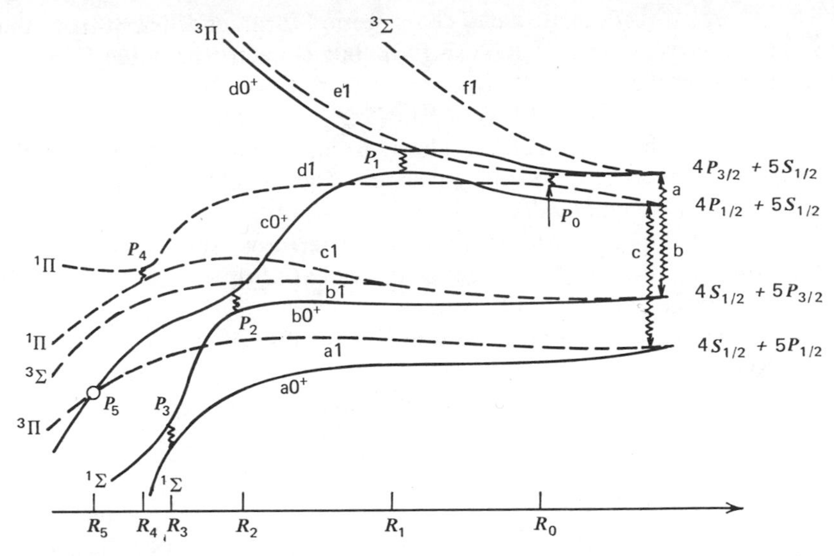

Figure 5.10 The pattern of molecular terms for the (KRb)* pair.

cross section is considerably lower than the experimental value. The discrepancy is analogous to that for similar partners and as yet has no explanation.

The excitation transfer $K^*(4\,^2P_{1/2}) \rightarrow Rb^*(5\,^2P_{3/2})$ is due to two parallel transitions at $R_2 \approx 13$ a.u. and $R_4 \approx 10$ a.u., which give comparable contributions to the cross section. The splitting of $b0^+$ and $c0^+$ terms at R_2 is proportional to $\Delta\varepsilon_k - \Delta\varepsilon_{Rb}$ and the splitting near R_4 depends on all types of interaction relevant to dissimilarity of partners: ΔE, $\Delta\varepsilon_k - \Delta\varepsilon_{Rb}$, and

Table 5.8
Excitation-Transfer Cross Sections (in Å²) for $K^*\ (4\,^2P_{1/2}) + Rb\ (5\,^2S_{1/2})$ Collisions at 370°K

	FINAL CHANNEL	σ_{theor} [a]	σ_{expt}
a	$K^*\ (4\,^2P_{3/2}) + Rb\ (5\,^2S_{1/2})$	67	260[b]
b	$K\ (4\,^2S_{1/2}) + Rb^*\ (5\,^2P_{3/2})$	10	40[b]
			5.3[c]
c	$K\ (4\,^2S_{1/2}) + Rb^*\ (5\,^2P_{1/2})$	4	2.7[c]
			2.3[d]

[a] Reference 49.
[b] Reference 94.
[c] Reference 95.
[d] Reference 96.

$Y^{(K,Rb)} - Y^{(Rb,K)}$. The calculated cross section is rather different from that reported by Hrycyshyn and Krause [94], but close to the value found by Stacey and Zare [95].

Excitation transfer $K^*(4\,^2P_{1/2}) \rightarrow Rb^*(5\,^2P_{3/2})$ proceeds via two pseudo-crossings at R_2 and $R_3 \approx 12$ a.u. The Coriolis coupling between crossing terms $a1$ and $c0^+$ at R_5 is of low efficiency. The calculated cross section is close to the experimental values (94, 96).

The calculated ratio of the excitation-transfer cross sections $\sigma(4\,^2P_{1/2} \rightarrow 5\,^2P_{3/2})/\sigma(4\,^2P_{1/2} \rightarrow 5\,^2P_{1/2})$ is in good agreement with experimental results reported by Stacey and Zare [95, 96].

Thus it seems that the overall interpretation of excitation transfer in the K*–Rb system is basically correct. This leaves some hope that other excitation-transfer processes occurring in collisions of atoms can be explained using a similar approach.

VIII. CONCLUSION

The problem of inelastic atomic collisions at low energies consists essentially of two parts. The first is the determination of adiabatic molecular terms, the second is the calculation of a multichannel scattering matrix from given potentials and coupling operators. Contribution of the theory of nonadiabatic transitions to the second part of the problem depends on the existence of distinctively separated regions of coupling which can be characterized by corresponding matrices of nonadiabatic transitions. Under semiclassical conditions the probability amplitudes oscillate rapidly, and this is the main practical difficulty in obtaining the scattering matrix. It seems more resonable to make approximations not while solving the equations but at an earlier stage when potentials and coupling within the nonadiabatic region are replaced by a simplified model allowing rather transparent treatment, either by perturbation technique or by obtaining the analytical solution. This approach can be adopted in view of our limited knowledge of interatomic interactions, in particular for excited states.

We believe that for a considerable time the uncertainty that arises in using models will be still less than that due to errors in potentials. Only in exceptional cases when exact reconstruction of potentials from scattering data is attempted, as is the case in collision spectroscopy, will more refined treatment of nonadiabatic coupling be needed.

IX. SUPPLEMENT

Since this review was first submitted for publication, a number of papers have appeared on the subject. Some comments on these papers in connection with the topics discussed in different sections are in order.

The semiclassical approach to the scattering problem is widely used now. Leaving aside papers on specific processes, we mention only those which are concerned with the general formulation of the method [97, 98] or which discuss the optimal choice of a trajectory within "average approximation" [99, 104] and within the Feynman path-integral formalism [101]. Construction of the S-matrix by matching of quantal adiabatic evolution matrices to semiclassical matrices of nonadiabatic transitions (equation 7) is discussed in [102, 103]. In common with different types of non-adiabatic coupling, reference is made to the paper [104] (analysis of the Stueckelberg method), paper [105] (analysis of the rotational coupling) and to the review paper [106] on the sudden approximation (often also called the exponential approximation).

In connection with the two-state approximation we mention paper [107] giving detailed analysis of the Landau–Zener–Stueckelberg model in which transition probabilities have been calculated for a wide set of parameters of the model, and also papers devoted to extensions of the model [108–110]. The exponential model with the Hamiltonian given by equation (26) was discussed in [111] and extended in [112].

In particular, in [111] phases ϕ and ψ entering equation (19) were calculated. As for approximate solutions of two-state problems, one more, differing from equations (29), (30), (31) and (32) has been suggested recently [113].

Ab initio calculations of systems M^*–X were performed recently for LiHe [114], NaHe [114, 115] and CsX [116]. Besides, there is indirect evidence on potentials from the broadening of spectral lines of Na [117, 118], extreme-wing line broadening in Cs [119] and on oscillations in continuum spectra of RbX and CsX pairs [120]. All data are in line with the term pattern shown in Fig. 5.2.

There are also papers on calculations of M*M systems: NaLi [121] and Li_2, Na_2, K_2, LiNa, LiK, NaK [122]. Comparison of the asymptotic method of calculation with results of *ab initio* approach [123] helps to estimate the accuracy of the former. The asymptotic method was found to give a different term pattern for different M*M pairs, significant, e.g., for the possibility of crossing of ${}^1\Sigma_g^+$ and ${}^1\Sigma_u^+$ terms of M*M. New information on potential curves will probably emerge from studies of radiation by M* excited in collisions with atoms X [124] and M [125].

The recent trend in theoretical investigation of intramultiplet mixing and depolarization was represented by attempts to consider both types of processes in terms of the same collision model. We mention here only works published in 1972 and 1973 on processes

$$\mathrm{M}^*({}^2P_{j,m}) + \mathrm{X} \rightarrow \mathrm{M}^*({}^2P_{j,m'}) + \mathrm{X} \tag{102}$$

Earlier works have been reviewed by Krause [126] and Lijnse [127].

If a collision $M^* + X$ proceeds under almost adiabatic conditions (i.e., if $\xi_1, \xi_2 \gg 1$) then joint consideration of $j \to j'$ (mixing) and $j, m \to j, m'$ (depolarization) processes is not really necessary as depolarization cross sections are by far larger than the mixing cross section. For such a case, depolarizing collisions can be considered first without taking into account the mixing collisions, and then $j \to j'$ transitions can be calculated assuming complete randomization of j over its projections. It should be pointed out here, that mixing and depolarization show a different dependence on the bending of the trajectory corresponding to relative motion of atoms M and X:

(i) If it is assumed as in earlier papers (references [38, 46]) that small impact parameters do not contribute to $P(\frac{1}{2} \to \frac{3}{2})$ then there is no question at all about the effect of the trajectory distortion. However, in this case calculated cross sections are considerably smaller than the experimental ones (see Figure 5.8).

(ii) To account for the bending of trajectories one can use the "broken path" corresponding to specular reflection from a rigid sphere with radius R^*. However this will introduce strong artificial non-adiabaticity because the radial component of velocity at the sphere is discontinuous, and the tangential component has a cusp. If $R^* < R_1$ this artificial effect increases with decreasing difference $\Delta V(R^*)$, and hence with increasing R^*. Thus in terms of this model, $P(\frac{1}{2} \to \frac{3}{2})$ increases with R^* until R^* reaches R_1 and then drops as the region $R \sim R_1$, responsible for the transition, can not be reached at all. This model will give larger cross sections $\sigma(\frac{1}{2} \to \frac{3}{2})$ compared to those calculated by the theory of almost adiabatic perturbations. This qualitative picture corresponds roughly to results of the paper [128] which thus might be criticized for improper use of "broken path" trajectories.

(iii) The broken path approximation using two straight-line segments intersecting at the classical scattering angle can, however, be used to describe the relative motion in resonant channels, say for depolarization of the state $^2P_{1/2}$. The total angle of rotation of the molecular axis in the region $R < R_1$ is important here rather than the angular velocity.

(v) For nonresonant channels one must use trajectories smoothly bent on a sphere $R = R_2$. It follows from equation (59) that the probability $P(\frac{1}{2} \to \frac{3}{2})$ is sensitive to the steepness of $U_\pi(R)$ near R_2. At a fixed steepness parameter α_2, the cross section $\sigma(\frac{1}{2} \to \frac{3}{2})$ decreases with increasing R_2, behavior just opposite to the case of specular reflection (compare ii).

In short, under almost adiabatic conditions a trajectory can not be chosen independent of the interaction responsible for the transition. Consequently one must be very cautious with regard to the semiclassical method when, as is usually the case, the interaction and trajectories are considered without any correlation.

Let us discuss now theoretical works on specific processes. The depolarization of the $^2P_{1/2}$ state of K^* was studied in [129]. It was found, in line with section VI, that the semi-empirical values of C_π which bring theoretical and experimental cross sections into agreement should be considerably larger than those calculated by Baylis (reference 41). As for depolarization of the $^2P_{1/2}$ state of Rb* and Cs*, Gallagher's experimental cross-sections (reference 81) should be reinterpreted [130] taking into account the spin of nuclei. Corrected cross sections are approximately three times less than those cited in the section VI. Calculations by Hidalgo and Geltman [128] for Rb*–Xe collisions can reproduce experimental findings if parameters of their collision model are properly adjusted. This approach is not much better than the use of the simple model leading to equation (78). The simple model approach is appropriate at the moment when considerable discrepancies exist among experimental results. Thus, for Cs*($^2P_{1/2}$) Guiry and Krause [131] give cross sections $\sigma^{(1/2)}_{1/2,-1/2}$ varying from 6 to 36A^2 (from He to Xe) which are approximately two times larger than those of Gallagher corrected by Bulos and Happer [130].

In contrast to depolarization of $^2P_{1/2}$ states, which can not be considered in the approximation of energetically isolated state, resonant processes in $^2P_{3/2}$ states can be discussed within this approximation. The matching method, which can take into account the distortion of trajectories, can provide an acceptable description of disorientation and disalignment [132]. Probably it is important to take into account both the bending of trajectories and adiabatic interaction between $^2P_{3/2}$ and $^2P_{1/2}$ states (i.e., to go beyond the isolated state approximation). An attempt of this kind was made in [129]. Though the theory [129] reproduces qualitatively the variation of disorientation and disalignment cross sections of $K^*(^2P_{3/2})$ colliding with noble gases, and shows the importance of exchange forces for light noble gases, discrepancies between theory and experiment [133] still remain. This completes the comment on works dealing separately with j–j' and j, m–j, m' transitions.

A unified description of $j \to j'$ and $j, m \to j, m'$ transitions was formulated in several papers for quasiresonant processes in Na*–He collisions. In the semiclassical approximation with rectilinear trajectories all cross sections (transfer of population and orientation between $^2P_{1/2}$ and $^2P_{3/2}$ states, relaxation of orientation in $^2P_{1/2}$ and $^2P_{3/2}$ states, and relaxation of alignment in $^2P_{3/2}$ state) were calculated by Masnou and Roueff [134]. The quantal study of these processes was completed recently by Reid [135]. It has been noted [132, 136, 137] that because of quasiresonance, the semiclassical solution can also be obtained by reducing the problem to depolarization of a spinless P-state, and that the bending of trajectories plays a non-negligible role. For such a case the best semiclassical approach would probably be based on matching adiabatic evolution matrices A (which describe deflection of wave packet moving over different adiabatic molecular terms) with matrices of non-adiabatic transitions N (which describe non-adiabatic coupling along one classical trajectory). It would be of interest to find out to what extent does the best semiclassical approach reproduce the exact quantal result just obtained [135]. As for comparison with experiment, the semiclassical calculations [134] satisfactorily agree with recent data [138, 139].

There has been lately no major development in the theory of intramultiplet mixing and excitation transfer for alkali–alkali collisions. The only important comment was made by Lewis [140] who pointed that at distances 5–6 Å the ionic term M^+M^- may cross the bunch of covalent terms of MM*. This crossing can provide additional channels for intramultiplet mixing. If the transition probability at crossing is taken to be maximally possible this crossing enhances mixing and essentially diminishes the discrepancy between theory and experiment reflected in Table 5.7 for Rb*–Rb and Cs*–Cs collisions. However the importance of ionic-covalent mixing can be estimated properly only after extensive *ab initio* calculations of excited molecular terms of MM* system are made. At present, there is no direct evidence of drastic perturbations of covalent terms at small interatomic distances.

Finally, two theoretical papers should be mentioned. The first [141] concerns orientation and alignment relaxation and transfer in resonant transitions $^2S_{1/2}$–$^2P_{1/2}$, $^2P_{3/2}$, and the second [142] deals with non-resonant excitation transfer for spinless S–P states. All results obtained refer to the simple dipole–dipole interaction, proportional to R^{-3}.

ACKNOWLEDGMENTS

This review is partly based on work done in cooperation with Dr. E. I. Dashevskaya, Dr. M. Ya. Ovchinnikova, Dr. E. P. Gordeev, Dr. A. I. Reznikov, Dr. A. I. Voronin, and Dr. A. A. Zembekov, of the Academy of Sciences of U.S.S.R., Moscow. Also, some questions were discussed with Professor R. S. Berry and Professor J. C. Light of the University of Chicago and with Professor L. Krause of University of Windsor where this work was initiated. Recognition is given to Professor Wm. E. Baylis, of the University of Windsor who assisted in the proofreading of this review. The author wishes to thank them all for their contribution to this paper.

REFERENCES

1. N. F. Mott and H. S. W. Massey, *The Theory of Atomic Collisions*, Clarendon Press, Oxford, 1965.

2. D. Allab, M. Barat, and I. Baudon, *J. Phys.*, **1,** 195 (1968).
3. F. T. Smith, in *Atomic Physics: Heavy Particle Collisions*, edited by F. Bederson, V. W. Cohen, and F. M. J. Pichanick, Plenum, New York, 1969, p. 353.
4. F. T. Smith, in *Lectures in Theoretical Physics: Atomic Collisions Processes*, edited by S. Geltman, K. T. Mahanthappa, and W. E. Brittin, Gordon and Breach, New York, 1969, Vol. XIc, p. 95.
5. E. E. Nikitin and M. Ja. Ovchinnikova, *Usp. Fiz. Nauk*, **104,** 379 (1971).
6. D. R. Bates, *Commun. Atom. Mol. Phys.*, **1,** 127 (1969).
7. E. E. Nikitin, in *Fast Reactions and Primary Processes in Chemical Kinetics*, edited by S. Claesson, Almqvist and Wiksell, Stockholm, 1967.
8. G. Herzberg, Spectra of Diatomic Molecules, 2nd ed., Van Nostrand Company, Inc., New York, 1950.
9. L. D. Landau and E. M. Lifshitz, *Quantum Mechanics. Non-Relativistic Theory*, 2nd ed., Pergamon, New York, 1965.
10. L. D. Landau, *Physik. Z. Sowjetunion*, **2,** 46 (1932).
11. V. L. Pokrovskii and I. M. Khalatnikov, *Zh. Eksperim. Teor. Fiz.*, **40,** 1713 (1961).
12. E. C. Kemble, *Phys. Rev.*, **48,** 549 (1935).
13. J. Heading, *An Introduction to Phase-Integral Methods*, Methnen's Monographs on Physical Subjects, London and New York, 1962.
14. N. Fröman and P. O. Fröman, *JWKB Approximation*, North-Holland, Amsterdam, 1965.
15. E. C. G. Stuckelberg, *Helv. Phys. Acta*, **5,** 369 (1932).
16. W. R. Thorson and J. B. Delos, *Phys. Rev.* (to be published).
17. L. D. Landau, *Phys. Z. Sowjetunion*, **1,** 88 (1932).
18. P. Pechukas and J. C. Light, *J. Chem. Phys.*, **44,** 3897 (1966).
19. D. Chang and J. C. Light, *J. Chem. Phys.*, **50,** 2517 (1969).
20. R. D. Levine, *Chem. Phys. Lett.*, **2,** 76 (1968).
21. C. Zener, *Proc. Roy. Soc.* (London), **A137,** 696 (1932).
22. N. Rosen and C. Zener, *Phys. Rev.*, **40,** 502 (1932).
23. E. E. Nikitin, *Opt. Spectry.*, **13,** 761 (1962).
24. E. E. Nikitin, *Discussions Faraday Soc.*, **33,** 14 (1962).
25. W. D. Ellison and S. Borovitz, in *Atomic Collision Processes*, edited by M. R. C. McDowell, North-Holland, Amsterdam, 1964.
26. M. Kunicke and Yu. N. Demkov, *Vest. Leningr. Univ.*, **16,** 39 (1969).
27. E. E. Nikitin, "Theory of Non-Adiabatic Transitions. Recent Development on the Landau–Zener Model," in *Chemische Elementazprozesse* edited by H. Hartmann, Springer-Verlag, Berlin, 1968.
28. L. P. Kotova, *Zh. Eksperim. Theor. Fiz.*, **55,** 1375 (1968).
29. M. S. Child, *Mol. Phys.*, **16,** 313 (1969).
30. M. S. Child, *Mol. Phys.*, **20,** 171 (1971).
31. E. E. Nikitin, *Commun. Atom. Mol. Phys.*, **1,** 166 (1970).
32. E. E. Nikitin, *Advan. Quantum Chem.*, **5,** 135 (1970).
33. E. F. Gurnee and J. H. Magee, *J. Chem. Phys.*, **26,** 1237 (1957).
34. L. Vainstein, L. Presnyakov, and I. Sobelman, *Zh. Eksperim. Teor. Fiz.*, **43,** 518 (1962).

35. B. G. Skinner, *Proc. Phys. Soc.* (London), **77,** 551 (1961).
36. H. Nakamura, *J. Phys. Soc. Japan*, **20,** 2272 (1965).
37. Y. Mori and H. Fujita, *J. Phys. Soc. Japan*, **20,** 432 (1965).
38. E. E. Nikitin, *J. Chem. Phys.*, **43,** 744 (1965).
39. S. B. Schneiderman and H. H. Michels, *J. Chem. Phys.*, **42,** 3706 (1965).
40. M. Krauss, *J. Res. Nat. Bur. Std. U.S.*, **72A,** 553 (1968).
41. W. E. Baylis, *J. Chem. Phys.*, **51,** 2665 (1969).
42. B. M. Smirnov, *Atomnye Stolknoveniya i Elementarnye Processy v Plazme*, Atomizdat, Moskva, 1968.
43. S. Ja. Umanski and A. I. Voronin, *Theoret. Chim. Acta*, **12,** 166 (1968).
44. E. I. Dashevskaya, E. E. Nikitin, and A. I. Reznikov, *J. Chem. Phys.*, **53,** 1175 (1970).
45. E. P. Gordeev, E. E. Nikitin, and M. Ja. Ovchinnikova, *Opt. Spectry.*, **30,** 189 (1971).
46. J. Callaway and E. Bauer, *Phys. Rev.*, **140,** A1072 (1965).
47. E. I. Dashevskaya, A. I. Voronin, and E. E. Nikitin, *Can. J. Phys.*, **47,** 1237 (1969).
48. B. Kockel and N. Grün, *Z. Naturforsch.*, **24a,** 731 (1969).
49. E. I. Dashevskaya, E. E. Nikitin, A. I. Voronin, and A. A. Zembekov, *Can. J.* Phys. **48,** 981 (1970).
50. E. I. Dashevskaya, E. E. Nikitin, and A. I. Reznikov, *Opt. Spectry.*, **29,** 1016 (1970).
51. E. E. Nikitin, *Opt. Spectry.*, **22,** 269 (1967).
52. E. I. Dashevskaya and E. E. Nikitin, *Opt. Spectry.*, **22,** 866 (1967).
53. J. Callaway and P. S. Laghos, *Phys. Lett.*, **26A,** 394 (1968).
54. L. Kumar and J. Callaway, *Phys. Lett.*, **28A,** 385 (1968).
55. R. H. G. Reid and A. Dalgarno, *Phys. Rev. Lett.*, **22,** 1029 (1969).
56. R. H. G. Reid and A. Dalgarno, *Chem. Phys. Lett.*, **6,** 85 (1970).
57. F. Masnou-Seeuws, *J. Phys. B: Atom. Mol. Phys.*, **3,** 1437 (1970).
58. J. Pitre and L. Krause, *Can. J. Phys.*, **45,** 2671 (1967).
59. L. Krause, *Appl. Opt.*, **5,** 1375 (1966).
60. W. Berdowski and L. Krause, *Phys. Rev.*, **165,** 158 (1968).
61. A. Gallagher, *Phys. Rev.*, **172,** 88 (1968).
62. H. P. Hooymayers and C. Th. J. Alkemade, *Chem. Phys. Lett.*, **4,** 277 (1969).
63. H. I. Mandelberg, in *Abstract of Papers Submitted to the Conference on Heavy Particle Collisions*, Belfast, ed. by D. R. Bates, Queen's University of Belfast, Northern Ireland, p. 177, 1968.
64. E. P. Gordeev, E. E. Nikitin, and M. Ja. Ovchinnikova, *Can. J. Phys.*, **47,** 1819 (1969).
65. G. Grawert, *Z. Phys.*, **225,** 283 (1969).
66. C. H. Wang and W. J. Tomlinson, *Phys. Rev.*, **181,** 115 (1969).
67. A. N. Okunevich and V. I. Perel, *Zh. Eksperim. Teor. Fiz.*, **58,** 666 (1970).
68. M. Elbel, *Can. J. Phys.*, **48,** 3047 (1970).
69. M. B. Hidalgo and S. Geltman, *J. Phys. B: Atom. Mol. Phys.*, **5,** 265 (1972).
70. R. M. Herman, *Phys. Rev.*, **136,** A1577 (1964).

71. M. Elbel and F. Naumann, *Z. Phys.* **204,** 501 (1967); **208,** 104 (1967).
72. M. Elbel, *Phys. Lett.*, **28A,** 4 (1968).
73. M. A. Bouchiat, in *Optical Pumping and Atomic Line Shape*, edited by T. Skalinski, Panstwowe Wydawrictwo Naukowe, Warszawa, 1969, p. 131.
74. M. I. Dyakonov and V. I. Perel, *Zh. Eksperim. Teor. Fiz.*, **48,** 345 (1965).
75. A. Omont, *J. Phys. Rad.*, **26,** 26 (1965).
76. W. Gough, *Proc. Phys. Soc.* (London), **90,** 287 (1967).
77. F. A. Franz and J. R. Franz, *Phys. Rev.*, **148,** 82 (1966).
78. J. Fricke and J. Haas, in ref. 73, p. 185.
79. E. I. Dashevskaya, *Chem. Phys. Lett.*, **11,** 184 (1971).
80. E. I. Dashevskaya and E. Kobseva, *Opt. Spectry.*, **30,** 807 (1971).
81. A. Gallagher, *Phys. Rev.*, **157,** 68 (1967).
82. P. A. Zhitnikov, P. P. Kuleshov, A. N. Okunevich, and B. M. Sebastyanov, *Zh. Eksperim. Teor. Fiz.*, **58,** 831 (1970).
83. O. B. Firsov, *Zh. Eksperim. Teor. Fiz.*, **42,** 1001 (1951).
84. T. Watanabe, *Phys. Rev.*, **138,** A1573 (1965); **139,** AB1 (1965); **140,** AB5 (1965).
85. A. Omont, *Compt. Rendu. Acad. Sci.*, **262,** 190 (1966).
86. A. P. Kasantsev, *Zh. Eksperim. Teor. Fiz.*, **51,** 1751 (1966).
87. T. Watanabe, *Advances in Chemistry Series*, No. 82, Radiation Chemistry II, ed. by the staff of Industrial and Engineering Chemistry, Wash. *Am. Chem. Soc.* 1968.
88. E. E. Nikitin, *Chem. Phys. Lett.*, **2,** 402 (1968).
89. D. R. Bates, *Discussions Faraday Soc.*, **33,** 7 (1962).
90. A. Gallagher, *Phys. Rev.*, **179,** 105 (1969).
91. Yu. A. Vdovin and N. A. Dobrodeev, *Zh. Eksperim. Teor. Fiz.*, **55,** 1047 (1968).
92. Yu. A. Vdovin, V. M. Galitski, and N. A. Dobrodeev, *Zh. Eksperim. Teor. Fiz.*, **56,** 1344 (1969).
93. J. Pitre and L. Krause, *Can. J. Phys.*, **46,** 125 (1968).
94. E. S. Hrycyshyn and L. Krause, *Can. J. Phys.*, **47,** 215 (1969).
95. V. Stacey and R. N. Zare, *Phys. Rev.* A, **1,** 1125 (1970).
96. M. H. Ornstein and R. N. Zare, *Phys. Rev.*, **181,** 214 (1969).
97. J. B. Delos, W. R. Thorson, and S. K. Knudsen, *Phys. Rev.*, **A6,** 709 (1972).
98. J. B. Delos, W. R. Thorson, and S. K. Knudsen, *Phys. Rev.*, **A6,** 720 (1972).
99. T. A. Green and M. E. Riley, *Electron. and Atom. Collisions. Abstr. pap.*, VIII ICPEAC, Beograd, 1973, p. 159.
100. M. E. Riley, *ibid*, p. 163.
101. A. P. Penner and R. Wallace, *Phys. Rev.*, **A7,** 1007 (1973).
102. E. E. Nikitin, in: *Physics of Ionized Gases 1972*, ed. M. V. Kurepa, Institute of Physics, Beograd, 1972, p. 117.
103. J. E. Bayfield, E. E. Nikitin, and A. I. Reznikov, *Chem. Phys. Lett.*, **19,** 471 (1973).
104. D. S. F. Crothers, *Adv. in Physics*, **20,** 405 (1971).
105. A. Russek, *Phys. Rev.*, **A4,** 1918 (1971).
106. R. D. Levine, *Mol. Phys.*, **22,** 497 (1971).
107. J. B. Delos and W. R. Thorson, *Phys. Rev.*, **A6,** 728 (1972).

108. J. B. Delos and W. R. Thorson, *Phys. Rev. Lett.*, **28,** 647 (1972).
109. M. Ya. Ovchinnikova, *Zh. Eksp. Teor. Fiz.*, **64,** 129 (1973).
110. E. E. Nikitin, M. Ya. Ovchinnikova, and A. I. Shushin, *Electron. and Atom. Collisions. Abstr. pap.*, VIII ICPEAC, Beograd, 1973, p. 207.
111. E. E. Nikitin and A. I. Reznikov, *Phys. Rev.*, **A6,** 522 (1972).
112. H. Nakamura, *Mol. Phys.*, **25,** 577 (1973).
113. D. S. F. Crothers, *J. Phys. B: Atom and Mol. Phys.*, **6,** 1418 (1973).
114. M. Krauss, P. Maldonado, and A. C. Wahl, *J. Chem. Phys.*, **54,** 4944 (1971).
115. C. Bottcher, *Chem. Phys. Lett.*, **18,** 457 (1973).
116. J. Pascale and J. Vandeplanque, *Electron. and Atom. Collisions. Abstr. pap.*, VIII ICREAC, Beograd, 1973, p. 629.
117. E. Roueff, *J. Phys. B: Atom. and Mol. Phys.*, **5,** L79 (1972).
118. E. L. Lewis and L. F. McNamara, *Phys. Rev.*, **A5,** 2643 (1972).
119. R. E. M. Hedges, D. L. Drummond, and A. Gallagher, *Phys. Rev.*, **A6,** 1519 (1972).
120. C. G. Carrington, D. Drummond, A. G. Gallagher, and A. V. Phelps, *Chem. Phys. Lett.*, **22,** 511 (1973).
121. P. J. Bertoncini, G. Das, and A. C. Wahl, *J. Chem. Phys.*, **52,** 5112 (1970).
122. A. C. Roach, *J. Mol. Spectrosc.*, **42,** 27 (1272).
123. A. I. Reznikov and S. Ya. Umanski, *Teor. Eksp. Kkim.*, **7,** 587 (1971).
124. V. Kempter, B. Kübler, and W. Mecklenbrauck, *Electron. and Atom. Collisions. Abstr. pap.*, VIII ICPEAC, Beograd 1973, p. 170. *J. Phys. B.*
125. V. Kempter, W. Koch, B. Kübler, W. Mecklenbrauck, and C. Schmidt, *Electron. and Atom. Collisions. Abstr. pap.*, VIII ICPEAC, Beograd 1973, p. 618; *Chem. Phys. Lett.*, **21,** 164 (1973).
126. L. Krause, in: *The Physics of Electronic and Atomic Collisions*, VII ICPEAC invited papers and progress reports, ed. T. R. Govers and F. J. deHeer, North Holland Publ. Co., 1972, p. 65.
127. P. L. Lijnse, Review of Literature on Quenching, Excitation and Mixing Collision Cross-Sections for the First Resonance Doublets of the Alkalies. Rep. 398, Fysisch Laboratorium, Rijksuniversiteit Utrecht, Netherlands, 1972.
128. M. B. Hidalgo and S. Geltman, *J. Phys. B: Atom. and Mol. Phys.*, **5,** 265 (1972).
129. E. P. Gordeev, E. E. Nikitin, and M. Ya. Ovchinnikova, *Optika i Spektroskopiya*, **30,** 189 (1971).
130. B. R. Bulos and W. Happer, *Phys. Rev.*, **A4,** 849 (1971).
131. J. Guiry and L. Krause, *Phys. Rev.*, **A6,** 273 (1972).
132. E. I. Dashevskaya and N. A. Mokhova, *Optika i Spektroskopiya*, **33,** 817 (1972).
133. W. Berdowski and L. Krause, *Phys. Rev.*, **A4,** 984 (1971).
134. F. Masnou and E. Roueff, *Chem. Phys. Lett.*, **16,** 593 (1972).
135. R. H. G. Reid, *J. Phys. B: Atom. and Mol. Phys.*, **6,** 2018 (1973).
136. E. I. Dashevskaya and N. A. Mokhova, *Chem Phys. Lett.*, **20,** 454 (1973).
137. M. Elbel, *Z. Phys.*, **248,** 375 (1971).
138. M. Elbel and W. Schneider, *Z. Phys.*, **241,** 244 (1970).

139. W. Schneider, *Z. Phys.*, **248**, 387 (1971).
140. E. L. Lewis, *Phys. Lett.*, **A35**, 387 (1971).
141. C. G. Carrington, D. N. Stacey, and J. Cooper, *J. Phys. B: Atom. and Mol. Phys.*, **6**, 417 (1973).
142. J. C. Gay and A. Omont, *J. Phys.*, **35**, 9 (1974).

CHAPTER SIX

EXCITATION AND DEEXCITATION PROCESSES RELEVANT TO THE UPPER ATMOSPHERE

J. William McGowan

Physics Department and the Centre for Interdiciplinary Studies in Chemical Physics
The University of Western Ontario, London, Canada

Ralph H. Kummler

Department of Chemical Engineering and Material Sciences
Wayne State University, Detroit, Michigan

Forrest R. Gilmore

R and D Associates, Santa Monica, California

Contents

I. GENERAL CONSIDERATIONS

This chapter, with reference to the upper atmosphere, discusses the internal degrees of freedom of atoms, ions, and molecules, including the mechanisms of production of excited species and the ways in which the energy is contained in excitation and distributed through collision. Emphasis is placed on the description and interrelation of such interactions in all areas that involve atoms, ions, and molecules in those excited states likely to be important in atmospheric processes. In the laboratory, as well as in the quiescent and disturbed atmospheres, it is often difficult to recognize and separate the effects of excited species, and for a long time it was expedient to assume that the excited-state reactions were subsidiary to ground-state reactions. However, it is now recognized that in both the laboratory and the atmosphere small numbers of excited species can have large effects on total reaction rates and upon steady-state conditions.

The experimental methods used in excited-state studies are not described specifically within this chapter. Some of these techniques are referred to in Table 2 of ref. 1. Others can be found in the associated chapters in this review. Special mention should be made of material contained in a review on excited nitrogen by Wright and Winkler [2]. Elsewhere in the present chapter, Section II examines the lifetimes and energies contained in the various excited atmospheric species, and in Section III some excitation and deexcitation results for the more important atmospheric species are presented. In Section IV a more complete list of pertinent rate constants and cross sections is given.

The bulk of the material in the chapter was prepared in advance of 1973. However, an attempt has been made to update the material in all tables and in many instances include the most recent references.

II. LIFETIMES OF, AND ENERGY STORED IN, EXCITED STATES

It is necessary to distinguish between the effective lifetime, τ_c, which allows for collisional deexcitation, and the collision-free radiative lifetime of the excited species, τ_0. These two lifetimes are simply related through the reactive collision frequency (kn):

$$\frac{1}{\tau_c} = \frac{1}{\tau_0} + \Sigma_i k_i n_i, \tag{1}$$

where k_i is the rate constant (cm³/sec) by which the excited atom, ion, or molecule inelastically scatters from another atom, ion, electron, or molecule of number density n_i (cm^{-3}).

Much of what is known about the radiative lifetimes of the longer-lived species is summarized in Table 6.1. These values are believed to be the best presently available, although some are controversial.

Table 6.1
Radiative Lifetimes and Transitions for Principal Atmospheric Species

SPECIES AND STATE	MEAN RADIATIVE LIFETIME (μsec)	PRINCIPAL TRANSITION; λ (IN Å); NAME OF TRANSITION	APPROX. ΔE_{if} (eV)	REF.
Atoms and Atomic ions				
$C(^3P)$	Ground state			
(^1D)	3.2×10^9	$^3P \leftarrow {}^1D$; 9823, 9850	1.26	3
(^1S)	2.0×10^6	$^1D \leftarrow {}^1S$; 8727	1.42	3
$(2p^3\,{}^5S^0)$	3×10^4	$^3P \leftarrow {}^5S$; 2965, 2967	4.18	3
$C^+(^2P^0)$	Ground state			
(^4P)	Mod. long	$^2P^0 \leftarrow {}^4P$; $\sim$2330	5.35	4
$N(^4S^0)$	Ground state			
$(^2D^0_{3/2})$	6.1×10^{10}	$^4S^0 \leftarrow {}^2D^0_{3/2}$; 5200 (nebular)	2.38	3, 5
$(^2D^0_{1/2})$	1.4×10^{11}	$^4S^0 \leftarrow {}^2D^0_{5/2}$; 5201	2.38	3, 5
$(^2P^0)$	1.3×10^7	$^2D \leftarrow {}^2P^0$; 10,396; 10,404	1.19	3, 5
$(3s\,{}^4P)$	2.5×10^{-3}	$^4S^0 \leftarrow {}^4P$; 1200, 1201	10.31	6
$N^+(^3P)$	Ground state			
(^1D)	2.5×10^8	$^3P \leftarrow {}^1D$; 6584, 6548	1.89	3, 5
(^1S)	9×10^5	$^1D \leftarrow {}^1S$; 5755	2.15	3
$O(^3P)$	Ground state			
(^1D)	1.48×10^8	$^3P \leftarrow {}^1D$; 6300, 6364 (red lines)	1.96	3, 7, 8
(^1S)	8×10^5	$^1D \leftarrow {}^1S$; 5577 (green line)	2.22	9
$(3s\,{}^5S^0)$	6×10^2	$^3P \leftarrow {}^5S^0$; 1356, 1359	9.13	3
$(3s\,{}^3S^0)$	1.8×10^{-3}	$^3P \leftarrow {}^3S^0$; 1302, 1305, 1306	9.51	10

$O^+(^4S^0)$	Ground state			
$(^2D^0_{3/2})$	5.9×10^9	$^4S^0 \leftarrow {}^2D^0_{3/2}$; 3726 (nebular)	3.33	3, 5
$(^2D^0_{5/2})$	2.1×10^{10}	$^4S^0 \leftarrow {}^2D^0_{5/2}$; 3729	3.32	3, 5
$(^2P^0_{1/2})$	5.4×10^6	$^2D^0 \leftarrow {}^2P^0_{1/2}$; 7319, 7330 (auroral)	1.69	3, 5
$(^2P^0_{3/2})$	4.2×10^6	$^2D^0 \leftarrow {}^2P^0_{3/2}$; 7319, 7330	1.69	3, 5
Diatomic molecules and molecular ions[a]				
$N_2(X\,^1\Sigma_g^+)$	Ground state			
$(A\,^3\Sigma_u^+)$	1.3×10^6 (F_2); 2.7×10^6 (F_1, F_3)	$A \rightarrow X$ (Vegard–Kaplan)	6.2	5, 11, 12
$(B\,^3\Pi_g)$	8.0	$B \rightarrow A$; 10,510 (first positive)	1.2	13, 14
$(W\,^3\Delta_u)$	1×10^3 $(v = 2)$	$W \rightarrow X$	7.4	17, 19,
		$W \rightarrow B$	0.003	
$(B'\,^3\Sigma_u^-)$	10 est.	$B' \rightarrow B$(Y bands)	0.8	4
$(a'\,^1\Sigma_u^-)$	$\geq 4 \times 10^4$	$a' \rightarrow X$ (Wilkinson)	8.4	4
$(a\,^1\Pi_g)$	1.4×10^2	$a \rightarrow X$; 1450 (Lyman–Birge–Hopfield)[b]	8.6	16, 18, 20
$(w\,^1\Delta_u)$	10^2 est.	$w \rightarrow a$; 36,400	0.3	4
$(C^3\pi_u)$	4.0×10^{-2}	$C \rightarrow B$; 3371 (second positive)	3.7	22, 23, 248
$(E\,^3\Sigma_g^+)$	2.0×10^2	$E \rightarrow A,C$	5.7, 0.8	15, 16
$N_2^+(X\,^2\Sigma_g^+)$	Ground state			
$(A\,^2\Pi_u)$	17	$A \rightarrow X$; 11,036 (Meinel)	1.0	249–251
$(B\,^2\Sigma_u^+)$	5.9×10^{-2}	$B \rightarrow X$; 3914 (first negative)	3.2	21–23, 248
$(^4\Sigma_u^+)$	Mod. long	$^4\Sigma_u^+ \rightarrow X$	$\sim$6	4, 24
$NO(X\,^2\Pi)$	Ground state			
$(a\,^4\Pi)$	$\sim 1.6 \times 10^5$ $(\Omega = \frac{5}{2})$	$a \rightarrow X$	4.7	25, 26
$(A\,^2\Sigma^+)$	0.2	$A \rightarrow X$; 2265 (γ bands)	5.5	27, 28, 252
$(B\,^2\Pi)$	3.6	$B \rightarrow X$; (β bands)	5.6	49, 253
$NO^+(X\,^1\Sigma^+)$	Ground state			

Table 6.1 (Continued)

SPECIES AND STATE	MEAN RADIATIVE LIFETIME (μsec)	PRINCIPAL TRANSITION; λ (IN Å); NAME OF TRANSITION	APPROX. ΔE_{if} (eV)	REF.
Diatomic molecules and molecular ions[a] *(cont'd)*				
$(a\ ^3\Sigma^+)$	Long	$a \rightarrow X$	6.4	29, 30
$(b\ ^3\Pi)$	1.4×10^2	$b \rightarrow a$	0.9	31
$(w\ ^3\Delta)$	10^2 est.	$w \rightarrow b$	0.3	4
$NO_2(^2B_1)$	0.55 to 0.9	$A\ ^2B_1 \rightarrow X\ ^2A_1$		32, 33
$O_2(X\ ^3\Sigma_g^-)$	Ground state			
$(a\ ^1\Delta_g)$	3.9×10^9	$a \rightarrow X$; 12,680 (infrared atmospheric)	0.98	7, 8, 34
$(b\ ^1\Sigma_g^+)$	1.2×10^7	$b \rightarrow X$; 7619 (atmospheric)	1.63	7, 35, 36
$(c\ ^1\Sigma_u^-)$	Long	$c \rightarrow X$; 2856 (Herzberg II)	4.0	37
$(C\ ^3\Delta_u)$	Long	$C \rightarrow X$ (Herzberg III)	~4.2	4
		$C \rightarrow a$		
$(A\ ^3\Sigma_u^+)$	3×10^4	$A \rightarrow X$; 2856 (Herzberg I)	4.3	38
		$A \rightarrow b$; 4586 (Broida–Gaydon)		
$(B\ ^3\Sigma_u^-)$	4.2×10^{-2}	$B \rightarrow X$; 2030 (Schumann–Runge)	6.1	39
$O_2^+(X\ ^2\Pi_u)$	Ground			
$(a\ ^4\Pi_u)$	Long	$a \rightarrow X$; 6026	4.0	4
$(A\ ^2\Pi_u)$	0.7	$A \rightarrow X$ (second negative)	5.0	40, 41
$(b\ ^4\Sigma_g^-)$	1.1	$b \rightarrow a$; 6026 (first negative)	2.1	40, 41, 254
$CO(X\ ^1\Sigma^+)$	Ground state			
$(a\ ^3\Pi)$	$\sim 10^4$ (dep on J)	$a \rightarrow X$ (Cameron bands)	6.0	44, 255
$(a'\ ^3\Sigma^+)$	10 ($v = 4$)	$a' \rightarrow a$ (Asundi bands)	~0.9	42, 43
		$a' \rightarrow X$ (Birge–Hopfield)	~6.9	

$(d\,^3\Delta)$	6.0	$d \rightarrow a$ (triplet bands)	1.5	42
$CO^+(X\,^2\Sigma^+)$	Ground state			
$(A\,^2\Pi)$	3.8	$A \rightarrow X$ comet tail	2.6	41, 249
$CH(X\,^2\Pi)$	Ground state			
$(A\,^2\Delta)$	0.5	$A \rightarrow X$	2.9	46, 47
$(B\,^2\Sigma^-)$	0.4	$B \rightarrow X$	3.2	47
$CN(X\,\Sigma^+)$	Ground state			
$(A\,^2\Pi_i)$	7 ($v = 1$)	$A \rightarrow X$ (red bands)	1.1	48
$(B^2\Sigma^+)$	0.08	$B \rightarrow X$; 3883 (violet bands)	3.2	45, 256

[a] The quoted lifetime is for the $v = 0$ level, and the wavelength and energy are for the (0, 0) transition unless otherwise stated or unknown.
[b] It is possible that the quoted lifetime is actually that of the w state, which cascades to the a state, while the true lifetime of the a state is about 1×10^{-5} sec, as obtained by Jeunehomme (49).

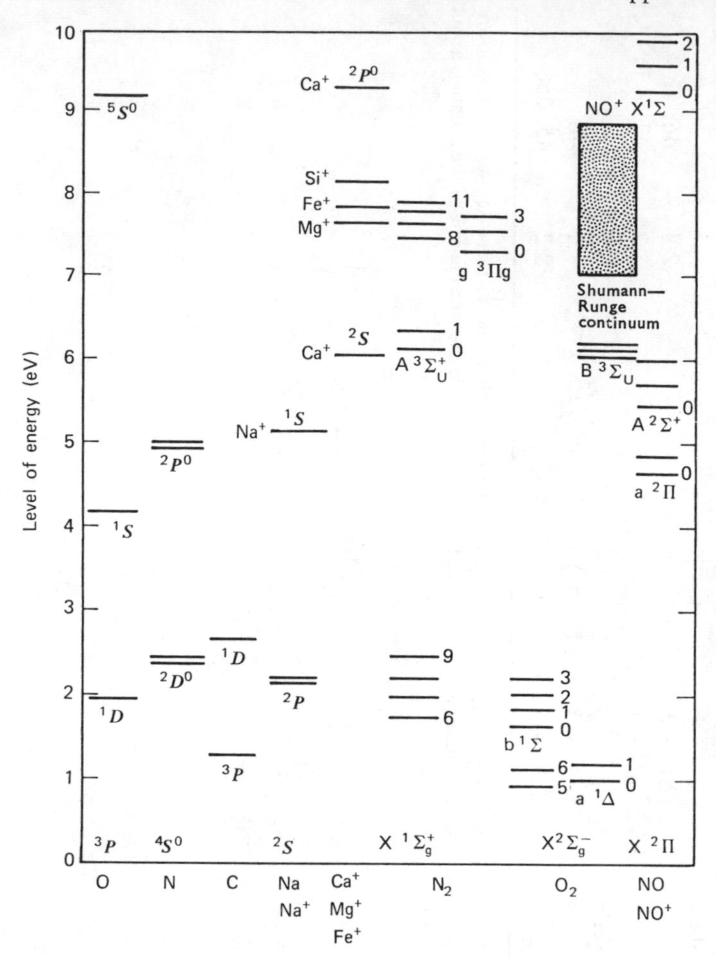

Figure 6.1 Energy levels of pertinent atoms, molecules, and metallic ions.

Figures 6.1 and 6.2 comprise energy-level diagrams for those levels of interest in upper-atmospheric chemistry. In energy-transfer reactions, energy resonance or near resonance may play some part, depending upon the details of the potential-energy surface associated with the interaction considered [50–54]. A recent compilation of lifetimes by Anderson [55] should be referred to.

III. THE EXCITATION AND DEEXCITATION OF SPECIFIC STATES

In this section we consider some of the literature that is relevant to an understanding of the role played by specific excited atoms, ions, and molecules

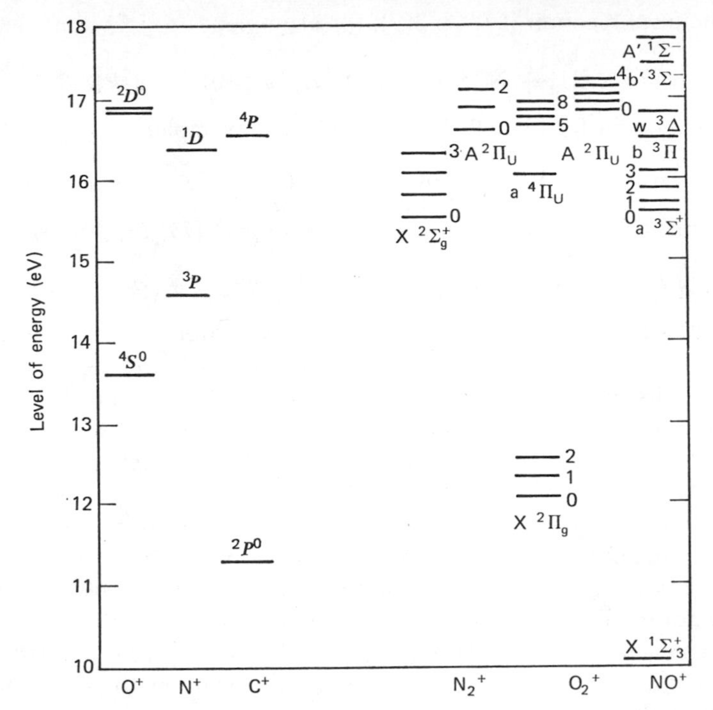

Figure 6.2 Energy levels of pertinent ions above those of corresponding ground-state neutral species.

in the quiescent and disturbed atmospheres and in the laboratory. This subject is also covered, at least in part, in papers by Hunten and McElroy [56], Bates [57], Muschlitz [58], Gilmore, Bauer, and McGowan [53], and Donovan and Hudson [59]; in review articles included in a special supplement on chemical lasers [60]; in the aforementioned review of active nitrogen [2]; and in books by Chamberlain [61] and others [5, 62–66].

A. Nitrogen

1. Vibrational Excitation and Deexcitation of the Ground Electronic State

An appreciable fraction of the energy deposited in the atmosphere finally comes to rest in vibrational excitation of nitrogen, either through radiative decay of higher-lying excited states [39], for example,

$$N_2(A\,^3\Sigma_u, v') \rightarrow N_2(X\,^1\Sigma_g^+, v'') + h\nu, \tag{2}$$

through chemical reaction [67, 68, 257], for example,

$$NO(X\,^2\Pi) + N(^4S^0) \rightarrow N_2(X\,^1\Sigma_g^+, v > 0) + O(^3P), \tag{3}$$

through energy transfer [56, 69–72, 258, 259], for example,

$$O(^1D) + N_2(X\,^1\Sigma_g^+, v = 0) \rightarrow O(^3P) + N_2(X\,^1\Sigma_g^+, v \leq 1), \tag{4}$$

or through resonance excitation under electron impact [73, 74], for example,

$$N_2(X\,^1\Sigma_g^+, v) + e \rightarrow N_2^-(^2\Pi, v') \rightarrow N_2(X\,^1\Sigma_g^+, v'') + e. \tag{5}$$

The latter process is very efficient, with a cross section approaching 6×10^{-16} cm^2 at its maximum [75]. These cross sections have been integrated over a Boltzmann distribution by Abraham and Fisher [76, 77], and the results are presented in Figure 6.3. [See also Ali [78] for similar results.]

A number of chemical reactions has been reported in which vibrationally excited products have been identified, but reaction (3) is the only one thus far studied in detail [67, 68, 277] which gives $N_2^\ddagger$ as a known product.

The quenching of $O(^1D)$ by N_2, reaction [4], can be a very efficient source of vibrationally excited nitrogen [260] even though the curve-crossing mechanism may not permit more than one or two quanta of N_2 vibrational energy per $O(^1D)$ quench [258, 259]. Hunten and McElroy [56] have pointed out that whereas $O(^1D)$ is readily quenched by energy transfer to N_2, $O(^1S)$ is not, even though it is nearly resonant with $v = 16$.

The vibrational excitation of nitrogen through inelastic resonance collisions of electrons has been examined experimentally by Schulz [73], by Golden [79], and by others. The cross section for this reaction has been measured and calculated with good agreement, but the reverse quenching (or superelastic) cross sections have been obtained only from the theoretical study by Chen [74]. Green et al. [80, 81] have used these and similar excitation data to calculate in detail the fractions of the energy from 30-keV incident electrons going into vibrational and into electronic excitation. (Note an error in the vibrational excitation in the latter work.) Quenching has also been observed experimentally [82].

The quenching of vibrationally excited molecules through energy transfer to the vibrational levels of other molecules (vibration–vibration, VV) or to kinetic energy (vibration–translation, VT) is thought to be well understood [83–86]. The $N_2(1 \rightarrow 0)$ transition at room temperature requires nearly 10^{10} collisions to transfer the one quantum of vibrational energy to kinetic energy [84]. For relaxation by atom exchange [87–90]

$$N\text{–}N^\ddagger + N \rightarrow N + N\text{–}N, \tag{6}$$

or by resonance energy transfer [84], the number of required collisions is

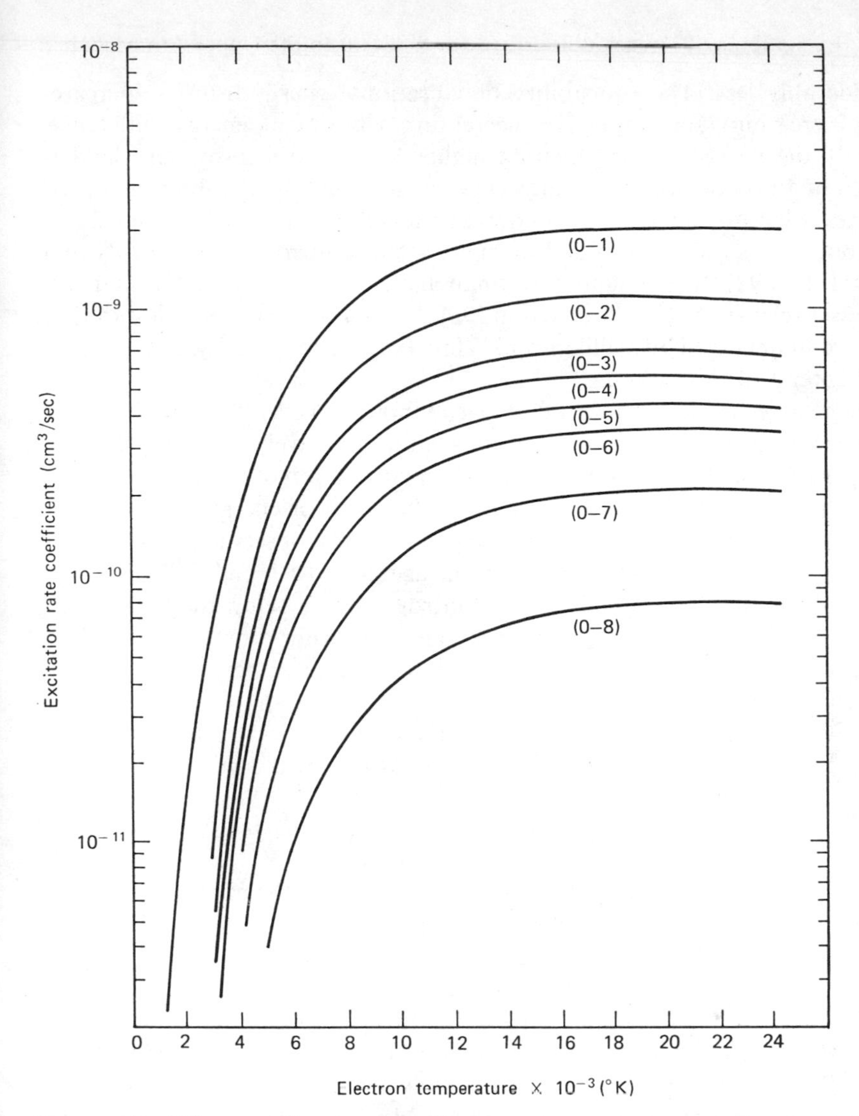

Figure 6.3 Nitrogen vibrational excitation rate constants as a function of the electron kinetic temperature for 300°K N_2 kinetic temperature [76, 77].

considerably less. The probability of vibrational energy transfer increases with increasing temperature or decreasing vibrational energy difference. Usually the process of deactivating higher levels is a step-by-step, ladder-descending process, so that a molecule put in a high-lying vibrational level may react before all its energy is equipartitioned. For most conditions $\Delta v = \pm 1$, but for very high levels and/or high kinetic temperatures it is likely that $\Delta v > 1$ [91, 92]. Taylor and Bitterman have reviewed those VT and VV processes relevant to CO_2 laser action [93]. For many more molecules, the VT data are summarized by Millikan and White [94] and are discussed thoroughly in Chapter 3 of this volume.

The near-resonance transfer of nitrogen vibrational energy to the electronic states of atoms and molecules depends quite sensitively on the potential-energy surfaces. Hunten [95] has argued that excitation of the sodium D line in low aurorae (<100 km) is likely to be the result of energy transfer from vibrationally excited nitrogen. This reaction has been observed in the laboratory [96], and the cross section has been measured as 10^{-15} cm^2 [97], although it is doubtful that vibrationally excited nitrogen exists below the turbopause [70, 260–262]. The cross sections for energy transfer from excited alkali metal atoms to the vibrational modes of nitrogen have been calculated by Fisher and Smith [98].

The degree of vibrational excitation significantly affects certain reaction rates, the most important of which probably is the exothermic reaction:

$$N_2^{\ddagger}(X\,^1\Sigma_g^+, v) + O^+(^4S) \rightarrow NO^+(X) + N(^4S), \tag{7}$$

which depends sensitively upon the vibrational level of the nitrogen molecule [99, 100]. From the work of Schmeltekopf et al. [100], Thomas and Norton [101], O'Malley [102], Walker et al. [71, 72], Jamshidi et al. [260], Varnum [264] and Brieg et al. [266], one can infer the following: (a) The electron density in the F region is affected by the nitrogen vibrational temperature through the conversion of atomic ions with low electron–ion recombination coefficients, mainly by reaction (7); (b) A decrease in electron density for a highly disturbed atmosphere is to be expected when the effective vibrational temperature is high. This probably occurs, for example, in the red arc associated with the sunspot maximum, in some aurorae, and in the atmosphere disturbed by a nuclear burst [103, 104].

In Table 6.2 some of the energy-transfer rate constants of interest to this chapter are listed. The values quoted are usually for bulk reaction coefficients and are normally related to the specific value of $v = 1$, except for the reaction $N_2^{\ddagger} + O^+$, treated separately in Table 6.3.

Data for vibrational–translational energy transfer are usually presented as a relaxation-time–pressure product $p\tau$, where τ refers to the e-folding time

Table 6.2
Energy Transfer from $N_2^{\ddagger}$

LEVEL, v	COLLISION PARTNER	PROBABLE PRODUCT	RATE CONSTANT (cm^3/sec^{-1})	TEMP. RANGE (°K)	REF.
1	N_2	Kinetic energy	$1{\cdot}3 \times 10^{-11}\, Te^{-220/T^{1/3}}$	300–5000	105
		$N_2^{\ddagger}$ (resonant VV)	3×10^{-13}	300	
1	O_2	Kinetic energy $O_2^{\ddagger}$ (nonresonant VV)	$\dfrac{2{\cdot}5 \times 10^{-7}\, e^{-263T^{-1/3}}}{(1 - e^{-3390/T})}$	1000–10,000	93[a]
1	O	Kinetic energy	$\dfrac{6{\cdot}21 \times 10^{-14}\, Te^{-51{\cdot}5/T^{1/3}}}{(1 - e^{-3990/T})}$ or $\dfrac{3{\cdot}43 \times 10^{-12}\, Te^{-152/T^{1/3}}}{(1 - e^{-3990/T})}$	3000–4500	106, 107
			$1.2 \times 10^{-13} e^{-23/T^{1/3}}$	300–3000	260, 265, 266
1	CO_2	$CO_2^{\ddagger}$ (near resonant VV)	$6 \times 10^{-13}\left(\dfrac{T}{300}\right)^{-1/2}$	300–1200	93
≥ 4	N_2O	$N_2O^{\ddagger}$	$2{\cdot}5 \times 10^{-14}$	300	54, 68
≥ 1	Ar	Kinetic energy	$2{\cdot}5 \times 10^{-16}$	300	54, 68
1	NO	$NO^{\ddagger}$ (nonresonant VV)	$1{\cdot}5 \times 10^{-16}$	300	94, 105

[a] Note that the rate constants for VT from Ref. 93 must be divided by $(1 - e^{-\theta_v T})$ to obtain rate equations of the conventional form.

Table 6.3
Rate Constants (Units of 10^{-12} cm³/sec) for the Reaction $N_2^{\ddagger} + O^+ \rightarrow NO^+ + N$ for a Range of Vibrational and Translation Temperatures [102]

TRANSLATION TEMPERATURE, T_{tr} (°K)	VIBRATIONAL TEMPERATURE, T_v (°K)										
	30*	1000	1500	2000	2500	3000	3500	4000	5000	6000	7000
300	1.3[a]	1.4[a]	2.1[a]	4.2	7.8	12.6	18.2	24.0	35.7	46.2	55.1
1000	0.7[a]	0.9	2.0	4.9	9.5	15.4	22.1	29.0	42.5	54.3	64.2
1500	0.6	0.9	2.3	5.5	10.6	16.9	24.0	31.2	45.0	57.0	67.0
2000	0.6	1.0	2.7	6.4	11.9	18.6	26.0	33.4	47.4	59.5	69.4
2500	0.8	1.3	3.4	7.6	13.6	20.7	28.3	36.0	50.2	62.3	72.1
3000	1.1	1.8	4.3	9.0	15.5	23.1	31.0	38.9	53.4	65.5	75.2
3500	1.6	2.5	5.4	10.7	17.7	25.7	34.0	42.1	56.7	68.9	78.5
4000	2.3	3.4	6.8	12.6	20.2	28.6	37.2	45.4	60.3	72.4	82.0
5000	4.4	6.0	10.4	17.3	25.8	35.0	44.0	52.6	67.7	79.7	89.1
6000	7.5	9.6	14.9	22.9	32.3	42.1	51.5	60.3	75.4	87.3	96.3
7000	11.7	14.2	20.5	29.5	39.6	49.8	59.5	68.3	83.3	94.8	103.5

[a] For these values, the dominant contribution comes from a low-energy mechanism.

of the vibrational energy, ε, according to the classical equation:

$$\frac{d\varepsilon}{dt} = \frac{1}{\tau}(\varepsilon_{\text{equil}} - \varepsilon), \tag{8}$$

at constant translational temperature in the absence of sources. The rate constant for deactivation of the first vibrational level k_{10} is related to τ by (see ref. 108, for example):

$$k_{10}M = [\tau(1 - e^{-h\nu/kT})]^{-1}, \tag{9}$$

where M is the number density of collision partner (at pressure p), and $h\nu$ is the vibrational-energy-level spacing. The data of Table 6.2 have been used in conjunction with the CIRA 1965 Atmosphere [109] to calculate the loss rates for $N_2^{\ddagger}$ pertinent to the atmosphere, as presented in Figure 6.4.

2. Excitation and Deexcitation of the $(A\ ^3\Sigma_u^+)$ State

The $(A\ ^3\Sigma_u^+)$ state at 6.2 eV is the lowest electronic metastable state of the nitrogen molecule, and as yet no significant reactions in the upper atmosphere have been ascribed to it, even though it is found to have a lifetime near 2 sec [11]. It can be populated through electron impact [20, 110–113] and by cascade from the higher-lying triplet states such as the $(B\ ^3\Pi_g)$ at 7.4 eV and the $(C\ ^3\Pi_u)$ at 11.0 eV.

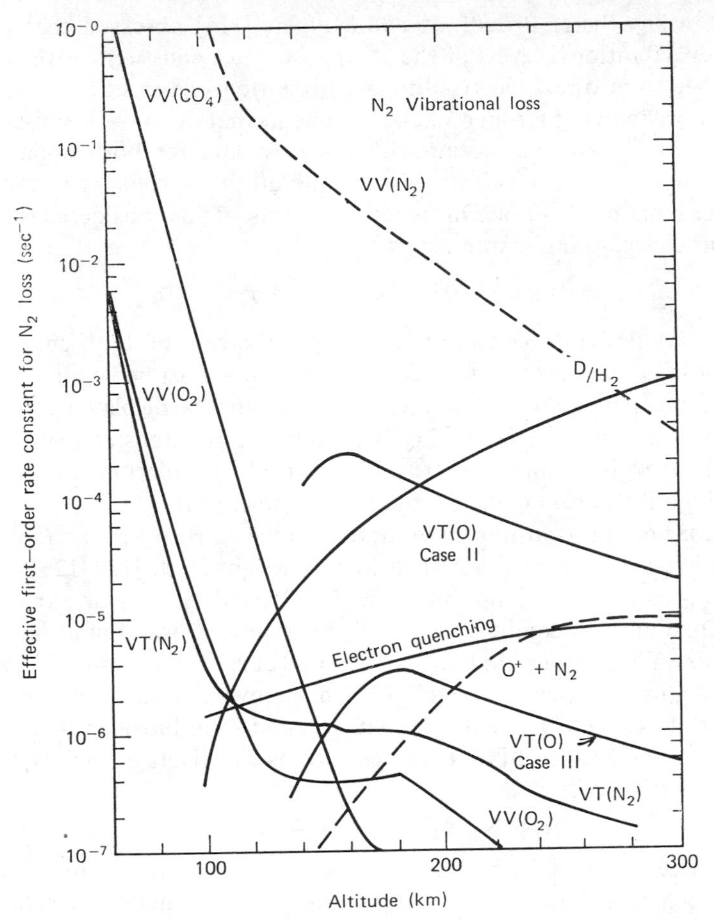

Figure 6.4 Loss frequencies (sec^{-1}) for N_2 vibration as a function of altitude.

Repeated measurements of the rotational temperature of the Vegard–Kaplan bands (A–X) of nitrogen in aurorae have indicated a rotational temperature of this diffuse radiation to be 800°K [114] or higher [115, 116]. To be consistent with these measurements, it was argued on the basis of the older 13-sec lifetime [117] that the radiation had to be emitted from an altitude in excess of 220 km, that is, where the kinetic temperature is 800°K. It was also postulated that there was no radiation from the *A* state at lower altitudes because the *A* state was effectively quenched by either atomic or molecular oxygen, which would consequently have to have large quenching rates. However, for the same auroral displays, measurements of the rotational temperatures of the N_2^+ first negative system (B–A) invariably gave temperatures between 200 and 600°K, implying that the aurora was considerably lower in the atmosphere. Broadfoot and Hunten [114] observed that the distribution of vibrational levels of the *A* state was not consistent with that obtained either from direct electron-impact excitation of the state or from cascade. It is now known [5] from rocket experiments that the A → X emission comes from 120–170 km, consistent with the new data for both radiative lifetimes and quenching coefficients. Hence, the altitude profile is resolved but the source remains in doubt. One reaction that should be considered is the near-resonant charge-transfer interaction:

$$N_2^+(X) + NO \rightarrow N_2(A) + NO^+, \tag{10}$$

which has a reaction rate constant [118, 119] in excess of 10^{-10} cm^3/sec. Although the concentrations of the reactants are known to be small in the undisturbed atmosphere, they will be increased in an auroral display. Furthermore, it has been observed [120, 121] that some charge-transfer reactions that proceed through an intermediate complex lead to molecular products that have a high (or poorly defined) rotational temperature.

Cross sections for electron-impact excitation to the *A*, *B*, *W*, *B'*, *C*, *E*, and *D* states of nitrogen have been obtained by Cartwright et al. [122–125] and were used to predict absolute photon fluxes in 11 band systems of nitrogen under nighttime auroral conditions. Significant radiation was found in the 1–10-μm region. The strongest radiation was found at 2.75, 3.33, and 4.23 μm.

Noxon [126], in agreement with Zipf [127], has shown that the deexcitation rate constant of $N_2(A\,^3\Sigma_u^+)$ in collision with ground-state nitrogen is small. It has also been observed [128] that nitrogen atoms effectively quench N_2(A) through atom exchange:

$$N + N_2(A) \rightarrow N_2(X) + N. \tag{11}$$

Hunten and McElroy [56] found that only oxygen atoms could provide the necessary quenching in the atmosphere. A summary of the available quenching data is presented in Table 6.4. Metallic species like Na, Fe, Hg, Ba, and so on are also effective quenchers through Penning ionization [58, 134, 135].

$$N_2(A) + M \rightarrow N_2(X) + M^+ + e. \tag{12}$$

3. High-Lying States of Nitrogen

No attempt is made here to review in detail information on metastable states of nitrogen above the $(A\,^3\Sigma_u^+)$ state, except to note that a metastable state at approximately 8.5 eV which was observed by Cermak [136] may lead to associative ionization:

$$N_2^* + NO \rightarrow N_2NO^+ + e. \tag{13}$$

The product N_2NO^+ has not been identified specifically in recent *D*-region mass-spectral studies, but may be important as an intermediate.

Table 6.4
Quenching Data for $N_2(A\,^3\Sigma_u^+)$

QUENCHANT	RATE CONSTANT (cm^3/sec)	REF.
N_2	$<3 \times 10^{-19}$	126
	$<1.2 \times 10^{-18}$	127
O_2	$<4 \times 10^{-10}$	56
	3.8×10^{-12}	129
O	$\leq 3 \times 10^{-11}$	56
N	5×10^{-11}	130
	5×10^{-12}	131
	5×10^{-11}	132
NO	7×10^{-11}	133

4. Excitation and Deexcitation of N_2^+ States, Particularly $(A\,^2\Pi_u)$, $(B\,^2\Sigma_u^+)$, $(^4\Sigma_u^+)$, and $(^4\Delta_u)$

The 3914-Å band associated with the (0, 0)(B → X) transition in N_2^+ is one of the strongest in the aurora, twilight glow, and dayglow. The primary sources of the N_2^+(B) state are photoionization by solar radiation,

$$N_2 + h\nu(>18.7 \text{ eV}) \rightarrow N_2^+(B\,^2\Sigma_u^+) + e; \tag{14}$$

ionization by energetic electrons,

$$N_2 + e \rightarrow N_2^+(B\,^2\Sigma_u^+) + 2e; \tag{15}$$

excitation by slow electrons [137],

$$N_2^+ + e \rightarrow N_2^+(B\,^2\Sigma_u^+) + e; \tag{16}$$

and resonance scattering of sunlight by ground-state ions,

$$N_2^+(X) + h\nu(>3.2\ eV) \rightleftarrows N_2^+(B\,^2\Sigma_u^+). \tag{17}$$

Only the second of these reactions [reaction (15)] is likely to be active in aurorae, where the secondary electrons are produced by heavy-particle bombardment of the atmosphere. Similar mechanisms can produce the $(A\,^2\Pi_u)$ state of the ion, although (as is shown later) charge transfer from $O^+(^2D)$ with N_2 also leads to A-state ions and possibly to an asymmetry in the subsequent Meinel radiation.

Wallace and McElroy [138] have discussed the relative importance of the above reactions in producing excited N_2^+. They show that above 100–150 km, resonance scattering is the major source of 3914-Å radiation. At low densities, reactive collisions involving either the A or B states of N_2^+ are slower than the excitation by radiation. However, at higher densities reactions such as

$$N_2^+(B\,^2\Sigma_u^+) + O_2(X\,^3\Sigma_g^-) = N_2(X\,^1\Sigma_g^+) + O_2^+(b\,^4\Sigma_g^-) \tag{18}$$

and

$$N_2^+(A\,^2\Pi_u) + O_2(X\,^3\Sigma_g^-) = N_2(X\,^1\Sigma_g^+) + O_2^+(a\,^4\Pi_u) \tag{19}$$

may be significant. Both reactions are nearly energy resonant and conserve spin. However, even these reactions cannot greatly affect electron density in laboratory afterglows or the upper atmosphere, since both parent and productions are diatomic with similar recombination rates. The major results of the above charge-transfer reactions would be to produce additional $O_2^+(b \rightarrow a)$ first-negative radiation and metastable $O_2^+(a\,^4\Pi_u)$ ions. Laboratory experiments [139, 140] have demonstrated that reaction (19) and perhaps reaction (18) have cross sections well in excess of the corresponding process for ground-state ions. However, the charge transfer of excited N_2^+ ions with N_2 has a smaller cross section than do ground-state ions [24, 140]. This reflects the fact that some $(N_2^+)^*$ ions formed under electron impact around the equilibrium internuclear distance of the neutral molecule relax to larger internuclear distances, so that charge transfer is no longer energy resonant.

The study of collision-induced dissociation of $(N_2^+)^*$ has proved to be an effective means of studying highly excited molecular ions. In particular, McGowan and Kerwin [24] have identified the metastable $(^4\Sigma_u^+)$ and $(^4\Delta)$ states of N_2^+ which form N_3^+ by

$$N_2^+(^4\Sigma_u \text{ or } ^4\Delta) + N_2(X\,^1\Sigma_g^+) = N_3^+ + N. \tag{20}$$

Higher excited states of N_2^+ (and O_2^+) have been identified in some laboratory experiments [24] where the reaction time was near 3 μsec. In other experiments [139] where the time between formation and collision was nearer 10 μsec, these states were not observed. Therefore, the higher states may decay with a lifetime of less than 10^{-5} sec and be of less importance in the atmosphere.

Studies of electron-impact ionization of molecular nitrogen and oxygen near threshold [141, 142] have demonstrated that a Franck–Condon distribution of vibrational levels is not obtained because many ions are formed indirectly, via autoionizing states. The importance of autoionization can be seen in the case of $N_2^+(X)$, where the Franck–Condon factors for transitions from $N_2(X, v'' = 0)$ to $N_2^+(X, v')$ decrease by nearly an order of magnitude for each successive vibrational level of the ion, while for electrons with energy not far above the ionization potential, Fineman et al. [143] have demonstrated that the populations of the $v' = 0$ and $v' = 1$ levels are nearly equal.

5. Excitation and Deexcitation of $N(^2D)$

The (^2D) states of atomic nitrogen, which are the upper states of the 5200-Å dayglow, are probably excited [138] by dissociative recombination,

$$N_2^+(X\,^2\Sigma_g^+) + e \rightarrow N(^2D) + N, \tag{21}$$

$$NO^+(X\,^1\Sigma_g^+) + e \rightarrow N(^2D) + O, \tag{22}$$

and by ion–atom interchange,

$$N_2^+(B\,^2\Sigma^+) + O(^3P) \rightarrow NO^+(X\,^1\Sigma^+) + N(^2D). \tag{23}$$

$N(^2D)$ may play a vital role in the NO balance of the D region via its major loss reaction [144, 228]

$$N(^2D) + O_2 \rightarrow NO + O, \tag{24}$$

which has a rate constant $k = 6 \times 10^{-12}$ cm^3/sec. By comparison, the quenching of $N(^2D)$ by molecular nitrogen is slow: $k = (3 \pm 3) \times 10^{-15}$ cm^3/sec [145].

Henry et al. (10) have calculated electron-impact excitation cross sections for $N(^2D)$ and $N(^2P)$. Ali [78] has integrated these cross sections over a Boltzmann distribution to obtain the rate constants illustrated in Figure 6.5.

B. Oxygen

1. The Vibrationally Excited Ground State, $O_2(X\,^3\Sigma_g^-)$

The cross section for dissociative attachment of electrons to oxygen,

$$O_2 + e \rightarrow O^- + O, \tag{25}$$

is strongly dependent upon the molecular oxygen temperature [142, 146, 147]. As the equilibrium oxygen temperature is increased from 300 to 2100°K, both the energy at maximum cross section and the threshold are shifted downward. O'Malley [148], through a multiparameter data fit, interpreted the results as being due to vibrational excitation of the molecule. Chen, however, demonstrated theoretically that rotational excitation must also be important [149].

Other methods for the production and destruction of the higher vibrational levels of $O_2(X\,^3\Sigma_g^-)$ have been reviewed by Dalgarno [88], by Hunten and McElroy [56], and by Dalgarno and McElroy [121].

Rate constants for molecular oxygen deactivation are given in Table 6.5; these data are then employed, in conjunction with the CIRA 1965 Atmosphere [109], to calculate the effective first-order rate constants for vibrationally excited $O_2(v = 1)$ deactivation, illustrated in Figure 6.6. At high energies

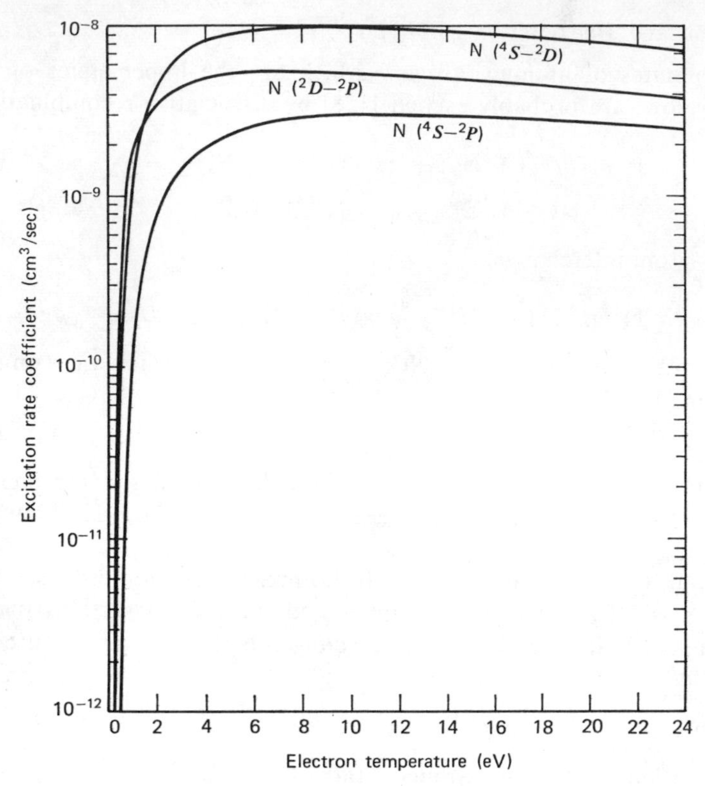

Figure 6.5 Excitation of atomic nitrogen by electron impact.

Table 6.5
Deactivation of $O_2^{\ddagger}(v = 1)$

REACTION	RATE CONSTANT, $k_{(1,0)}$ (cm^3/sec)	T RANGE (°K)	REF.
$O_2^{\ddagger} + M \rightarrow O_2 + M$	$\dfrac{2.5 \times 10^{-12} T \exp[-(2.95 \times 10^6/T)]^{1/3}}{(1 - e^{-2270/T})}$	800–3200	93
$M = N_2$ or O_2			
$O_2^{\ddagger} + O \rightarrow O_2 + O$	$3.3 \times 10^{-13} T^{1/2} \exp\left(\dfrac{-483}{T}\right)$	300–1700	90[a]
	$1.7 \times 10^{-10} \exp\left(\dfrac{-4000}{T}\right)$	2000–4000	151
$O_2^{\ddagger} + H_2O \rightarrow O_2 + H_2O^{\ddagger}$	$10^{-(12 \pm 1)} \left(\dfrac{T}{300}\right)^{-1/2}$		93

[a] Assuming that reaction occurs as fast as isotopic exchange.

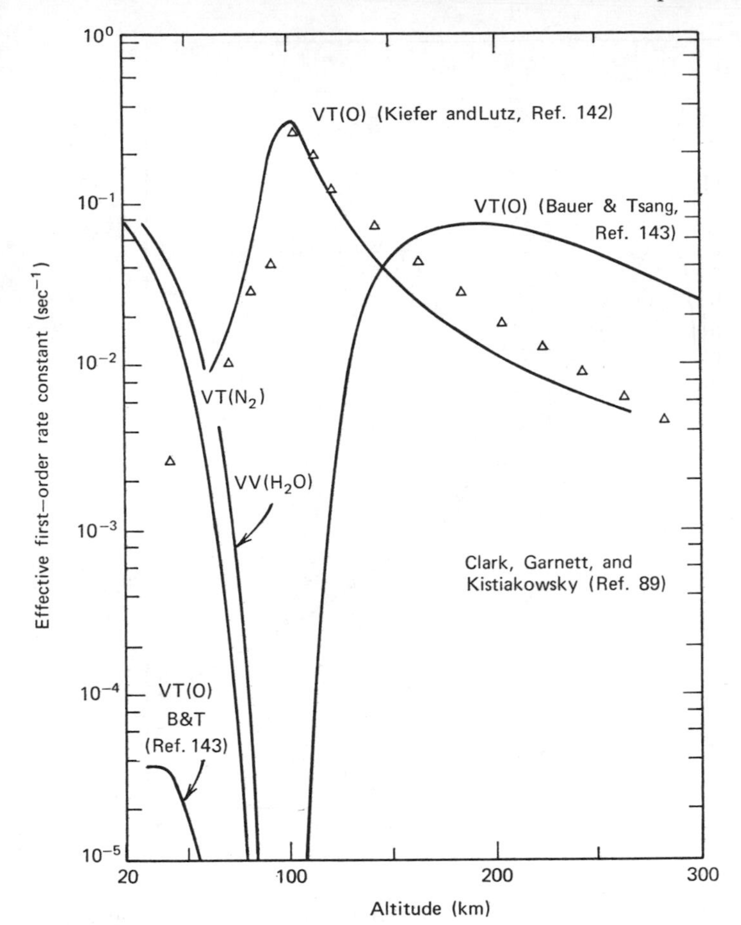

Figure 6.6 Deactivation of $O_2^{\ddagger}$ ($v = 1$).

(10–20 eV), vibrational excitation by ion–molecule collision, for example,

$$O^+ + O_2 \rightarrow O^+ + O_2^{\ddagger}(v \geq 1),$$

has been observed.

As shown by Schulz [152, 153], $O_2^{\ddagger}$ may also be excited by electron impact. The cross sections of Schulz have been integrated over a Boltzmann distribution by Ali [78], whose results are displayed in Figure 6.7.

2. $O_2(a\,^1\Delta_g)$ and $(b\,^1\Sigma_g^+)$

Perhaps the most abundant metastable state in the upper atmosphere is the $(a\,^1\Delta_g)$ state of oxygen, which is produced in the D region by the

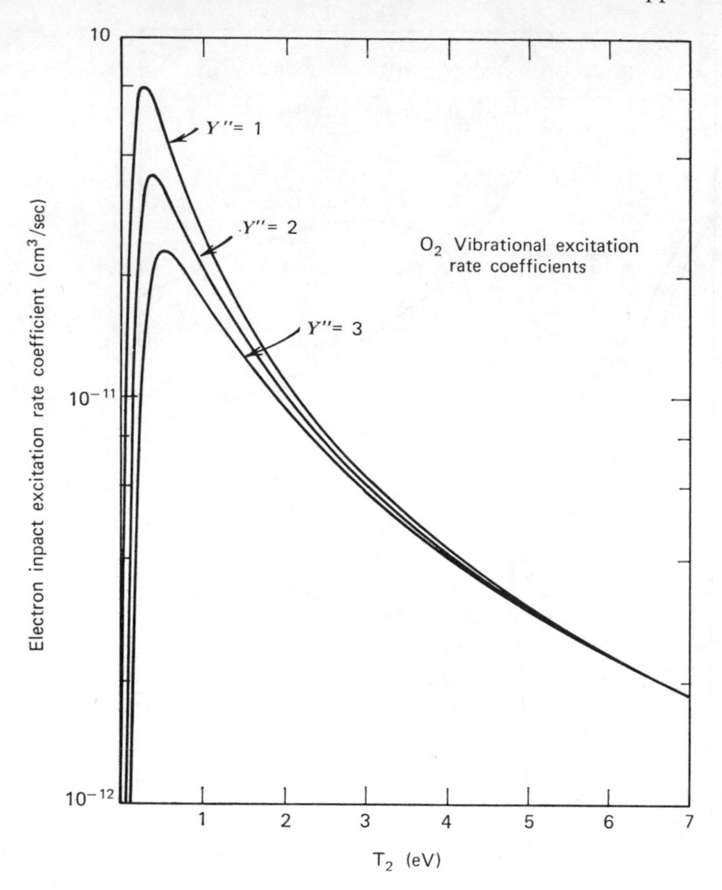

Figure 6.7 The rate coefficient of O_2 vibrational excitation by electron impact as reported by A. W. Ali [78].

photodissociation of ozone [154]:

$$O_3 + h\nu \text{ (Hartley continuum)} \rightarrow O_2(a\,{}^1\Delta_g) + O({}^1D). \tag{27}$$

(Table 6.6 gives the quantum yields for the ozone photolysis as a function of wavelength [270].) This produces a density of approximately 10^{10} $(a\,{}^1\Delta_g)$ molecules cm^{-3} in the normal atmosphere at altitudes in the vicinity of 50 km [155, 156]. Another mechanism that may contribute, particularly in the disturbed atmosphere [157], is

$$O({}^3P) + O_3 \rightarrow O_2 + O_2(a\,{}^1\Delta_g), \tag{28}$$

Table 6.6
Summary of Evaluated Photochemical Data

REACTION	QUANTUM YIELD, $\phi(\lambda)$	WAVE-LENGTH λ, nm	WAVELENGTH RANGE nm, FOR ABSORPTION COEFFICIENTS
$O_3 + h\nu$ (vis) $\rightarrow O + O_2$	1	450–750	440–850
$O_3 + h\nu$ (uv) $\rightarrow O(^1D) + O_2(^1\Delta)$	1	250–310	200–360
	0	>310	
$\rightarrow O(^1D) + O_2(^3\Sigma_g^-)$	0	<350	
$\rightarrow O(^3P) + O_2$ (Singlet)	0	<310	
	~1	310–350	
$\rightarrow$ O(total) $+ O_2$	1	250–350	
$\rightarrow O(^1D) + O_2(^1\Sigma_g^+)$	0	250–350	
$\rightarrow O(^3P) + O_2(^3\Sigma_g^-)$	0	250–350	

with a reported rate constant $k \approx 4.5 \times 10^{-15}$ cm^3/sec. Collisional energy transfer from $O(^1D)$ may produce the $(b\,^1\Sigma_g^+)$ state of oxygen [158, 159]:

$$O(^1D) + O_2(X\,^3\Sigma_g^-) \rightarrow O(^3P) + O_2(b\,^1\Sigma_g^+, v \leq 2). \qquad (29)$$

Noxon [159] observed the decay of 6300-Å radiation and the rise of 7618-Å radiation, thereby measuring the rate constant $k = 9 \times 10^{-11}$ cm^3/sec. This reaction is probably important in aurorae [160]. Electronically excited oxygen may also be produced efficiently by the impact of relatively low-energy electrons. Cartwright et al. [123] have measured $(a\,^1\Delta_g)$, $(b\,^1\Sigma_g^+)$, $(c\,^1\Sigma_u^-)$, $(A\,^3\Sigma_u^+)$, and $(B\,^3\Sigma_u^-)$ electron excitation cross sections. They find that excitation of the $(c\,^1\Sigma_u^-)$ and of the $(B\,^3\Sigma_u^-)$ states yields dissociation into two $O(^3P)$ atoms and $O(^3P) + O(^1D)$, respectively, but produces little direct radiation. The cross sections for excitation of $(a\,^1\Delta_g)$ and $(b\,^1\Sigma_g^+)$ are shown in Figure 6.8.

The $O_2(a\,^1\Delta_g)$ state is quite stable. Its behavior is documented in a monograph [161]. Data on $O_2(^1\Delta_g)$ quenching have shown excellent agreement, according to the review of Clark and Wayne [162]. The most reliable values (excluding the data of ref. 163) are given in Table 6.7. It is obvious from these data that molecular oxygen is the dominant quenching partner in the atmosphere, a fact confirmed by Evans' interpretation of $O_2(^1\Delta_g)$ emission at 1.27 μm in the atmosphere [156].

It was suggested by Megill and Hasted [169] in relation to polar-cap absorption events, and by Kummler and Bortner [170] with respect to

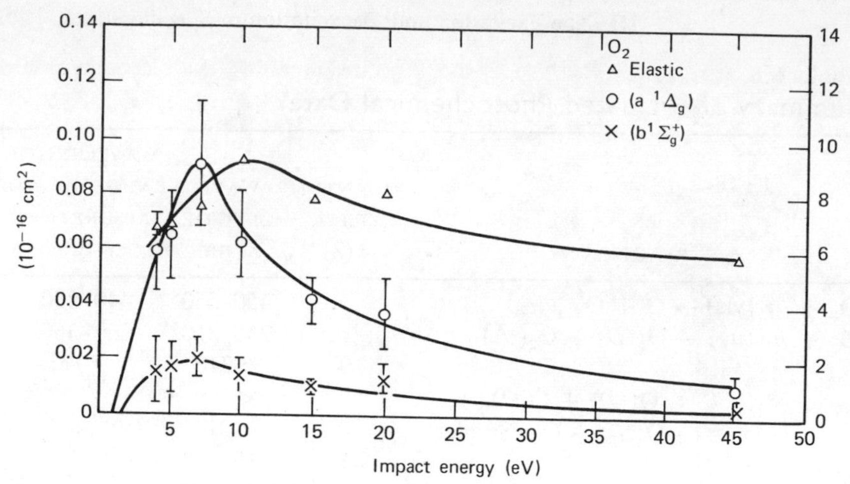

Figure 6.8 Cross section for electron-impact excitation of O_2 (from Cartwright, ref. 123).

Table 6.7
Quenching Data for $O_2(^1\Delta_g)$.

QUENCHING SPECIES	RATE CONSTANT (cm³/sec)	REF.
O_2	2.4×10^{-18}	162
	2.2×10^{-18}	164
	2.2×10^{-18}	165
	2.0×10^{-18}	166
	$2.2 \left(\frac{T}{300}\right)^{0.8} \times 10^{-18}$	270
N_2	$<1.1 \times 10^{-19}$	156
CO_2	3.9×10^{-18}	162
H_2O	1.5×10^{-17}	162
Ar	$\leq 2.1 \times 10^{-19}$	162
O	$\leq 1.3 \times 10^{-16}$	167
N	$(2.8 \pm 2) \times 10^{-15}$	167
O_3	3×10^{-15}	168

disturbed atmospheres, that $O_2(a\,^1\Delta_g)$ might be effective in detaching electrons from O_2^- by

$$O_2^- + O_2(a\,^1\Delta_g) \rightarrow 2O_2 + e, \tag{30}$$

in a deexcitation–detachment reaction analogous to a Penning ionization reaction. Subsequent investigations have verified the rapidity of the reaction [171, 172], which has a rate constant $k \approx 2 \times 10^{-10}$ cm³/sec, and have indicated that

$$O^- + O_2(^1\Delta_g) \rightarrow O_3 + e \tag{31}$$

is rapid as well: $k \approx 3 \times 10^{-10}$ cm³/sec.

Deexcitation–detachment reactions are particularly significant for the disturbed atmosphere, in which anomalously high 1.27 mμ intensities are found, presumably due to $O_2(^1\Delta_g)$ in some aurorae [173–175].

Another channel through which high concentrations of $O_2(a\,^1\Delta_g)$ may be quenched is radiative O_2 dimer formation [161, 176],

$$2\ O_2(a\,^1\Delta_g) \rightleftarrows O_4^* \rightarrow 2\ O_2(X\,^3\Sigma_g^-) + h\nu\ (6340,\ 7030\ \text{A}). \tag{32}$$

The same collision leads also to the excitation of the $(b\,^1\Sigma_g^+)$ state [176, 177],

$$2\ O_2(a\,^1\Delta_g) \rightarrow O_2(b\,^1\Sigma_g^+) + O_2(X\,^3\Sigma_g^-). \tag{33}$$

Reaction (29) is the dominant source of $O_2(b\,^1\Sigma_g^+)$ in the atmosphere, and molecular nitrogen is the major quenching agent. There is good agreement between laboratory results ($k_{N_2} = 1.5\text{–}2.5 \times 10^{-15}$ cm³/sec) and the study of Wallace and Hunten [5, 178] in the upper atmosphere ($k_{N_2} = 1.5 \times 10^{-15}$ cm³/sec). There is still considerable disagreement over the relative effectiveness of other quenchants, but molecular oxygen appears to be an order of magnitude less effective than nitrogen (5) ($k_{O_2} < 10^{-16}$ cm³/sec), although the results of Welge [179] ($k_{O_2} \approx 4.5 \times 10^{-16}$ cm³/sec) indicate that data taken at high pressure may be suspect. The recent review of Hamson et al. [270] supports these values.

3. Production and Destruction of the (^1S) and (^1D) States of Atomic Oxygen

The (^1D) state of atomic oxygen yields the forbidden red-line radiations (6300 and 6364 Å) that are prominent in the aurora and dayglow, twilight, and nightglow. In the dayglow, the major source of $O(^1D)$ is photodissociation in the Schumann–Runge continuum [180]:

$$O_2 + h\nu(>7.1\ \text{eV}) \rightarrow O(^3P) + O(^1D). \tag{34}$$

Evidence suggests that dissociative recombination may be the dominant source of $O(^1D)$ at night; a recombination rate constant $k = 2.2 \times 10^{-7}$ cm³/sec at 300°K is consistent with other available data [138]. Zipf [181, 182] studied excited-state formation from the dissociative recombination of O_2^+ in afterglows. His results are given in Table 6.8. Note that the total of α is equal to twice the rate constant since each recombination electron leads to two product atoms.

Table 6.8
Dissociative Recombination of O_2^+ with Electrons [182]

$O_2^+ + e \rightarrow O^* + O^{**}$	PRODUCT RATIO	RATE CONSTANT[a] (cm^3/sec) FOR PRODUCTION AT 300°K
Total $O^*(^1S)$	0.1	2.1×10^{-8}
Total $O^*(^1D)$	0.9	1.9×10^{-7}
Total $O(^3P)$	1	2.1×10^{-7}

[a] The rate constant is defined here in terms of the individual species production and not in terms of O_2^+ disappearance.

In 1931 Chapman [183] suggested that $O(^1S)$ may be formed in the nightglow by the reaction

$$3\,O(^3P) \rightarrow O_2 + O(^1S). \tag{35}$$

Young and Black [184] found the three-body rate constant $k = 1.5 \times 10^{-34}$ cm^6/sec, which is consistent with the value required by Chapman's theory. They also found the rate constant $k = 3 \times 10^{-33}$ cm^6/sec for the reaction

$$N(^4S) + 2\,O(^3P) \rightarrow NO + O(^1S). \tag{36}$$

The production and destruction of both $O(^1D)$ and $O(^1S)$ through electron collisions,

$$O(^3P) + e \rightleftarrows O(^1D \text{ or } ^1S) + e, \tag{37}$$

have been studied extensively by Seaton [185] and applied to atmospheric problems by Hunten and McElroy [56]. The latter authors conclude that below 100 km neither excitation nor quenching through electron impact is important in comparison with other excitation and quenching mechanisms. This is not true, however, for the lowest excited states of N, N^+, and O^+: For these cases superelastic quenching is likely to remain important even below 100 km. In studies of higher-density plasma afterglows in helium, Ingraham and Brown [186] showed that superelastic collisions play an important role in heating the electrons in the afterglow. Metastable–metastable collisions leading to Penning ionization are also active. Additional

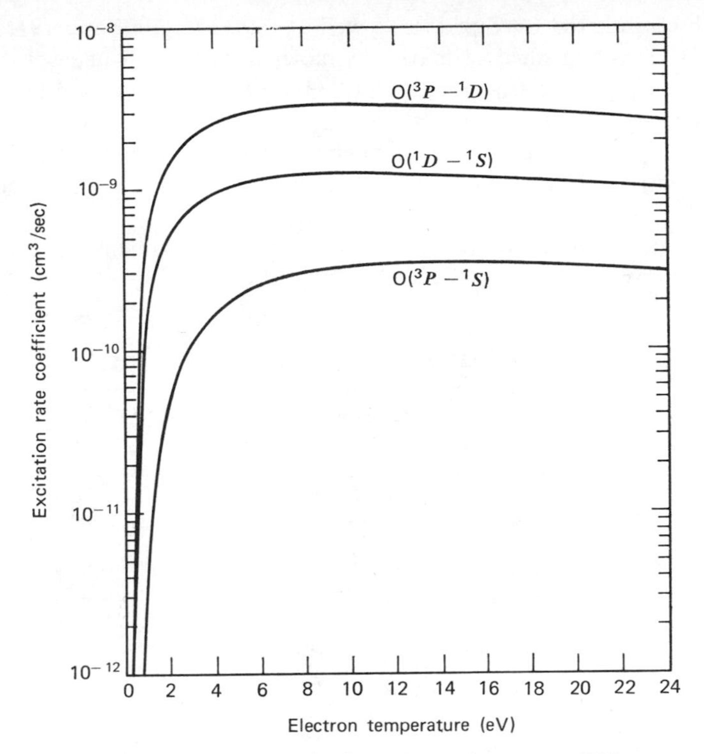

Figure 6.9 Electron-impact excitation of atomic oxygen [78].

atomic excitation cross sections have been calculated by Henry et al. [187]. The latter values have been integrated over a Boltzmann distribution of electron energies by Ali [78], and these results are illustrated in Figure 6.9.

Data from the flowing-afterglow experiments of Young and Black [184] suggest that the deactivation process

$$O(^1S) + O(^3P) \rightarrow 2\,O(^1D) \text{ or } O(^1D) + O(^3P) \tag{38}$$

has a rate coefficient $k = 1.8 \times 10^{-13}$ cm³/sec, which would make it important in the nightglow [56]. Evans and Vallance-Jones [188] found in auroral measurements that the effective lifetime for the green-line ($^1S \rightarrow {}^1D$) transition was shorter than the predicted lifetime. This decrease in the effective lifetime is confirmed by the decreased intensity of the green line relative to that of the 3914-Å line of N_2^+. It seems unlikely that reaction (38) could account for the observed decrease in lifetime, nor does the stimulated ($^1S \rightarrow {}^1D$) emission reported at high pressures (>100 torr) [189].

From the data gathered by Zipf [5], as well as refs. 145, 190, and 191, it appears that $O(^1S)$ is quenched primarily by molecular oxygen whenever the density ratio n_{O_2}/n_0 is large ($k_{O_2} = 4.3 \times 10^{-12} \exp(-830/T)$ cm³/sec [276]. Quenching by molecular nitrogen is slow ($k_{N_2} < 10^{-16}$ cm³/sec) [190], and the only other possible atmospheric quenchers are water vapor ($k_{H_2O} = 3 \times 10^{-10}$ cm³/sec) [192, 193] and carbon dioxide ($k_{CO_2} = 3 \times 10^{-13}$ cm³/sec) [145, 191].

The recommended values of the Climatic Impact Assessment Program for $O(^1S)$ quenching are given in Table 6.9.

Table 6.9
Rate Constants for $O(^1S)$ Deactivation [275]

REACTION	TEMP. RANGE/K	REACTION RATE CONSTANT k/cm³ molecule⁻¹ s⁻¹	RELIABILITY OF $\log k$
$O(^1S) + O(^3P) \rightarrow ?$	300	7.5×10^{-12}	± 0.2
$O(^1S) + O_2 \rightarrow ?$	200–377	$4.3 \times 10^{-12} \exp(-850/T)$	± 0.15
$O(^1S) + O_3 \rightarrow ?$	300	5.8×10^{-10}	± 0.7
$O(^1S) + NO \rightarrow ?$	200–291	$3.2 \times 10^{-11} (T)^{0.5}$	± 0.05
$O(^1S) + NO_2 \rightarrow ?$	300	5×10^{-10}	± 0.2
$O(^1S) + N_2 \rightarrow ?$	200–380	$< 5 \times 10^{-17}$	
$O(^1S) + N_2O \rightarrow ?$	300	1.4×10^{-11}	± 0.1
$O(^1S) + NH_3 \rightarrow ?$	300	5×10^{-10}	± 0.2
$O(^1S) + H_2O \rightarrow ?$	300	$> 10^{-10}$	± 1
$O(^1S) + CO_2 \rightarrow ?$	200–450	$3.1 \times 10^{-11} \exp(-1320/T)$	± 0.15
$O(^1S) + CH_4 \rightarrow ?$	300	2×10^{-14}	± 0.4

Considerable controversy has centered upon the rates at which $O(^1D)$ is quenched by collisions with various atoms and molecules. Much of the older data and results are discussed in detail by Hunten and McElroy [56], who conclude that molecular nitrogen is the most efficient quencher, with a rate constant $k \approx 8 \times 10^{-11}$ cm³/sec. This value is supported by the work of Carleton et al. [194] and by that of McGrath and McGarvey [195], who found that the rate for nitrogen as a quencher exceeds that for oxygen. This subject has been reviewed by Zipf [5], who concurs in the view that nitrogen is the major atmospheric quenchant. Relevant quenching data are summarized in Table 6.10, together with the recommended quenching constant for nitrogen. Clearly the contribution to $O(^1D)$ quenching by oxygen is not negligible.

Hunt [205] and Hampson [206] have demonstrated how very small concentrations of $O(^1D)$ can greatly affect the lower D region. Hampson, for example, showed that in the stratosphere the free radicals OH and HO_2 are present only because of the action of $O(^1D)$ on water vapor. Hunt showed

Table 6.10
Quenching of $O(^1D)$ by Various Gases, Relative to Molecular Nitrogen; $k_{N_2} \approx 8 \times 10^{-11}$ cm³/sec [a]

Quencher	Source of data (ref.)									
	129	196	197, 198	158	159	199	200, 201	202	203	196, 204
He			0.077							
Ar		0.046		<0.036					<0.002	
Kr		0.25	0.35							
Xe	2.5	3.2	0.32							
N_2	1.00	1.00	1.00	1.00	1.00	1.00	1.00	1.00	1.00	1.00
CO_2		4.20	3.80	4.7	0.3	15.0			4.35	0.038
H_2O		4.25	9.2	3.6		28.0			5.0	
NO_2		6.75								
O_2	7.5			0.82	0.60	2.1	1.0	0.25	1.0	

[a] References 5, 56, 194, and 195.

that an ozone concentration in agreement with measured values could be derived only if several $O(^1D)$ reactions involving hydrogen and water were taken into consideration, for example,

$$O(^1D) + H_2 \rightarrow OH + H \text{ (assumed } k = 10^{-11} \text{ cm}^3/\text{sec)} \quad (39)$$

with a rate constant of 3×10^{-10} [267], and

$$O(^1D) + H_2O \rightarrow 2\,OH,$$

with a rate constant of 6×10^{-10} [268] or 3.5×10^{-10} [269, 270]. In addition, $O(^1D)$ plays an important stratospheric role through

$$O(^1D) + CH_4 \rightarrow CH_3 + OH \; (k = 3 \times 10^{-10} \text{ from [271]})$$

and

$$O(^1D) + N_2O \rightarrow N_2 + O_2$$
$$\rightarrow 2\,NO \text{ (both } k \simeq 1 \times 10^{-10} \text{ from [271])}$$

Hunt [205] concluded that the concentration of $O(^1D)$ from 40 to 100 km lies between 10^3 and 10^4 cm^{-3}. This value is probably too large because his assumed rate constants are wrong, but the general conclusion persists that even small concentrations of $O(^1D)$ are extremely important, even when turbulent transport is taken into account [207, 208].

$O(^1D)$ is also extremely important in tropospheric chemistry as noted by Kummler et al. [272, 273] and by Chameides and Walker [274]. A review of $O(^1D)$ reaction rate constants has been performed for the Climatic Impact Assessment Program by Garvin and Hampson [275]. Their recommended values are given in Table 6.11.

Table 6.11
Reaction Rate Constants for O(^1D) [275]

REACTION	TEMP. RANGE/K	REACTION RATE CONSTANT k/cm^3 molecule^{-1} s^{-1}	RELIABILITY OF log k
$O(^1D_2) + O_2 \rightarrow O_2(^1\Sigma_g^+) + O(^3P)$	298	7.4×10^{-11} cm^3 molecule^{-1} s^{-1}	± 0.1
$O(^1D_2) + O_3 \rightarrow O_2(^3\Sigma_2^-) + O_2$ (?) (a)	298	5.3×10^{-10} $(k_a + k_b) k_a/k_b \sim 1$	± 0.3
$\rightarrow O_2 + 2\,O(^3P)$ (b)	298		
$O(^1D_2) + NO \rightarrow NO + O(^3P)$	298	1.7×10^{-10}	± 0.3
$O(^1D_2) + NO_2 \rightarrow NO + O_2$	298	2.8×10^{-10}	± 0.1
$O(^1D_2) + N_2 \rightarrow N_2 + O(^3P)$	298	5.4×10^{-11}	± 0.15
$O(^1D_2) + N_2 + M \rightarrow N_2O + M$	298	2.8×10^{-36} cm^6 molecule^{-2} s^{-1}	
$O(^1D_2) + N_2O \rightarrow N_2 + O_2$ (a)	298	1.1×10^{-10} cm^3 molecule^{-1} s^{-1}	± 0.1
$\rightarrow 2\,NO$ (b)	298	1.1×10^{-10}	± 0.1
$O(^1D_2) + NH_3 \rightarrow NH_2 \pm HO$	298	(Probably about 3×10^{-10} No measurements available)	
$O(^1D_2) + H_2 \rightarrow HO + H$	298	2.9×10^{-10}	± 0.1
$O(^1D_2) + H_2O \rightarrow 2\,HO$	298	3.5×10^{-10}	± 0.1
$O(^1D_2) + H_2O_2 \rightarrow HO + HO_2$	298	$>3 \times 10^{-10}$	
$O(^1D_2) + CO \rightarrow CO + O(^3P)$	298	7.7×10^{-11}	± 0.1
$O(^1D_2) + CO_2 \rightarrow CO_2 + O(^3P)$	298	1.8×10^{-10}	± 0.1
$O(^1D_2) + CH_4 \rightarrow CH_3 + HO$ (a) $\rightarrow CH_2O + H_2$ (b)	298	4.0×10^{-10} $(k_a + k_b) k_a/k_b = 10$	± 0.1
$O(^1D_2) + C_2H_6 \rightarrow C_2H_5 + HO$ (a) $\rightarrow CH_3 + CH_2OH$ (b)	298	4.8×10^{-10} $(k_a + k_b)$	± 0.1

4. Production and Destruction of $O^+(^2D)$ and $O^+(^2P)$

The production of both $O^+(^2D)$ and $O^+(^2P)$ occurs in the dayglow through direct photoionization,

$$O(^3P) + h\nu \rightarrow O^+(^4S, {}^2D, \text{ or } {}^2P) + e, \tag{40}$$

and through ionization by photoelectrons,

$$O(^3P) + e \rightarrow O^+(^4S\ {}^2D, \text{ or } {}^2P) + 2e. \tag{41}$$

Both excited states can be destroyed through superelastic collisions with electrons, or through superelastic, energy-transfer, or reactive collisions with neutral particles. The rate for deactivation or ions through superelastic electron collision is large [185], giving a quenching frequency in the dayglow of 0.1 sec^{-1}, while the transition probabilities for such states are much smaller.

In aurora the N_2^+ Meinel radiation is probably affected by the near-resonant reaction [209, 210]:

$$O^+(^2D) + N_2(X\,^1\Sigma_g, v = 0) \rightarrow O(^3P) + N_2^+(A\,^2\Pi_u, v = 1). \tag{42}$$

Similarly, the cross sections for the near-resonant reactions,

$$O^+(^2D) + O_2(X\,^3\Sigma_g^-) \rightarrow O(^3P) + O_2^+(a\,^4\Pi_u, v = 7) \tag{43}$$

$$\rightarrow O(^3P) + O_2^+(A\,^2\Pi_u, v = 0), \tag{44}$$

are large [24, 139] ($k \approx 3 \times 10^{-10}$ cm^3/sec).

Wallace and McElroy (138) tried to match theoretically the observed dayglow emission of the 3914-Å (B → X) transition of N_2^+. Using the excitation mechanisms discussed above, as well as reactions (14), (15), and (17), and the loss reactions:

$$N_2^+ + e = N + N, \tag{45}$$

$$N_2^+ + O = NO^+ + N, \tag{46}$$

and

$$N_2^+ + O_2 = N_2 + O_2^+; \tag{47}$$

a fit of the data gives the values of k_{45}, k_{46}, and k_{47} shown in the second column of Table 6.12.

Table 6.12
Calculated Reaction Coefficients

	WITH NO $O^+(^2D)$ SOURCE (cm^3/sec)	WITH $O^+(^2D)$ REACTION INCLUDED (cm^3/sec)
k_{45}	6×10^{-8}	3.5×10^{-10}
k_{46}	7×10^{-11}	1×10^{-10}
k_{47}	9×10^{-11}	5×10^{-11}

Table 6.13
Excitation and Deexcitation Rate Coefficients or Cross Sections[a]

REACTION	(1) REACTION COEFFICIENT (cm^3 sec^{-1}) OR (2) CROSS SECTION (cm^2) OR (3) REACTION COEFFICIENT (cm^6 sec^{-1})	TEMPERATURE OR ENERGY	TYPE OF EXPERIMENT	REF.
Electron-impact excitation and deexcitation				
A major review of electron-impact excitation rates is being performed under the sponsorship of the Data Center at the Joint Institute for Laboratory Astrophysics.			Beams and fast-flowing systems[b]	216
Application to the atmosphere see				215
$e + H(^1S) \rightarrow e + H(^2S)$	(2) 6×10^{-17} (max)	30–50 eV	[b]	217, 218
$\rightarrow e + H(^2S)$	(2) 1.0×10^{-17} (max)	12 eV	[b]	219
$e + Na(3\,^2S) \rightarrow e + Na(3\,^2P)$	(2) 2×10^{-15} (max)	10 eV	[b]	220
$e + He(1\,^1S) \rightarrow e + He^*(2\,^3S)$	(2) 3×10^{-18} (max)	20.6 eV		221
$e + He^*(2\,^1S) \rightarrow e + He^*(2\,^3S)$	(2) 300×10^{-14}	0.025 eV	Static afterglow	186, 222
$e + N^*(^2D) \rightarrow e + N(^4S)$	(1) 5×10^{-10}	1000°K	Theory	185
$e + N^*(^2P) \rightarrow e + N(^4S)$	(1) 2×10^{-9}	1000°K	Theory	185
$e + O^*(^1D) \rightarrow e + O(^3P)$	(1) 1.5×10^{-9}	1000°K	Theory	185
$e + O^*(^1S) \rightarrow e + O(^3P)$	(1) 1.8×10^{-9}	1000°K	Theory	185
$e + N_2 \rightarrow e + N_2^{\ddagger}$	(2) 6×10^{-16}	2.3 eV	[b], Theory	73, 75
$e + N_2 \rightarrow e + N_2^*(C, v = 0)$	(2) 1.6×10^{-17} (max)	17 eV	[b]	223
$e + CO \rightarrow e + CO_2^{\ddagger}$	(2) 8×10^{-16} (max)	1.75 eV	[b]	73
$e + H_2 \rightarrow e + H_2^{\ddagger}$	(2) 6×10^{-17} (max)	2 eV	[b]	73
$e + N_2 \rightarrow 2e + N_2^+(B)$	(2) 9.5×10^{-8} (max)	90 eV	[b]	224
$e + N_2O^{\ddagger} \rightarrow N_2O^-$	(2) 8.3×10^{-8}	2.3 eV	[b]	225
Two-body reactions				
$O^*(^1S) + O(^3P) \rightarrow O(^1D) + O(^1D$ or $^3P)$	(1) 1.8×10^{-13}	300°K	Flowing afterglow	184
$O^*(^1S) + N_2O \rightarrow O + N_2O$	(1) 6×10^{-13}	300°K	Flowing afterglow	184

$O^*(^1S) + O_2 \rightarrow O + O_2$	(1) 1×10^{-13}	300°K	Flowing afterglow	184
$O^*(^1S) + CO_2 \rightarrow O + CO_2$ or $O_2 + CO$	(1) 2.5×10^{-14}	300°K	Flowing afterglow	184
$O^*(^1S) + N_2 \rightarrow O + N_2$	(1) $<10^{-17}$	300°K	Flowing afterglow	184
$O^*(^1D) + N_2 \rightarrow O(^3P) + N_2$	(1) 8×10^{-11}	300°K	Atmospheric, flowing afterglow	6, 8
$O^*(^1D) + O_2 \rightarrow O(^3P) + O_2$	(1) 9×10^{-11}	300°K	Atmospheric, flowing afterglow	
$O^+(^2D) + N_2 \rightarrow O(^3P) + N_2^+$	(1) $\sim 3 \times 10^{-10}$		[b]	139
$O^+(^2D) + O_2 \rightarrow O(^3P) + O_2^+$	(1) $\sim 3 \times 10^{-10}$		[b]	139
$O + O_3 \rightarrow O_2 + O_2(a\,^1\Delta_g)$	(1) $4.5 \times 10^{(-15\pm 2)}$	300°K	Flowing afterglow	157
$O_2^\ddagger(X, v \geq 17) + O_3 \rightarrow 2O_2 + O^*(^1D)$	(1) Large	300°K	Flash photolysis	226
$O_2^*(a\,^1\Delta_g) + O_2 \rightarrow O_2(X) + O_2$	(1) 2.4×10^{-18}	300°K	Flowing afterglow	154, 161, 162,
$O_2^*(a\,^1\Delta_g) + O_2^- \rightarrow 2O_2 + e$	(1) 2×10^{-10}	300°K	Flowing afterglow	169, 171
$O_2^*(a\,^1\Delta_g) + O^- \rightarrow O_3 + e$	(1) 3×10^{-10}	300°K	Flowing afterglow	171
$O_2^*(a\,^1\Delta_g) + N \rightarrow NO + O$	(1) $(2.8 \pm 2) \times 10^{-15}$	300°K	Flowing afterglow	167
$O_2^*(b\,^1\Sigma_g^+) + O_3 \rightarrow 2O_2 + O$	(1) 6×10^{-13}	300°K	Flowing afterglow	227
$O_2^+ + e \rightarrow O^* + O^{**}$	(1) See Table 6-7	300°K	Flowing afterglow	181
$N^*(^2D) + O_2 \rightarrow NO + O$	(1) $4 \times 10^{-13} T^{1/2}$	236–365°K	Flash photolysis	228
$N + NO \rightarrow N_2^\ddagger(X, v \approx 8) + O(^3P)$	(1) 2.2×10^{-11}	300°K	Flowing afterglow	126
$N + N_2(X) \rightarrow N + N_2^*(A)$	(1) $1.9 \times 10^{-6} T^{-3/2} \exp(-E_{XA}/kT)$		Shocks	132
$N_2 + N_2(X) \rightarrow N_2^* + N_2(A)$	(1) $k(N_2) \geq 0.01k(N)$		Active discharge, shocks	132
$A^\ddagger + B \rightarrow A + B^\ddagger$	Considerable literature is available on this type of reaction. The general agreement of vibrational-energy transfer associated with lower vibrational levels is encouraging.		Flames, flash photolysis, shocks, theory	84, 86, 93
$A + BC^\ddagger \rightarrow AB + C$	Considerable literature on this subject of a nonatmospheric nature has been generated by Polanyi at the University of Toronto. (Cf. the listed references.)			229, 230
$NO^*(A, v = x) + N_2(X, v = 0) \rightarrow NO^*(A, v = x - 1) + N_2^\ddagger(X, v = 1)$	(1) Very large	300°K	Flash photolysis, theory	84, 86
$N_2^\ddagger(X, v) + O^+ \rightarrow NO^+ + N$	(1) See Table 6-3	300°K	Flowing afterglow	99
$N_2^\ddagger(X, v \geq 7) + Na(3\,^2S) \rightarrow N_2 + Na^*$	(1) $\sim 10^{-10}$	300°K	Flowing afterglow	97

Table 6.13 (Continued)

REACTION	(1) REACTION COEFFICIENT ($cm^3\ sec^{-1}$) OR 2. CROSS SECTION (cm^2) OR (3) REACTION COEFFICIENT (cm^6/sec)	TEMPERATURE OR ENERGY	TYPE OF EXPERIMENT	REF.
$N_2^{\ddagger}(X, v \geq 4) + M = N_2(X, v < 4) + M$	(1) See Table 6-2	$\sim 300°K$	Flowing afterglow	56, 68, 96, 97, 99
$N_2^*(A) + N_2 \rightarrow N_2(X) + N_2$	(1) $<3 \times 10^{-19}$	300°K	High-pressure chemical reactions	126
$N_2^*(A) + O_2 \rightarrow N_2 + (O_2 \text{ or } 2O)$	(1) Large	$\sim 300°K$	Atmospheric	56
$N_2^*(A) + O \rightarrow N_2 + O \text{ or } NO + N$	(1) Large	$\sim 300°K$	Atmospheric	56
$N_2^*(A) + N \rightarrow N + N_2^{\ddagger}(X)$	(1) $\geq 2 \times 10^{-10}$	$\sim 300°K$	Flowing afterglow	128
$N_2^*(a, v) + N_2 \rightarrow N_2^*(a, v' < v) + N_2$	(1) Large	1000°K	Theory	231
$N_2^* + Ba \rightarrow N_2(X) + Ba^{+*}(6\,^2P_{3/2}) + e$	(1) Large	300°K	[b], static afterglow	228, 232, 233
$OH^{\ddagger}(X, v') + OH^{\ddagger}(X, v) \rightarrow OH^*(A) + OH(X)$	(1) $\sim 10^{-10}$	$\sim 340°K$	Flames	234
$N_2^{+*}(A\,^2\Pi_u) + O_2 \rightarrow N_2 + O^{\ddagger}$	(1) Large		[b]	120
$N_2^{+*}(A) + N_2 \rightarrow N_2^+ + N_2$	(1) 1×10^{-9}	300°K	[b]	235
$N_2^{+*}(B) + N_2 \rightarrow N_2^+ + N_2$	(1) 6×10^{-10}	300°K	Static afterglow	236
$N_2^{+*}(B) + O_2 \rightarrow N_2^+ + O_2$	(1) 2×10^{-9}	300°K	Static afterglow	236
$N_2^+ + CO \rightarrow N_2 + CO^{+*}(A)$	(2) $>10^{-16}$	8 eV	[b]	116
$He^*(2\,^3S) + N_2 \rightarrow He(^1S) + N_2^{+*}(B) + e$	(1) 1.4×10^{-10}	300°K	[b]	96, 237
$He^*(2\,^3S) + O_2 \rightarrow He(^1S) + O_2^{+*} + e$	(1) 5.0×10^{-10}		Flowing afterglow	237
$He^+ + N_2(X) \rightarrow He + N_2^*(C, v = 3)$	(2) Large	$2\text{–}10^3$ eV	[b]	238
$H^+ + Cs \rightarrow H^*(2p) + Cs^+$	(2) 38×10^{-16}	200 eV	[b]	239
Three-body reactions				
$N + N + N_2 \rightarrow N_2^*(B) + N_2$	(3) 1.4×10^{-33}	300°K	Flowing afterglow	240

$N + O + N_2 \rightarrow NO^*(B) + N_2$	(3) 1×10^{-34}	300°K	Flowing afterglow	184
$O + O + N_2 \rightarrow O_2^*(A) + N_2$	(3) 2.1×10^{-37}	300°K	Flowing afterglow	184
$O + O + N_2 \rightarrow O_2^*(b) + N_2$	(3) 1.7×10^{-37}	300°K	Flowing afterglow	184
$O + O + O_2 \rightarrow O_2^*(A, v = 9, 10) + O_2$	(3) $1 \times 10^{-33}(v = 9)$	1000°K	Theory	241
	(3) $5.5 \times 10^{-33}(v = 10)$	1000°K	Theory	241
$O + O + CN \rightarrow O_2 + CN^*(A)$	(3) 10^{-31}–10^{-30}	300°K	Flash photolysis	242
$O + O + Na \rightarrow O_2 + Na^*(^2P)$	(3) 1.5×10^{-29}	1250–1500°K	Flames	243
$H + H + Na \rightarrow H_2 + Na^*(^2P)$	(3) 5×10^{-31}	1250–1500°K	Flames	243
$N + N + O \rightarrow N_2 + O^*(^1S)$	(3) $10^{-33} + (5 \times 10^{-33})P_{N_2}$ (P_{N_2} in torr)	~300°K	Flowing afterglow	184
$N + O + O \rightarrow NO + O^*(^1S)$	(3) 3×10^{-33}	300°K	Flowing afterglow	184
$O + O + O \rightarrow O_2 + O^*(^1S)$	(3) 1.5×10^{-34}	300°K	Flowing afterglow	184
$O + O + O \rightarrow O_2 + O^*(^1D)$	(3) Large	300°K	Theory	244
$H + H_2 + O_2 \rightarrow H_2O + OH^*(A)$	(3) 5×10^{-37}	1000–1900°K	Static afterglow	245
$H^+ + e + e \rightarrow H^*(n) + e$	Cross sections for recombination into the higher values of n are very large and can be estimated from detailed balance from the calculations for ionization.		Theory	246, 247

[a] This list, which was revised last in December 1970, is believed to include the major portion of all simple atomic and diatomic chemical excitation and energy-transfer reaction rates of interest for the quiescent and disturbed atmosphere.
[b] Including electron, ion, and neutral inelastic and reactive scattering.

At twilight the first negative bands have a high rotational temperature, suggesting a source of $N_2^+(B)$ that gives high rotational temperatures, such as the charge transfer of $O^+(^2D)$ with N_2 might provide. If the rate coefficient for this charge-transfer reaction is taken as $k = 2 \times 10^{-9}$ cm^3/sec, Dalgarno and McElroy [121] found that the reaction rate constants k_{46} and k_{47} are not greatly changed, but that k_{41} must be increased by a factor of 6 to give the measured 3914-Å emission data (see Table 6.12, column 3). It is interesting to note that this change is brought about by $O^+(^2D)$ ion densities as low as 100 cm^{-3}.

C. Metallic Ions

It has been demonstrated [211] that metallic ions are present in the *D* region. Since these species are atomic and will tend therefore to recombine slowly with electrons (though faster with negative ions), it is important to study their reaction rates with the various molecular atmosphere species. Ground-state metallic ions, owing to their low energy, will not readily enter into charge-transfer or ion-molecule reactions. However, ions that are in excited states may quickly be converted to molecular ions through ion-molecular reactions, or transfer their charge directly to molecular species. In atmospheric detonations of nuclear devices, where tremendous energy and much metallic debris are deposited in the atmosphere, a large fraction of the metallic ions that are formed may be excited. For Na^+, Fe^+, Al^+, and a few other metallic ions, Fogel [212] and Layton [213] have shown that excitation has a marked effect upon the cross sections measured at kilovolt energies. As the ionizing electron energy is increased in the ion source, so that excited ions are added to the beam, the total measured cross section varies markedly.

Observation of metallic ion populations have also led to speculation about vibrational temperatures. Without an elevated oxygen vibrational temperature at 110 km, Ferguson et al. [214] could not explain the observed atmospheric Si^+/SiO^+ ratio. However, as noted by Bauer et al. [70], it is difficult to reconcile an elevated oxygen vibrational temperature with known O_2^* deactivation rates by atomic oxygen.

IV. REACTION RATES FOR REACTIONS INVOLVING EXCITED SPECIES

The reaction rates and cross sections listed in Table 6.13 represent a considerable portion of the information available on excited-state production and destruction processes involving atmospheric species. The temperature dependence of most reaction coefficients is not given; the temperature at which the measurements were made is given instead. Also included are the general types of experiments used, with appropriate references, in obtaining the data quoted.

During the compilation of Table 6.13, it became clear that most of the information on energy transfer and the excitation of long-lived excited atoms, ions, and molecules is limited to reactions where the products are only inferred and for which temperature dependences are as yet unmeasured. Clearly, the detailed study of both reaction products and energy dependences is essential to future progress in this field.

REFERENCES

1. J. Wm. McGowan, General Atomic Division (of General Dynamics Corporation) Report GA-7590, 1967.
2. A. Wright and C. Winkler, *Active Nitrogen*, Academic Press, New York, 1968.
3. W. L. Wiese, M. Smith, and B. M. Glennon, "Atomic Transition Probabilities," Vol. I, National Standard Reference Data System Report NSRDS NBS-4, 1966.
4. F. R. Gilmore, unpublished work, 1971.
5. E. C. Zipf, Jr., *Can. J. Chem.*, **47,** 1863 (1969).
6. G. M. Lawrence and B. D. Savage, *Phys. Rev.*, **141,** 67 (1966).
7. R. Nichols, *Ann. Geophys.*, **20,** 144 (1964).
8. R. H. Garstang, *Monthly Notices Roy. Astron. Soc.*, **111,** 115 (1951).
9. C. Nicolaides, O. Sinanoglu, and P. Westhaus, *Phys. Rev. A*, **4,** 1400 (1971).
10. G. M. Lawrence, *Can. J. Chem.*, **47,** 1856 (1969); *Phys. Rev. A*, **2,** 397 (1970).
11. D. Shemansky, *J. Chem. Phys.*, **51,** 689 (1969).
12. D. Shemansky and N. P. Carleton, *J. Chem. Phys.*, **51,** 682 (1969).
13. M. Jeunehomme, *J. Chem. Phys.*, **45,** 1805 (1966).
14. M. Hollstein, D. C. Lorents, J. R. Peterson, and J. R. Sheridan, *Can. J. Chem.*, **47,** 1858 (1969).
15. R. S. Freund, *J. Chem. Phys.*, **50,** 3734 (1969).
16. W. L. Borst and E. C. Zipf, *Phys. Rev. A*, **3,** 979 (1971).
17. H. L. Wu and W. Benesch, *Phys. Rev.*, **172,** 31 (1968).
18. D. E. Shemansky, *J. Chem. Phys.*, **51,** 5487 (1969).
19. R. Covey and W. Bensch, *Bull APS*, **18,** 575 (1973).
20. J. Olmstead, A. S. Newton, and K. Street, Jr., *J. Chem. Phys.*, **42,** 2321 (1965).
21. C. E. Head, *Phys. Lett.*, **34A,** 92 (1971).
22. J. E. Hesser, *J. Chem. Phys.*, **48,** 2518 (1968).
23. A. W. Johnson and R. G. Fowler, *J. Chem. Phys.*, **53,** 65 (1970)
24. J. Wm. McGowan and L. Kerwin, *Can. J. Phys.*, **42,** 2086 (1964).
25. R. Frosch and G. Robinson, *J. Chem. Phys.*, **41,** 367 (1964).
26. H. Lefebvre-Brion and F. Guerin, *J. Chem. Phys.*, **49,** 1446 (1968).
27. J. Jeunehomme, *J. Chem. Phys.*, **45,** 4433 (1966).
28. H. Bubert and F. W. Froben, *Chem. Phys. Lett.*, **8,** 242 (1971).

29. O. Edqvist et al., *Arkiv. Fys.*, **40,** 439 (1970).
30. R. F. Mathis, B. R. Turner, and J. A. Rutherford, *J. Chem. Phys.*, **49,** 2051 (1968).
31. W. B. Maier and R. F. Holland, *J. Chem. Phys.*, **54,** 2693 (1971).
32. S. Schwartz and H. Johnston, *J. Chem. Phys.*, **51,** 2186 (1969).
33. L. Keyser, S. Levine, and F. Kaufman, *J. Chem. Phys.*, **54,** 355 (1971).
34. R. M. Badger, A. C. Wright, and R. F. Whitlock, *J. Chem. Phys.*, **43,** 4345 (1965).
35. J. H. Miller, R. W. Boese, and L. P. Giver, *J. Quant. Spectr. Radiative Transfer*, **9,** 1507 (1969).
36. D. E. Burch and D. A. Gryvak, *Appl. Opt.*, **8,** 1493 (1969).
37. V. Degen, *Can. J. Phys.*, **46,** 783 (1968).
38. W. R. Jarmain and R. W. Nicholls, *Proc. Phys. Soc.* (London), **90,** 545 (1967).
39. R. W. Nicholls, *Ann. Geophys.*, **20,** 144 (1964).
40. M. Jeunehomme, *J. Chem. Phys.*, **44,** 4253 (1966).
41. E. H. Fink and K. H. Welge, *Z. Naturforsch.*, **23a,** 358 (1968).
42. T. Wentink Jr., E. P. Marram, L. Isaacson, and R. J. Spindler, "Ablative Material Spectroscopy," AFWL-TR67-30, Nov. 1967.
43. L. Isaacson, E. P. Marram, and T. Wentink Jr., *Appl. Opt.*, **8,** 235 (1969).
44. T. C. James, *J. Chem. Phys.*, **55,** 4118 (1971).
45. R. G. Bennett and E. W. Dalby, *J. Chem. Phys.*, **36,** 344 (1962).
46. R. G. Bennett and E. W. Dalby, *J. Chem. Phys.*, **32,** 1716 (1960).
47. E. H. Fink and K. H. Welge, *J. Chem. Phys.*, **46,** 4315 (1967).
48. M. Jeunehomme, *J. Chem. Phys.*, **42,** 4086 (1965).
49. M. Jeunehomme, Air Force Weapons Laboratory Report AFWL-TR-66-143, 1967.
50. K. J. Laidler, *Chemical Kinetics of the Excited States*, Oxford U. P., London, 1955.
51. D. R. Bates, editor, *Atomic and Molecular Processes*, Academic, New York, 1962.
52. U. Fano and W. Lichten, *Phys. Rev. Lett.*, **14,** 627 (1965).
53. F. R. Gilmore, E. Bauer, and J. Wm. McGowan, *J. Quant. Spectr. Radiative Transfer*, **9,** 157 (1969).
54. E. Bauer, E. R. Fisher, and F. R. Gilmore, *J. Chem. Phys.*, **51,** 4173 (1969).
55. R. Anderson, *Atomic Data*, **3,** 227 (1971).
56. D. M. Hunten and M. B. McElroy, *Rev. Geophys.*, **4,** 303 (1966).
57. D. R. Bates, *Discussions Faraday Soc.*, **37,** 21 (1964).
58. E. E. Muschlitz, in *Molecular Beams*, edited by J. Ross. Interscience, New York, 1966, p. 171.
59. R. J. Donovan and D. Husain, *Chem. Rev.*, **70,** 489 (1970).
60. K. E. Shuler and W. R. Bennett, editors, *Applied Optics Supplement 2: Chemical Lasers*, Vol. 4, 1965.
61. J. W. Chamberlain, *Physics of the Aurora and Airglow*, Academic, New York, 1961.
62. E. B. Armstrong and A. Dalgarno, editors, *The Airglow and Aurorae*, Pergamon, London, 1956.

63. J. A. Ratcliff, editor, *Physics of the Upper Atmosphere*, Academic, New York, 1960.
64. M. Zelikoff, editor, *The Threshold of Space*, Pergamon, New York, 1957.
65. R. D. Cadle, editor, *Chemical Reactions in the Lower and Upper Atmosphere*, Interscience, New York, 1961.
66. C. O. Hines et al., editors, *Physics of the Earth's Upper Atmosphere*, Prentice-Hall, Englewood Cliffs, N.J., 1965.
67. L. F. Phillips and H. I. Schiff, *J. Chem. Phys.*, **36,** 1509, 3283 (1962).
68. J. E. Morgan, L. F. Phillips, and H. I. Schiff, *Discussions Faraday Soc.*, **33,** 118 (1962).
69. E. Bauer and E. R. Fisher, private communication, 1970.
70. E. Bauer, R. H. Kummler, and M. H. Bortner, *Appl. Opt.*, **10,** 1861 (1971).
71. J. Walker, *Planet. Space Sci.*, **16,** 321 (1968).
72. J. C. Walker, R. S. Stolarksi, and A. F. Nagy, *Ann. Geophys.*, **25,** 831 (1969).
73. G. J. Schulz, *Phys. Rev.*, **125,** 229 (1962); **135,** A988 (1964).
74. J. C. Y. Chen, *J. Chem. Phys.*, **40,** 3513 (1964); *Phys. Rev.*, **146,** 61 (1966).
75. A. G. Engelhardt, A. V. Phelps, and C. G. Risk, *Phys. Rev.*, **135,** A1566 (1964).
76. G. Abraham and E. R. Fisher, Wayne State University Report RIES 70-01, 1970.
77. E. R. Fisher and R. H. Kummler, Spring Symposium on Quantum Electronics, Sun Valley, Idaho, 1971.
78. A. W. Ali, U.S. Naval Research Laboratory, Plasma Dynamics Technical Note 24, 1970.
79. D. E. Golden, *Bull. Am. Phys. Soc.*, **12,** 222 (1967).
80. A. E. S. Green and C. A. Barth, *J. Geophys. Res.*, **70,** 1083 (1965).
81. A. E. S. Green, editor, *The Middle Ultraviolet*, Wiley, New York, 1966, p. 165.
82. P. D. Burrow and P. Davidovitz, *Phys. Rev. Lett.*, **21,** 1789 (1968).
83. D. Rapp and T. E. Sharp, *J. Chem. Phys.*, **38,** 2641 (1963).
84. A. B. Callear, p. 145 of ref. 60.
85. K. F. Herzfeld and T. A. Litovitz, *Absorption and Dispersion of Ultrasonic Waves*, Academic, New York, 1959.
86. K. Takayanagi, in *Advances in Atomic and Molecular Physics*, edited by D. R. Bates and I. Estermann, Academic, New York, 1965, Vol. 1, p. 149.
87. D. R. Bates, *J. Atmos. Terrest. Phys.*, **6,** 171 (1955).
88. A. Dalgarno, *Planet. Space Sci.*, **10,** 19 (1963).
89. T. C. Clark, S. H. Garnett, and G. B. Kistiakowsky, *J. Chem. Phys.*, **52,** 4694 (1970).
90. S. H. Garnett, G. B. Kistiakowsky, and B. V. O'Grady, *J. Chem. Phys.*, **51,** 84 (1969).
91. E. Bauer and F. W. Cummings, *J. Chem. Phys.*, **36,** 618 (1962).
92. C. E. Treanor, *J. Chem. Phys.*, **43,** 532 (1965).
93. R. Taylor and S. Bitterman, *Rev. Mod. Phys.*, **41,** 26 (1969).
94. R. Millikan and D. White, *J. Chem. Phys.*, **39,** 3209 (1963).
95. D. M. Hunten, *J. Atmos. Terrest. Phys.*, **27,** 538 (1965).
96. W. L. Starr, *J. Chem. Phys.*, **43,** 73 (1965).

97. W. L. Fite, W. R. Henderson, H. F. Krause, and J. E. Mentall, Fifth International Conference on Physics of Electronic Atomic Collisions, Leningrad, USSR, 1967.
98. E. R. Fisher and G. Smith, *Appl. Opt.*, **10**, 1803 (1971).
99. A. L. Schmeltekopf, G. I. Gilman, F. C. Fehsenfeld, and E. E. Ferguson, *Planet. Space Sci.*, **15**, 401 (1967).
100. A. L. Schmeltekopf, E. E. Ferguson, and F. C. Fehsenfeld, *J. Chem. Phys.*, **48**, 2966 (1968).
101. L. Thomas and R. B. Norton, *J. Geophys. Res.*, **15**, 401 (1967).
102. T. F. O'Malley, *J. Chem. Phys.*, **52**, 3269 (1970).
103. R. Whitten and A. Dalgarno, *Planet. Space Sci.*, **15**, 1419 (1967).
104. R. D. Sears, private communication, 1968.
105. E. R. Fisher and R. H. Kummler, *J. Chem. Phys.*, **49**, 1075 (1968).
106. M. H. Bortner and R. H. Kummler, General Electric Company Report DASA 2407, 1970.
107. D. Breshears and R. Bird, *J. Chem. Phys.*, **48**, 4768 (1968).
108. W. G. Vincenti and C. H. Kruger, *Physical Gas Dynamics*, Wiley, New York, 1965, p. 202.
109. *COSPAR International Reference Atmosphere*, North-Holland, Amsterdam, 1965.
110. W. Lichten, *Phys. Rev.*, **120**, 848 (1960).
111. H. F. Winters, *J. Chem. Phys.*, **43**, 926 (1965).
112. G. J. Schulz, *Phys. Rev.*, **116**, 1141 (1959).
113. S. N. Foner and R. L. Hudson, *J. Chem. Phys.*, **37**, 1662 (1962).
114. A. L. Broadfoot and D. M. Hunten, *Can. J. Phys.*, **42**, 1212 (1964).
115. L. Wallace, *J. Atmos. Terrest. Phys.*, **17**, 46 (1959).
116. W. Petrie, *Phys. Rev.*, **86**, 790 (1952).
117. T. Wentink and L. Isaacson, *J. Chem. Phys.*, **46**, 822 (1967).
118. B. R. Turner, J. A. Rutherford, and R. F. Stebbings, *J. Geophys. Res.* **71**, 4521 (1966).
119. P. D. Goldan, A. L. Schmeltekopf, F. C. Fehsenfeld, H. I. Schiff, and E. E Ferguson, *J. Chem. Phys.*, **44**, 4095 (1966).
120. J. Peterson and D. C. Lorents, private communication.
121. A. Dalgarno and M. B. McElroy, *Planet. Space Sci.*, **14**, 1321 (1966).
122. D. C. Cartwright, W. Williams, and S. Trajmar, I.U.G.G. Meeting, Moscow, USSR, 1971.
123. D. C. Cartwright, S. Trajmar, and W. Williams, I.U.G.G. Meeting, Moscow, USSR, 1971.
124. D. C. Cartwright, Aerospace Corp. Report TR-0059 (9260-01)-6, 1970.
125. D. C. Cartwright, *Phys. Rev. A*, **2**, 1331 (1970).
126. J. F. Noxon, *J. Chem. Phys.*, **36**, 926 (1962).
127 E. C. Zipf, Jr., *Bull. Am. Phys. Soc.*, **9**, 185 (1964).
128. R. A. Young, *Can. J. Chem.*, **44**, 1171 (1966).
129. R. A. Young, G. Black, and T. G. Slanger, *J. Chem. Phys.*, **50**, 303 (1969).
130. R. A. Young and G. A. St. John, *J. Chem. Phys.*, **48**, 895 (1968).
131. B. A. Thrush, *J. Chem. Phys.*, **47**, 3691 (1967).

132. K. Wray, *J. Chem. Phys.*, **44,** 623 (1966).
133. R. A. Young and G. A. St. John, *J. Chem. Phys.*, **48,** 898 (1968).
134. C. Kenty, *J. Chem. Phys.*, **23,** 1555 (1955).
135. V. Cermak, *J. Chem. Phys.*, **44,** 1318 (1966).
136. V. Cermak, *J. Chem. Phys.*, **43,** 4527 (1965).
137. A. R. Lee and N. P. Carleton, *Phys. Lett.*, **27A,** 195 (1968).
138. L. Wallace and M. B. McElroy, *Planet. Space Sci.*, **14,** 677 (1966).
139. R. F. Stebbings, B. R. Turner, and J. Rutherford, *J. Geophys. Res.*, **71,** 771 (1966).
140. R. C. Amme and N. G. Utterback, in *Atomic Collision Processes*, edited by M. R. C. MacDowell, North-Holland, Amsterdam, 1964, p. 847.
141. J. Wm. McGowan, E. M. Clarke, H. D. Hanson, and R. F. Stebbings, *Phys. Rev. Lett.*, **14,** 620 (1964).
142. W. L. Fite and R. T. Brackmann, *Proceedings of the Sixth International Conference on Ionization in Gases*, North-Holland, New York, 1963, Vol. 1, p. 21.
143. M. A. Fineman et al., *Proceedings of the Fourth International Conference on the Physics of Electronic and Atomic Collisions*, Science Bookcrafters, New York, 1965, p. 452.
144. C. Lin and F. Kaufman, *J. Chem. Phys.*, **55,** 3760 (1971).
145. G. Black, T. G. Slanger, G. St. John, and R. A. Young, *J. Chem. Phys.*, **51,** 116 (1969).
146. R. F. Stebbings et al., General Atomic Division (of General Dynamics Corporation) Report DASA 1708, 1965.
147. W. L. Fite, R. T. Brackmann, and W. R. Henderson, *Proceedings of the Fourth International Conference on Electronic and Atomic Collisions*, Science Bookcrafters, New York, 1965, p. 100.
148. T. F. O'Malley, *Phys. Rev.*, **155,** 59 (1967).
149. J. C. Y. Chen and J. L. Preacher, *Phys. Rev.*, **163,** 103 (1967).
150. J. H. Kiefer and R. W. Lutz, *Eleventh Symposium (International) on Combustion*, The Combustion Institute, Pittsburgh, Pa., 1967, p. 67.
151. S. H. Bauer and S. C. Tsang, *Phys. Fluids*, **6,** 182 (1963).
152. G. Schulz and J. T. Dowell, *Phys. Rev.*, **128,** 174 (1962).
153. D. Spence and G. Schulz, *Phys. Rev. A*, **2,** 1802 (1970).
154. I. T. N. Jones and R. P. Wayne, *J. Chem. Phys.*, **51,** 3617 (1969).
155. A. Vallance Jones and R. L. Gattinger, *Planet. Space Sci.*, **11,** 961 (1963).
156. W. F. Evans, D. M. Hunten, E. J. Llewellyn, and A. Vallance Jones, *J. Geophys. Res.*, **73,** 2885 (1968).
157. R. Flugge and D. Headrick, Cornell Aeronautical Laboratory Report DASA 2551, 1970.
158. R. A. Young, G. Black, and T. G. Slanger, *J. Chem. Phys.*, **49,** 4758 (1968).
159. J. Noxon, *J. Chem. Phys.*, **52,** 1852 (1970).
160. L. Wallace and J. W. Chamberlain, *Planet. Space Sci.*, **2,** 60 (1959).
161. A. M. Trozzolo, editor, *International Conference on Singlet Molecular Oxygen and Its Role in Environmental Sciences*, Ann. N.Y. Acad. Sci. **171,** Art. 1 (1970).

162. I. D. Clark and R. P. Wayne, *Proc. Roy. Soc.* (London), **A314,** 111 (1969).
163. A. Winer and K. Bayes, *J. Phys. Chem.*, **70,** 302 (1966).
164. I. D. Clark and R. P. Wayne, *Chem. Phys. Lett.*, **3,** 93 (1969).
165. F. Findlay, C. Fortin, and D. Snelling, *Chem. Phys. Lett.*, **3,** 204 (1969).
166. R. P. Steer, R. A. Ackerman, and J. N. Pitts, Jr., *J. Chem. Phys.*, **51,** 843 (1969).
167. I. D. Clark and R. P. Wayne, *Chem. Phys. Lett.*, **3,** 405 (1969).
168. R. J. McNeal and G. R. Cook, *J. Chem. Phys.*, **47,** 5385 (1967).
169. L. R. Megill and J. B. Hasted, *Planet. Space Sci.*, **13,** 339 (1965).
170. R. H. Kummler and M. H. Bortner, General Electric Company TIS Report R67SD20, 1967.
171. F. C. Fehsenfeld, D. L. Albritton, J. A. Burt, and H. I. Schiff, *Can. J. Chem.*, **47,** 1793 (1969).
172. R. H. Kummler and M. H. Bortner, p. 237 of ref. 161.
173. J. Noxon, *J. Geophys. Res.*, **75,** 1879 (1970).
174. L. Megill, A. Despain, D. Baker, and K. Baker, *J. Geophys. Res.*, **75,** 4775 (1970).
175. H. I. Schiff, J. Haslett, and L. Megill, *J. Geophys. Res.*, **75,** 4363 (1970).
176. S. J. Arnold, N. Finlayson, and E. A. Ogryzlo, *J. Chem. Phys.*, **44,** 2529 (1966).
177. R. A. Young and G. Black, *J. Chem. Phys.*, **42,** 3750 (1965).
178. L. Wallace and D. M. Hunten, *J. Geophys. Res.*, **73,** 4813 (1968).
179. S. V. Filseth, A. Zia, and K. H. Welge, *J. Chem. Phys.*, **52,** 5502 (1970).
180. K. Watanabe, *Advan. Geophys.*, **5,** 153 (1958).
181. E. C. Zipf, Jr., *Bull. Am. Phys. Soc.*, **12,** 225 (1967).
182. E. C. Zipf, Jr., *Bull. Am. Phys. Soc.*, **15,** 418 (1970).
183. S. Chapman, *Proc. Roy. Soc.* (London), **A132,** 353 (1931).
184. R. A. Young and G. Black, *J. Chem. Phys.*, **44,** 3741 (1966).
185. M. J. Seaton, p. 289 of ref. 62.
186. J. C. Ingraham and S. C. Brown, *Phys. Rev.*, **138,** A1015 (1965).
187. R. Henry, P. Burke, and A. L. Sinfailam, *Phys. Rev.*, **178,** 218 (1969).
188. W. F. J. Evans and A. Vallance Jones, *Can. J. Phys.*, **43,** 697 (1965).
189. R. F. Hampson, Jr., and H. Okabe, *J. Chem. Phys.*, **52,** 1930 (1970).
190. I. D. Clark and R. P. Wayne, *Proc. Roy. Soc.* (London), **A316,** 539 (1970).
191. S. V. Filseth, F. Stuhl, and K. H. Welge, *J. Chem. Phys.*, **52,** 239 (1970).
192. F. Stuhl and K. H. Welge, *Can. J. Chem.*, **47,** 1879 (1969).
193. R. A. Young, G. Black, and T. G. Slanger, *J. Chem. Phys.*, **50,** 309 (1969).
194. N. P. Carleton, F. J. LeBlanc, and O. Oldenberg, *Bull. Am. Phys. Soc.*, **11,** 503 (1966).
195. W. D. McGrath and J. J. McGarvey, *Planet. Space Sci.*, **15,** 427 (1967).
196. K. F. Preston and R. J. Cvetanovich, *J. Chem. Phys.*, **45,** 2888 (1966).
197. H. Yamazaki and R. J. Cvetanovich, *J. Chem. Phys.*, **40,** 582 (1964).
198. H. Yamazaki and R. J. Cvetanovich, *J. Chem. Phys.*, **41,** 3703 (1964).
199. J. O. Sullivan and P. Warneck, *J. Chem. Phys.*, **46,** 953 (1967).
200. W. B. DeMore and O. F. Raper, *Astrophys. J.*, **139,** 1381 (1964).
201. W. B. DeMore and O. F. Raper, *J. Chem. Phys.*, **44,** 1780 (1964).
202. T. Izod and R. P. Wayne, *Chem. Phys. Lett.*, **4,** 208 (1969).
203. W. B. DeMore, *J. Chem. Phys.*, **52,** 4309 (1970).

204. I. D. Clark, *Chem. Phys. Lett.*, **5,** 317 (1970).
205. B. G. Hunt, *J. Geophys. Res.*, **71,** 1385 (1966).
206. J. Hampson, Canadian Armaments Research and Development Establishment Report TN 1627/64, 1964.
207. E. Hesstvedt, *Geofys. Norveg.*, **27,** 1 (1967).
208. J. Anderson, Ph.D. dissertation, University of Colorado, 1970.
209. A. Ohmholt, *J. Atmos. Terrest. Phys.*, **10,** 320 (1957).
210. D. M. Hunten, *Ann. Geophys.*, **14,** 167 (1958).
211. R. S. Narcisi and A. D. Bailey, *J. Geophys. Res.*, **7,** 3687 (1965).
212. M. Ya. Fogel, *Sov. Phys. Usp.*, **3,** 390 (1960).
213. J. K. Layton, *J. Chem. Phys.*, **47,** 1869 (1967).
214. E. E. Ferguson, F. C. Fehsenfeld, and J. Whitehead, *J. Geophys. Res.*, **S.P.75,** 4366 (1970).
215. A. Dalgarno, *Can. J. Chem.*, **47,** 1723 (1969).
216. B. Moiseiwitsch and S. Smith, *Rev. Mod. Phys.*, **40,** 238 (1968).
217. W. L. Fite, R. F. Stebbings, and R. T. Brackmann, *Phys. Rev.*, **116,** 356 (1959).
218. S. Geltman and P. G. Burke, *J. Phys. B*, **3,** 1062 (1970).
219. D. Hils, H. Kleinpoppen, and H. Koschmieder, *Proc. Phys. Soc.* (London), **89,** 35 (1966).
220. I. P. Zapesochnyi and L. L. Shimon, *Opt. Spectry.*, **19,** 268 (1965).
221. H. K. Holt and R. Krotkov, *Phys. Rev.*, **144,** 82 (1966).
222. A. V. Phelps, *Phys. Rev.*, **99,** 1307 (1955).
223. D. T. Stewart and E. Gabathuler, *Proc. Phys. Soc.* (London), **A72,** 287 (1958).
224. D. T. Stewart, *Proc. Phys. Soc.* (London), **A69,** 437 (1956).
225. E. L. Chaney and L. G. Christophorou, *J. Chem. Phys.*, **51,** 883 (1969).
226. N. Basco and R. A. W. Norrish, *Discussions Faraday Soc.*, **33,** 99 (1962).
227. R. March, S. Furnival, and H. I. Schiff, *Photochem. Photobiol.*, **4,** 971 (1965).
228. T. G. Slanger, B. Wood, and G. Black, *J. Geophys. Res.*, **76,** 8430 (1971).
229. K. Anlauf, D. Maylotte, J. Polanyi, and R. Bernstein, *J. Chem. Phys.*, **51,** 5716 (1969).
230. J. Polanyi and D. Tardy, *J. Chem. Phys.*, **51,** 5717 (1969).
231. E. Bauer and F. W. Cummings, *J. Chem. Phys.*, **36,** 618 (1962).
232. A. B. King and C. Gatz, *J. Chem. Phys.*, **37,** 1566 (1962).
233. C. Kenty, *J. Chem. Phys.*, **37,** 1567 (1962).
234. H. P. Broida, *J. Chem. Phys.*, **36,** 444 (1962).
235. W. F. Sheridan, O. Oldenberg, and N. P. Carleton, in *Atomic Collision Processes*, edited by M. R. C. McDowell, North-Holland, Amsterdam, 1964, p. 440.
236. M. N. Hirsch, P. N. Eisner, and J. A. Selvin, G. C. Dewey Corp. Report R-173-4 (AD607071), 1964.
237. M. Cher and C. Hollingsworth, *Can. J. Chem.*, **47,** 1937 (1969).
238. R. F. Stebbings, J. Rutherford, and B. R. Turner, *Planet. Space Sci.*, **13,** 1125 (1965).
239. B. L. Donnally, T. Clapp, W. Sawyer, and M. Schultz, *Phys. Rev. Lett.*, **12,** 502 (1964).
240. I. M. Campbell and B. A. Thrush, *Chem. Commun.*, **12,** 250 (1965).

241. E. Bauer and M. Salkoff, *J. Chem. Phys.*, **33**, 1202 (1960).
242. D. W. Setser and B. A. Thrush, *Proc. Roy. Soc.* (London), **A288**, 275 (1965).
243. R. A. Carabetta and W. E. Kaskan, Eleventh Int'l Symposium on Combustion, The Combustion Inst., Pittsburgh, Pa. p. 321 (1967). Contract No. DA-31-124-ARO-D-214, 1965.
244. D. R. Bates, *Earth is a Planet*, Univ. of Chicago Press, Chicago, 1960, p. 576.
245. F. E. Belles and M. R. Lauver, *J. Chem. Phys.*, **40**, 415 (1964).
246. K. Omidvar, *Phys. Rev.*, **140**, A26 (1965).
247. M. J. Seaton, *Monthly Notices Roy. Astron. Soc.*, **127**, 177 (1964).
248. L. W. Dotchin, E. L. Chapp, and D. J. Pegg, *J. Chem. Phys.*, **59**, 3960 (1973).
249. R. F. Holland and W. B. Maier II, *J. Chem. Phys.*, **56**, 5229 (1972).
250. J. R. Peterson and J. T. Mosely, *J. Chem. Phys.*, **58**, 172 (1973).
251. D. C. Cartwright, *J. Chem. Phys.*, **58**, 178 (1973).
252. E. M. Weinstock, R. N. Zare, and L. A. Melton, *J. Chem. Phys.*, **56**, 3456 (1972).
253. M. Jeunehomme and A. B. F. Duncan, *J. Chem. Phys.*, **41**, 1692 (1964).
254. A. R. Fairbairn, *J. Chem. Phys.*, **60**, 521 (1974).
255. C. E. Johnson and R. I. VanDyke, Jr., *J. Chem. Phys.*, **56**, 1506 (1972).
256. J. H. Moore, Jr., and D. W. Robinson, *J. Chem. Phys.*, **48**, 4870 (1968).
257. G. Black, R. Sharpless, and T. Slanger, *J. Chem. Phys.*, **58**, 4792 (1973).
258. E. R. Fisher and E. Bauer, *J. Chem. Phys.*, **57**, 1966 (1972).
259. T. Slanger and G. Black, *J. Chem. Phys.*, **60**, (1974).
260. E. Jamshidi, E. R. Fisher, and R. H. Kummler, *J. Geophys. Res.*, **78**, 6151 (1973).
261. R. H. Kummler and M. H. Bortner, *Space Res.*, **12**, 711 (1972).
262. J. B. Kumer and T. C. James, *J. Geophys. Res.*, **79**, 638 (1974).
263. T. E. Van Landt and T. F. O'Malley, *J. Geophys. Res.*, **78**, 6818 (1973).
264. W. Varnum, *Planet. Space Sci.*, **20**, 1865 (1972).
265. R. McNeal, M. Whitson, Jr., and C. Cook, *Chem. Phys. Lett.*, **16**, 507 (1972).
266. E. Brieg, M. Brenner, and R. NcNeal, *J. Geophys.* **78**, 1225 (1973).
267. P. J. Crutzen, *J. Geophys. Res.*, **76**, 7311 (1971).
268. G. Paraskevopoulos and R. J. Cvetanovic, *Chem. Phys. Lett.*, **9**, 603 (1971).
269. D. Garvin and R. F. Hampson, Proceedings of the Second Conference on the Climatic Impact Assessment Program, A. Broderick, editor, U.S. Department of Transportation Report DOT-TSC-OST-73-4, April (1973).
270. R. F. Hampson, W. Braun, R. Brown, D. Garvin, J. T. Herron, R. E. Huie, M. J. Kurylo, A. H. Laufor, J. D. McKinley, H. Okake, M. D. Scheer, and W. Tsang, *J. Phys. Chem.*, Ref. Data, **2**, 267 (1973).
271. M. Nicolet, Proceedings of the Second Conference on the Climatic Impact Assessment Program, A. Broderick, editor, U.S. Department of Transportation Report DOT-TSC-OST-73-4, April (1973).
272. R. H. Kummlėr, M. H. Bortner, and T. Baurer, *Environ. Sci. Technol.* **3**, 248 (1969).
273. R. H. Kummler and T. Baurer, *J. Geophys. Res.*, **78**, 5306 (1973).
274. W. Chameides and J. C. G. Walker, *J. Geophys. Res.* **78**, 8751 (1973).
275. D. Garvin and R. F. Hampson, National Bureau of Standards Report NBS1R **74**, 430 January (1974).
276. R. Atkinson and K. Welge, *J. Chem. Phys.* **57**, 3689 (1972).
277. G. Black, R. Sharpless, and T. Slanger, *J. Chem. Phys.* **58**, 4792 (1973).

CHAPTER SEVEN

APPLICATIONS TO LASERS

Robert H. Bullis

United Aircraft Research Laboratories, East Hartford, Connecticut

Contents

I. INTRODUCTION

One of the modern day examples of the high efficiency of energy transfer processes in gases is that associated with gas discharge lasers. Of the host of electrically excited laser systems that presently exist, the CO and CO_2 lasers stand out as having achieved high optical power outputs at high efficiencies. For the CO_2 laser, continuous-wave (cw) optical power outputs in excess of 25 kW at efficiencies of 17% have been reported [1], while experimental efficiencies in CO lasers of nearly 50% have been achieved [2]. This type of performance is a direct reflection of the highly efficient nature of the collisional coupling between electrons and the vibrationally excited states of N_2, CO, and CO_2. In this chapter the key processes influencing the performance of CO_2 gas discharge lasers will be described along with a brief description of the evolution of key scientific events that led to the discovery and development of this type of laser. The reader who is interested in pursuing the details of the collisional phenomena governing CO laser operation, which are similar in major respects to the phenomena occurring in CO_2 lasers, is referred to refs. 2–7.

Numerous energy transfer processes take place in typical CO_2–N_2–He laser discharges. In addition to direct excitation, ionization, and vibrational energy transfer processes, which are described in detail in Sections II and III, the ultimate performance of the CO_2 gas discharge laser has been found to be profoundly influenced by plasma chemical reactions taking place between minority species produced in the discharge. The nature of these reactions and their influence on discharge current-voltage and stability characteristics is developed in Sections IV and V. The basic collision phenomena governing CO_2–N_2–He gas discharge laser performance can best be understood by reference to Figure 7.1. The energy distributions within the combined CO_2–N_2–He molecular system as shown in this figure can be characterized by three temperatures. This model, which was developed by Fowler [8], is similar to models proposed by Moore, Wood, Hu, and Yardley [9] and Gordietz, Sobolev, Sokovikov, and Shelepin [10]. The energy distribution within the asymmetric stretch mode of CO_2 denoted by ($00n$) is characterized by temperature T_2, while that of the symmetric stretch mode of CO_2 denoted by ($l00$) is characterized by temperature T_1. Because the symmetric stretch and bending modes of CO_2 are coupled together collisionally [11, 12] and by the Fermi resonance [13] that exists between them, the energy distribution within the bending mode, which is denoted by ($0m0$), is also characterized by the temperature T_1. Lastly, the nitrogen vibrational energy distribution is characterized by temperature T_N. It is possible to represent the energy distribution within each of these modes by a temperature because each vibrational mode can be considered to be a simple harmonic oscillator with intermolecular vibration transfer resonant in character. Therefore, energy transfer within a mode proceeds at a rate that is faster than transfer to other modes [14]. Through collisional processes a population inversion is established between the asymmetric stretch (001) level and symmetric stretch (100) level of CO_2. In addition, electron and heavy particle collisions also serve to influence the populations of each of the remaining levels of the CO_2–N_2 system. The key energy transfer channels governing laser performance are denoted in Fig. 7.1 by numbers 1–7. These channels will be described in detail in Sections II and III. From this brief description it becomes obvious that collisional phenomena governing laser performance are quite complex. For this reason, it is not too surprising that from the time of the original discovery of stimulated emission in CO_2 at the milliwatt level by Patel [15], it required a period of several years to achieve increases in laser output to levels on the order of kilowatts [1, 16–18].

As indicated by Patel [19], the attractive feature that prompted him to investigate the feasibility of producing a molecular gas discharge laser with output in the infrared was the favorable features of the energetics of molecular systems in comparison to those of atomic systems. In an atomic system the upper laser level typically lies near the ionization limit. As a consequence, collisional processes occurring in a discharge are highly nonselective in populating this level because of the presence of numerous competing nearby levels. Further, the quantum efficiency, which is the ratio of the energy associated with the laser transition to the energy of the upper laser level, is typically small. In contrast, in molecular systems the upper laser level can be located potentially quite close to the ground state. In this situation there are fewer competing levels and the quantum efficiency can be quite high. The CO_2 system that Patel chose to investigate represents an excellent example of this situation, as can be partially recognized from the energy-level diagram of Figure 7.1.

Initial investigations which led to the observation of laser action near 10.6 μm were carried out in CO_2 gas discharges [15, 20]. Subsequent investigations were conducted in a mixing laser configuration with N_2 and CO_2 gas constituents [21]. In this configuration N_2 was excited in a discharge and then mixed with CO_2 in an optical cavity. In this type of laser the population inversion in CO_2 was produced by vibrational energy transfer collisions taking place between the N_2 ($v = 1$) level of the ground electronic state and the asymmetric (001) level of CO_2. This is represented as channel 2 in Figure 7.1. In a somewhat similar experiment Legay and Legay–Sommaire [22, 23] observed vibrational energy transfer from the $N_2(v = 1)$ level to the ground state of CO as well as to CO_2 and N_2O.

Despite the fact that successful laser action had been achieved in CO_2 and

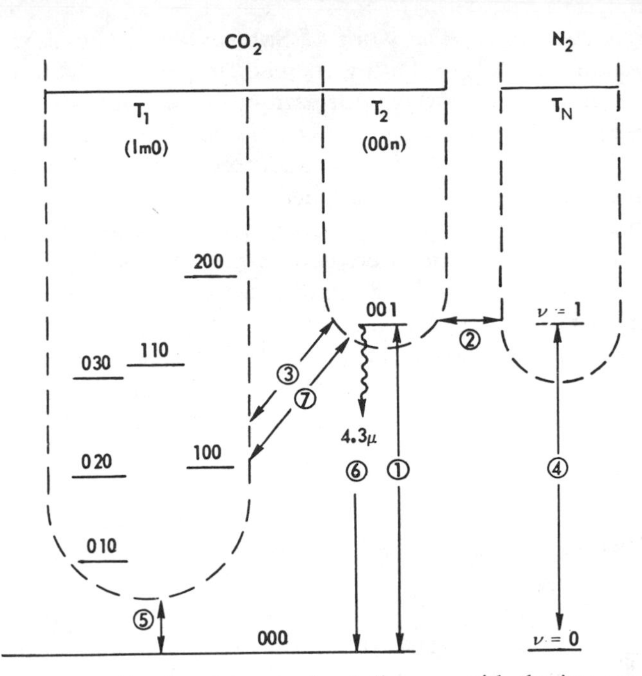

Figure 7.1 CO_2–N_2 vibrational energy level diagram with the important energy-transfer processes indicated [8].

CO_2–N_2 mixtures, little at this time was clearly understood about the actual processes producing the vibration excitation in either CO_2 or N_2. From pulsed CO_2 discharge measurements, Patel [24] concluded from observed time delays between discharge and laser output that recombination and cascade processes rather than direct electron impact excitation from the ground state of CO_2 were responsible for producing the observed population inversions. In contrast, Sobolev and Sokovikov [25, 26] proposed that direct electron excitation of N_2 and CO_2 was the key process of importance in producing the observed population inversions. The basis for their proposal was the low-energy electron collisional measurements in N_2 and CO reported by Schulz [27–29]. Furthermore, experimental evidence existed at this time which suggested that electron energy distribution functions in pure N_2 discharges typical of the laser systems under investigation were highly non-Maxwellian [30]. This occurs because of the large cross sections for vibrational excitation at energies in the 1–3 eV range, as discussed by Engelhardt, Phelps, and Risk [31]. Because of the complex nature of the electron collisional processes in N_2 and CO, the fact that electron energy distribution functions were expected to be highly non-Maxwellian, and because only meager data existed on CO_2 collisional processes, only qualitative estimates could be made of excitation rates. The work of Sobolev and Sokovikov [32] is an excellent example of the initial attempts made to predict excitation rates and laser performance based on this information. Furthermore, in this work the importance of other gases such as He as relaxants for the lower laser level was recognized. However, it was not until additional collisional information on CO_2 became available [33–37] that detailed analyses could be made of electron kinetic processes in laser plasmas. The first work to be reported in which detailed excitation rates were determined for laser mixtures was that of Nighan [38–40]. With the development of this detailed excitation rate information, which will be described in Section II, it was possible with the heavy particle kinetic analysis of Fowler [8], which will be described in Section III, to analyze the CO_2–N_2–He gas discharge laser in a highly quantitative fashion as described by Bullis, Nighan, Fowler, and Wiegand [41].

II. ELECTRON KINETIC PROCESSES

The work of Sobolev and Sokovikov [32] represents the first attempt to analyze in detail electron excitation processes in CO_2 laser discharges. The significance of this work is the recognition of the importance of direct electron vibrational pumping and that the electron energy distribution functions in typical laser discharges are highly non-Maxwellian. The work of Engelhardt, Phelps, and Risk [31] supported the conclusion that electron distribution functions would be highly non-Maxwellian in pure N_2 and CO discharges, as did the experimental probe measurements in N_2 discharges reported by Swift [30]. Early swarm information obtained by Haas [42] in N_2 suggested the presence of a large energy loss channel in N_2 at low energies. More detailed resolution of the electron energy-loss processes in N_2 and CO were obtained by Schulz using a trapped-electron experiment [27] and subsequently with high-resolution beam measurements [28, 29]. The large cross sections for vibrational excitation observed in these measurements were attributed to the presence of a temporary negative ion state. The analysis of swarm data by Frost and Phelps [43] and Engelhardt, Phelps, and Risk [31] served to identify the important collisional processes and energy loss channels for N_2. Similar developments that followed this original work in N_2 served to identify the collisional processes of significance in CO_2. Analysis of swarm data by Hake and Phelps [33] and Phelps [34] in conjunction with beam measurements by Boness and Hasted [36] and Boness and Schulz [37] aided in the development of the understanding of CO_2 collisional processes. Unfortunately, the description of CO_2 low-energy collisional processes based on this information was not as complete as the understanding of the N_2

system. More recent high-resolution beam measurements by Andrick, Danner, and Ehrhardt [35] have served to provide much needed additional information. With these cross section data and the techniques developed by Phelps et al. in analysis of swarm experiments through detailed solution of the Boltzmann equation, the techniques to analyze CO_2 molecular laser plasmas were available. Application of these techniques by Nighan [38–40] led to the development of a clear understanding of electron collisional processes in CO_2–N_2–He discharges.

A. Electron Distribution Functions

Plasmas typical of CO_2 laser discharges operate over a pressure range from 1 Torr to several atmospheres with degrees of ionization, that is, n_e/N (the ratio of electron density to neutral density) in the range from 10^{-8} to 10^{-6}. Under these conditions the electron energy distribution function is highly non-Maxwellian. As a consequence it is necessary to solve the Boltzmann transport equation based on a detailed knowledge of the electron collisional channels in order to establish the electron distribution function as a function of the ratio of the electric field to the neutral gas density, E/N, and species concentration. Development of the fundamental techniques for solution of the Boltzmann equation are presented in detail by Shkarofsky, Johnston, and Bachynski [44] and Holstein [45].

The Boltzmann equation for a spatially uniform gas in the presence of a steady electric field $\mathbf{E}$ is of the form [44, 45]

$$\frac{-e\mathbf{E}}{m} \cdot \nabla_v f(\mathbf{v}) = \left(\frac{\delta f(\mathbf{v})}{\delta \mathbf{t}}\right)_c, \tag{1}$$

where e is electronic charge, m is the electron mass, v is the electron velocity, and $f(\mathbf{v})$ is the electron velocity distribution function. The right-hand side of this equation represents the rate of change of $f(\mathbf{v})$ produced by electron collisions of all kinds. When the electron velocity distribution is nearly spherically symmetric, $f(\mathbf{v})$ can be separated into two parts as follows:

$$f(\mathbf{v}) = f_0(v) + \mathbf{f}_1(v) \cdot \frac{\mathbf{v}}{v}, \tag{2}$$

where $f_0(v)$ is isotropic, and $\mathbf{f}_1(v)$ is the small anisotropic portion of the distribution. Using this expansion, equation (1) may be resolved into two coupled equations following the development of refs. 44 and 45:

$$\frac{-e\mathbf{E}}{3mv^2} \cdot \frac{d}{dv} [v^2{}_1(\mathbf{f}v)] = \left(\frac{\delta f_0(v)}{\delta t}\right)_c, \tag{3}$$

$$\frac{-e\mathbf{E}}{m} \frac{df_0(v)}{dv} = \left(\frac{\delta \mathbf{f}_1(v)}{\delta t}\right)_c. \tag{4}$$

A necessary condition for the two-term expansion of the distribution function of equation (2) to be valid is that the electron collision frequency for momentum transfer must be larger than the total electron collision frequency for excitation for all values of electron energy. Under these conditions electron-heavy particle momentum-transfer collisions are of major importance in reducing the asymmetry in the distribution function. In many cases as pointed out by Phelps in ref. 34, this condition is not met in the analysis of N_2, CO, and CO_2 transport data primarily because of large vibrational excitation cross sections. The effect on the accuracy of the determination of distribution functions as a result is a factor still remaining to be assessed.

Following the development of refs. 44 and 45, equation (4) reduces to

$$\left(\frac{\delta \mathbf{f}_1(v)}{\delta t}\right)_c = -\sum \nu_{ej}(v)\mathbf{f}_1(v), \tag{5}$$

where $\nu_{ej}(v)$ is the electron momentum-transfer collision frequency for the jth species of heavy particle,

$$\nu_{ej}(v) = N_j Q_{ej}(v)v, \tag{6}$$

where N_j is the density of the jth species and $Q_{ej}(v)$ is the momentum-transfer cross section. By transforming to a new independent variable $u = mv^2/2e$, that is, electron energy in eV, and using equations (3)–(5), the isotropic portion of the distribution function becomes

$$-\frac{E^2}{3}\frac{d}{du}\left\{u\frac{df_0}{du}\left[\sum_j N_j Q_{ej}(u)\right]^{-1}\right\} = \left(\frac{\delta f_0}{\delta t}\right)_c. \tag{7}$$

Here the distribution function has been normalized in a fashion such that the fraction of electrons with energies between u and $u + du$ is $u^{1/2}f_0(u)\,du$. This results in

$$\int_0^\infty u^{1/2} f_0(u)\,du = 1. \tag{8}$$

The right-hand side of equation (7) represents the combined effect on the isotropic part of the distribution function of elastic and inelastic electron–neutral, electron–electron, and electron–ion collisions. Fortunately, for the CO_2 laser cases of interest the significant collisional cross sections of importance as outlined above have been determined from the analyses of swarm data of refs. 31 and 33 for the energy range from 0.5 to 3.0 eV. The effects of elastic and rotational excitation collisions are negligible in comparison to energy-transfer collisions resulting in vibrational and electronic excitation. Furthermore, as Nighan [38] has shown, the thermalizing influence of electron–electron collisions is unimportant because of the low degree of ionization typically present in CO_2 laser discharge plasmas. The effect on the isotropic portion of the distribution function of collisional processes in which an appreciable fraction of the total electron energy can be transferred in a single collision, such as is typical of vibrational and electronic collisions, has been analyzed by Holstein [45] and is expressed in the following manner:

$$\left(\frac{\delta f}{\delta t}\right)_c = \sum_{j,k} N_j[(u + u_{jk})f(u + u_{jk})Q_{jk}(u + u_{jk}) - uf(u)Q_{jk}(u)], \tag{9}$$

where $f(u)$, following the development of Nighan [40], represents the isotropic portion of the distribution function, $Q_{jk}(u)$ is the electron cross section for excitation or deexcitation of the kth vibrational or electronic level of the jth neutral species, and u_{jk} is the energy loss or gain for this species. From this development a working equation for $f(u)$ can be obtained by combining equations (7) and (9) and integrating once. The total result is then divided by the total gas density to cast the equation in the following form:

$$-\frac{(E/N)^2}{3} u \frac{df}{du}\left[\sum_j \delta_j Q_{ej}(u)\right]^{-1} = \sum_{j,k} \delta_j \int_u^{u+u_{jk}} uf(u)Q_{jk}(u)\, du, \tag{10}$$

where δj is the fractional concentration of the jth neutral species, that is, N_j/N. For a given value of E/N and mixture, $f(u)$ is determined from equation (10). It should be noted that $f(u)$ is not dependent on absolute particle density, but is dependent on fractional mixture composition, that is, δj. As indicated previously, the input data needed to obtain a meaningful solution of equation (10) are the cross sections for the various collisional channels in N_2 and CO_2. Shown in Figure 7.2 is the structure of the electron–vibrational excitation cross sections for N_2, CO, and CO_2 obtained from refs. 31 and 33–35. A detailed discussion of each of the collisional channels considered in this analysis is contained in ref. 40 and the above indicated references. It is important to note that the magnitudes of the vibrational excitation cross sections are quite large ($1–4 \times 10^{-16}$ cm^2) in the 1–5 eV range. Furthermore, the cross section for direct vibrational excitation of the CO_2 asymmetric stretch mode is of significance in 0.5–1.0 eV range. This is especially true when this cross section is weighted by the magnitude of the energy exchange associated with this process (0.29 eV).

Computed values of the distribution function over the energy range of interest to laser applications for a CO_2–N_2–He mixture having a composition by density of 1–1–8, respectively [46], are presented in Figure 7.3 [40]. Since a Maxwellian distribution would appear as a straight line in this figure it is clear, as expected, that the electron energy distributions are non-Maxwellian. As pointed out in a more recent analysis by Lowke, Phelps, and Irwin [47] that parallels the approach described herein, the fact that the distribution function is non-Maxwellian does not significantly influence the determination of plasma parameters such as the electron drift velocity. However, as shown

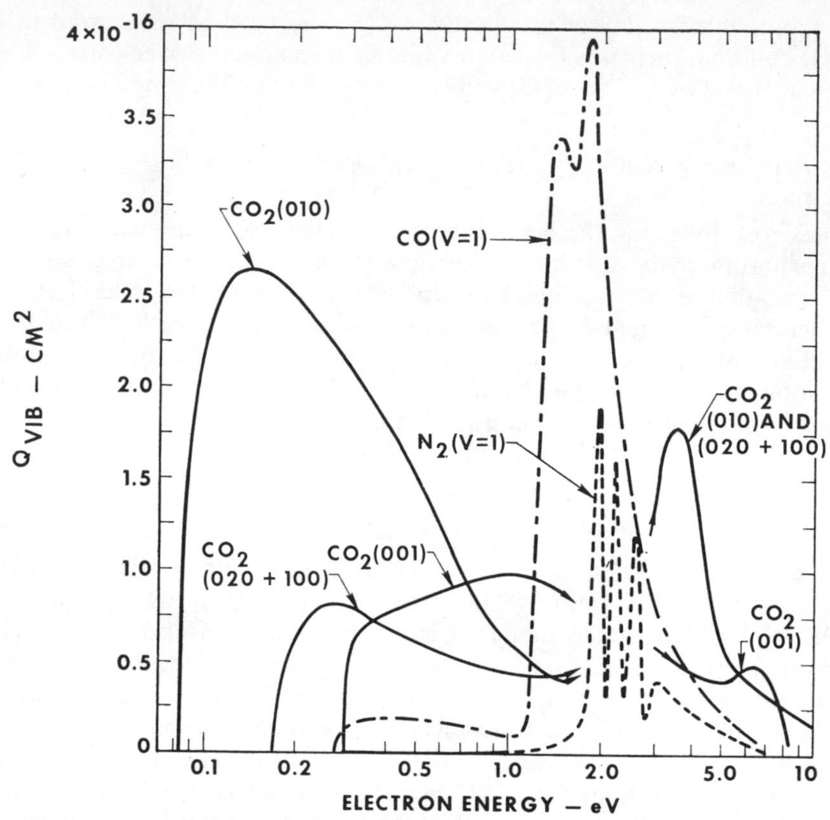

Figure 7.2 Vibration cross sections for CO_2, N_2, and CO from refs. 31 and 33–35. Reference [47] contains in tabulated form the CO_2 cross-section data of ref. 35.

in ref. 47, collisional processes that critically depend on the characteristics of the electron population in the tail of the distribution, such as the ionization coefficient, α, are particularly sensitive to the precise nature of the distribution function for E/N values of importance to laser applications. This sensitivity of the ionization coefficient is clearly demonstrated by the results presented in Figure 9 of ref. 47. Further, calculations by Nighan [48] of the effective N_2 vibrational excitation rate, ν_{eff}/N, exhibit a similar sensitivity to choice of distribution function as well as laser mixture, as shown in Figure 7.4. In particular, the assumption of a Maxwellian distribution in the low-energy range typical of electron beam augmented laser operating conditions results in a more severe overestimate of the effective excitation rate than at higher energies, as shown in this figure. The influence of fractional mixture concentration can also be of significance as can be seen by comparing the pure N_2 rates to those for a 1–1–8 laser mixture.

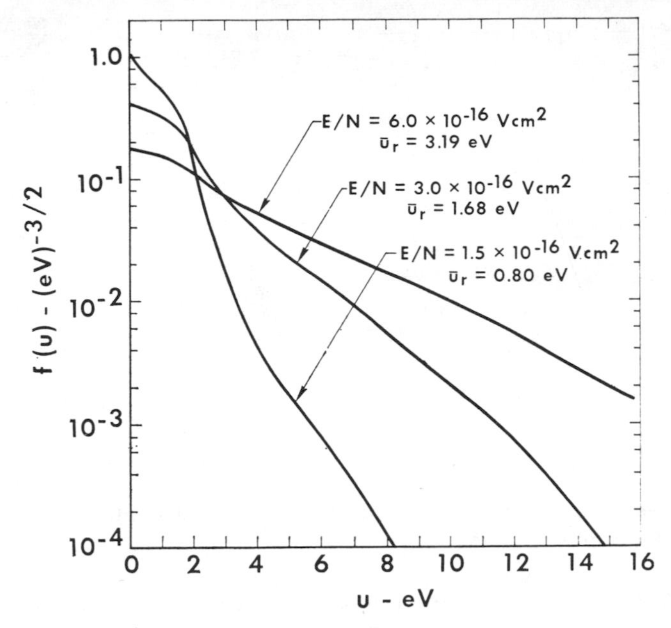

Figure 7.3 Computed electron energy distributions for a 1–1–8 laser mixture (41).

Since it is estimated that the vibrational temperatures T_N and T_2 can reach several thousand degrees, the population of vibrationally excited molecules becomes significant in typical laser discharges. Furthermore, the vibrational populations are strongly influenced by the extraction of laser optical power from the discharge. Therefore, it is expected that the electron kinetics will be influenced not only by the presence of vibrationally excited molecules in the discharge, but that these kinetics will also be a function of the operating conditions of the laser.

It is difficult to obtain a precise assessment of the influence of superelastic collisions on laser electron kinetics since the theoretical superelastic cross sections of Chen [49] for N_2 are the only information available on this type of processes for laser species. Nighan [39], using Chen's N_2 data, has calculated the effect a change in vibrational population has on the electron energy distribution function in N_2. Shown in Figure 7.5 are electron energy distribution functions calculated for two typical N_2 vibrational populations corresponding to the vibrational temperatures indicated. As can be seen from this figure, the major change occurs in the high-energy tail of the distribution function, with little change occurring in the reduced average energy $\bar{u}_r$, which is defined such that

$$\bar{u}_r = \frac{2}{3}\int_0^\infty u^{3/2} f(u)\, du, \tag{11}$$

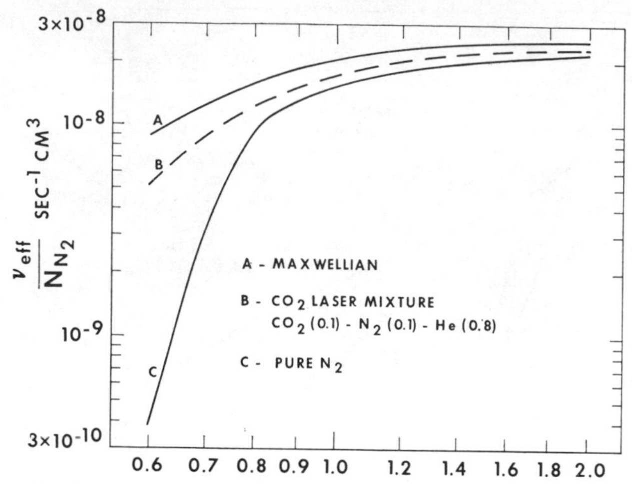

Figure 7.4 Dependence of effective electron–N_2 vibrational excitation rate on reduced average electron energy [48].

where for a Maxwellian distribution $\bar{u}_r = kT_e/e$ and T_e is the electron temperature. Changes of the type shown in Figure 7.5 are most expected to influence the electronic excitation and ionization processes in the discharge. For vibrational populations corresponding to vibrational temperatures of less than 1000°K, the electron energy distribution function was found to be unaffected by changes in vibrational population. The distribution function was found, however, to be more sensitive to changes in vibrational temperature for reduced average electron energies below 1.5 eV, as shown in ref. 40. Using the distributions shown in Figure 7.5, fractional power transfer to the N_2 vibrational levels was calculated using the technique outlined in the following section. It was found that the fractional power transfer increased from 50% to 56.5% of the total as the vibrational temperatures decreased from 3000 to 1000°K [39]. The changes in the distribution produced by changes in the vibrational population also suggest that under optical power extraction conditions, an increase in the discharge E/N will be required to maintain the same level of ionization. Probe measurements by Bletzinger and Garscadden [50] of both electron distribution functions and discharge electric fields under various discharge operation conditions serve to confirm the above findings.

Since equation (10) suggests that the two parameters of importance in the determination of the distribution function are mixture composition and E/N, a series of calculations were conducted [40] to establish the influence of mixture composition on reduced average energy $\bar{u}_r$ for a fixed E/N. Shown in Figure 7.6 is the rather profound change in E/N ratio that is required to yield

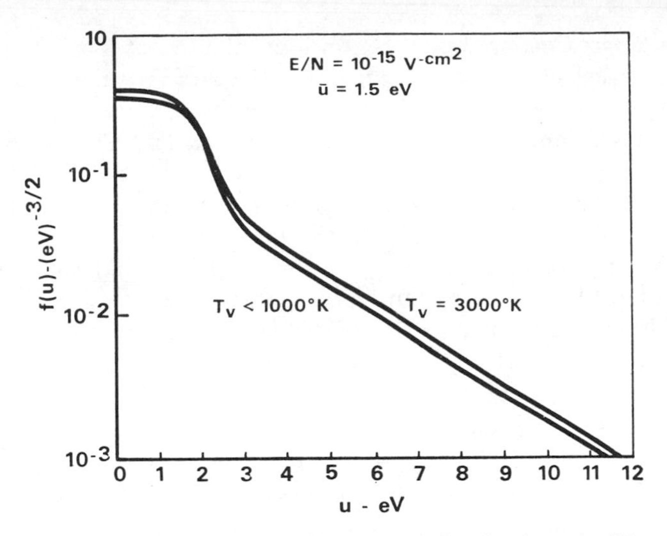

Figure 7.5 Dependence of electron energy distributions in N_2 on vibrational temperature [39].

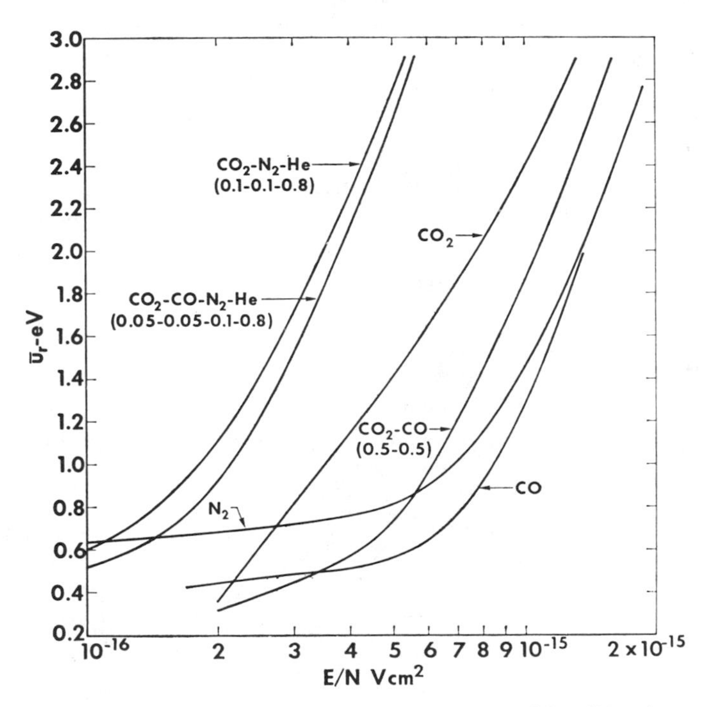

Figure 7.6 Variation of reduced average energy, $\bar{u}_r$, with E/N ratio and discharge mixture [40].

a typical $\bar{u}_r$ of approximately 1.0 eV for various mixtures of interest in laser discharges. These calculations have been carried out for mixtures containing significant concentrations of CO, which has been found to be an important discharge generated impurity species [51]. In pure N_2 and CO the lack of sensitivity to changes in E/N for values below 5×10^{-16} V cm^2 as shown in Figure 7.6 is a direct reflection of the inability of the electrons to penetrate the barrier produced by the large vibrational cross section in these gases in the 1.5–2.0 eV range (Figure 7.2). These results in general indicate that extreme caution must be exercised in interpretation of data from laser systems operating with different mixtures. In particular, one common misunderstanding is the concept that the presence of helium [26] influences the electron distribution function. Since direct ionization processes in both CO_2 and N_2 occur at 13.8 and 15.6 eV, respectively, while the first excitation potential of helium is 19.6 eV, it is not surprising to find that helium does not contribute in a significant way to ionization processes in the ranges of interest to most CO_2 laser applications. The presence of helium does, however, alter the total neutral species concentration and as a result the E/N ratio required to yield a given average energy.

B. Fractional Power Transfer

Probably one of the more important results obtained from a knowledge of the distribution functions for laser conditions has been the development of fractional power transfer information. From this type of information, which was first presented in ref. 31 for N_2, it is possible to determine directly the conditions required to optimize laser operating characteristics for maximum efficiency. Fractional power transfer information is obtained by performing an energy balance for the electrons by integrating the electron kinetic equation over all electron energies. This is accomplished by multiplying equation (7) by $e(2e/m)^{1/2}u\,du$, dividing by total gas density N, and integrating over all electron energies. In this development $[\delta f(u)/\delta t]_c$ is given by equation (9). This results in the following expression, as developed by Nighan [40]:

$$-e\left(\frac{2e}{m}\right)^{1/2}\frac{(E/N)^2}{3}\int_0^\infty u\,\frac{df}{du}\left[\sum_j \delta_j Q_{ej}(u)\right]^{-1} du$$

$$= e\left(\frac{2e}{m}\right)^{1/2}\sum_{j,k}\delta_j u_{jk}\int_0^\infty uf(u)Q_{jk}(u)\,du. \quad (12)$$

The rate of energy deposition from the electric field, which is the electron-joule heating expressed on a per-electron-per-neutral-particle basis, is represented by the left-hand side of this equation. The term on the right-hand side of equation (10), which is similar in form to the right-hand side of equation (12), corresponds to the net rate at which electrons exchange energy through vibrational and electronic excitation and/or deexcitation collisions with molecules. Therefore, by introducing terms for electron drift velocity v_D and normalized electron–molecule collision frequency ν_{jk}/N_j the energy balance can be simplified in the following manner:

Defining electron drift velocity as

$$v_D = -\left(\frac{2e}{m}\right)^{1/2} \frac{E/N}{3} \int_0^\infty u \frac{df}{du} \left[\sum_j \delta_j Q_{ej}(u)\right]^{-1} du \tag{13}$$

and the normalized collision frequency as

$$\frac{\bar{\nu}_{jk}}{N_j} = \left(\frac{2e}{m}\right)^{1/2} \int_0^\infty u f(u) Q_{jk}(u)\, du, \tag{14}$$

the energy balance equation becomes

$$e v_D \frac{E}{N} = \sum_{j,k} \delta_j e u_{jk} \bar{\nu}_{jk}/N_j. \tag{15}$$

A major point of importance is that the electron energy balance, and therefore the average electron energy, in weakly ionized plasmas is independent of electron density. As a consequence, the energy balance can be treated on a per electron basis. From equation (15) the total power density can be established by multiplying by the product of electron and neutral densities.

Direct application of equation (15), which is the electron energy conservation equation, to the CO_2–N_2–He laser mixture for density concentrations in the ratio of 1–1–8, respectively, results in the fractional power transfer information shown in Figure 7.7 [41]. Power transfer information for similar collisional channels has been combined for ease of interpretation in this figure. For reduced average energies ranging between 0.4 and 3.0 eV, electron energy-transfer processes change from vibrational excitation in both N_2 and CO_2 at 0.4 eV to predominantly electronic excitation processes at 3.0 eV. For reduced average energies of 0.8 eV, the major electron energy loss is to the vibrational levels of N_2, with a significant fraction of energy being transferred directly to the CO_2 (001) upper laser level. Losses to the combined bending and symmetric stretch modes, although of importance, do not represent a significant energy loss channel at this energy. As the electron energy is increased, electron energy loss to vibrational levels decreases while the influence of electronic excitation loss processes becomes more important. Since a dc discharge must operate in a manner in which excitation and ionization processes balance charged-particle loss processes, it is not too surprising

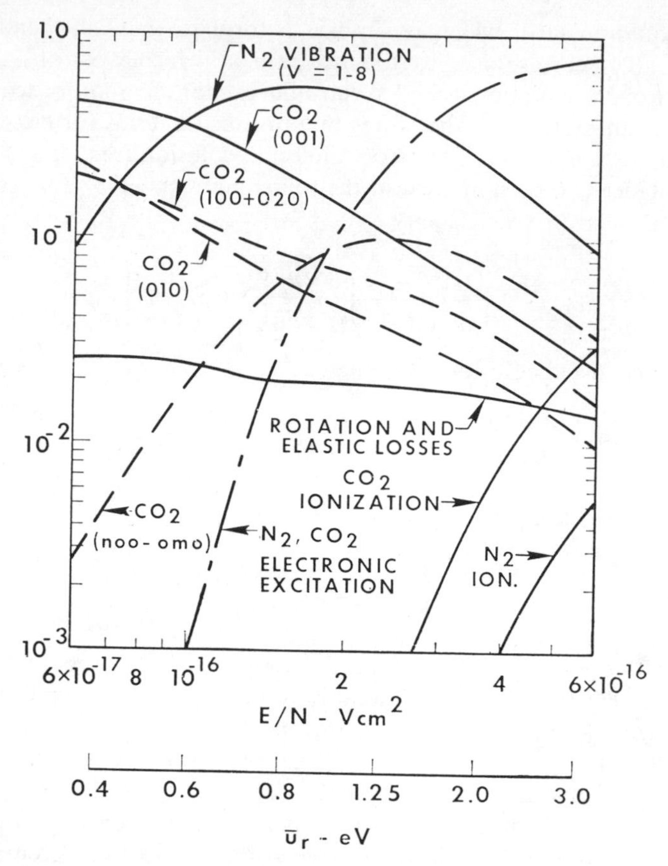

Figure 7.7 Fraction power transfer for a 1–1–8 laser mixture [41].

to find typical operating E/N values for the 1–1–8 mixture to be in the range of 1.5–2 × 10^{-16} V cm^2, which corresponds to $\bar{u}_r$ values in the range of 0.8–1.2 eV. Under these operating conditions direct vibrational excitation of the CO_2 asymmetric stretch mode plays a role in producing the population inversion. Unfortunately, the electron energy transfer processes to the electronic excitation channels of N_2 and CO_2 under these conditions not only result in the production of sufficient ionization to sustain the discharge, but also produce significant minority species chemistry. In particular, the dissociative attachment channel in CO_2 has been found to be one of the major reaction channels influencing discharge minority species chemistry and discharge stability [51–53]. By decreasing the discharge operating E/N, it is possible to maintain a relatively high level of energy transfer to desirable vibrational channels while reducing the detrimental effects of CO_2 dissociative attachment reactions. In this case direct excitation of the CO_2 (001) asymmetric stretch level begins to dominate over N_2 vibrational excitation. In this situation the ionization in the discharge has to be maintained by an auxiliary source. Recognition of the possibility that efficient energy transfer to the vibrational levels required to produce a population inversion could be maintained at reduced average electron energies led to the development of high-energy electron beams for use as an auxiliary ionization source to sustain laser discharges [54, 55].

The fractional power transfer information contained in Figure 7.7 can also be utilized to explain in a qualitative fashion the observed improvement in laser output and efficiency reported by Bletzinger and Garscadden [50, 56] and Clark and Wada [57] when xenon is added to a laser mixture. As reported in these references, experimental determination of average electron energies and electron densities by electrostatic probe and microwave techniques indicates that the presence of Xe in the discharge reduces the average electron energy while increasing the discharge current density at lower discharge voltage. For the conditions reported in these measurements the enhancement in laser performance can be produced by the following factors. An increase in electron density directly influences the total rate of power transfer to favorable laser levels in N_2 and CO_2 since the total power deposition is the product of the fractional power transfer data of Figure 7.7 times the electron density–neutral gas density product. In addition, although xenon has an ionization potential comparable to that of CO_2, the ionization rate is approximately a factor of 3 greater when added under the conditions typical of the experiments in refs. 50, 56–58. As a consequence, enhancements in the ionization rate, as observed, can be achieved at lower discharge applied voltage. A decrease in applied discharge voltage which reduces E/N also results in a larger fractional power transfer to the $N_2(v = n)$ and CO_2 (001) levels since reductions in average electron energy from 2.0 to 0.8 eV result in almost a 60% increase in fractional power transfer to these levels, as shown in Figure 7.7.

III. MOLECULAR KINETIC PROCESSES

The development of an understanding of the electron kinetic processes taking place in the CO_2–N_2–He laser discharge has served to point out the highly efficient nature of electron energy transfer processes. However, a recognition of the importance of heavy particle energy transfer and relaxation processes was required before significant advances could be made in the development of compact high-power CO_2 gas discharge lasers. In particular, recognition of the critical dependence of many of the heavy particle processes on neutral gas temperature was one of the major factors required before significant improvements in laser performance could be achieved [41]. Prior to this development, cw kilowatt laser outputs were achieved by producing extraordinary long laser discharges; for example, a 54 m long discharge was utilized to produce a cw output of 2.5 kW [19].

In contrast to the relatively limited number of experimental approaches utilized to determine electron collisional information for CO_2 laser species, many different types of experiments have been employed in the determination of heavy particle rates as a function of temperature, for temperatures slightly below room temperature up to several thousand degrees. At room temperature, measurements have been obtained using sound absorption and/or dispersion as well as impact-tube and spectrophone techniques. High temperature rate data have been obtained primarily from shock tube experiments in which electron beam, infrared emission, *schlieren*, and interferometric diagnostic techniques are employed. For example, as many as 36 separate experiments have been conducted to determine the relaxation rate of the CO_2 bending mode in pure CO_2 [59]. The reader is referred to the review by Taylor and Bitterman [59] of heavy-particle processes of importance to laser applications for a detailed description and interpretation of available experimental and theoretical data.

Also of significance is the fact that the laser itself has been utilized to determine relaxation rates of interest through the use of Q-switching techniques. Kovacs, Flynn, and Javan [60] and Flynn, Kovacs, Rhodes, and Javan [61] first reported use of this technique to determine rates of importance to the CO_2 laser systems. Yardley and Moore [62] initially employed laser induced fluorescence techniques to determine V–V rates in methane. This technique was subsequently employed quite successfully in the determination of many important rates for the CO_2 laser system [63–65]. The review by Moore [66] presents a critique of the potential of this experimental technique and an interpretation of results.

Until the work of Fowler [8, 67] on heavy particle modeling of the CO_2 laser, most analyses either assumed a Maxwellian distribution to establish electron pumping rates or based estimates on incomplete knowledge of electron collisional processes (processes 1 and 4, Figure 7.1). The energy level diagram used in this modeling of the CO_2–N_2–He laser heavy particle kinetics is shown in Figure 7.1. Electron pumping rates employed in this model were obtained from the work of Nighan, as described in Section II. Heavy particle information was obtained from the available information reported in the literature as summarized in refs. 59 and 66. As discussed in refs. 8 and 67, there are seven important energy transfer processes in the CO_2–N_2 system, as indicated in Figure 7.1. These seven basic collisional and radiative processes that influence the CO_2 and N_2 vibrational level populations are as follows:

Process 1

Electron excitation and deexcitation of the CO_2 asymmetric stretch mode [40],

$$e + CO_2[l, m, (n-1)] \rightleftarrows e + CO_2(l, m, n).$$

Process 2

Vibrational energy exchange between N_2 and the asymmetric stretch mode of CO_2 [59],

$$N_2(v) + CO_2[l, m, (n-1)] \rightleftarrows N_2(v-1) + CO_2(l, m, n).$$

Process 3

Vibrational energy exchange between the asymmetric stretch and the coupled symmetric stretch and bending modes of CO_2 [59],

$$M + CO_2[l, m, (n-1)] \rightleftarrows M + CO_2[(l-1), (m-1), n].$$

Process 4

Electron excitation and deexcitation of N_2 [40],

$$e + N_2(v) \rightleftarrows e + N_2(v').$$

Process 5

Excitation and deexcitation by both electrons and heavy particles of the bending mode of CO_2 [40, 59],

$$M + CO_2[l, (m-1), n] \rightleftarrows M + CO_2(l, m, n).$$

Process 6

Absorption and spontaneous emission of 4.3-μm radiation by CO_2,

$$CO_2[l, m, (n-1)] + \text{photon}(4.3\ \mu\text{m}) \rightleftarrows CO_2(l, m, n).$$

Process 7

Absorption and stimulated emission of 10.6-μm radiation by CO_2 at the $P(20)$ rotational line,

$$CO_2(100) + 2\ \text{photons}(10.6\ \mu\text{m}) \rightleftarrows CO_2(001) + \text{photon}(10.6\ \mu\text{m}).$$

In the modeling by Fowler [8] the rate constants for the above processes were averaged over all the appropriate vibrational levels by using the harmonic oscillator approximation [68]. This results in rate constants dependent on the characteristic vibrational temperatures.

The two parameters that characterize the performance of a CO_2–N_2–He laser plasma are the small-signal gain, α_0, and the saturation intensity, I_s. The small-signal gain–saturation intensity product, $\alpha_0 I_s$, represents a measure of the optical power density available for extraction from the laser medium. Small-signal gain is the fractional increase in the intensity an infinitesimal 10.6-μm beam experiences in transversing a centimeter path length of active laser medium under conditions in which the beam intensity is below the saturation intensity of the medium. The saturation intensity of the medium, I_s, is the value of the radiation intensity that reduces the small-signal gain to one-half its original value of α_0. For most conditions typical of present CO_2 laser operation, the small-signal gain is the product of the cross section for stimulated emission (process 7) times the difference in the vibrational populations of the asymmetric stretch (001) level and the symmetric stretch (100) level of CO_2.

For the operating range of importance to laser applications, both Doppler and collisional broadening effects [8] influence the line width of the laser transition. As a result the cross section for stimulated emission (process 7) is a function of gas temperature, pressure, and mixture composition. This cross section as determined by Fowler [8], over a range of pressures and temperatures of interest to laser application for a 1–1–8 mixture of CO_2–N_2–He is shown in Figure 7.8. The sensitivity of the cross section to pressure at high pressures is predominantly due to collisional broadening effects, while the sensitivity to temperature is primarily due to changes in rotational relaxation processes with temperature.

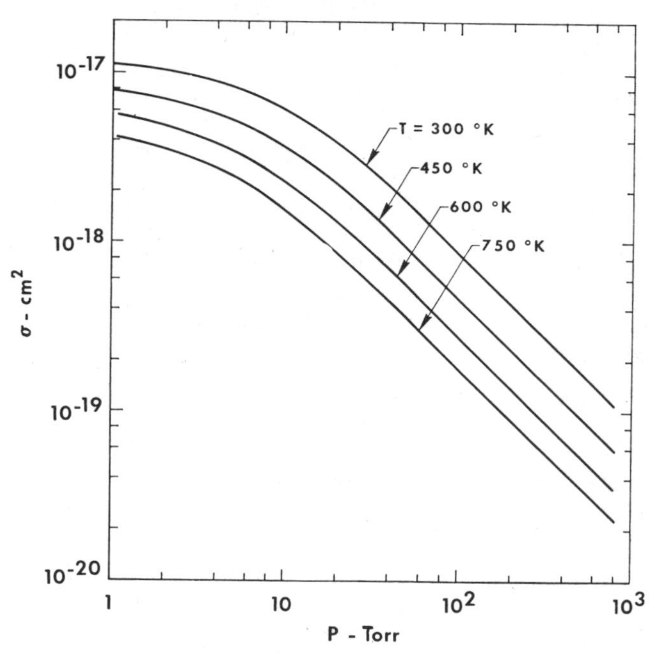

Figure 7.8 Dependence of stimulated emission cross section for the $P(20)$ line of the (001)–(100) laser transition in CO_2 on gas temperature and pressure for 1–1–8 laser mixture [8].

In order to determine the influence of various discharge processes on laser characteristics, gas temperature, electron density, and average electron energy were taken as independent modeling parameters [8]. In reality this cannot be achieved under ordinary dc discharge conditions. However, by using this approach it is possible clearly to identify the discharge parameters that have major influence on laser performance characteristics.

A. Small-Signal Gain

Shown in Figure 7.9 [8] is the influence variations in gas temperature and electron density have on small-signal gain, α_0. This calculation was carried out for a 1–1–8, CO_2–N_2–He mixture for a fixed-discharge E/N typical of laser operating conditions. The conditions depicted in Figure 7.9 correspond to a

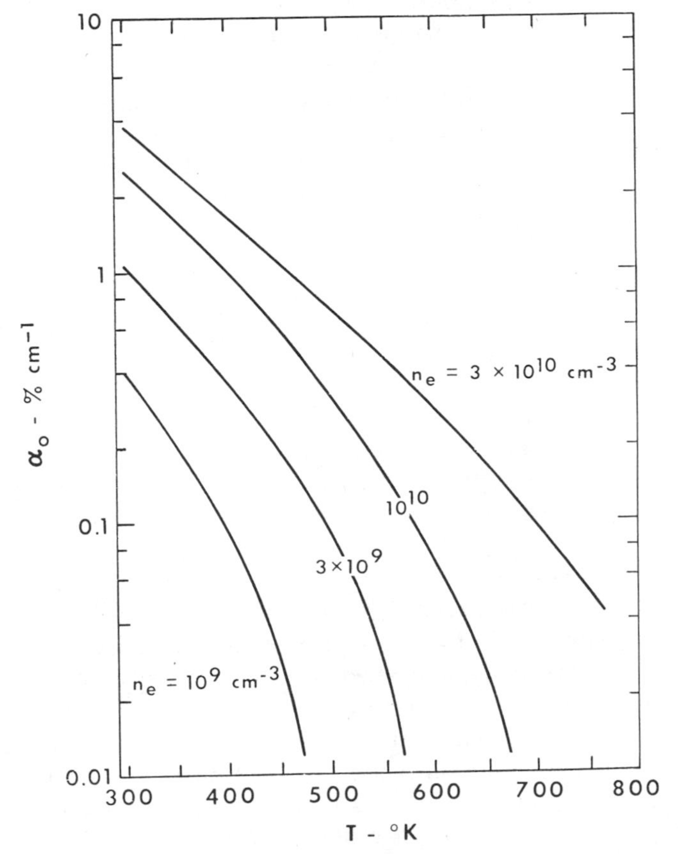

Figure 7.9 Variation of small signal gain, α_0 with gas temperature and electron density for a 1–1–8 laser mixture and reduced average electron energy of 1.5 eV [8].

discharge pressure of 10 Torr and reduced average energy, $\bar{u}_r$, of 1.5 eV. In this calculation it was assumed that only the original neutral species were present in the discharge, which is a condition that does not occur in actual operation. The impact of plasma chemical reactions on discharge species concentrations and discharge stability will be treated in Sections IV and V. The conditions presently under consideration are most nearly achieved in high-flow-velocity, open-cycle lasers in which the gas passes through the discharge only once and has a residence time in the discharge region on the order of 10^{-4}–10^{-3} sec.

For a fixed gas temperature, increases in electron density result in significant increases in gain, as shown in Figure 7.9. This occurs because electron density increases in the discharge enhance the rate of pumping of the $CO_2(001)$ and $N_2(v = n)$ levels (processes 1 and 4), and as a consequence, increase temperatures T_2 and T_N. This result could be partially anticipated from the fractional power transfer information of Figure 7.7, which shows that the power transfer to the CO_2 asymmetric stretch and $N_2(v = n)$ levels is significantly greater than the transfer to the levels in the symmetric and bending modes of CO_2. Further, since the gas temperature is held constant, atom–molecule collisions (process 5) involving He as a relaxant [59] are highly effective in maintaining the combined bending and symmetric stretch mode of vibration at a low effective temperature. As a result, the difference in the populations of the (001) and (100) levels is enhanced, and therefore, the gain increases.

For a constant electron density and increasing gas temperature, the gain in the medium is drastically reduced. As shown in Figure 7.9, the medium will actually become absorbing even at high electron densities for temperatures above approximately 800°K. The dramatic decreases in gain with increases in gas temperature result from the fact that all the quantities that determine gain are adversely affected by increases in gas temperature. The stimulated emission cross section, as shown in Figure 7.8, decreases with increasing gas temperature, while the collisional relaxation rate of the upper laser level (process 3) is increased. Furthermore, under constant pressure conditions the total density of neutral atoms is reduced. Lastly, one of the most important factors to be considered is the fact that since the gas temperature is quite close to the combined bending–symmetric stretch mode temperature, T_1, increases in gas temperature will result in an increased (100) level population despite the effectiveness of He relaxation processes (process 5). Recognition of the strong influence gas temperature has on laser gain [41] and the development of techniques to enhance discharge cooling rates over those typically achievable with wall-cooled discharges resulted in the achievement of kilowatt outputs from extremely small discharges [1].

Experimentally determined gains usually range from 0.1 to 0.5% cm^{-1} [69].

Furthermore, gains on the order of several percent per centimeter are completely consistent with results reported in ref. 16 in a somewhat different laser configuration. A good assessment of the ability of the model developed by Fowler to predict actual laser conditions can be obtained from the information presented in Section IV.

One additional point of importance that should be noted is that helium plays an important role as a relaxant in determining the population of the (100) level and, therefore, gain. The importance of He as a lower level relaxant can best be appreciated by noting the early work of Moeller and Rigden [70] in which a factor of 5 enhancement in laser performance was achieved through addition of helium to the system. In subsequent work, Witteman [71] demonstrated the effectiveness of water vapor in place of He as a relaxant for the lower laser level.

B. Saturation Intensity

As in the case of gain, the saturation intensity, I_s, of the medium is strongly influenced by collisional processes. Therefore, it would be expected that variations in gas temperature and electron density would also influence this laser parameter. Basically, the saturation intensity can be looked upon as a measure of the ability of the applied radiation field to perturb the steady-state conditions in the medium, which are governed by the collisional and energy transfer processes 1–7. In the absence of N_2, the magnitude of the saturation intensity is determined predominantly by the sum of all the excitation and deexcitation processes from the asymmetric stretch mode, which are governed by processes 1, 3, and 6. At a fixed electron density this sum increases with gas temperature, therefore the saturation intensity also increases with gas temperature, as shown in Figure 7.10. The addition of N_2 to the system influences the rate of process 3 and also the energy transfer between the $N_2(v = n)$ and the CO_2 asymmetric level (process 2). Because of the presence of the latter energy transfer channel, the saturation intensity is larger than would be predicted if the only influence of N_2 on the system was to affect the rate of process 3.

As indicated previously in consideration of the gain of the medium, increasing the discharge electron density at constant gas temperature results in an elevation of temperatures T_2 and T_N. As a consequence, the losses from the asymmetric stretch mode increase. Further, at higher electron densities the vibrational energy content of N_2 is increased, allowing this level to act more effectively as an energy storage level for the CO_2 asymmetric stretch level. As a consequence, the saturation intensity increases with electron density, as shown in Figure 7.10. Since the rate of transfer from the CO_2

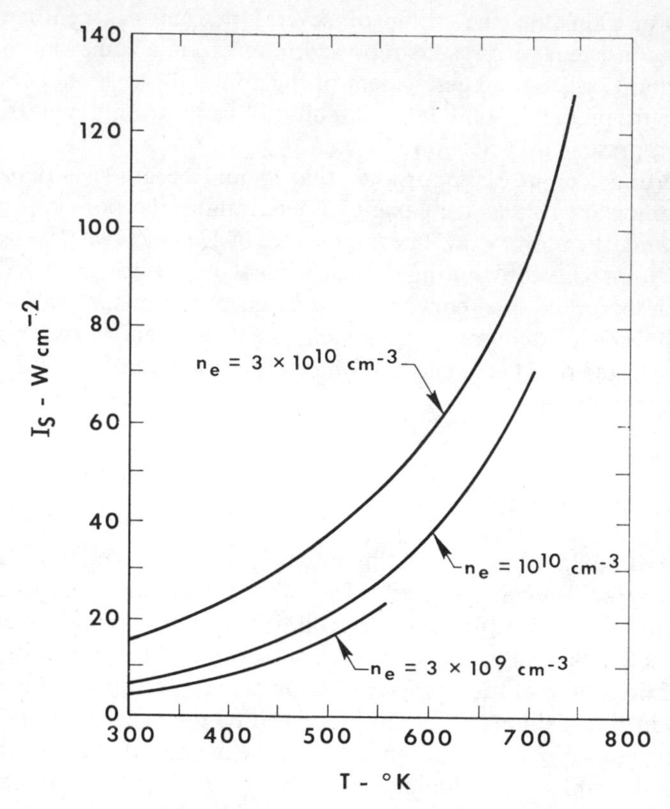

Figure 7.10 Dependence of saturation intensity I_s, on gas temperature and electron density for a 1–1–8 laser mixture and reduced average electron energy of 1.5 eV [8].

bending and symmetric stretch modes to the asymmetric stretch mode (reverse of process 3) increases with gas temperature, the influence of increasing electron density as the gas temperature increases is minimized, as shown in Figure 7.10. Furthermore, it is important to note that the influence of increases in gas temperature and electron density on saturation intensity are not as profound as the effect these two discharge parameters have on gain, as shown in Figure 7.9.

C. Optical Power Density

The gain-saturation intensity product, $\alpha_0 I_s$, is a measure of the optical power density available for extraction from the laser discharge medium.

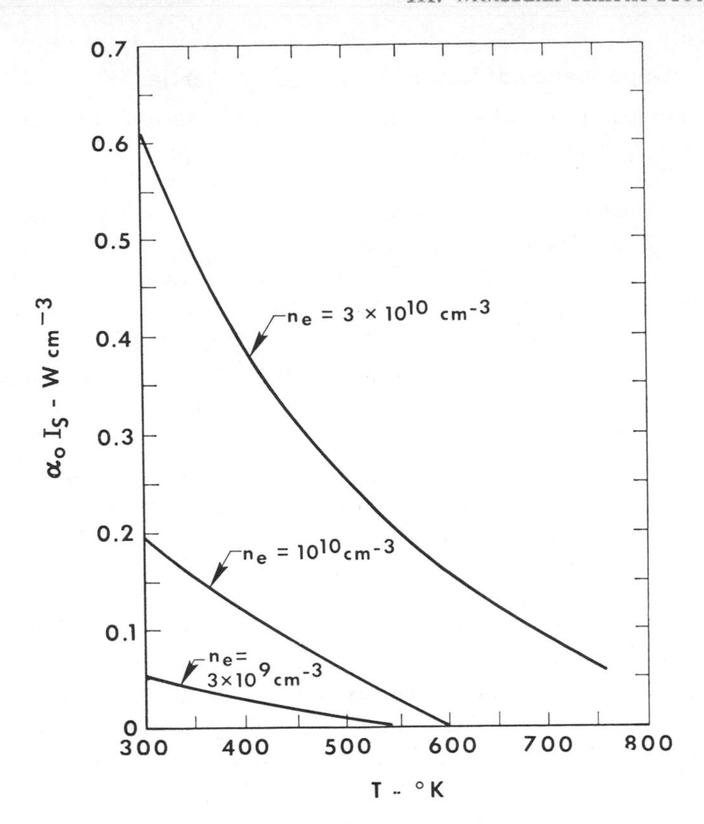

Figure 7.11 Variation of optical power density with electron density and gas temperature for a 1–1–8 laser mixture and reduced average electron energy of 1.5 eV [8].

Shown in Figure 7.11 is the influence that changes in gas temperature and electron density have on this parameter. As noted in the above discussions, the gain of the medium exhibits a marked decrease with increases in gas temperature. As a consequence, even though the saturation intensity increases with increasing gas temperature, the product of these two parameters decreases. As indicated in Figure 7.11, operation of the laser discharge at gas temperatures above 700°K will result in minimal optical power available for extraction even at elevated electron densities. Under certain conditions at temperatures in the 700°K range the medium actually becomes absorbing. Recognition of the importance to laser performance of maintaining low gas temperatures as indicated previously resulted in the development of discharges with enhanced heat removal rates achieved by employing convective cooling techniques [1, 16–18, 72].

D. Pressure Dependence of Laser Discharge Properties

In addition to using convective flow techniques to circumvent the problems associated with gas heating, pulsed techniques have proven to be quite effective [73, 74]. With pulsed techniques, optical power from a pulsed discharge is extracted before gas heating brings about a significant gas temperature rise. Furthermore, the pulsed discharge laser, developed by Beaulieu [73], which is known as a TEA laser, is operated at atmospheric pressure. Operation at elevated pressure has decided performance advantages, as can be seen from the following laser kinetics considerations.

The performance improvements to be achieved by increasing the operating gas pressure or density of a laser discharge can be appreciated by considering the influence that increases in gas pressure have on gain and saturation intensity, and also by considering the factors influencing discharge power density. With the exception of absorption and emission of 4.3-μm radiation from the asymmetric stretch level, the magnitudes of all the energy transfer processes of the system increase linearly with electron density and pressure. Therefore, the values of T_N, T_2, and T_1 are independent of pressure for conditions of constant fractional ionization, n_e/N, fixed gas composition, and gas temperature. Under these conditions the factors determining small-signal gain that are influenced by pressure increases are the cross section for stimulated emission (Figure 7.8) and the total CO_2 density. In the low pressure range the small-signal gain increases linearly with pressure, because the cross section for stimulated emission is nearly constant up to approximately 30 Torr as shown in Figure 7.8. Above this pressure the stimulated emission cross section decreases inversely with pressure. As a result, at higher pressures the small-signal gain remains constant. The saturation intensity for these same conditions has a nearly linear dependence on pressure at low pressure because of the influence of pressure on the stimulated emission rate from the laser transition (process 7) and the other collisional processes of the system. At approximately 30 Torr the saturation intensity variation due to combined collisional processes becomes quadratic. Inclusion of process 6 results in higher losses from the asymmetric stretch level at lower pressures because of less trapping of 4.3-μm radiation [75]. This results in a decrease in small-signal gain and an increase in saturation intensity over that determined by exclusion of this process. Shown in Figure 7.12 [8] is the small-signal gain and saturation intensity dependence on pressure. The product, $\alpha_0 I_s$, based on this information varies quadratically across the entire pressure range, which points up one of the major motivations for developing laser discharges operating at higher pressures.

Achievement of atmospheric pressure operation with a dc discharge under conditions suitable for laser operation is extremely difficult. This has led to

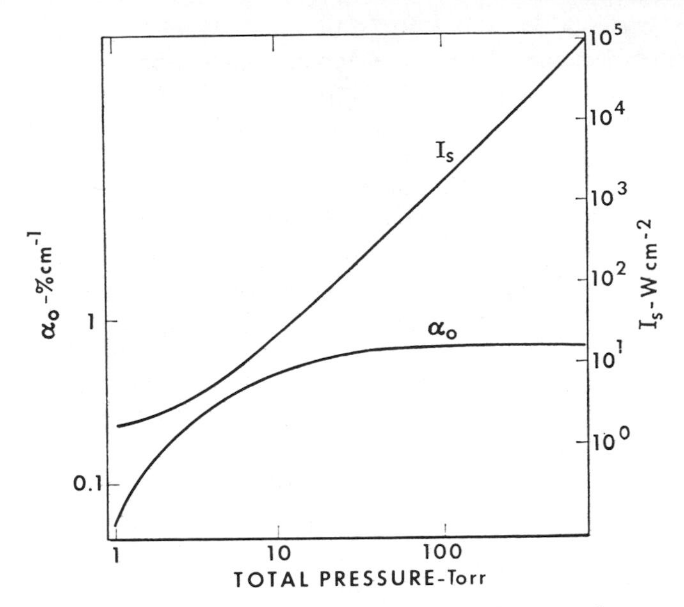

Figure 7.12 Dependence of saturation intensity and gain on total gas pressure [8].

the development of several novel approaches for achieving high pressure operation. In addition to the pulsed TEA laser work of Beaulieu [73], auxiliary ionization techniques, such as the use of electron beams [54, 55] and photo-ionization as developed by Levine and Javan [77], have permitted stable discharge operation at high pressures by decoupling the ionization and excitation processes in the discharge. Pulsed techniques with high repetition rate carefully controlled pulses, as developed by Hill [76], have also been successfully employed to achieve quasi-cw operation at high pressures. In all these systems enhancements in the small-signal gain–saturation intensity product are realized only if the discharge power density scales quadratically with the neutral density. This is most easily recognized by examining the factors influencing electron-joule heating in the discharge.

Electron-joule heating is the product of the electron current density, J, and the electric field intensity, E; it has the following form as developed by Nighan [78]:

$$JE = n_e e v_D E = \left[\left(\frac{n_e}{N e v_D}\right)\left(\frac{E}{N}\right)\right] N^2; \tag{16}$$

where n_e is the electron density, e is the electronic charge, v_D is the electron drift velocity as defined by equation (13), and N is the total neutral particle density. In CO_2 lasers the highly selective nature of the excitation and relaxation processes as described in this section and in Section II, to a first approximation, limit laser discharge operation to a relatively narrow fractional ionization range ($n_e/N \sim 10^{-8}$–10^{-7}). In addition, the discharge E/N, once the gas mixture is established, is set for maximum vibrational excitation in the case of auxiliary ionization lasers or by the combined requirements of producing sufficient ionization to maintain the discharge while maintaining a reasonable level of vibrational excitation in pure dc discharges. Therefore, the E/N ratio for a given system, once the mixture composition is established, is nearly constant. This results in a nearly constant value for the electron drift velocity since the drift velocity is directly proportional to E/N for a given discharge mixture. Thus to a first approximation the quantity in square brackets in equation (16) is nearly a constant, and discharge power density is proportional to the square of the neutral density. As will be discussed in the following sections, several factors, including plasma chemical processes and discharge instabilities, strongly influence achievement of the ultimate potential of electric discharge lasers predicted on the basis of small-signal gain, saturation intensity and power deposition considerations.

IV. DISCHARGE PROCESSES

A. Laser Discharge Characteristics

The results of Sections II and III illustrate the profound sensitivity of laser characteristics to various discharge parameters. Several measurements of laser discharge properties tend to confirm these results qualitatively [79–81]. The first detailed evaluation of the predictions of the above work was carried out by Wiegand [41] and Wiegand, Fowler, and Benda [69]. In this work a flowing gas wall dominated discharge was analyzed to establish gas discharge laser parameters. Modeling of this discharge system was conducted by developing a modified Schottky analysis [82] of the positive cylindrical column of the discharge. From this analysis, which was similar to the work of Ecker and Zoller [83], it is possible to calculate the radial variation of electron density and gas temperature. The input information for this model was electron transport data (mobility and diffusion coefficients) and various electron–molecule excitation and ionization rates. These rates were obtained by taking appropriate averages over the distribution functions established from the work of Section II for fixed gas mixture and discharge E/N. Since no optical power was extracted from the discharge and because heavy particle V–T transfer rates are fast for typical 1–1–8 gas mixtures [59], the principal energy loss from the discharge was thermal conduction to the discharge-tube walls. The other required inputs to this analysis are the temperature-dependent thermal conductivity of the gas mixture, which is independent of pressure, and the mobility of the dominant positive ion.

From detailed analysis of the charged-particle processes taking place in the CO_2 laser discharge, direct electron impact ionization of ground state CO_2 has been found to dominate charged-particle production in typical 1–1–8 mixtures [69]. However, because of significant dissociation of CO_2 [51], sufficient O_2 is produced to enter into charge exchange reactions with CO_2 to make O_2^+ the dominant positive ion. Multistage ionization processes involving N_2 metastables have been found to become important only when the concentration of N_2 is significantly greater than CO_2. Further, O_2^+ will probably remain the dominant positive ion as long as other species with lower ionization potentials are not introduced into the discharge or produced by minority species plasma chemical reactions. The loss process for the positive ions in this system is wall recombination with electrons. It should be noted, however, that in many laser discharges, such as *e*-beam augmented lasers, dissociative recombination is the dominant loss process.

For the discharge conditions outlined above the radial electron density profile was found to be nearly parabolic. This resulted in computed values of electron density on the discharge axis in the range between 10^9 and 10^{10} cm^{-3}. The gas temperature for these conditions exhibits a similar nearly parabolic radial profile with the value on the discharge axis ranging from 400 to 700°K over the range of discharge conditions tested [69]. Shown in Figure 7.13 are both the experimental and analytically determined temperature profiles. Experimentally, the gas temperature was determined in these measurements by using a calibrated electrically floating thermocouple. The same good agreement was established throughout the entire discharge operating range

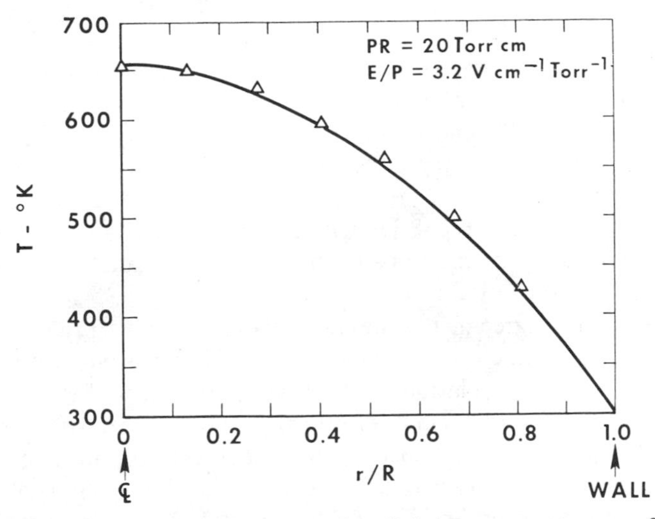

Figure 7.13 Calculated and measured radial gas temperature profiles in a wall-dominated laser discharge operating at 10.5 Torr [69].

for all mixture compositions in which He dominated the neutral-species concentrations, confirming the validity of the thermal conduction model. This information is used to establish discharge current-voltage characteristics. These characteristics are most conveniently expressed in the form of the current-pressure product, *IP*, and the ratio of electric field to pressure, *E*/*P*. Shown in Figure 7.14 is the agreement obtained between theory represented as the solid line and the experimentally determined points. It is important

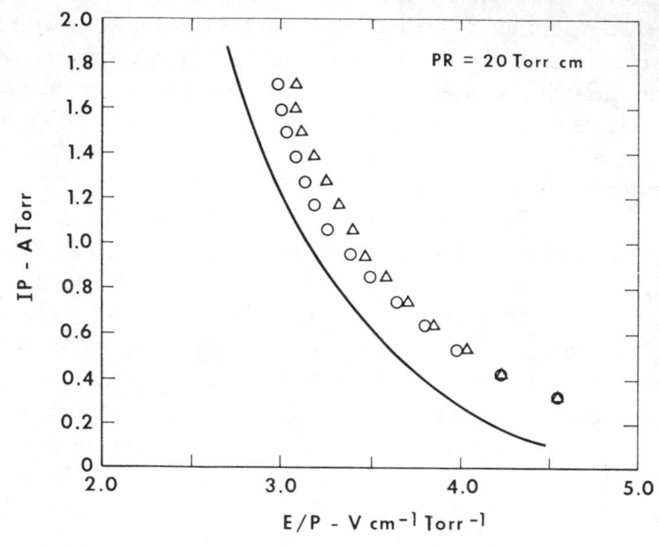

Figure 7.14 Computed and measured discharge characteristics [69].

to note that this type of agreement could not be achieved unless the essential physical mechanisms governing discharge characteristics were included in the positive column analysis. The most important of these is the reasoning leading to the assumption that single-step direct ionization of CO_2 is the dominant channel for the production of positive ions. The negative-discharge characteristic that is typical of the 1–1–8 laser discharge mixture is due to the influence increases in gas temperature have on *E*/*N*, which result in exponential changes in discharge ionization rate.

The relatively good agreement obtained between various predicted and measured discharge characteristics serves to establish the validity of the model for establishing correlations with laser parameters. In particular, the most critical prediction from the electron and heavy particle models of Sections II and III is influence of gas temperature on laser gain. Measurements of radial gain profile were obtained by using techniques similar to those described by Smith and McCoy [84]. Shown in Figure 7.15

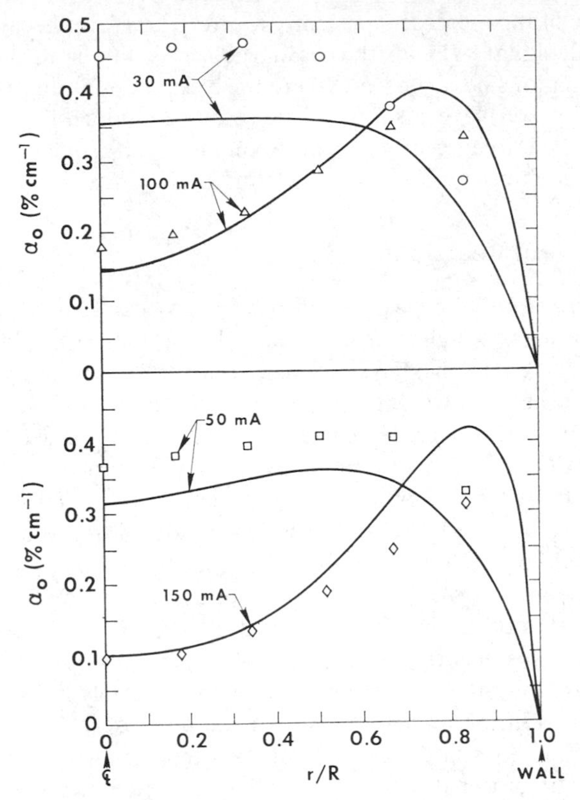

Figure 7.15 Variation of computed and measured radial gain profiles [69].

is the variation of measured and predicted small-signal gain with radial position and total discharge current. For low discharge currents (30 mA) the radial gas temperature profile is parabolic with a low center-line temperature. As a consequence, the experimentally measured and theoretical gain profiles are nearly parabolic. In this case the population of the upper laser level (001) is significantly greater than the population of lower laser level (100). Since little gas heating is taking place, the gain profile should follow closely the radial electron density profile, which is also parabolic. Increases in discharge current density produce significant gas heating on the center line of the discharge. As a result, even though the CO_2 asymmetric stretch level and $N_2(v = n)$ levels are pumped harder, the center-line gain decreases. The gain decreases because increased gas temperature produces a decrease in the stimulated emission cross section (Figure 7.8) and also an increase in the collisional relaxation rate of $CO_2(001)$ asymmetric stretch level (process 3). Further, the population of the symmetric stretch level (100) is increased because of the increased gas temperature. Therefore, because of gas heating, the small-signal gain on the discharge center line is lower than in cooler discharge regions near the wall where the electron pumping rate of the $CO_2(001)$ and $N_2(v = n)$ levels is lower. The results contained in Figure 7.15 vividly illustrate the influence of gas heating on laser performance and serve to confirm the predictions of the laser modeling of Sections II and III.

B. Plasma Chemical Reactions

The first experimental evidence that discharge processes were producing new chemical species through plasma chemical reactions was obtained in seal-off laser discharges. In these systems laser output was found to decrease significantly with time. The work of Witteman [85] served to point out that the major cause for the observed decreases in laser output was due to dissociation of CO_2 into CO and O. Subsequent work by Wiegand et al. [51] confirmed the observations of Witteman in a more quantitative fashion. In these measurements it was found that because of dissociation processes, CO concentrations could build up to levels of 70% of the total CO_2 concentration for discharge residence times on the order of seconds. As predicted by Nighan [39], CO_2 dissociation exceeding 10% was found to result in degradation of laser performance. This occurs not only because of the loss of CO_2 from the discharge, but also because the presence of CO adversely influences the electron energy distribution and molecular kinetic processes.

In addition to the formation of CO, many other minority species are produced in operating laser discharges [52]. The presence of these minority species, although in concentrations of less than 1% of the total discharge mixture, has been found to have an important influence on discharge stability, as described in Section V. Analysis of the plasma chemical processes occurring in laser discharges has been reported by Wiegand and Nighan [52]. In this work the temporal evolution of charged and neutral species in a CO_2–N_2–He convection laser discharge dominated by volume processes has been determined. The model employed in this investigation, which parallels work by Niles [86] for atmospheric systems, considers 40 neutral and charged species in 300 reaction channels. The major reaction processes treated are typically ionization, dissociative attachment, dissociation, recombination, and detachment. Dissociation in the discharge, it is found, leads to the formation of CO and O and to somewhat smaller concentrations of N. These dissociation products subsequently lead to the formation of minority neutral and charged molecular species. The production of significant negative ion concentrations, in particular, appears to be quite important to discharge stability. Although in high-velocity convective-flow discharge lasers concentrations of minority species do not build up to levels comparable to the original constituents (CO_2–N_2–He), the concentrations of these species do become equivalent to the electron density concentration in the discharge. Of more significance is the fact that negative ion concentrations produced as a result of plasma chemical reactions have been found to exceed electron concentrations. While the presence of these minority species does not influence electron–molecule vibrational excitation processes, they do have a major influence on discharge stability [53, 87].

For typical laser conditions, only the electrons in the tail of the distribution enter into ionization [88], dissociation [51, 89], and dissociative attachment [90] reactions. Since the rates for these reactions are a strong function of E/N, the rates must be determined by using known cross sections [51, 88–90] and electron distribution functions determined by techniques outlined in Section II. Heavy particle rate constants are obtained from refs. 86 and 91.

Shown in Figure 7.16 [52] is the calculated evolution of neutral species for typical laser discharge conditions. As indicated, the initial neutral species

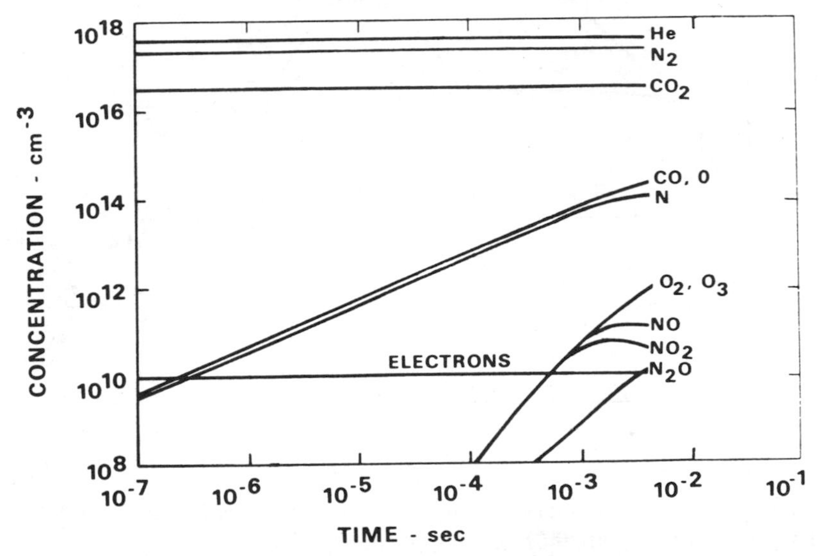

Figure 7.16 Calculated temporal evolution of neutral minority species in 20 Torr 1–1–8 laser mixture [52].

concentrations remain essentially constant while minority species concentrations increase to approximately 1% of the original neutral concentration in one discharge residence time. The principal dissociation products for this system are CO, O, and N, as indicated in the figure. It should also be noted that numerous secondary minority species, namely, O_2 and O_3 and the oxides of nitrogen, reach concentrations in excess of the electron concentration.

Because of rapid exchange and rearrangement reactions, the final ion species are directly related to the minority species concentrations. As can be seen from Figure 7.16, dissociation of CO_2 into CO and O, and N_2 into N results in significant concentrations of these species on a time scale of 10^{-6} sec. Shown in Figure 7.17 is the temporal evolution of negative ion species. The principal source of negative ions in typical laser mixtures is the formation of O^- by dissociative attachment of electrons to CO_2. However, clustering of O^- with CO_2 as shown in Figure 7.17*a* results in the formation CO_3^-. A

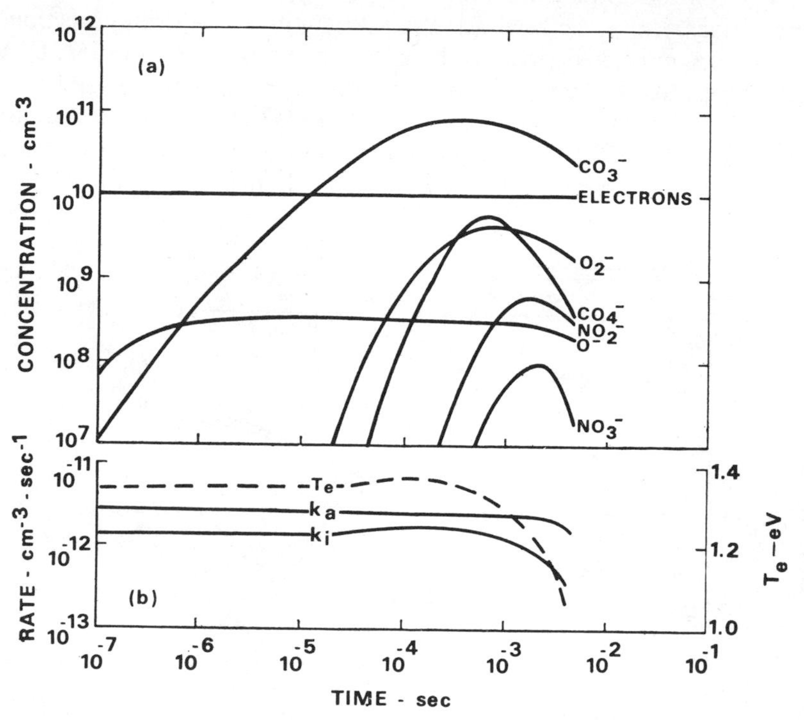

Figure 7.17 Calculated temporal evolution of negative ions for 20 Torr 1–1–8 laser discharge operating at 300°K gas temperatures [52].

multitude of additional negative ions are also formed from minority species collisional processes. As seen from the data in Figure 7.17, it is important to note that under certain conditions negative ion concentrations can exceed electron concentrations by almost an order of magnitude. From comparison of direct electron impact ionization rates for CO_2 and N_2 and O^- formation by dissociative attachment, it is found that the loss of electrons would exceed the production rate by direct ionization for electron temperatures typical of this example, namely, 1.3 eV. Therefore, as shown in Figure 7.17*b*, the electron temperature must exceed 1.3 eV to maintain a constant discharge electron current. Further, the balance for charge neutrality must be between positive and negative ion concentrations rather than the electron concentration under these conditions. As a result, this forces positive ion concentrations to higher levels to maintain neutrality, which results in higher recombination losses. This requires a higher discharge electron temperature in order to increase the ionization rate to compensate for these additional losses. This increase in effective temperature occurs as shown in Figure 7.17*b* in the 10^{-5}–10^{-4} sec time range. As CO and O concentrations build to sufficient levels in the discharge, these species free electrons from negative ions by associative detachment reactions [91, 92]. This process reduces the net effect of attachment. As detachment processes increase in importance, the negative ion density peaks and subsequently decreases. Under these conditions as shown in Figure 7.17*b*, the discharge can be maintained at significantly lower effective electron temperatures. This actually occurs under conditions where the direct electron–molecule attachment rate is greater than the ionization rate, which is a condition of major importance to discharge stability, as shown in ref. 53 and in the following section.

V. PLASMA STABILITY

The development of instabilities in discharges has been the subject of many investigations over the past several decades. In the case of laser discharges the formation of plasma instabilities has limited the achievement of the ultimate predicted potential of the electric discharge laser. Furthermore, efforts to enhance laser performance by operation at higher pressures have aggravated discharge stability problems. Fortunately, in the CO_2 laser discharge situation sufficient collisional information has become available, as indicated in Sections II–IV, to allow a critical evaluation of the role various collisional processes play in influencing discharge stability.

From the development of a comprehensive model of the factors leading to instabilities in molecular discharges, Haas [87] has found that these systems are prone to several modes of instability. Of these modes, the conditions leading to the development of ionization and neutral particle energy transfer modes of instabilities have been found to be most easily satisfied under laser conditions. Ionization instabilities are produced when conditions enhance the temporal amplification of an imbalance between electron production and loss processes. The characteristic growth time for this type of instability that has been found to be independent of pressure and power density is in the 10^{-6}–10^{-4} sec range [53, 78, 87]. In contrast, neutral particle energy transfer instability modes that occur due to amplification of disturbances in translational and/or vibrational energy density occur on the 10^{-4}–10^{-2} sec time range [53, 78, 87]. The conditions leading to the development of these types of instabilities are dependent on discharge power density and pressure. All of the instability modes of importance to laser applications have also been found to depend critically on the response of the electron density to disturbances in the medium properties [53, 87]. On this basis it is easy to understand the motivation for developing auxiliary ionization systems, such as electron beams, to uncouple the ionization and excitation processes [54, 55]. With *e*-beam techniques the ionization processes, and therefore the electron density, can be maintained uniform throughout the discharge volume independent of other plasma processes.

From the development of the plasma stability model by Haas [87] and by Nighan, Wiegand, and Haas [53], it has been found that a fluctuation in electron temperature can be related to a fluctuation in electron density in the following manner:

$$\frac{T_{ek}}{T_e} = \left(\frac{-2\cos^2\phi}{1 + \hat{\nu}_u - \hat{\nu}_m \cos 2\phi}\right)\frac{n_{ek}}{n_e}, \tag{17}$$

where T_{e_k} and n_{e_k} are the perturbed amplitudes of the electron temperature and density, respectively, ν_u is the total electron energy exchange collision frequency, ν_m is the momentum transfer collision frequency, and ϕ is the angle between the direction of the steady state unperturbed current density flow and the direction of the wave propagation vector $\mathbf{k}$ for a disturbance. In equation (17) $\hat{\nu}_m = \partial \ln \nu_m / \partial \ln T_e$, where T_e is the generalized electron temperature, $\frac{2}{3}\bar{u}_r$. Similarly, $\hat{\nu}_u = \partial \ln \nu_u / \partial \ln T_e$. From equation (17) the amplitudes of the disturbances in electron temperature and electron density are in phase at either 0° or 180°. Due to the dependences of ν_u and ν_m for the laser mixture the quantity $1 + \hat{\nu}_u - \hat{\nu}_m \cos 2\phi$ is always positive. Thus, electron density disturbances are always out of phase with electron temperature fluctuations. This means that an increase in electron density due to a fluctuation results in a decrease in electron temperature. This finding is quite contrary to the usual intuitive notions of plasma processes, and serves to point up the quite complicated nature of the collisional processes influencing plasma stability. Furthermore, as a consequence of the nearly instantaneous electron kinetics, the amplitudes of the electron temperature and density fluctuations are not dependent on ionization, recombination, or attachment processes [53, 87]. These processes will, however, determine whether or not a disturbance will grow.

Furthermore, from the development of the stability criteria in refs. 53 and 87, the growth of a mode is dependent on collisional processes dominating the production and loss of electrons and ions. These processes for the CO_2 laser are outlined in Section IV. From this model it is found that a disturbance in electron density will be amplified if the following condition is met [53, 87]:

$$\left(\frac{-2\cos^2\phi}{1+\hat{\nu}_u-\hat{\nu}_m\cos 2\phi}\right)\left(1-\frac{k_a\hat{k}_a}{k_i\hat{k}_i}\right)\frac{\hat{k}_i}{\tau_i} - \frac{1}{\tau_i}\left(\frac{n_n\tau_i}{n_p\tau_r^i}+\frac{n_e\tau_i}{n_n\tau_a}+\frac{n_e\tau_i}{n_p\tau_r^e}+\frac{n_n\tau_i}{n_e\tau_d}\right) > 0, \quad (18)$$

where n_e, n_m, and n_p are the number densities of electrons, negative ions, positive ions, and k_i and k_a are the mixture weighted electron rate coefficients for direct ionization and attachment averaged over the electron distribution function. In equation (18), $\tau_i = (Nk_i)^{-1}$, $\tau_a = (Nk_a)^{-1}$, $\tau_r^e = (n_p k_r^e)^{-1}$, $\tau_r^i = (n_p k_r^i)^{-1}$, and $\tau_d = (Nk_d)^{-1}$ are the characteristic times for ionization, attachment, electron–positive ion recombination, and positive ion–negative ion molecular detachment, and N is the total density.

In equation (18) it is important to note that only steady-state unperturbed quantities influence the growth rate of instabilities in the discharge. The first term in this equation represents the effects of changes in electron production and loss produced by fluctuations in electron temperature. The second term represents the effect of fluctuations in electron and negative ion density on electron production and loss. Since this second term contains only positive quantities, the effects of electron and negative ion density fluctuations are stabilizing.

Since $1 + \hat{\nu}_u - \hat{\nu}_m \cos 2\phi$ is always positive for typical laser conditions, the requirement for ionization instability is that $k_a\hat{k}_a/k_i\hat{k}_i$ be greater than unity. In addition, the angular dependence of the growth rate indicates that disturbances will favor growth in the direction of discharge current flow. Thus to achieve an unstable condition in the discharge it is required that the attachment rate must increase with electron temperature ($\hat{k}_a > 0$). For conditions when $k_a\hat{k}_a/k_i\hat{k}_i > 1$, the loss of electrons resulting from electron–molecule attachment dominates over electron production from ionization during a fluctuation. Therefore, attachment dominance over ionization during a disturbance favors an inverse relationship between electron temperature and electron density, which is the same result obtained in equation (17). This type of amplification mechanism can lead to instabilities if the electron temperature fluctuation is large enough to overcome the stabilizing influence of the second term in equation (18).

Shown in Figure 7.18 [53] is the electron temperature dependence of the term $k_a\hat{k}_a/k_i\hat{k}_i$ computed from the collisional data of Sections II and IV for several different laser mixtures. It is quite important to note that this function is strongly temperature dependent over a very small temperature range. Furthermore, this term exceeds unity for electron temperatures precisely

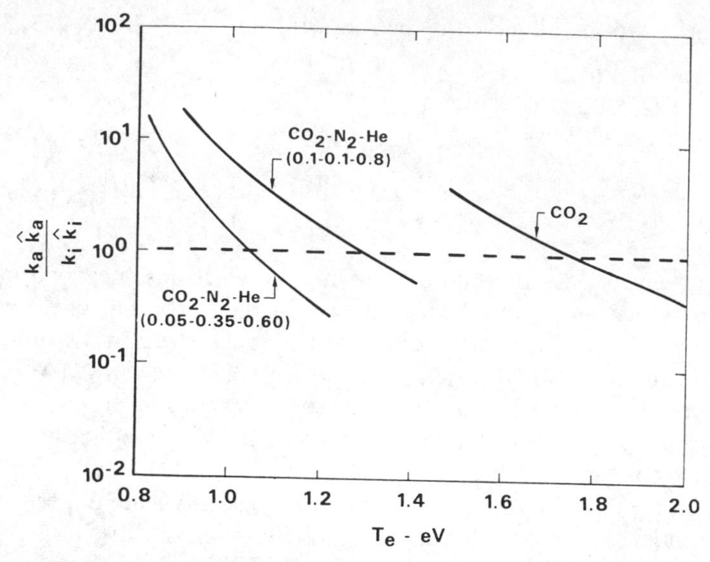

Figure 7.18 Calculated electron temperature and gas mixture dependence of stability parameter $k_a\hat{k}_a/k_i\hat{k}_i$ [53].

in the laser operating range. These results indicate that the necessary condition for ionization instability exists in typical CO_2 laser discharges.

For the effect of negative ion formation to exceed the stabilizing influence of electron and negative ion density fluctuations, negative ion concentrations must be significant when compared to electron concentrations. From the plasma chemical calculations of Section IV, this condition was shown to exist for discharge residence times of 10^{-5} sec or greater, as indicated in Figure 7.17. Therefore, the influence of significant negative ion concentrations must be considered in determining discharge stability conditions. Using the kinetics and chemistry models described in Sections II and IV, the conditions for instability [equation (18)] were determined as a function of CO_2 dissociation. Shown in Figure 7.19 [53] is the computed stability boundary determined from equation (18). The variation of the computed stability boundary shown in Figure 7.19 indicates the dependence of discharge stability on the steady-state values of electron temperature and negative ion concentration. The results presented in Figure 7.19 show that large negative ion concentrations do not necessarily lead to discharge instabilities, but that the critical factor is the effect of negative ion formation on electron production and loss processes during a fluctuation.

The theoretical modeling predictions of Fig. 7.19 have been confirmed [53] in a discharge system in which carefully controlled neutral species are introduced into the discharge to produce operating conditions comparable to

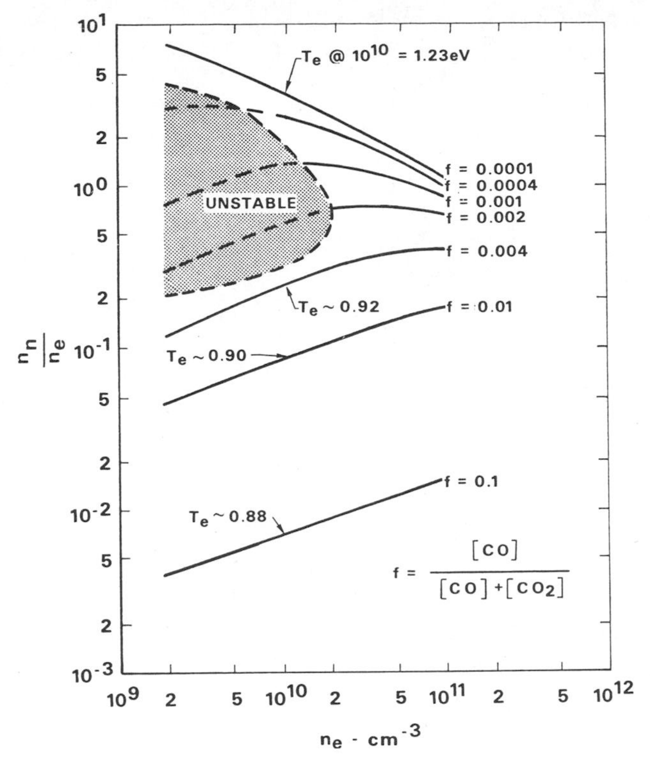

Figure 7.19 Calculated ionization stability boundary for 1–7–12 laser mixture [53].

the conditions depicted in the figure. From sidelight emission measurements and floating probe measurements the ionization mode of discharge instability has been found to lead to the development of a striated plasma as predicted by theory. While this type of instability has been found to be benign, these experimental results besides suggesting that the theoretical stability and plasma chemistry models are providing a good qualitative picture of minority species processes in the discharge, also indicate that plasma stability phenomena are basically collisionally dominated. Investigation of the other potential instability modes identified to be of importance to CO_2 laser applications, i.e., thermal and vibrational has revealed that these modes are also influenced by charged particle production and loss processes. In addition, these modes are suspected to be the major cause of the catastrophic collapse observed in CO_2 laser discharges which limits power deposition. With this recognition of the important role played by minority species collisional processes in influencing the various plasma instability modes and overall gas discharge laser performance the basic understanding is available to develop new techniques and approaches for improvement of gas discharge laser performance.

VI. SUMMARY

From the material contained in this chapter there can be no doubt of the importance and highly efficient nature of collisional energy transfer processes in CO_2 molecular gas discharge lasers. From the level of understanding of the details of the electron and heavy particle collisional processes taking place in these discharges, there is no question that further laser performance improvements will be achieved in the immediate future. Furthermore, as the understanding of laser processes improves it is expected new wavelength gas discharge laser systems with performance characteristics comparable to the CO_2 laser will emerge. Lastly, development of the CO_2 laser has also provided the opportunity to investigate in greater detail than heretofore possible the factors influencing the stability of gas discharges, which is a subject that has been of major interest for decades. It is important to note that the key factor that allowed development of the present level of understanding of the CO_2 gas discharge laser was the availability of basic collisional information.

ACKNOWLEDGMENTS

The author would like to acknowledge the many helpful discussions with W. L. Nighan, W. J. Wiegand, and M. C. Fowler that took place during the preparation of this manuscript.

REFERENCES

1. C. O. Brown and J. W. Davis, *Appl. Phys. Lett.*, **21,** 480 (1972).
2. M. L. Bhaumik, W. B. Lacina, and M. M. Mann, *IEEE J. Quantum Electron.*, **QE-8,** 150 (1972).
3. J. W. Rich, *J. Appl. Phys.*, **42,** 2719 (1971).
4. W. L. Nighan, *Appl. Phys. Lett.*, **20,** 96 (1972).
5. R. E. Center and G. E. Caledonia, *Appl. Phys. Lett.*, **19,** 211 (1971).
6. J. T. Yardley, *Appl. Opt.*, **10,** 1760 (1971).
7. N. N. Sobolev and V. V. Sokovikov, *Sov. J. Quantum Electron.*, **2,** 305 (1973).
8. M. C. Fowler, *J. Appl. Phys.*, **43,** 3480 (1972).
9. C. B. Moore, R. E. Wood, B. Hu, and J. T. Yardley, *J. Chem. Phys.*, **46,** 4222 (1967).

10. B. F. Gordietz, N. N. Sobolev, V. V. Sokovikov, and L. A. Shelepin, *IEEE J. Quantum Electron.*, **QE-4,** 796 (1968).
11. C. K. Rhodes, M. J. Kelly, and A. Javan, *J. Chem. Phys.*, **48,** 5730 (1968).
12. R. D. Sharma, *J. Chem. Phys.*, **49,** 5195 (1968).
13. G. Herzberg, *Molecular Spectra and Molecular Structure*, Vol. II, Van Nostrand, Princeton, N.J., 1945, pp. 215–218.
14. P. K. Cheo, *J. Appl. Phys.*, **38,** 3563 (1967).
15. C. K. N. Patel, W. L. Faust, and R. A. McFarlance, *Bull. Am. Phys. Soc.*, **9,** 500 (1964).
16. C. O. Brown, *Appl. Phys. Lett.*, **17,** 388 (1970).
17. W. B. Tiffany, R. Targ, and J. D. Foster, *Appl. Phys. Lett.*, **15,** 91 (1969).
18. A. E. Hill, *Appl. Phys. Lett.*, **18,** 194 (1971).
19. C. K. N. Patel, *Scientific American*, **219,** 22 (1968).
20. C. K. N. Patel, *Phys. Rev. Lett.*, **12,** 588 (1964).
21. C. K. N. Patel, *Phys. Rev. Lett.*, **13,** 617 (1964).
22. F. Legay and N. Legay-Sommaire, *Compt. Rend.*, **257,** 2644 (1964).
23. F. Legay and N. Legay-Sommaire, *Compt. Rend.*, **259,** 99 (1964).
24. C. K. N. Patel, *Phys. Rev.*, **136,** A1187 (1964).
25. N. N. Sobolev and V. V. Sokovikov, *Zh. Eksperim. Teor. Fiz. Pis'ma*, **4,** 303 (1966).
26. N. N. Sobolev and V. V. Sokovikov, *Zh. Eksperim. Teor. Fiz. Pis'ma*, **5,** 122 (1967).
27. G. J. Schulz, *Phys. Rev.*, **116,** 1141 (1959).
28. G. J. Schulz, *Phys. Rev.*, **125,** 229 (1962).
29. G. J. Schulz, *Phys. Rev.*, **135,** A988 (1964).
30. J. D. Swift, *Br. J. Appl. Phys.*, **16,** 837 (1965).
31. A. G. Engelhardt, A. V. Phelps, and C. G. Risk, *Phys. Rev.*, **135,** A1566 (1964).
32. N. N. Sobolev and V. V. Sokovikov, *Sov. Phys. Usp.*, **10,** 153 (1967).
33. R. D. Hake Jr. and A. V. Phelps, *Phys. Rev.*, **158,** 70 (1967).
34. A. V. Phelps, *Rev. Mod. Phys.*, **40,** 399 (1968).
35. A. Andrick, D. Danner, and H. Ehrhardt, *Phys. Lett.*, **29A,** 346 (1969).
36. M. J. W. Boness and J. B. Hasted, *Phys. Lett.*, **21,** 526 (1966).
37. M. J. W. Boness and G. J. Schulz, *Phys. Rev. Lett.*, **21,** 1031 (1968).
38. W. L. Nighan and J. H. Bennett, *Appl. Phys. Lett.*, **14,** 240 (1969).
39. W. L. Nighan, *Appl. Phys. Lett.*, **15,** 355 (1969).
40. W. L. Nighan, *Phys. Rev. A*, **2,** 1989 (1970).
41. R. H. Bullis, W. L. Nighan, M. C. Fowler, and W. J. Wiegand, *AIAA J.*, **10,** 407 (1972).
42. R. Haas, *Z. Physik*, **148,** 177 (1957).
43. L. S. Frost and A. V. Phelps, *Phys. Rev.*, **127,** 1621 (1962).
44. I. P. Shkarofsky, T. W. Johnston, and M. O. Bachynski, *The Particle Kinetics of Plasmas*, Addison-Wesley, Reading, Mass., 1966.
45. T. Holstein, *Phys. Rev.*, **70,** 367 (1946).
46. Unless otherwise indicated all mixture compositions referred to in the text are by density with CO_2–N_2–He the sequence of species.
47. J. J. Lowke, A. V. Phelps, and B. W. Irwin, *J. Appl. Phys. 44*, 4664 (1973).

48. W. L. Nighan (unpublished).
49. J. C. Y. Chen, *J. Chem. Phys.*, **40,** 3513 (1964).
50. P. Bletzinger and A. Garscadden, *Proc. IEEE*, **59,** 675 (1971).
51. W. J. Wiegand, M. C. Fowler, and J. A. Benda, *Appl. Phys. Lett.*, **16,** 237 (1970).
52. W. J. Wiegand and W. L. Nighan, *Appl. Phys. Lett.*, **22,** 583 (1973).
53. W. L. Nighan, W. J. Wiegand, and R. A. Haas, *Appl. Phys. Lett.*, **22,** 579 (1973).
54. C. A. Fenstermacher, M. J. Nutter, W. T. Leland, and K. Boyer, *Appl. Phys. Lett.*, **20,** 56 (1972).
55. J. D. Dougherty, E. R. Pugh, and D. H. Douglas-Hamilton, *Bull. Am. Phys. Soc.*, **17,** 339 (1972).
56. P. Bletzinger and A. Garscadden, *Appl. Phys. Lett.*, **12,** 289 (1968).
57. P. O. Clark and J. Y. Wada, *IEEE J. Quantum Electron.*, **QE-4,** 263 (1968).
58. W. J. Wiegand (private communication).
59. R. L. Taylor and S. Bitterman, *Rev. Mod. Phys.*, **41,** 26 (1969).
60. M. A. Kovacs, G. W. Flynn, and A. Javan, *Appl. Phys. Lett.*, **8,** 62 (1966).
61. G. W. Flynn, M. A. Kovacs, C. K. Rhodes, and A. Javan, *Appl. Phys. Lett.*, **8,** 63 (1966).
62. J. T. Yardley and C. B. Moore, *J. Chem. Phys.*, **45,** 1066 (1966).
63. L. O. Hocker, M. A. Kovacs, C. K. Rhodes, G. W. Flynn, and A. Javan, *Phys. Rev. Lett.*, **17,** 233 (1966).
64. G. W. Flynn, L. O. Hocker, A. Javan, M. A. Kovacs, and C. K. Rhodes, *IEEE J. Quantum Electron.*, **QE-2,** 378 (1966).
65. J. T. Yardley and C. B. Moore, *J. Chem. Phys.*, **46,** 4491 (1967).
66. C. B. Moore, *Acct. Chem. Res.*, **2,** 103 (1969).
67. M. C. Fowler, *Appl. Phys. Lett.*, **18,** 175 (1971).
68. K. F. Herzfeld and T. A. Litovitz, *Absorption and Dispersion of Ultrasonic Waves*, Academic, New York, 1959, p. 86.
69. W. J. Wiegand, M. C. Fowler, and J. A. Benda, *Appl. Phys. Lett.*, **18,** 365 (1971).
70. G. Moeller and J. D. Rigden, *Appl. Phys. Lett.*, **7,** 274 (1965).
71. W. J. Witteman, *Phys. Lett.*, **18,** 125 (1965).
72. A. E. Hill, *Appl. Phys. Lett.*, **16,** 423 (1970).
73. A. J. Beaulieu, *Appl. Phys. Lett.*, **16,** 504 (1970).
74. J. D. Daugherty, *IEEE J. Quantum Electron.*, **QE-8,** 594 (1972).
75. M. A. Kovacs and A. Javan, *J. Chem. Phys.*, **50,** 4111 (1969).
76. A. E. Hill, *Appl. Phys. Lett.*, **22,** 670 (1973).
77. J. S. Levine and A. Javan, *Appl. Phys. Lett.*, **22,** 55 (1973).
78. W. L. Nighan, in *Proceedings of the Eleventh International Conference on Phenomena in Ionized Gases*, 1973 (Invited Papers), edited by L. Pekárek and L. Láska (Czechoslovakia Academy of Sciences, Prague, 1973).
79. A. K. McQuillan and A. I. Carswell, *Appl. Phys. Lett.*, **17,** 158 (1970).
80. P. K. Cheo and H. G. Cooper, *IEEE J. Quantum Electron.*, **QE-3,** 79 (1967).
81. T. A. Cool and J. A. Shirley, *Appl. Phys. Lett.*, **14,** 70 (1969).
82. Gordon Francis, in *Encyclopedia of Physics*, Vol. 22, edited by S. Flugge, Springer-Verlag, Berlin, 1956, p. 114.
83. G. Ecker and O. Zoller, *Phys. Fluids*, **7,** 1996 (1964).

84. D. C. Smith and J. C. McCoy, *Appl. Phys. Lett.*, **15,** 282 (1969).
85. W. J. Witteman, *Appl. Phys. Lett.*, **11,** 337 (1967).
86. F. E. Niles, *J. Chem. Phys.*, **52,** 408 (1970).
87. R. A. Haas, *Phys. Rev. A*, **8,** 1017 (1973).
88. D. Rapp and P. Englander-Golden, *J. Chem. Phys.*, **43,** 1464 (1965).
89. H. F. Winters, *J. Chem. Phys.*, **44,** 1472 (1966).
90. D. Rapp and D. D. Briglia, *J. Chem. Phys.*, **43,** 1480 (1965).
91. E. W. McDaniel, V. Cermak, A. Dalgarno, E. E. Ferguson, and L. Friedman, *Ion-Molecule Reactions*, Wiley, New York, 1970, Chap. 6.
92. F. C. Fehsenfeld, E. E. Ferguson, and A. L. Schmeltekopf, *J. Chem. Phys.*, **45,** 1844 (1966).

AUTHOR INDEX

SUBJECT INDEX